VIDEO PROCESSING
AND
COMMUNICATIONS

PRENTICE HALL SIGNAL PROCESSING SERIES
Alan V. Oppenheim, *Series Editor*

BRACEWELL *Two Dimensional Imaging*
BRIGHAM *The Fast Fourier Transform and Its Applications (AOD)*
BUCK, DANIEL, & SINGER *Computer Explorations in Signals and Systems Using MATLAB, 2/E*
CASTLEMAN *Digital Image Processing*
COHEN *Time-Frequency Analysis*
CROCHIERE & RABINER *Multirate Digital Signal Processing (AOD)*
JOHNSON & DUDGEON *Array Signal Processing (AOD)*
KAY *Fundamentals of Statistical Signal Processing, Vols. I & II*
KAY *Modern Spectral Estimation (AOD)*
LIM *Two-Dimensional Signal and Image Processing*
MCCLELLAN *et al. Computer-Based Exercises in Signal Processing using MATLAB5*
MENDEL *Lessons in Estimation Theory for Signal Processing, Communications, and Control, 2/e*
NIKIAS & PETROPULU *Higher Order Spectra Analysis*
OPPENHEIM & SCHAFER *Digital Signal Processing*
OPPENHEIM & SCHAFER *Discrete-Time Signal Processing*
OPPENHEIM & WILLSKY, with NAWAB *Signals and Systems, 2/e*
ORFANIDIS *Introduction to Signal Processing*
PHILLIPS & NAGLE *Digital Control Systems Analysis and Design, 3/e*
QUATIERI *Discrete-Time Speech Signal Processing: Principles and Practice*
RABINER & JUANG *Fundamentals of Speech Recognition*
RABINER & SCHAFER *Digital Processing of Speech Signals*
STEARNS & DAVID *Signal Processing Algorithms in MATLAB*
TEKALP *Digital Video Processing*
VAIDYANATHAN *Multirate Systems and Filter Banks*
VETTERLI & KOVACEVIC *Wavelets and Subband Coding*
WANG, OSTERMANN, & ZHANG *Video Processing and Communications*
WIDROW & STEARNS *Adaptive Signal Processing*

VIDEO PROCESSING
AND
COMMUNICATIONS

Yao Wang

Jörn Ostermann

Ya-Qin Zhang

PRENTICE HALL
Upper Saddle River, New Jersey 07458

Library of Congress Cataloging-in-Publication Data

Wang, Yao, 1962–
    Video processing and communications / Yao Wang, Jörn Ostermann, Ya-Qin Zhang.
       p.   cm.
    Includes bibliographical references and index.
    ISBN 0-13-017547-1
    1. Image transmission.   2. Digital video   I. Ostermann, Jörn.   II. Zhang, Ya-Qin, 1969–
III. Title.

    TK5105.2 .W36   2001
    621.388′33—dc21

                                                                                    200103699

Vice President and Editorial Director, ECS: *Marcia J. Horton*
Publisher: *Tom Robbins*
Editorial Assistant: *Jessica Romeo*
Vice President and Director of Production and Manufacturing, ESM: *David W. Riccardi*
Executive Managing Editor: *Vince O'Brien*
Managing Editor: *David A. George*
Production Editor: *Kathy Ewing*
Director of Creative Services: *Paul Belfanti*
Creative Director: *Carole Anson*
Art Director: *Jayne Conte*
Art Editor: *Greg Dulles*
Manufacturing Manager: *Trudy Pisciotti*
Manufacturing Buyer: *Lynda Castillo*
Marketing Manager: *Holly Stark*
Marketing Assistant: *Karen Moon*

The author and publisher of this book have used their best efforts in preparing this book. These efforts include the development, research, and testing of the theories and programs to determine their effectiveness. The author and publisher make no warranty of any kind, expressed or implied, with regard to these programs or the documentation contained in this book. The author and publisher shall not be liable in any event for incidental or consequential damages in connection with, or arising out of, the furnishing, performance, or use of these programs.

MATLAB is a registered trademark of The MathWorks, Inc., 3 Apple Hill Drive, Natick, MA 07160-2098.

Printed in the United States of America
10 9 8 7 6 5 4 3 2 1

ISBN   0-13-017547-1

Pearson Education Ltd., *London*
Pearson Education Australia Pty. Ltd., *Sydney*
Pearson Education Singapore, Pte. Ltd.
Pearson Education North Asia Ltd., *Hong Kong*
Pearson Education Canada, Inc., *Toronto*
Pearson Educacíon de Mexico, S.A. de C.V.
Pearson Education—Japan, *Tokyo*
Pearson Education Malaysia, Pte. Ltd.

To Our Beloved Ones:

Hai and Yana
Dorota, Olivia, Oktavia, and Orry
Jenny, Sophie, and Brandon

VIDEO PROCESSING
AND
COMMUNICATIONS

Contents

PREFACE   *xxi*

GLOSSARY OF NOTATIONS   *xxv*

**1 VIDEO FORMATION, PERCEPTION,
AND REPRESENTATION**   *1*

    1.1   Color Perception and Specification   2
           1.1.1   Light and Color, 2
           1.1.2   Human Perception of Color, 3
           1.1.3   The Trichromatic Theory of Color Mixture, 4
           1.1.4   Color Specification by Tristimulus Values, 5
           *1.1.5   Color Specification by Luminance and Chrominance
                    Attributes, 6*

    1.2   Video Capture and Display   7
           1.2.1   Principles of Color Video Imaging, 7
           1.2.2   Video Cameras, 8
            1.2.3   Video Display, 10
           1.2.4   Composite versus Component Video, 11
           1.2.5   Gamma Correction, 11

    1.3   Analog Video Raster   12
           1.3.1   Progressive and Interlaced Scan, 12
           1.3.2   Characterization of a Video Raster, 14

1.4   Analog Color Television Systems   16
       1.4.1   Spatial and Temporal Resolution, 16
       1.4.2   Color Coordinate, 17
       1.4.3   Signal Bandwidth, 19
       1.4.4   Multiplexing of Luminance, Chrominance, and Audio, 19
       1.4.5   Analog Video Recording, 21

1.5   Digital Video   22
       1.5.1   Notation, 22
       1.5.2   ITU-R BT.601 Digital Video, 23
       1.5.3   Other Digital Video Formats and Applications, 26
       1.5.4   Digital Video Recording, 28
       1.5.5   Video Quality Measure, 28

1.6   Summary   30

1.7   Problems   31

1.8   Bibliography   32

**2   FOURIER ANALYSIS OF VIDEO SIGNALS AND
       FREQUENCY RESPONSE OF THE HUMAN
       VISUAL SYSTEM   33**

2.1   Multidimensional Continuous-Space Signals and Systems   33

2.2   Multidimensional Discrete-Space Signals and Systems   36

2.3   Frequency Domain Characterization of Video Signals   38
       2.3.1   Spatial and Temporal Frequencies, 38
       2.3.2   Temporal Frequencies Caused by Linear Motion, 40

2.4   Frequency Response of the Human Visual System   42
       2.4.1   Temporal Frequency Response and Flicker Perception, 43
       2.4.2   Spatial Frequency Response, 45
       2.4.3   Spatiotemporal Frequency Response, 46
       2.4.4   Smooth Pursuit Eye Movement, 48

2.5   Summary   50

2.6   Problems   51

2.7   Bibliography   52

3   VIDEO SAMPLING   53

3.1   Basics of the Lattice Theory   54

3.2   Sampling over Lattices   59
       3.2.1   Sampling Process and Sampled-Space Fourier Transform, 60
       3.2.2   The Generalized Nyquist Sampling Theorem , 61
       3.2.3   Sampling Efficiency, 63

          3.2.4   Implementation of the Prefilter and Reconstruction Filter, 65

          3.2.5   Relation between Fourier Transforms over Continuous, Discrete, and Sampled Spaces, 66

   3.3   Sampling of Video Signals   67

          3.3.1   Required Sampling Rates, 67

          3.3.2   Sampling Video in Two Dimensions: Progressive versus Interlaced Scans, 69

          3.3.3   Sampling a Raster Scan: BT.601 Format Revisited, 71

          3.3.4   Sampling Video in Three Dimensions, 72

          3.3.5   Spatial and Temporal Aliasing, 73

   3.4   Filtering Operations in Cameras and Display Devices   76

          3.4.1   Camera Apertures, 76

          3.4.2   Display Apertures, 79

   3.5   Summary   80

   3.6   Problems   80

   3.7   Bibliography   83

4   VIDEO SAMPLING RATE CONVERSION   84

   4.1   Conversion of Signals Sampled on Different Lattices   84

          4.1.1   Up-Conversion, 85

          4.1.2   Down-Conversion, 87

          4.1.3   Conversion between Arbitrary Lattices, 89

          4.1.4   Filter Implementation and Design, and Other Interpolation Approaches, 91

   4.2   Sampling Rate Conversion of Video Signals   92

          4.2.1   Deinterlacing, 93

          4.2.2   Conversion between PAL and NTSC Signals, 98

          4.2.3   Motion-Adaptive Interpolation, 104

   4.3   Summary   105

   4.4   Problems   106

   4.5   Bibliography   109

5   VIDEO MODELING   111

   5.1   Camera Model   112

          5.1.1   Pinhole Model, 112

          5.1.2   CAHV Model, 114

          5.1.3   Camera Motions, 116

   5.2   Illumination Model   116

          5.2.1   Diffuse and Specular Reflection, 116

　　　　　5.2.2   *Radiance Distribution under Differing Illumination and Reflection Conditions, 117*

　　　　　5.2.3   *Changes in the Image Function Due to Object Motion, 119*

　　　5.3   Object Model   120

　　　　　5.3.1   *Shape Model, 121*

　　　　　5.3.2   *Motion Model, 122*

　　　5.4   Scene Model   125

　　　5.5   Two-Dimensional Motion Models   128

　　　　　5.5.1   *Definition and Notation, 128*

　　　　　5.5.2   *Two-Dimensional Motion Models Corresponding to Typical Camera Motions, 130*

　　　　　5.5.3   *Two-Dimensional Motion Corresponding to Three-Dimensional Rigid Motion, 133*

　　　　　5.5.4   *Approximations of Projective Mapping, 136*

　　　5.6   Summary   137

　　　5.7   Problems   138

　　　5.8   Bibliography   139

6   TWO-DIMENSIONAL MOTION ESTIMATION   141

　　　6.1   Optical Flow   142

　　　　　6.1.1   *Two-Dimensional Motion versus Optical Flow, 142*

　　　　　6.1.2   *Optical Flow Equation and Ambiguity in Motion Estimation, 143*

　　　6.2   General Methodologies   145

　　　　　6.2.1   *Motion Representation, 146*

　　　　　6.2.2   *Motion Estimation Criteria, 147*

　　　　　6.2.3   *Optimization Methods, 151*

　　　6.3   Pixel-Based Motion Estimation   152

　　　　　6.3.1   *Regularization Using the Motion Smoothness Constraint, 153*

　　　　　6.3.2   *Using a Multipoint Neighborhood, 153*

　　　　　6.3.3   *Pel-Recursive Methods, 154*

　　　6.4   Block-Matching Algorithm   154

　　　　　6.4.1   *The Exhaustive Block-Matching Algorithm, 155*

　　　　　6.4.2   *Fractional Accuracy Search, 157*

　　　　　6.4.3   *Fast Algorithms, 159*

　　　　　6.4.4   *Imposing Motion Smoothness Constraints, 161*

　　　　　6.4.5   *Phase Correlation Method, 162*

　　　　　6.4.6   *Binary Feature Matching, 163*

　　　6.5   Deformable Block-Matching Algorithms   165

　　　　　6.5.1   *Node-Based Motion Representation, 166*

　　　　　6.5.2   *Motion Estimation Using the Node-Based Model, 167*

6.6   Mesh-Based Motion Estimation   169
      6.6.1   Mesh-Based Motion Representation, 171
      6.6.2   Motion Estimation Using the Mesh-Based Model, 173

6.7   Global Motion Estimation   177
      6.7.1   Robust Estimators, 177
      6.7.2   Direct Estimation, 178
      6.7.3   Indirect Estimation, 178

6.8   Region-Based Motion Estimation   179
      6.8.1   Motion-Based Region Segmentation, 180
      6.8.2   Joint Region Segmentation and Motion Estimation, 181

6.9   Multiresolution Motion Estimation   182
      6.9.1   General Formulation, 182
      6.9.2   Hierarchical Block Matching Algorithm, 184

6.10   Application of Motion Estimation in Video Coding   187

6.11   Summary   188

6.12   Problems   189

6.13   Bibliography   191

7   THREE-DIMENSIONAL MOTION ESTIMATION   194

7.1   Feature-Based Motion Estimation   195
      7.1.1   Objects of Known Shape under Orthographic Projection, 195
      7.1.2   Objects of Known Shape under Perspective Projection, 196
      7.1.3   Planar Objects, 197
      7.1.4   Objects of Unknown Shape Using the Epipolar Line, 198

7.2   Direct Motion Estimation   203
      7.2.1   Image Signal Models and Motion, 204
      7.2.2   Objects of Known Shape, 206
      7.2.3   Planar Objects, 207
      7.2.4   Robust Estimation, 209

7.3   Iterative Motion Estimation   212

7.4   Summary   213

7.5   Problems   214

7.6   Bibliography   215

8   FOUNDATIONS OF VIDEO CODING   217

8.1   Overview of Coding Systems   218
      8.1.1   General Framework, 218
      8.1.2   Categorization of Video Coding Schemes, 219

8.2   Basic Notions in Probability and Information Theory   221
      8.2.1   *Characterization of Stationary Sources, 221*
      8.2.2   *Entropy and Mutual Information for Discrete Sources, 222*
      8.2.3   *Entropy and Mutual Information for Continuous
              Sources, 226*

8.3   Information Theory for Source Coding   227
      8.3.1   *Bound for Lossless Coding, 227*
      8.3.2   *Bound for Lossy Coding, 229*
      8.3.3   *Rate-Distortion Bounds for Gaussian Sources, 232*

8.4   Binary Encoding   234
      8.4.1   *Huffman Coding, 235*
      8.4.2   *Arithmetic Coding, 238*

8.5   Scalar Quantization   241
      8.5.1   *Fundamentals, 241*
      8.5.2   *Uniform Quantization, 243*
      8.5.3   *Optimal Scalar Quantizer, 244*

8.6   Vector Quantization   248
      8.6.1   *Fundamentals, 248*
      8.6.2   *Lattice Vector Quantizer, 251*
      8.6.3   *Optimal Vector Quantizer, 253*
      8.6.4   *Entropy-Constrained Optimal Quantizer Design, 255*

8.7   Summary   257

8.8   Problems   259

8.9   Bibliography   261

9   **WAVEFORM-BASED VIDEO CODING**   **263**

9.1   Block-Based Transform Coding   263
      9.1.1   *Overview, 264*
      9.1.2   *One-Dimensional Unitary Transform, 266*
      9.1.3   *Two-Dimensional Unitary Transform, 269*
      9.1.4   *The Discrete Cosine Transform, 271*
      9.1.5   *Bit Allocation and Transform Coding Gain, 273*
      9.1.6   *Optimal Transform Design and the KLT, 279*
      9.1.7   *DCT-Based Image Coders and the JPEG Standard, 281*
      9.1.8   *Vector Transform Coding, 284*

9.2   Predictive Coding   285
      9.2.1   *Overview, 285*
      9.2.2   *Optimal Predictor Design and Predictive Coding Gain, 286*
      9.2.3   *Spatial-Domain Linear Prediction, 290*
      9.2.4   *Motion-Compensated Temporal Prediction, 291*

9.3   Video Coding Using Temporal Prediction and Transform Coding   293
        9.3.1   Block-Based Hybrid Video Coding, 293
        9.3.2   Overlapped Block Motion Compensation, 296
        9.3.3   Coding Parameter Selection, 299
        9.3.4   Rate Control, 302
        9.3.5   Loop Filtering, 305

9.4   Summary   308

9.5   Problems   309

9.6   Bibliography   311

10   CONTENT-DEPENDENT VIDEO CODING   314

10.1   Two-Dimensional Shape Coding   314
        10.1.1 Bitmap Coding, 315
        10.1.2 Contour Coding, 318
        10.1.3 Evaluation Criteria for Shape Coding Efficiency, 323

10.2   Texture Coding for Arbitrarily Shaped Regions   324
        10.2.1 Texture Extrapolation, 324
        10.2.2 Direct Texture Coding, 325

10.3   Joint Shape and Texture Coding   326

10.4   Region-Based Video Coding   327

10.5   Object-Based Video Coding   328
        10.5.1 Source Model F2D, 330
        10.5.2 Source Models R3D and F3D, 332

10.6   Knowledge-Based Video Coding   336

10.7   Semantic Video Coding   338

10.8   Layered Coding System   339

10.9   Summary   342

10.10   Problems   343

10.11   Bibliography   344

11   SCALABLE VIDEO CODING   349

11.1   Basic Modes of Scalability   350
        11.1.1 Quality Scalability, 350
        11.1.2 Spatial Scalability, 353
        11.1.3 Temporal Scalability, 356
        11.1.4 Frequency Scalability, 356

11.1.5   *Combination of Basic Schemes, 357*
11.1.6   *Fine-Granularity Scalability, 357*

11.2   Object-Based Scalability   359

11.3   Wavelet-Transform-Based Coding   361
11.3.1   *Wavelet Coding of Still Images, 363*
11.3.2   *Wavelet Coding of Video, 367*

11.4   Summary   370

11.5   Problems   370

11.6   Bibliography   371

12   STEREO AND MULTIVIEW SEQUENCE PROCESSING   374

12.1   Depth Perception   375
12.1.1   *Binocular Cues—Stereopsis, 375*
12.1.2   *Visual Sensitivity Thresholds for Depth Perception, 375*

12.2   Stereo Imaging Principle   377
12.2.1   *Arbitrary Camera Configuration, 377*
12.2.2   *Parallel Camera Configuration, 379*
12.2.3   *Converging Camera Configuration, 381*
12.2.4   *Epipolar Geometry, 383*

12.3   Disparity Estimation   385
12.3.1   *Constraints on Disparity Distribution, 386*
12.3.2   *Models for the Disparity Function, 387*
12.3.3   *Block-Based Approach, 388*
12.3.4   *Two-Dimensional Mesh-Based Approach, 388*
12.3.5   *Intra-Line Edge Matching Using Dynamic Programming, 391*
12.3.6   *Joint Structure and Motion Estimation, 392*

12.4   Intermediate View Synthesis   393

12.5   Stereo Sequence Coding   396
12.5.1   *Block-Based Coding and MPEG-2 Multiview Profile, 396*
12.5.2   *Incomplete Three-Dimensional Representation
         of Multiview Sequences, 398*
12.5.3   *Mixed-Resolution Coding, 398*
12.5.4   *Three-Dimensional Object-Based Coding, 399*
12.5.5   *Three-Dimensional Model-Based Coding, 400*

12.6   Summary   400

12.7   Problems   402

12.8   Bibliography   403

13   VIDEO COMPRESSION STANDARDS   405

13.1   Standardization   406
    13.1.1   Standards Organizations, 406
    13.1.2   Requirements for a Successful Standard, 409
    13.1.3   Standard Development Process, 411
    13.1.4   Applications for Modern Video Coding Standards, 412

13.2   Video Telephony with H.261 and H.263   413
    13.2.1   H.261 Overview, 413
    13.2.2   H.263 Highlights, 416
    13.2.3   Comparison, 420

13.3   Standards for Visual Communication Systems   421
    13.3.1   H.323 Multimedia Terminals, 421
    13.3.2   H.324 Multimedia Terminals, 422

13.4   Consumer Video Communications with MPEG-1   423
    13.4.1   Overview, 423
    13.4.2   MPEG-1 Video, 424

13.5   Digital TV with MPEG-2   426
    13.5.1   Systems, 426
    13.5.2   Audio, 426
    13.5.3   Video, 427
    13.5.4   Profiles, 435

13.6   Coding of Audiovisual Objects with MPEG-4   437
    13.6.1   Systems, 437
    13.6.2   Audio, 441
    13.6.3   Basic Video Coding, 442
    13.6.4   Object-Based Video Coding, 445
    13.6.5   Still Texture Coding, 447
    13.6.6   Mesh Animation, 447
    13.6.7   Face and Body Animation, 448
    13.6.8   Profiles, 451
    13.6.9   Evaluation of Subjective Video Quality, 454

13.7   Video Bit Stream Syntax   454

13.8   Multimedia Content Description Using MPEG-7   458
    13.8.1   Overview, 458
    13.8.2   Multimedia Description Schemes, 459
    13.8.3   Visual Descriptors and Description Schemes, 461

13.9   Summary   465

13.10   Problems   466

13.11   Bibliography   467

14   ERROR CONTROL IN VIDEO COMMUNICATIONS   472

14.1   Motivation and Overview of Approaches   473

14.2   Typical Video Applications and Communication Networks   476
          14.2.1   *Categorization of Video Applications, 476*
          14.2.2   *Communication Networks, 479*

14.3   Transport-Level Error Control   485
          14.3.1   *Forward Error Correction, 485*
          14.3.2   *Error-Resilient Packetization and Multiplexing, 486*
          14.3.3   *Delay-Constrained Retransmission, 487*
          14.3.4   *Unequal Error Protection, 488*

14.4   Error-Resilient Encoding   489
          14.4.1   *Error Isolation, 489*
          14.4.2   *Robust Binary Encoding, 490*
          14.4.3   *Error-Resilient Prediction, 492*
          14.4.4   *Layered Coding with Unequal Error Protection, 493*
          14.4.5   *Multiple-Description Coding, 494*
          14.4.6   *Joint Source and Channel Coding, 498*

14.5   Decoder Error Concealment   498
          14.5.1   *Recovery of Texture Information, 500*
          14.5.2   *Recovery of Coding Modes and Motion Vectors, 501*
          14.5.3   *Syntax-Based Repair, 502*

14.6   Encoder–Decoder Interactive Error Control   502
          14.6.1   *Coding-Parameter Adaptation Based on Channel Conditions, 503*
          14.6.2   *Reference Picture Selection Based on Feedback Information, 503*
          14.6.3   *Error Tracking Based on Feedback Information, 504*
          14.6.4   *Retransmission without Waiting, 504*

14.7   Error-Resilience Tools in H.263 and MPEG-4   505
          14.7.1   *Error-Resilience Tools in H.263, 505*
          14.7.2   *Error-Resilience Tools in MPEG-4, 508*

14.8   Summary   509

14.9   Problems   511

14.10   Bibliography   513

**15   STREAMING VIDEO OVER THE INTERNET AND
        WIRELESS IP NETWORKS   519**

15.1   Architecture for Video Streaming Systems   520

15.2   Video Compression   522

15.3   Application-Layer QoS Control for Streaming Video   522
         15.3.1   Congestion Control, 522
         15.3.2   Error Control, 525

15.4   Continuous Media Distribution Services   529
         15.4.1   Network Filtering, 529
         15.4.2   Application-Level Multicast, 531
         15.4.3   Content Replication, 532

15.5   Streaming Servers   533
         15.5.1   Real-Time Operating System, 534
         15.5.2   Storage System, 537

15.6   Media Synchronization   539

15.7   Protocols for Streaming Video   542
         15.7.1   Transport Protocols, 543
         15.7.2   Session Control Protocol: RTSP, 545

15.8   Streaming Video over Wireless IP Networks   546
         15.8.1   Network-Aware Applications, 548
         15.8.2   Adaptive Service, 549

15.9   Summary   554

15.10   Bibliography   555

**APPENDIX A: DETERMINATION OF SPATIAL–TEMPORAL
GRADIENTS**   **562**

A.1   First- and Second-Order Gradient   562

A.2   Sobel Operator   563

A.3   Difference of Gaussian Filters   563

APPENDIX B: GRADIENT DESCENT METHODS   **565**

B.1   First-Order Gradient Descent Method   565

B.2   Steepest Descent Method   566

B.3   Newton's Method   566

B.4   Newton-Ralphson Method   567

B.5   Bibliography   567

APPENDIX C: GLOSSARY OF ACRONYMS   **568**

APPENDIX D: ANSWERS TO SELECTED PROBLEMS   **575**

Preface

In the past decade or so, there have been fascinating developments in multimedia representation and communications. First of all, it has become very clear that all aspects of media are "going digital"; from representation to transmission, from processing to retrieval, from studio to home. Second, there have been significant advances in digital multimedia compression and communication algorithms, which make it possible to deliver high-quality video at relatively low bit rates in today's networks. Third, the advancement in VLSI technologies has enabled sophisticated software to be implemented in a cost-effective manner. Last but not least, the establishment of half a dozen international standards by ISO/MPEG and ITU-T laid the common groundwork for different vendors and content providers.

At the same time, the explosive growth in wireless and networking technology has profoundly changed the global communications infrastructure. It is the confluence of wireless, multimedia, and networking that will fundamentally change the way people conduct business and communicate with each other. The future computing and communications infrastructure will be empowered by virtually unlimited bandwidth, full connectivity, high mobility, and rich multimedia capability.

As multimedia becomes more pervasive, the boundaries between video, graphics, computer vision, multimedia database, and computer networking start to blur, making video processing an exciting field with input from many disciplines. Today, video processing lies at the core of multimedia. Among the many technologies involved, video coding and its standardization are definitely the key enablers of these developments. This book covers the fundamental theory and techniques for digital video processing, with a focus on video coding and communications. It is intended as a textbook for a graduate-level course on video processing, as well as a reference or self-study text for researchers and engineers. In selecting the topics to cover, we have tried to achieve a balance between providing a solid theoretical foundation and presenting complex system issues in real video systems.

SYNOPSIS

Chapter 1 gives a broad overview of video technology, from analog color TV system to digital video. Chapter 2 delineates the analytical framework for video analysis in the frequency domain, and describes characteristics of the human visual system. Chapters 3–12 focus on several very important sub-topics in digital video technology. Chapters 3 and 4 consider how a continuous-space video signal can be sampled to retain the maximum perceivable information within the affordable data rate, and how video can be converted from one format to another. Chapter 5 presents models for the various components involved in forming a video signal, including the camera, the illumination source, the imaged objects and the scene composition. Models for the three-dimensional (3-D) motions of the camera and objects, as well as their projections onto the two-dimensional (2-D) image plane, are discussed at length, because these models are the foundation for developing motion estimation algorithms, which are the subjects of Chapters 6 and 7. Chapter 6 focuses on 2-D motion estimation, which is a critical component in modern video coders. It is also a necessary preprocessing step for 3-D motion estimation. We provide both the fundamental principles governing 2-D motion estimation, and practical algorithms based on different 2-D motion representations. Chapter 7 considers 3-D motion estimation, which is required for various computer vision applications, and can also help improve the efficiency of video coding.

Chapters 8–11 are devoted to the subject of video coding. Chapter 8 introduces the fundamental theory and techniques for source coding, including information theory bounds for both lossless and lossy coding, binary encoding methods, and scalar and vector quantization. Chapter 9 focuses on waveform-based methods (including transform and predictive coding), and introduces the block-based hybrid coding framework, which is the core of all international video coding standards. Chapter 10 discusses content-dependent coding, which has the potential of achieving extremely high compression ratios by making use of knowledge of scene content. Chapter 11 presents scalable coding methods, which are well-suited for video streaming and broadcasting applications, where the intended recipients have varying network connections and computing powers. Chapter 12 introduces stereoscopic and multiview video processing techniques, including disparity estimation and coding of such sequences.

Chapters 13–15 cover system-level issues in video communications. Chapter 13 introduces the H.261, H.263, MPEG-1, MPEG-2, and MPEG-4 standards for video coding, comparing their intended applications and relative performance. These standards integrate many of the coding techniques discussed in Chapters 8–11. The MPEG-7 standard for multimedia content description is also briefly described. Chapter 14 reviews techniques for combating transmission errors in video communication systems, and also describes the requirements of different video applications, and the characteristics of various networks. As an example of a practical video communication system, we end the text with a chapter devoted to video streaming over the Internet and wireless network. Chapter 15 discusses the requirements and representative solutions for the major subcomponents of a streaming system.

SUGGESTED USE FOR INSTRUCTION AND SELF-STUDY

As prerequisites, students are assumed to have finished undergraduate courses in signals and systems, communications, probability, and preferably a course in image processing. For a one-semester course focusing on video coding and communications, we recommend covering the two beginning chapters, followed by video modeling (Chapter 5), 2-D motion estimation (Chapter 6), video coding (Chapters 8–11), standards (Chapter 13), error control (Chapter 14) and video streaming systems (Chapter 15). On the other hand, for a course on general video processing, the first nine chapters, including the introduction (Chapter 1), frequency domain analysis (Chapter 2), sampling and sampling rate conversion (Chapters 3 and 4), video modeling (Chapter 5), motion estimation (Chapters 6 and 7), and basic video coding techniques (Chapters 8 and 9), plus selected topics from Chapters 10–13 (content-dependent coding, scalable coding, stereo, and video coding standards) may be appropriate. In either case, Chapter 8 may be skipped or only briefly reviewed if the students have finished a prior course on source coding. Chapters 7 (3-D motion estimation), 10 (content-dependent coding), 11 (scalable coding), 12 (stereo), 14 (error-control), and 15 (video streaming) may also be left for an advanced course in video, after covering the other chapters in a first course in video. In all cases, sections denoted by asterisks (*) may be skipped or left for further exploration by advanced students.

Problems are provided at the end of Chapters 1–14 for self-study or as homework assignments for classroom use. Appendix D gives answers to selected problems. The website for this book (www.prenhall.com/wang) provides MATLAB scripts used to generate some of the plots in the figures. Instructors may modify these scripts to generate similar examples. The scripts may also help students to understand the underlying operations. Sample video sequences can be downloaded from the website, so that students can evaluate the performance of different algorithms on real sequences. Some compressed sequences using standard algorithms are also included, to enable instructors to demonstrate coding artifacts at different rates by different techniques.

ACKNOWLEDGMENTS

We are grateful to the many people who have helped to make this book a reality. Dr. Barry G. Haskell of AT&T Labs, with his tremendous experience in video coding standardization, reviewed Chapter 13 and gave valuable input to this chapter as well as other topics. Prof. David J. Goodman of Polytechnic University, a leading expert in wireless communications, provided valuable input to Section 14.2.2, part of which summarize characteristics of wireless networks. Prof. Antonio Ortega of the University of Southern

California and Dr. Anthony Vetro of Mitsubishi Electric Research Laboratories, then a Ph.D. student at Polytechnic University, suggested what topics to cover in the section on rate control, and reviewed Sections 9.3.3–4. Mr. Dapeng Wu, a Ph.D. student at Carnegie Mellon University, and Dr. Yiwei Hou from Fijitsu Labs helped to draft Chapter 15. Dr. Ru-Shang Wang of Nokia Research Center, Mr. Fatih Porikli of Mitsubishi Electric Research Laboratories, also a Ph.D. student at Polytechnic University, and Mr. Khalid Goudeaux, a student at Carnegie Mellon University, generated several images related to stereo. Mr. Haidi Gu, a student at Polytechnic University, provided the example image for scalable video coding. Mrs. Dorota Ostermann provided the brilliant design for the cover.

We would like to thank the anonymous reviewers who provided valuable comments and suggestions to enhance this work. We would also like to thank the students at Polytechnic University, who used draft versions of the text and pointed out many typographic errors and inconsistencies. Solutions included in Appendix D are based on their homeworks. Finally, we would like to acknowledge the encouragement and guidance of Tom Robbins at Prentice Hall. Yao Wang would like to acknowledge research grants from the National Science Foundation and New York State Center for Advanced Technology in Telecommunications over the past ten years, which have led to some of the research results included in this book.

Most of all, we are deeply indebted to our families, for allowing and even encouraging us to complete this project, which started more than four years ago and took away a significant amount of time we could otherwise have spent with them. The arrival of our new children Yana and Brandon caused a delay in the creation of the book but also provided an impetus to finish it. This book is a tribute to our families, for their love, affection, and support.

YAO WANG
Polytechnic University, Brooklyn, NY, USA
yao@vision.poly.edu

JÖRN OSTERMANN
AT&T Labs—Research, Middletown, NJ, USA
osterman@research.att.com

YA-QIN ZHANG
Microsoft Research, Beijing, China
yzhang@microsoft.com

Glossary of Notations

Mathematical notation	

$[\mathbf{A}]$	Boldface upper case letters in roman with square bracket represent matrices
$\mathbf{a}, \mathbf{A}$	Boldface lower or upper case letters in roman represent vectors
$[\mathbf{A}]^T, \mathbf{a}^T$	Transpose of matrices or vectors
$[\mathbf{A}]^{-1}$	Inverse of matrix $[\mathbf{A}]$
$\det[\mathbf{A}]$	Determinant of matrix $[\mathbf{A}]$
$\|\mathbf{a}\|$	l_2-norm of vector $\mathbf{a}$
$\mathcal{R}^K$	K-D real space
$\mathcal{C}^K$	K-D complex space
$\mathcal{Z}^K$	K-D integer space
$\mathcal{F}$	Upper case letters in script represent random variables
$\boldsymbol{\mathcal{F}}$	Boldface upper case letters in script represent random vectors, $\boldsymbol{\mathcal{F}} = [\mathcal{F}_1, \mathcal{F}_2, \ldots, \mathcal{F}_N]^T$, or a random sequence, $\boldsymbol{\mathcal{F}} = \{\mathcal{F}_1, \mathcal{F}_2, \ldots, \}$
$E\{\cdot\}$	Expectation operation
$\forall$	For all
Λ	Representing a lattice
Λ^*	Reciprocal lattice of Λ
$d(\Lambda)$	Density of lattice Λ.
$[\mathbf{V}]$	Generating matrix for Λ
$[\mathbf{U}]$	Generating matrix for Λ^*, $[\mathbf{U}] = ([\mathbf{V}]^T)^{-1}$
$\mathcal{V}(\Lambda)$	Voronoi cell of Λ

Video representation

$\mathbf{X} = [X, Y, Z]^T$	Coordinate of a point in the 3-D space: X, Y, and Z represent the horizontal, vertical, and depth positions, respectively
$\mathbf{x} = [x, y]^T$	Coordinate of a pixel in the 2-D image plane: x and y represent the horizontal and vertical positions, respectively
t	Time index, either continuous or discrete
$\mathbf{m} = [m, n]^T$	Discrete coordinate of a pixel in a 2-D digital image: m and n represent the column and row indices of pixels, respectively
k	Discrete time index (i.e., frame number)
$\psi(x, y, t)$	Image value at pixel location $\mathbf{x}$ and frame time t of a video sequence. The image value could be a scalar representing the luminance or a vector representing three color components. This notation is used to refer to a video signal in general, defined over either a continuous or discrete space
$\psi_1(x, y), \psi_2(x, y)$	Representing the anchor and target frames involved in 2-D motion estimation
$\psi_p(x, y, t)$	The predicted image of frame $\psi(x, y, t)$ in video coding
$\psi_r(x, y)$	The reference frame used to predict a frame
$\psi_v(x, y, t)$	Image function in view v in a multiview system
$\psi(m, n, k)$	Image value at pixel location (m, n) and frame time k in a digital video
$\nabla\psi$	Spatial gradient of $\psi(x, y, t)$, $\nabla\psi = [\frac{\partial\psi}{\partial x}, \frac{\partial\psi}{\partial y}]^T$
$\Delta_x, \Delta_y, \Delta_t$	Sampling interval in horizontal, vertical, and temporal directions, respectively
$f_{s,t}$	Sampling frequency in temporal direction or frame rate, $f_{s,t} = 1/\Delta_t$, measured in frames/s (fps) or Hz
$f_{s,x}, f_{s,y}$	Sampling frequencies in horizontal and vertical directions, $f_{s,x} = 1/\Delta_x$, $f_{s,y} = 1/\Delta_y$, commonly measured in pixels/picture-width and pixels/picture-height
R	Bit rate, specified in bits/s (bps) for a video sequence, bits/pixel (bpp) for an image, or bits/sample for a general discrete source

Frequency-domain representations

f_x, f_y	Horizontal and vertical frequencies, usually measured in cycles/degree (cpd)
f_θ	Angular frequencies, measured in cpd
f_t	Temporal frequency, measured in cycles/s (cps) or Hz
$\mathbf{f}$	Frequency index in a multidimensional space. For video signals, $\mathbf{f} = [f_x, f_y, f_t]^t$
$\Psi_c(f_x, f_y, f_t)$	Continuous-space Fourier transform (CSFT) of $\psi(x, y, t)$
$\Psi_s(f_x, f_y, f_t)$	Sampled-space Fourier transform (SSFT) of $\psi(x, y, t)$
$\Psi_d(f_x, f_y, f_t)$	Discrete-space Fourier transform (DSFT) of $\psi(m, n, k)$
$\Psi(f_x, f_y, f_t)$	Generally used to refer to the Fourier transform of a video signal; can be CSFT, SSFT, or DSFT

Motion/disparity/camera characterization

$\mathbf{D}(\mathbf{X}, t_1, t_2)$	3-D motion vector of point $\mathbf{X}$ from time t_1 to time t_2. When the underlying t_1 and t_2 are clear, we simply write $\mathbf{D}(\mathbf{X})$. The X-, Y-, and Z-components of $\mathbf{D}$ is denoted by D_x, D_y, D_z
$\mathbf{d}(\mathbf{x}, t_1, t_2)$	2-D motion vector of pixel $\mathbf{x}$ from time t_1 to time t_2. When the underlying t_1, t_2 are clear, we simply write $\mathbf{d}(\mathbf{x})$. The $x-$ and $y-$components of $\mathbf{d}$ are denoted by d_x and d_y. $\mathbf{d}(\mathbf{x}; \mathbf{a})$ represents the motion field as a function of motion parameter vector $\mathbf{a}$. The same notation $\mathbf{d}(\mathbf{x})$ is also used to represent the disparity vector between two views in a stereo sequence
$\mathbf{w}(\mathbf{x})$	The mapping function between two image frames, $\mathbf{w}(\mathbf{x}) = \mathbf{x} + \mathbf{d}(\mathbf{x})$. $\mathbf{w}(\mathbf{x}; \mathbf{a})$ represents the mapping function as a function of motion parameters $\mathbf{a}$
$\theta_x, \theta_y, \theta_z$	Camera or object rotation angles with respect to X, Y, and Z axes, respectively, of a predefined world coordinate
$[\mathbf{R}]$	The rotation matrix of a camera or object in 3-D
$\mathbf{T}$	The translation vector of a camera or object in 3-D, $\mathbf{T} = [T_x, T_y, T_z]^T$
F	Camera focal length
$\mathbf{C}$	Camera center in the world coordinate

1

Video Formation, Perception, and Representation

In this first chapter, we describe what a video signal is, how it is captured and perceived, how it is stored and transmitted, and what the important parameters are that determine the quality and bandwidth (which in turn determines the data rate) of the signal. We first present the underlying physics of color perception and specification (Section 1.1). We then describe the principles and typical devices for video capture and display (Section 1.2). As will be seen, analog video is captured, stored, and transmitted in a raster scan format, using either progressive or interlaced scans. As an example, we review the analog color television (TV) system (Section 1.4), and give insights as to how certain critical parameters (such as frame rate and line rate) are chosen, what the spectral content of a color TV signal is, and how different components of a signal can be multiplexed into a composite signal. Finally, Section 1.5 introduces the ITU-R BT.601 video format (formerly CCIR601), the digitized version of the analog color TV signal. We present some of the considerations that have gone into the selection of various digitization parameters. We also describe several other digital video formats, including high-definition TV (HDTV). The compression standards developed for different applications and their associated video formats are summarized.

The purpose of this chapter is to give the reader background knowledge about analog and digital video, and to provide insights into common video system design problems. As such, the presentation is intentionally more qualitative than quantitative. In later chapters, we will return to certain problems mentioned in this chapter and provide more rigorous descriptions and solutions.

1.1 COLOR PERCEPTION AND SPECIFICATION

A video signal is a sequence of two-dimensional (2-D) images, projected from a dynamic three-dimensional (3-D) scene onto the image plane of a video camera. The color value at any point in a video frame records the emitted or reflected light at a particular 3-D point in the observed scene. To understand what the color value means physically, we review basics of light physics in this section, and describe the attributes that characterize light and its color. We also describe the principle of human color perception, and different ways to specify a color signal.

1.1.1 Light and Color

Light consists of an electromagnetic wave, with wavelengths in the range of 380–780 nanometers (nm), to which the human eye is sensitive. The energy of light is measured by *flux,* with a unit of the watt, which is the rate at which energy is emitted. The *radiant intensity* of a light, which is directly related to our perception of the brightness of the light, is defined as the flux radiated into a unit solid angle in a particular direction, measured in watt/solid-angle. A light source usually can emit energy in a range of wavelengths, and its intensity can vary in both space and time. In this text, we use the notation $C(\mathbf{X}, t, \lambda)$ to represent the radiant intensity distribution of a light, which specifies the light intensity at wavelength λ, spatial location $\mathbf{X} = (X, Y, Z)$, and time t.

The perceived color of a light depends on its spectral content (i.e., its wavelength composition). For example, a light that has its energy concentrated near 700 nm appears red. A light that has equal energy across the entire visible band appears white. In general, a light that has a very narrow bandwidth is referred to as a *spectral color.* On the other hand, a white light is said to be *achromatic*.

There are two types of light sources: the *illuminating source,* which emits an electromagnetic wave, and the *reflecting source,* which reflects an incident wave.* Illuminating light sources include the sun, lightbulbs, television monitors, and so on. The perceived color of an illuminating light source depends on the wavelength range in which it emits energy. Illuminating light follows an *additive rule:* the perceived color of several mixed illuminating light sources depends on the sum of the spectra of all the light sources. For example, combining red, green, and blue lights in the right proportions creates the color white.

Reflecting light sources are those that reflect an incident light (which could itself be a reflected light). When a light beam hits an object, the energy in a certain wavelength range is absorbed, while the rest is reflected. The color of a reflected light depends on the spectral content of the incident light and the wavelength range that is absorbed. The most notable reflecting light sources are color dyes and paints. A reflecting light source follows a *subtractive rule:* the perceived color of several mixed reflecting light sources depends on the remaining, unabsorbed wavelengths. For example, if the incident light

---

*Illuminating and reflecting light sources are also referred to as primary and secondary light sources, respectively. We do not use those terms, to avoid confusion with the primary colors associated with light. In other sources, illuminating and reflecting lights are also called *additive colors* and *subtractive colors,* respectively.

is white, a dye that absorbs the wavelength near 700 nm (red) appears as cyan. In this sense, we say that cyan is the *complement* of red (i.e., white minus red). Similarly, magenta and yellow are complements of green and blue, respectively. Mixing cyan, magenta, and yellow dyes produces black, which absorbs the entire visible spectrum.

1.1.2 Human Perception of Color

The perception of light by a human being starts with the photoreceptors located in the retina (the rear surface of the interior of the eyeball). There are two types of receptors: *cones,* which function under bright light and can perceive color tone; and *rods,* which work under low ambient light and can only extract luminance information. Visual information from the retina is passed via optic nerve fibers to the brain area called the visual cortex, where visual processing and understanding is accomplished. There are three types of cones, which have overlapping pass-bands in the visible spectrum with peaks at red (near 570 nm), green (near 535 nm), and blue (near 445 nm) wavelengths, respectively, as shown in Figure 1.1. The responses of these receptors to an incoming light distribution $C(\lambda)$ can be described by the following equation:

$$C_i = \int C(\lambda)a_i(\lambda)\,d\lambda, \quad i = r, g, b, \tag{1.1.1}$$

where $a_r(\lambda), a_g(\lambda), a_b(\lambda)$ are referred to as the frequency responses or relative absorption functions of the red, green, and blue cones. The combination of these three types of receptors enables a human being to perceive any color. This implies that the perceived color depends only on three numbers, C_r, C_g, C_b, rather than the complete light spectrum $C(\lambda)$. This is known as the *tri-receptor theory of color vision,* first advanced by Young [17].

There are two attributes that describe the color sensation of a human being: *luminance* and *chrominance*. Luminance refers to the perceived brightness of the light,

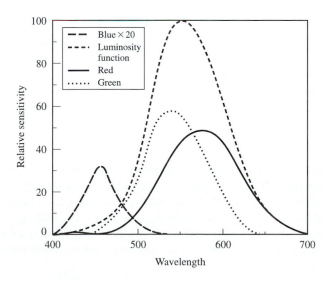

Figure 1.1 Frequency responses of the three types of cones in the human retina and the luminous efficiency function. The blue response curve is magnified by a factor of twenty in the figure. Reprinted from D. H. Pritchard, US color television fundamentals, *IEEE Trans. Consum. Electron.* (1977), CE-23:467–78. Copyright 1977 IEEE.

which is proportional to the total energy in the visible band. Chrominance describes the perceived color tone of a light, which depends on the wavelength composition of the light. Chrominance is in turn characterized by two attributes: *hue* and *saturation*. Hue specifies the color tone, which depends on the peak wavelength of the light, whereas saturation describes how pure the color is, which depends on the spread or bandwidth of the light spectrum. In this book, we use the word "color" to refer to both the luminance and chrominance attributes of a light, although it is customary to use the word color to refer to the chrominance aspect of a light only.

Experiments have shown that there exists a secondary processing stage in the human visual system (HVS), which converts the three color values obtained by the cones into one value that is proportional to the luminance and two other values that are responsible for the perception of chrominance. This is known as the *opponent color model* of the HVS [2, 8]. It has been found that the same amount of energy produces different sensations of brightness at different wavelengths, and this wavelength-dependent variation of brightness sensation is characterized by a *relative luminous efficiency function*, $a_y(\lambda)$, which is also shown in Figure 1.1. It is essentially the sum of the frequency responses of all three types of cones. We can see that the green wavelength contributes the most to the perceived brightness, the red wavelength the second-most, and the blue the least. The luminance (often denoted by Y) is related to the incoming light spectrum by

$$Y = \int C(\lambda) a_y(\lambda) \, d\lambda. \tag{1.1.2}$$

In the preceding equations, we have omitted the time and space variables, since we are only concerned with the perceived color or luminance at a fixed spatial and temporal location. We have also dropped the scaling factor commonly associated with each equation, which depends on the desired unit for describing the color intensities and luminance.

1.1.3 The Trichromatic Theory of Color Mixture

A very important finding in color physics is that most colors can be produced by mixing three properly chosen primary colors. This is known as the *trichromatic theory of color mixture*, first demonstrated by Maxwell in 1855 [8, 16]. Let $C_k, k = 1, 2, 3$, represent the colors of three primary color sources, and C a given color. Then the theory essentially says that

$$C = \sum_{k=1,2,3} T_k C_k, \tag{1.1.3}$$

where the T_k are the amounts of the three primary colors required to match the color C. The T_k are known as *tristimulus values*. In general, some of the T_k can be negative. Assuming that only T_1 is negative, this means that one cannot match color C by mixing C_1, C_2, C_3, but one can match color $C + |T_1|C_1$ with $T_2C_2 + T_3C_3$. In practice, the primary colors should be chosen so that most natural colors can be reproduced using positive combinations of primary colors. The most popular primary set for illuminating light sources contains red, green, and blue colors, known as the *RGB primary*. The most common primary set for reflecting light sources contains cyan, magenta, and yellow, known as the *CMY primary*. In fact, RGB and CMY primary sets are complement of each other, in that mixing two colors in one set will produce one color in the other set. For example, mixing red with green will yield yellow. This complementary information is best illustrated by a color wheel, which can be found in many image processing books, for example, [8, 3].

For a chosen primary set, one way to determine tristimulus values for any color is by first determining the *color matching functions, $m_i(\lambda)$*, for primary colors, C_i, $i = 1, 2, 3$. These functions describe the tristimulus values of a spectral color with wavelength λ, for various λ in the entire visible band, and can be determined by visual experiments with controlled viewing conditions. Then the tristimulus values for any color with a spectrum $C(\lambda)$ can be obtained [8] by:

$$T_i = \int C(\lambda) m_i(\lambda) \, d\lambda, \quad i = 1, 2, 3. \tag{1.1.4}$$

To produce all visible colors with positive mixing, the matching functions associated with the primary colors must be positive.

The preceding theory forms the basis for color capture and display. To record the color of an incoming light, a camera must have three sensors that have frequency responses similar to the color matching functions of a chosen primary set. This can be accomplished by optical or electronic filters with the desired frequency responses. Similarly, to display a color picture, the display device must emit three optical beams of the chosen primary colors with appropriate intensities, as specified by the tristimulus values. In practice, electronic beams that strike phosphors with the red, green, and blue colors are used. All present display systems use an RGB primary, although the standard spectra specified for the primary colors may be slightly different. Likewise, a color printer can produce different colors by mixing three dyes with the chosen primary colors in appropriate proportions. Most color printers use the CMY primary. For a more vivid and wide-range color rendition, some color printers use four primaries, by adding black (K) to the CMY set. This is known as the *CMYK primary,* which can render the color black more truthfully.

1.1.4 Color Specification by Tristimulus Values

Tristimulus Values   The tristimulus representation of a color specifies the proportions, the T_k in Equation (1.1.3), of the three primary colors needed to create the desired color. In order to make the color specification independent of the absolute energy of the primary colors, these values should be normalized so that $T_k = 1$, $k = 1, 2, 3$, for a reference white color (equal energy in all wavelengths) with a unit energy. When we use an RGB primary, the tristimulus values are usually denoted by R, G, and B.

Chromaticity Values   The preceding tristimulus representation mixes the luminance and chrominance attributes of a color. To measure only the chrominance information (the hue and saturation) of a light, the chromaticity coordinate is defined as

$$t_k = \frac{T_k}{T_1 + T_2 + T_3}, \quad k = 1, 2, 3. \tag{1.1.5}$$

Since $t_1 + t_2 + t_3 = 1$, two chromaticity values are sufficient to specify the chrominance of a color.

Obviously, the color value of an imaged point depends on the primary colors used. To standardize color description and specification, several standard primary color systems have been specified. For example, the CIE,* an international body of color scientists, defined a *CIE RGB primary system,* which consists of colors at 700 (R_0), 546.1 (G_0), and 435.8 (B_0) nm.

Color Coordinate Conversion   One can convert the color values based on one set of primaries to the color values for another set of primaries. Conversion of an (R, G, B) coordinate to a (C, M, Y) coordinate, for example, is often required for printing color images stored in the (R, G, B) coordinate. Given the tristimulus representation of one primary set in terms of another primary, we can determine the conversion matrix between the two color coordinates. The principle of color conversion and the derivation of the conversion matrix between two sets of color primaries can be found in [8].

1.1.5 Color Specification by Luminance and Chrominance Attributes

The RGB primary commonly used for color display mixes the luminance and chrominance attributes of a light. In many applications, it is desirable to describe a color in terms of its luminance and chrominance content separately, to enable more efficient processing and transmission of color signals. Towards this goal, various three-component color coordinates have been developed, in which one component reflects the luminance and the other two collectively characterize hue and saturation. One such coordinate is the CIE XYZ primary, in which Y directly measures the luminance intensity. The (X, Y, Z) values in this coordinate are related to the (R, G, B) values in the CIE RGB coordinate [8] by

$$\begin{bmatrix} X \\ Y \\ Z \end{bmatrix} = \begin{bmatrix} 2.365 & -0.515 & 0.005 \\ -0.897 & 1.426 & -0.014 \\ -0.468 & 0.089 & 1.009 \end{bmatrix} \begin{bmatrix} R \\ G \\ B \end{bmatrix}. \tag{1.1.6}$$

In addition to separating the luminance and chrominance information, another advantage of the CIE XYZ system is that almost all visible colors can be specified with nonnegative tristimulus values, which is a very desirable feature. The problem is that the X, Y, Z colors so defined are not realizable by actual color stimuli. As such, the XYZ primary is not directly used for color production; rather it is mainly introduced

*CIE stands for Commission Internationale de L'Eclariage (International Commission on Illumination).

for the purpose of defining other primaries and for numerical specification of color. As will be seen, the color coordinates used for transmission of color TV signals, such as YIQ and YUV, are all derived from the XYZ coordinate.

There are other color representations in which the hue and saturation of a color are explicitly specified, in addition to the luminance. One example is the HSI coordinate, where "H" stands for hue, "S" for saturation, and "I" for intensity (equivalent to luminance).* Although this color coordinate clearly separates different attributes of a light, it is nonlinearly related to the tristimulus values and is difficult to compute. Gonzalez and Woods [3] present a comprehensive coverage of various color coordinates and their conversions.

1.2 VIDEO CAPTURE AND DISPLAY

1.2.1 Principles of Color Video Imaging

Having defined what light is and how it is perceived and characterized, we are now in a position to understand the meaning of a video signal. In short, a video[†] records the emitted and/or reflected light intensity, $C(\mathbf{X}, t, \lambda)$, from the objects in a scene that is observed by a viewing system (a human eye or a camera). In general, this intensity changes both in time and space. Here, we assume that there are some illuminating light sources in the scene. Otherwise, there will be neither injected nor reflected light and the image will be totally dark. When the scene is observed by a camera, only those wavelengths to which the camera is sensitive will be visible. Let the *spectral absorption function* of the camera be denoted by $a_c(\lambda)$; then the light intensity distribution in the 3-D world that is "visible" to the camera is:

$$\bar{\psi}(\mathbf{X}, t) = \int_0^\infty C(\mathbf{X}, t, \lambda) a_c(\lambda) \, d\lambda. \tag{1.2.1}$$

The image function captured by the camera at any time t is the projection of the light distribution in the 3-D scene onto a 2-D image plane. Let $\mathcal{P}(\cdot)$ represent the camera projection operator, so that the projected 2-D position of the 3-D point $\mathbf{X}$ is given by $\mathbf{x} = \mathcal{P}(\mathbf{X})$. Furthermore, let $\mathcal{P}^{-1}(\cdot)$ denote the inverse projection operator, so that $\mathbf{X} = \mathcal{P}^{-1}(\mathbf{x})$ specifies the 3-D position associated with a 2-D point $\mathbf{x}$. Then the projected image is related to the 3-D image by

$$\psi(\mathcal{P}(\mathbf{X}), t) = \bar{\psi}(\mathbf{X}, t) \quad \text{or} \quad \psi(\mathbf{x}, t) = \bar{\psi}(\mathcal{P}^{-1}(\mathbf{x}), t). \tag{1.2.2}$$

The function $\psi(\mathbf{x}, t)$ is what is known as a video signal. We can see that it describes the radiant intensity at the 3-D position $\mathbf{X}$ that is projected onto $\mathbf{x}$ in the image plane at time t. In general, the video signal has a finite spatial and temporal range. The spatial

---

*The HSI coordinate is also known as HSV, where "V" stands for the value of the intensity.

†Throughout the text, when we refer to "a video" without further qualification, we are referring to a recorded sequence of moving images.

range depends on the camera viewing area, while the temporal range depends on the duration for which the scene is captured. A point in the image plane is called a *pixel* (meaning picture element) or simply *pel*.* For most camera systems, the projection operator $\mathcal{P}(\cdot)$ can be approximated by a perspective projection. (This is discussed in more detail in Section 5.1.)

If the camera absorption function is the same as the relative luminous efficiency function of a human being, $a_c(\lambda) = a_y(\lambda)$, then a luminance image is formed. If the absorption function is nonzero over a narrow band, then a monochrome (or monotone) image is formed. To perceive all visible colors, according to the trichromatic color vision theory (see Section 1.1.2), three sensors are needed, each with a frequency response similar to the color matching function for a selected primary color. As described, most color cameras use red, green, and blue sensors for color acquisition.

If the camera has only one luminance sensor, $\psi(\mathbf{x}, t)$ is a scalar function that represents the luminance of the projected light. In this book, we use the term *grey scale* to refer to such an image. The term *black and white* will be used strictly to describe an image that has only two colors: black and white. On the other hand, if the camera has three separate sensors, each tuned to a chosen primary color, the signal is a vector function that contains three color values at every point. Instead of specifying these color values directly, one can use other color coordinates (each consisting of three values) to characterize light, as explained in the previous section.

Note that for special purposes, one may use sensors that work in a frequency range that is invisible to the human being. For example, in X-ray imaging, the sensor is sensitive to the spectral range of the X ray. On the other hand, an infrared camera is sensitive to the infrared range, which can function at very low ambient light. These cameras can "see" things that cannot be perceived by the human eye. Yet another example is the range camera, in which the sensor emits a laser beam and measures the time it takes for the beam to reach an object and then be reflected back to the sensor. Because the round trip time is proportional to the distance between the sensor and the object surface, the image intensity at any point in a range image describes the distance or range of its corresponding 3-D point from the camera.

1.2.2 Video Cameras

All the analog cameras of today capture video in a frame-by-frame manner, with a certain time spacing between the frames. Some cameras (such as TV cameras and consumer video camcorders) acquire a frame by scanning consecutive lines with a certain line spacing. Similarly, all the display devices present a video as a consecutive set of frames; and with TV monitors, the scan lines are played back sequentially as separate lines. Such capture and display mechanisms are designed to take advantage of the fact that the HVS cannot perceive very high frequency changes in time and space. (This property of the HVS will be discussed more extensively in Section 2.4.)

There are basically two types of video cameras: (1) tube-based cameras such as vidicons, plumbicons, or orthicons; and (2) solid-state sensors such as charge-coupled

---

*Strictly speaking, the notion of pixel or pel is only defined in digital imagery, in which each image or frame in a video recording is represented by a finite 2-D array of pixels.

devices (CCDs). The lens of a camera focuses the image of a scene onto a photosensitive surface of the imager of the camera, which converts optical signals into electrical signals. The photosensitive surface of a tube imager is typically scanned line by line (known as a raster scan) with an electron beam or by other electronic methods, and the scanned lines in each frame are then converted into an electrical signal that represents variations of light intensity as variations in voltage. Different lines are therefore captured at slightly different times in a continuous manner. With progressive scan, the electronic beam scans every line continuously; whereas with interlaced scan, the beam scans every other line in one half of the frame time (a field) and then scans the other half of the lines. (We discuss raster scan in more detail in Section 1.3.) With a CCD camera, the photosensitive surface comprises a 2-D array of sensors, each corresponding to one pixel, and the optical signal reaching each sensor is converted to an electronic signal. The sensor values captured in each frame time are first stored in a buffer, and are then read out sequentially, one line at a time, to form a raster signal. Unlike the tube-based cameras, all the read-out values in the same frame are captured at the same time. With an interlaced scan camera, alternate lines are read out in each field.

To capture color, there are usually three types of photosensitive surfaces or CCD sensors, each with a frequency response that is determined by the color matching function of the chosen primary color, as described in Section 1.1.3. To reduce the cost, most consumer cameras use a single CCD chip for color imaging. This is accomplished by dividing the sensor area for each pixel into three or four sub-areas, each sensitive to a different primary color. The three captured color signals can either be converted to one luminance signal and two chrominance signals and sent out as a component color image, or multiplexed into a composite signal. This subject is explained further in Section 1.2.4.

Many cameras of today are CCD-based because they can be made much smaller and lighter than the tube-based cameras, to acquire the same spatial resolution. Advancement in CCD technology has made it possible to capture in a very small chip a very high resolution image array. For example, $\frac{1}{3}$-in. CCDs with 380 K pixels are commonly found in consumer-use camcorders, whereas a $\frac{2}{3}$-in. CCD with two million pixels has been developed for HDTV. The tube-based cameras are more bulky and costly, and are only used in special applications, such as those requiring very high resolution or high sensitivity under low ambient light. In addition to the circuitry for color imaging, most cameras also implement color coordinate conversion (from RGB to luminance and chrominance) and compositing of luminance and chrominance signals. For digital output, analog-to-digital (A/D) conversion is also incorporated. Figure 1.2 shows the typical processing involved in a professional video camera. The camera provides outputs in both digital and analog form, and in the analog case, includes both component (CN) and composite (CS) formats. To improve the image quality, digital processing is introduced within the camera. The A/D for each of the three CCDs and preprocessing (including point and area processing, data detection and data correction) are performed at the CCD output rate, $f_{s,1}$. These are followed by image enhancement and nonlinear filtering at twice the CCD output rate. The rate conversion of $2f_{s,1}$ to $f_{s,2}$ is required to match the standard camera output rate $f_{s,2}$ ($f_{s,2} = 13.5$ MHz for the ITU-R BT.601

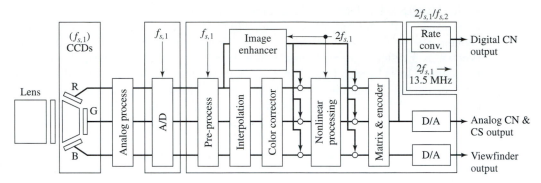

Figure 1.2 Schematic block diagram of a professional color video camera. Reprinted from Y. Hashimoto, M. Yamamoto, and T. Asaida, Cameras and display systems, *IEEE* (July 1995), 83(7):1032–43. Copyright 1995 IEEE.

digital video discussed in Section 1.5.2). For an excellent exposition of video camera and display technologies, see [5].

1.2.3 Video Display

To display a video, the most common device is the cathode ray tube (CRT). In a CRT monitor, an electron gun emits an electron beam across the screen, line by line, exciting phosphors with intensities proportional to the intensity of the video signal at corresponding locations. To display a color image, three beams are emitted by three separate guns, exciting red, green, and blue phosphors with the desired intensity combination at each location. To be more precise, each color pixel consists of three elements arranged in a small triangle, known as a triad.

The CRT can produce an image having a very large dynamic range, so that the displayed image can be very bright, sufficient for viewing during daylight or from a distance. However, the depth of a CRT needs to be about the same as the width of the screen, for the electrons to reach the side of the screen. A large screen monitor is thus too bulky, unsuitable for applications requiring thin and portable devices. To circumvent this problem, various flat-panel displays have been developed. One popular device is the liquid crystal display (LCD). The principle idea behind the LCD is to change the optical properties and consequently the brightness or color of the liquid crystal by an applied electric field. The electric field can be generated or adapted by an array of transistors, such as in LCDs using active-matrix thin-film transistors (TFTs), or by using plasma. Plasma technology eliminates the need for TFTs and makes large-screen LCDs possible. There are also new designs for flat CRTs. A more comprehensive description of video display technologies can be found in [5].

The raster scan and display mechanisms just described apply only to video cameras and displays. With motion picture cameras, the color pattern seen by the camera at any frame instant is completely recorded on the film. For display, consecutive recorded frames are played back using an analog optical projection system.

1.2.4 Composite versus Component Video

Ideally, a color video should be specified by three functions or signals, each describing one color component, in either a tristimulus color representation, or a luminance-chrominance representation. A video in this format is known as *component video*. Mainly for historical reasons, various *composite video* formats also exist, wherein the three color signals are multiplexed into a single signal. These composite formats were invented when the color TV system was first developed, and there was a need to transmit the color TV signal in such a way that a black-and-white TV set could extract from it the luminance component. The construction of a composite signal relies on the property that the chrominance signals have a significantly smaller bandwidth than the luminance component. By modulating each chrominance component to a frequency that is at the high end of the luminance component, and adding the resulting modulated chrominance signals to the original luminance signal, one creates a composite signal that contains both luminance and chrominance information. To display a composite video signal on a color monitor, a filter is used to separate the modulated chrominance signals from the luminance signal. The resulting luminance and chrominance components are then converted to red, green, and blue color components. With a grey-scale monitor, the luminance signal alone is extracted and displayed directly.

All present analog TV systems transmit color TV signals in a composite format. The composite format is also used for video storage on some analog tapes (such as VHS). In addition to being compatible with a grey-scale signal, the composite format eliminates the need for synchronizing different color components when processing a color video. A composite signal also has a bandwidth that is significantly lower than the sum of the bandwidths of three component signals, and therefore can be transmitted or stored more efficiently. These benefits are achieved, however, at the expense of image quality: there often exist noticeable artifacts caused by cross talk between color and luminance components.

As a compromise between data rate and image quality, *S-video* was invented, which consists of two components: the luminance component, and a single chrominance component that is the multiplex of two original chrominance signals. Many advanced consumer-level video cameras and displays enable recording or display of video in S-video format. Component format is used only in professional video equipment.

1.2.5 Gamma Correction

We have said that the video frames captured by a camera reflect the color values of the imaged scene. In reality, the output signals from most cameras are not linearly related to the actual color values, rather the relation takes a nonlinear form:*

$$v_c = B_c^{-\gamma_c}, \tag{1.2.3}$$

---

*A more precise relation is $B_c = cv_c^{\gamma_c} + B_0$, where c is a gain factor, and B_0 is the cutoff level of light intensity. If we assume that the output voltage value is properly shifted and scaled, then the presented equation is valid.

where B_c represents the actual light brightness, and v_c the camera output voltage. The value of γ_c ranges from 1.0 for most CCD cameras to 1.7 for a vidicon camera [6]. Similarly, most of the display devices also suffer from such a nonlinear relation between the input voltage value v_d and the displayed color intensity B_d, namely,

$$B_d = v_d^{\gamma_d}. \tag{1.2.4}$$

The CRT displays typically have a γ_d of 2.2–2.5 [6]. In order to present true colors, one must apply an inverse power function on the camera output. Similarly, before sending real image values for display, one must precompensate the "gamma effect" of the display device. These processes are known as *gamma correction*.

In TV broadcasting, ideally, at the TV broadcaster side, the RGB values captured by the TV cameras should first be corrected based on the camera gamma value, and then converted to the color coordinates used for transmission (YIQ for NTSC, and YUV for PAL and SECAM). At the receiver side, the received YIQ or YUV values should first be converted to the RGB values, and then compensated for the monitor gamma values. In reality, however, in order to reduce the processing to be done in the millions of receivers, the broadcast video signals are pre-gamma corrected in the RGB domain. Let v_c represent the R, G, or B signal captured by the camera; the gamma-corrected signal for display, v_d, is obtained by

$$v_d = v_c^{\gamma_c/\gamma_d}. \tag{1.2.5}$$

In most TV systems, a ratio of $\gamma_c/\gamma_d = 2.2$ is used. This is based on the assumption that a CCD camera with $\gamma_c = 1$ and a CRT display with $\gamma_d = 2.2$ are used [6]. These gamma-corrected values are converted to the YIQ or YUV values for transmission. The receiver simply applies a color coordinate conversion to obtain the RGB values for display. Notice that this process applies display gamma correction before the conversion to the YIQ/YUV domain, which is not strictly correct. But the distortion is insignificant and not noticeable by the average viewer [6].

1.3 ANALOG VIDEO RASTER

As we have described, the analog TV systems of today use raster scan for video capture and display. As this is the most popular analog video format, in this section we describe the mechanism of raster scan in more detail, including both progressive and interlaced scan. As an example, we also explain the video formats used in various analog TV systems.

1.3.1 Progressive and Interlaced Scan

Progressive Scan   In a raster scan, a camera captures a video sequence by sampling it in both temporal and vertical directions. The resulting signal is stored in a continuous one dimensional (1-D) waveform. As shown in Figure 1.3(a), the electronic or optic beam of an analog video camera continuously scans the imaged region from the top to bottom and then back to the top. The resulting signal consists of a series of frames

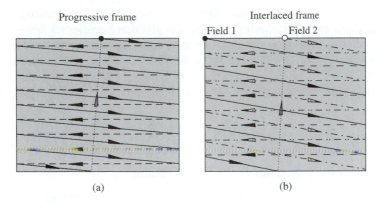

Figure 1.3   Progressive (a) and interlaced (b) raster scan formats.

separated by a regular frame interval, Δ_t, and each frame consists of a consecutive set of horizontal scan lines, separated by a regular vertical spacing. Each scan line is actually slightly tilted. Also, the bottom line is scanned about one frame interval later than the top line of the same frame. However, for analysis purposes, we often assume that all the lines in a frame are sampled at the same time, and each line is perfectly horizontal. The intensity values captured along contiguous scan lines over consecutive frames form a 1-D analog waveform, known as a *raster scan*. With a color camera, three 1-D rasters are converted into a composite signal, which is a color raster.

Interlaced Scan   The raster scan format just described is more accurately known as *progressive scan* (also known as sequential or noninterlaced scan), in which the horizontal lines are scanned successively. In the *interlaced scan,* each frame is scanned in two fields and each field contains half the number of lines in a frame. The time interval between two fields, that is, the field interval, is half of the frame interval, while the line spacing in a field is twice that desired for a frame. The scan lines in two successive fields are shifted by half of the line spacing in each field, as illustrated in Figure 1.3(b). Following the terminology used in the MPEG standard, we call the field containing the first line and following alternating lines in a frame the *top field,* and the field containing the second line and following alternating lines the *bottom field.** In certain systems, the top field is sampled first, while in other systems, the bottom field is sampled first. It is important to remember that two adjacent lines in a frame are separated in time by the field interval. This fact leads to the infamous zig-zag artifacts in interlaced video images that contain fast-moving objects with vertical edges. The motivation for using the interlaced scan is to trade off the vertical resolution for an enhanced temporal resolution, given the total number of lines that can be recorded

---

*A more conventional definition is to call the field that contains all even lines the even field, and the field containing all odd lines the odd field. This definition depends on whether the first line is numbered 0 or 1, and is therefore ambiguous.

within a given time. A more thorough comparison of progressive and interlaced scans in terms of their sampling efficiency is given in Section 3.3.2.

The interlaced scan just introduced should more precisely be called 2:1 interlace. In general, one can divide a frame into $K \geq 2$ fields, each separated in time by Δ_t / K. This is known as K:1 interlace and K is called the *interlace order*. In a digital video in which each line is represented by discrete samples, the samples on the same line may also appear in different fields. For example, the samples in a frame may be divided into two fields using a checker-board pattern. The most general definition of the interlace order is the ratio of the number of samples in a frame to the number of samples in each field.

1.3.2 Characterization of a Video Raster

A raster is described by two basic parameters: the *frame rate* (frames/second or fps or Hz), denoted by $f_{s,t}$, and the *line number* (lines/frame or lines/picture-height), denoted by $f_{s,y}$. These two parameters define the temporal and vertical sampling rates of a raster scan. From these parameters, one can derive another important parameter, the *line rate* (lines/second), denoted by $f_l = f_{s,t} f_{s,y}$.* We can also derive the temporal sampling interval or frame interval, $\Delta_t = 1/f_{s,t}$, the vertical sampling interval or line spacing, $\Delta_y = $ picture-height$/f_{s,y}$, and the line interval, $T_l = 1/f_l = \Delta_t/f_{s,y}$, which is the time used to scan one line. Note that the line interval T_l includes the time for the sensor to move from the end of a line to the beginning of the next line, which is known as the *horizontal retrace time* or just *horizontal retrace,* to be denoted by T_h. The actual scanning time for a line is $T_l' = T_l - T_h$. Similarly, the frame interval Δ_t includes the time for the sensor to move from the end of the bottom line in a frame to the beginning of the top line of the next frame, which is called *vertical retrace time* or just *vertical retrace,* to be denoted by T_v. The number of lines that is actually scanned in a frame time, known as the *active lines,* is $f_{s,y}' = (\Delta_t - T_v)/T_l = f_{s,y} - T_v/T_l$. Normally, T_v is chosen to be an integer multiple of T_l.

A typical waveform of an interlaced raster signal is shown in Figure 1.4(a). Notice that portions of the signal during the horizontal and vertical retrace periods are held at a constant level above the level corresponding to black. These are called *sync signals.* The display devices start the retrace process upon detecting these sync signals.

Figure 1.4(b) shows the spectrum of a typical raster signal. It can be seen that the spectrum contains peaks at the line rate f_l and its harmonics. This is because adjacent scan lines are very similar, so that the signal is nearly periodic with a period of T_l. The width of each harmonic lobe is determined by the maximum vertical frequency in a frame. The overall bandwidth of the signal is determined by the maximum horizontal spatial frequency.

The frame rate is one of the most important of the parameters that determine the quality of a video raster. For example, the TV industry uses an interlaced scan with a frame rate of 25–30 Hz, with an effective temporal refresh rate of 50–60 Hz, while the

---

*The frame rate and line rate are also known as the vertical sweep frequency and the horizontal sweep frequency, respectively.

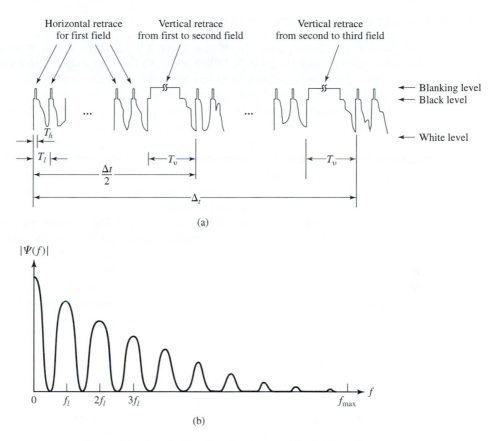

Figure 1.4   A typical interlaced raster scan: (a) waveform, (b) spectrum.

motion picture industry uses a frame rate of 24 Hz.* On the other hand, in the computer industry, 72 Hz has become a de facto standard. The line number used in a raster scan is also a key factor affecting the video quality. In analog TVs, a line number of about 500–600 is used, while for computer display, a much higher line number is used (e.g., the SVGA display has 1024 lines). These frame rates and line numbers are determined based on the visual temporal and spatial thresholds in different viewing environments, as described in Section 2.4. Higher frame rates and line numbers are necessary in computer applications to accommodate a significantly shorter viewing distance and higher frequency contents (line graphics and texts) in the displayed material.

The width-to-height ratio of a video frame is known as the *image aspect ratio* (IAR). For example, an IAR of 4:3 is used in standard-definition TV (SDTV) and computer display, while a higher IAR is used in wide-screen movies (up to 2.2) and HDTVs (IAR = 16:9) for a more dramatic visual sensation.

---

*To reduce the visibility of flicker, a rotating blade is used to create an illusion of 72 fps.

1.4 ANALOG COLOR TELEVISION SYSTEMS

In this section, we briefly describe analog color TV systems, which provide good examples of many concepts we have talked about so far. One major constraint in designing a color TV system is that it must be compatible with the previous monochrome TV system. First, the overall bandwidth of a color TV signal has to fit within that allocated for a monochrome TV signal (6 MHz per channel in the United States). Second, all the color signals must be multiplexed into a single composite signal in such a way that a monochrome TV receiver can extract from it the luminance signal. The successful design of color TV systems that satisfy the above constraints is one of the great technological innovations of the twentieth century. Figure 1.5 illustrates the main processing steps involved in color TV signal production, transmission, and reception. We briefly review each of the steps in the following.

There are three different systems worldwide: the NTSC system used in North America as well as some parts of Asia, including Japan and Taiwan; the PAL system used in most of Western Europe and Asia including China, and the Middle East; and the SECAM system used in the former Soviet Union, Eastern Europe, France, as well as the Middle East. We will compare these systems in terms of their spatial and temporal resolution, color coordinate, and multiplexing mechanism. The materials presented here are mainly from [8, 9]. More complete coverage on color TV systems can be found in [4, 15].

1.4.1 Spatial and Temporal Resolution

All three color TV systems use the 2:1 interlaced scan mechanism described in Section 1.3 for capturing as well as displaying video images. The NTSC system uses a field rate of 59.94 Hz, and a line number of 525 lines/frame. The PAL and SECAM systems both use a field rate of 50 Hz and a line number of 625 lines/frame. These frame rates are chosen not to interfere with the standard electric power systems in the involved

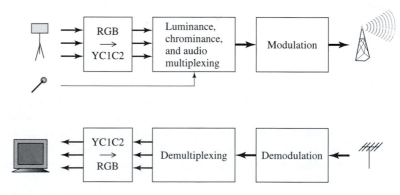

Figure 1.5   Analog color TV systems: video production, transmission, and reception.

TABLE 1.1   PARAMETERS OF ANALOG COLOR TV SYSTEMS

Parameters	NTSC	PAL	SECAM
Field rate	59.94	50	50
Line number/frame	525	625	625
Line rate (line/s)	15,750	15,625	15,625
Image aspect ratio	4:3	4:3	4:3
Color coordinate	YIQ	YUV	YDbDr
Luminance bandwidth (MHz)	4.2	5.0, 5.5	6.0
Chrominance bandwidth (MHz)	1.5 (I), 0.5 (Q)	1.3 (U, V)	1.0 (U, V)
Color subcarrier (MHz)	3.58	4.43	4.25 (Db), 4.41 (Dr)
Color modulation	QAM	QAM	FM
Audio subcarrier (MHz)	4.5	5.5, 6.0	6.5
Composite signal bandwidth (MHz)	6.0	8.0, 8.5	8.0

countries. They also turned out to be good choices, in that they match with the critical flicker fusion frequency of the human visual system (as described in Section 2.4). All systems have an IAR of 4:3. The parameters of the NTSC, PAL, and SECAM video signals are summarized in Table 1.1. For NTSC, the line interval is $T_l = 1/(30 * 525) = 63.5\,\mu s$. But the horizontal retrace takes $T_h = 10\,\mu s$, so that the actual time for scanning each line is $T_l' = 53.5\,\mu s$. The vertical retrace between adjacent fields takes $T_v = 1333\,\mu s$, which is equivalent to the time for 21 scan lines per field. Therefore, the number of active lines is $525 - 42 = 483$/frame. The actual vertical retrace only takes the time to scan nine horizontal lines. The remaining time (twelve scan lines) is for broadcasters wishing to transmit additional data in the TV signal (e.g., closed caption, teletext, etc.).*

1.4.2 Color Coordinate

The color coordinate systems used in the three TV systems are different. For video capture and display, all three systems use an RGB primary, but with slightly different definitions of the spectra of individual primary colors. For transmission of the video signal, in order to reduce the bandwidth requirement and to be compatible with monochrome TV systems, a luminance/chrominance coordinate is employed. In the following, we describe the color coordinates used in these systems.

The color coordinates used in the NTSC, PAL, and SECAM systems are all derived from the YUV coordinate used in PAL, which in turn originates from the XYZ coordinate. Based on the relation between the RGB primary and the XYZ primary, one can determine the Y value from the RGB value, which forms the luminance component. The two chrominance values, U and V, are proportional to color differences, $B - Y$ and $R - Y$, respectively, scaled to have the desired range. Specifically, the YUV coordinate

---

*The number of active lines cited in different references varies from 480 to 495. This number is calculated from the vertical blanking interval cited in [4].

is related to the PAL RGB primary values [8] by:

$$
\begin{bmatrix} Y \\ U \\ V \end{bmatrix} = \begin{bmatrix} 0.299 & 0.587 & 0.114 \\ -0.147 & -0.289 & 0.436 \\ 0.615 & -0.515 & -0.100 \end{bmatrix} \begin{bmatrix} \tilde{R} \\ \tilde{G} \\ \tilde{B} \end{bmatrix} \tag{1.4.1}
$$

and

$$
\begin{bmatrix} \tilde{R} \\ \tilde{G} \\ \tilde{B} \end{bmatrix} = \begin{bmatrix} 1.000 & 0.000 & 1.140 \\ 1.000 & -0.395 & -0.581 \\ 1.000 & 2.032 & 0.001 \end{bmatrix} \begin{bmatrix} Y \\ U \\ V \end{bmatrix}, \tag{1.4.2}
$$

where $\tilde{R}$, $\tilde{G}$, $\tilde{B}$ are normalized gamma-corrected values, so that $(\tilde{R}, \tilde{G}, \tilde{B}) = (1, 1, 1)$ corresponds to the reference white color defined in the PAL/SECAM system.

The NTSC system uses the YIQ coordinate, where the I and Q components are rotated versions (by 33°) of the U and V components. This rotation serves to make I correspond to colors in the orange-to-cyan range, and Q the green-to-purple range. Because the human eye is less sensitive to changes in the green-to-purple range than in the yellow-to-cyan range, the Q component can be transmitted with less bandwidth than the I component [9]. (This point will be elaborated further in Section 1.4.3.) The YIQ values are related to the NTSC RGB system by:

$$
\begin{bmatrix} Y \\ I \\ Q \end{bmatrix} = \begin{bmatrix} 0.299 & 0.587 & 0.114 \\ 0.596 & -0.275 & -0.321 \\ 0.212 & -0.523 & 0.311 \end{bmatrix} \begin{bmatrix} \tilde{R} \\ \tilde{G} \\ \tilde{B} \end{bmatrix} \tag{1.4.3}
$$

and

$$
\begin{bmatrix} \tilde{R} \\ \tilde{G} \\ \tilde{B} \end{bmatrix} = \begin{bmatrix} 1.000 & 0.956 & 0.620 \\ 1.000 & -0.272 & -0.647 \\ 1.000 & -1.108 & 1.700 \end{bmatrix} \begin{bmatrix} Y \\ I \\ Q \end{bmatrix} \tag{1.4.4}
$$

With the YIQ coordinate, $\tan^{-1}(Q/I)$ approximates the hue, and $\sqrt{I^2 + Q^2}/Y$ reflects the saturation. In an NTSC composite video, the I and Q components are multiplexed into one signal, so that the phase of the modulated signal is $\tan^{-1}(Q/I)$, whereas the magnitude equals $\sqrt{I^2 + Q^2}/Y$. Because transmission errors affect the magnitude more than the phase, the hue information is better retained than the saturation in a broadcast TV signal. This is desired, as the human eye is more sensitive to the color hue. The names "I" and "Q" come from the fact that the I signal is in-phase with the color modulation frequency, whereas the Q signal is in quadrature (i.e., one quarter of the way around the circle or 90 degrees out of phase) with the modulation frequency. (The color multiplexing scheme is explained in Section 1.4.4.)

Note that because the RGB primary set and the reference white color used in the NTSC system are different from those in the PAL/SECAM system, the same set of RGB values corresponds to slightly different colors in these two systems.

The SECAM system uses the YDbDr coordinate, where the Db and Dr values are related to the U and V values [6] by

$$D_b = 3.059U, \quad D_r = -2.169V. \tag{1.4.5}$$

1.4.3 Signal Bandwidth

The bandwidth of a video raster can be estimated from its line rate. First of all, the maximum vertical frequency results when white and black lines alternate in a raster frame, which is equal to $f'_{s,y}/2$ cycles/picture-height, where $f'_{s,y}$ represents the number of active lines. The maximum frequency that can be rendered properly by a system is usually lower than this theoretical limit. The attenuation factor is known as the *Kell factor,* denoted by K, which depends on the camera and display aperture functions. Typical TV cameras have a Kell factor of $K = 0.7$. The maximum vertical frequency that can be accommodated is related to the Kell factor by

$$f_{v,\max} = Kf'_{s,y}/2 \text{ (cycles/picture-height)}. \tag{1.4.6}$$

If we assume that the maximum horizontal frequency is identical to the vertical frequency for the same spatial distance, then $f_{h,\max} = f_{v,\max} \cdot \text{IAR}$ (cycles/picture-width). Because each line is scanned in T'_l seconds, the maximum frequency in the 1-D raster signal is

$$f_{\max} = f_{h,\max}/T'_l = \text{IAR} \cdot Kf'_{s,y}/2T'_l \text{ Hz}. \tag{1.4.7}$$

For the NTSC video format, we have $f'_{s,y} = 483$, $T'_l = 53.5\,\mu s$. Consequently, the maximum frequency of the luminance component is 4.2 megacycles/second or 4.2 MHz. Although the potential bandwidth of the chrominance signal could be just as high, usually it is significantly lower than the luminance signal. Furthermore, the HVS has been found to have a much lower threshold for observing changes in chrominance. Typically, the two chrominance signals are bandlimited to have much narrower bandwidth. As mentioned, the human eye is more sensitive to spatial variations in the orange-to-cyan color range, represented by the I component, than in the green-to-purple range, represented by the Q component. Therefore, the I component is bandlimited to about 1.5 MHz, and the Q component to about 0.5 MHz.* Table 1.1 lists the signal bandwidths of different TV systems.

1.4.4 Multiplexing of Luminance, Chrominance, and Audio

In order to make the color TV signal compatible with the monochrome TV system, all three analog TV systems use the composite video format, in which the three color components as well as the audio component are multiplexed into one signal. Here, we briefly describe the mechanism used by NTSC. First, the two chrominance components $I(t)$ and $Q(t)$ are combined into a single signal $C(t)$, using quadrature amplitude

---

*The actual numbers depend on the definition of bandwidth, varying from 1.3–1.6 MHz for the I component, from 0.5–0.6 MHz for the Q component. The numbers cited here are from [15].

modulation (QAM). The *color subcarrier frequency* f_c is chosen to be an odd multiple of half of the line rate, $f_c = 455\ f_l/2 = 3.58$ MHz. This is chosen to satisfy the following criteria: (1) It should be high enough where the luminance component has very low energy; (2) It should be midway between two line rate harmonics where the luminance component is strong; and (3) It should be sufficiently far away from the audio subcarrier, which is set at 4.5 MHz (286 f_l), the same as in the monochrome TV signal.

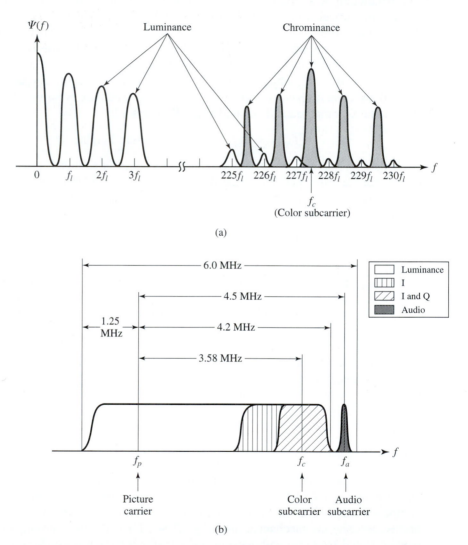

(a)

(b)

Figure 1.6 Multiplexing of luminance, chrominance, and audio signals in NTSC system: (a) interleaving between luminance and chrominance harmonics; (b) overall spectral composition of the NTSC composite signal.

Figure 1.6(a) shows how the harmonic peaks of the luminance and chrominance signals interleave with each other. Finally, the audio signal is frequency modulated (FM) using an *audio subcarrier frequency* of $f_a = 4.5$ MHz and added to the composite video signal, to form the final multiplexed signal. Because the I component has a bandwidth of 1.5 MHz, the modulated chrominance signal has a maximum frequency of up to 5.08 MHz. In order to avoid interference with the audio signal, the I signal is bandlimited in the upper sideband to 0.5 MHz. Notice that the lower sideband of the I signal will run into the upper part of the Y signal. For this reason, sometimes the I signal is bandlimited to 0.5 MHz on both sidebands. As mentioned before, the Q component is bandlimited to 0.5 MHz on both sides. Finally, the entire composite signal, with a bandwidth of about 4.75 MHz, is modulated onto a *picture carrier frequency*, f_p, using *vestigial sideband modulation* (VSB), so that the lower sideband only extends to 1.25 MHz below f_p and the overall signal occupies 6 MHz. This process is the same as in the monochrome TV system. The picture carrier f_p depends on the broadcasting channel. Figure 1.6(b) illustrates the spectral composition of the NTSC composite signal. The signal bandwidths and modulation methods in the three color TV systems are summarized in Table 1.1.

At a television receiver, the composite signal first must be demodulated to the baseband, and then the audio and three components of the video signals must be demultiplexed. To separate the video and audio signals, a low-pass filter can be used. This process is the same in a monochrome and a color TV. To further separate the chrominance signal from the luminance signal, ideally, a comb filter should be used, to take advantage of the interleaving of the harmonic frequencies in these two signals. Most high-end TV sets implement a digital comb filter with null frequencies at the harmonics corresponding to the chrominance component to accomplish this. Low-end TV sets, however, use a simple RC circuit to perform low-pass filtering with a cut-off frequency at 3 MHz, which would retain the unwanted I component in the extracted luminance signal, and vice versa. This will lead to cross-color and cross-luminance artifacts. *Cross-color* refers to the spurious colors created by the high-frequency luminance signal that is close to the color subcarrier frequency. *Cross-luminance* refers to false high-frequency edge patterns caused by the modulated chrominance information. (For a good illustration of the effects of different filters, see [1].) After extracting the chrominance signal, a corresponding color-demodulation method is used to separate the two chrominance components. Finally, the three color components are converted to the RGB coordinate for display.

1.4.5 Analog Video Recording

Along with the development of analog TV systems, various video tape recording (VTR) technologies have been developed, to allow professional video production (recording and editing) as well as for consumer recording (home video) and playback (VCR). Table 1.2 summarizes common videotape formats.

TABLE 1.2   ANALOG VIDEOTAPE FORMATS

Video format	Tape format	Horizontal lines	Luminance bandwidth	Applications
composite	VHS, 8mm	240	3.0 MHz	Consumer
	Umatic SP	330	4.0 MHz	Professional
S-video	S-VHS, Hi8	400	5.0 MHz	High-quality consumer
component	Betacam SP	480	4.5 MHz	Professional

1.5 DIGITAL VIDEO

1.5.1 Notation

A digital video can be obtained either by sampling a raster scan, or directly using a digital video camera. Presently, all digital cameras use CCD sensors. As with analog cameras, a digital camera samples the imaged scene as discrete frames. Each frame consists of output values from a CCD array, which is by nature discrete both horizontal and vertically. A digital video, is defined by the frame rate, $f_{s,t}$, the line number, $f_{s,y}$, as well as the number of samples per line, $f_{s,x}$. From these, one can find the temporal sampling interval or frame interval, $\Delta_t = 1/f_{s,t}$, the vertical sampling interval $\Delta_y =$ picture-height$/f_{s,y}$, and the horizontal sampling interval $\Delta_x =$ picture-width$/f_{s,x}$. In this book, we use $\psi(m, n, k)$ to represent a digital video, where integer indices m and n are the column and row indices, and k the frame number. The actual spatial and temporal locations corresponding to the integer indices are $x = m\Delta_x$, $y = n\Delta_y$, and $t = k\Delta_t$. For convenience, we use the notation $\psi(x, y, t)$ to describe a video signal in a general context, which could be either analog or digital. We will use $\psi(m, n, k)$ only when specifically addressing digital video.

In addition to the preceding parameters, another important parameter of digital video is the number of bits used to represent a pixel value (luminance only, or three color values), to be denoted by N_b. Conventionally, the luminance and/or each of the three color values is specified with eight bits or 256 levels. Therefore, $N_b = 8$ for monochrome video, whereas $N_b = 24$ for color video. The data rate, R, of digital video is determined by $R = f_{s,t} f_{s,x} f_{s,y} N_b$, with a unit of bits/second (or bps—usually measured in kilobits/second (kbps) or megabits/second (mbps)). In general, the spatial and temporal sampling rates can be different for the luminance and chrominance components of a digital video signal. In this case, N_b should reflect the equivalent number of bits used for each pixel in the luminance sampling resolution. For example, if the horizontal and vertical sampling rates for each chrominance component are both half of that for the luminance, then there are two chrominance samples for every four Y samples. If each sample is represented with eight bits, the equivalent number of bits per sample in the Y resolution is $(4*8 + 2*8)/4 = 12$ bits.

When displaying digital video on a monitor, each pixel is rendered as a rectangular region with a constant color that is specified for this pixel. The ratio of the width to the height of this rectangular area is known as the *pixel aspect ratio* (PAR). It is related to the IAR of the display area and the image dimension by

$$\text{PAR} = \text{IAR} \cdot f_{s,y}/f_{s,x}. \tag{1.5.1}$$

For proper display of a digitized video signal, one must specify either PAR or IAR, along with $f_{s,x}$ and $f_{s,y}$. The display device should conform to the PAR specified for this signal (or derived from the specified IAR). Otherwise, the object shape will be distorted. For example, the image of a person will become fatter and shorter, if the display PAR is larger than the specified PAR. In the computer industry, a PAR of 1.0 is normally used. On the other hand, in the TV industry, nonsquare pixels are used because of historical reasons; the rationale behind this is explained in Section 1.5.2.

1.5.2 ITU-R BT.601 Digital Video

Spatial Resolution of the BT.601 Signal   In an attempt to standardize the digital formats used to represent different analog TV video signals with a quality equivalent to broadcast TV, the International Telecommunications Union—Radio Sector (ITU-R) developed the BT.601 recommendation [35]. The standard specifies digital video formats with both 4:3 and 16:9 aspect ratios; we discuss only the version with aspect ratio 4:3.* To convert a raster scan to a digital video signal, one need only sample the 1-D waveform. If a total number of $f_{s,x}$ samples are taken per line, the equivalent sampling rate is $f_s = f_{s,x} f_{s,y} f_{s,t} = f_{s,x} f_l$ samples/second. In the BT.601 standard, the sampling rate f_s is chosen to satisfy two constraints: (1) the horizontal sampling resolution should match the vertical sampling resolution as closely as possible, that is, let $\Delta_x \approx \Delta_y$; and (2) the same sampling rate should be used for the NTSC and PAL/SECAM systems, and it should be a multiple of respective line rates in these systems. The first criterion calls for $f_{s,x} \approx \text{IAR} \cdot f_{s,y}$, or $f_s \approx \text{IAR} \cdot f_{s,y}^2 f_{s,t}$, which leads to $f_s \approx 11$ and 13 MHz for the NTSC and PAL/SECAM systems, respectively. A number that is closest to both numbers and yet satisfies the second criterion is then chosen, which is

$$f_s = 858 f_l \text{ (NTSC)} = 864 f_l(\text{PAL/SECAM}) = 13.5 \text{ MHz}. \tag{1.5.2}$$

The numbers of pixels per line are $f_{s,x} = 858$ for NTSC and 864 for PAL/SECAM. These two formats are known as 525/60 and 625/50 signals, respectively, and are illustrated in Figure 1.7. The numbers of active lines are, respectively, $f'_{s,y} = 480$ and 576 in the 525- and 625-line systems, but the numbers of active pixels/line are the same, both equal to $f'_{s,x} = 720$ pixels. The rest are samples obtained during the horizontal and vertical retraces, which fall in the nonactive area.

With the BT.601 signal, the pixel width-to-height ratio is not 1, that is, the physical area associated with a pixel is not a square. Specifically, $\text{PAR} = \Delta_x/\Delta_y = \text{IAR} \cdot f'_{s,y}/f'_{s,x} = 8/9$ for 525/60 and 16/15 for 625/50 signals. To display a BT.601 signal, the display device must have a proper PAR, otherwise the image will be distorted. For example, when displayed on a computer screen that has a PAR of 1, a 525/60

---

*The ITU-R was formerly known as the International Radio Consultative Committee (CCIR), and the 4:3 aspect ratio version of the BT.601 format was called the CCIR601 format.

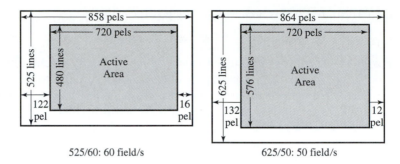

525/60: 60 field/s   625/50: 50 field/s

Figure 1.7   BT.601 video formats.

signal will appear stretched horizontally, whereas a 625/50 signal will appear stretched vertically. Ideally, one should resample the original signal so that $f'_{s,x} = \text{IAR} \cdot f'_{s,y}$. For example, the 525/60 and 625/50 signals should be resampled to have 640 and 768 active pixels/line, respectively.

    Color Coordinate and Chrominance Subsampling   Along with the image resolution, the BT.601 recommendation also defines a digital color coordinate, known as YCbCr. The Y, Cb, and Cr components are scaled and shifted versions of the analog Y, U, and V components, where the scaling and shifting operations are introduced so that the resulting components take values in the range of (0–255). For a more detailed explanation on the design of this color coordinate, readers are referred to [8]; here we present only the transformation matrix for deriving this coordinate from the digital RGB coordinate. Assuming that the RGB values are in the range of (0–255), the YCbCr values are related to the RGB values by:

$$\begin{bmatrix} Y \\ C_b \\ C_r \end{bmatrix} = \begin{bmatrix} 0.257 & 0.504 & 0.098 \\ -0.148 & -0.291 & 0.439 \\ 0.439 & -0.368 & -0.071 \end{bmatrix} \begin{bmatrix} R \\ G \\ B \end{bmatrix} + \begin{bmatrix} 16 \\ 128 \\ 128 \end{bmatrix}. \tag{1.5.3}$$

The inverse relation is:

$$\begin{bmatrix} R \\ G \\ B \end{bmatrix} = \begin{bmatrix} 1.164 & 0.000 & 1.596 \\ 1.164 & -0.392 & -0.813 \\ 1.164 & 2.017 & 0.000 \end{bmatrix} \begin{bmatrix} Y - 16 \\ C_b - 128 \\ C_r - 128 \end{bmatrix}. \tag{1.5.4}$$

    In the preceding relations, $R = 255\tilde{R}, G = 255\tilde{G}, B = 255\tilde{B}$ are the digital equivalent of the normalized RGB primaries, $\tilde{R}, \tilde{G},$ and $\tilde{B}$, as defined either in the NTSC or PAL/SECAM system. In the YCbCr coordinate, Y reflects the luminance and is scaled to have a range of (16–235); C_b and C_r are scaled versions of color differences $B - Y$ and $R - Y$, respectively. The scaling and shifting is designed so that they have a range of (16–240). The maximum value of C_r corresponds to red ($C_r = 240$ or $R = 255, G = B = 0$), whereas the minimum value yields cyan ($C_r = 16$ or $R = 0, G = B = 255$). Similarly, the maximum and minimum values of C_b correspond to blue ($C_b = 240$ or $R = G = 0, B = 255$) and yellow ($C_b = 16$ or $R = G = 255, B = 0$).

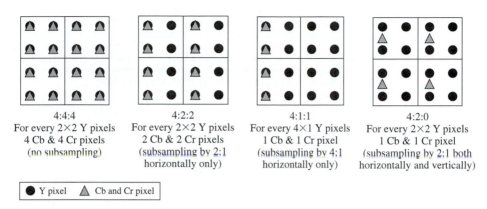

4:4:4
For every 2×2 Y pixels
4 Cb & 4 Cr pixels
(no subsampling)

4:2:2
For every 2×2 Y pixels
2 Cb & 2 Cr pixels
(subsampling by 2:1
horizontally only)

4:1:1
For every 4×1 Y pixels
1 Cb & 1 Cr pixel
(subsampling by 4:1
horizontally only)

4:2:0
For every 2×2 Y pixels
1 Cb & 1 Cr pixel
(subsampling by 2:1 both
horizontally and vertically)

● Y pixel   ▲ Cb and Cr pixel

Figure 1.8 BT.601 chrominance subsampling formats. Note that the two adjacent lines in any one component belong to two different fields.

The spatial sampling rate introduced previously refers to the luminance component, Y. For the chrominance components, C_b and C_r, usually only half of the sampling rate is used, that is, $f_{s,c} = f_s/2$. This leads to half the number of pixels in each line, but the same number of lines per frame. This is known as the 4:2:2 format, implying that there are two Cb samples and two Cr samples for every four Y samples. To further reduce the required data rate, BT.601 also defined the 4:1:1 format, in which the chrominance components are subsampled along each line by a factor of four, that is, there is one Cb sample and one Cr sample for every four Y samples. This sampling method, however, yields very asymmetric resolutions in the horizontal and vertical directions. Another sampling format has therefore been developed, which subsamples the Cb and Cr components by half in both the horizontal and vertical directions. In this format, there is also one Cb sample and one Cr sample for every four Y samples. But to avoid confusion with the previously defined 4:1:1 format, this format is designated as 4:2:0. For applications requiring very high resolutions, the 4:4:4 format is defined, which samples the chrominance components in exactly the same resolution as the luminance components. The relative positions of the luminance and chrominance samples for different formats are shown in Figure 1.8.*

In Chapter 4, we will discuss solutions for converting video signals with different spatial/temporal resolutions. The conversion between different color subsampling formats is considered in one of the exercise problems in that chapter.

The raw data rate of a BT.601 signal depends on the color subsampling factor. With the most common 4:2:2 format, there are two chrominance samples per two Y samples, each represented with 8 bits. Therefore, the equivalent bit rate for each Y sample is $N_b = 16$ bits, and the raw data rate is $f_s N_b = 216$ mbps. The raw data rate corresponding to the active area is $f_{s,t} f'_{s,y} f'_{s,x} N_b = 166$ mbps. With the 4:2:0 format, there are two chrominance samples per four Y samples, and the equivalent bit rate for

---

*For the 4:2:0 format, the Cr and Cb samples may also be positioned in the center of the four corresponding Y samples, as shown in Figure 13.14(a).

TABLE 1.3   DIGITAL VIDEO FORMATS FOR DIFFERENT APPLICATIONS

Video format	Y size	Color sampling	Frame rate	Raw data (mbps)
HDTV over air, cable, satellite, MPEG-2 video 20–45 mbps				
SMPTE 296M	1280×720	4:2:0	24P/30P/60P	265/332/664
SMPTE 295M	1920×1080	4:2:0	24P/30P/60I	597/746/746
Video production, MPEG-2, 15–50 mbps				
BT.601	$720 \times 480/576$	4:4:4	60I/50I	249
BT.601	$720 \times 480/576$	4:2:2	60I/50I	166
High-quality video distribution (DVD, SDTV), MPEG-2, 4–8 mbps				
BT.601	$720 \times 480/576$	4:2:0	60I/50I	124
Intermediate-quality video distribution (VCD, WWW), MPEG-1, 1.5 mbps				
SIF	$352 \times 240/288$	4:2:0	30P/25P	30
Videoconferencing over ISDN/Internet, H.261/H.263, 128–384 kbps				
CIF	352×288	4:2:0	30P	37
Video telephony over wired/wireless modem, H.263, 20–64 kbps				
QCIF	176×144	4:2:0	30P	9.1

each Y sample is $N_b = 12$ bits. Therefore the raw data rate is 162 mbps, with 124 mbps in the active area. For the 4:4:4 format, the equivalent bit rate for each Y sample is $N_b = 24$ bits, and the raw data rate is 324 mbps, with 249 mbps in the active area. The resolutions and data rates of different BT.601 signals are summarized in Table 1.3.

The BT.601 formats are used in high-quality digital video applications, with the 4:4:4 and 4:2:2 formats typically used for video production and editing, and the 4:2:0 for video distribution (e.g., movies on digital video disks (DVD), video-on-demand (VOD), etc.). The MPEG-2* video compression standard was primarily developed for compression of BT.601 4:2:0 signals, although it can also handle video signals in lower or higher resolutions. A typical 4:2:0 signal with a raw active data rate of 124 mbps can be compressed to about 4–8 mbps. We will introduce the MPEG-2 video coding algorithm in Section 13.5.

1.5.3 Other Digital Video Formats and Applications

In addition to the BT.601 format, several other standard digital video formats have been defined. Table 1.3 summarizes these video formats, along with their main applications, compression methods, and compressed bit rates. The CIF (Common Intermediate Format), specified by the International Telecommunications Union—Telecommunications Sector (ITU-T), has about half the resolution of BT.601 4:2:0 in both horizontal and

---

*"MPEG" stands for Motion Picture Expert Group of the International Standards Organization or ISO.

vertical dimensions and was developed for video conferencing applications; the QCIF, which has half the resolution of CIF in both horizontal and vertical dimensions, is used for videophone and similar applications. Both are noninterlaced. The ITU-T H.261 coding standard was developed to compress video signals in either of these formats to $p \times 64$ kbps, with $p = 1, 2, \ldots, 30$, for transport over ISDN (integrated service digital network) lines, which only allow transmission rates in multiples of 64 kbps. Typically, a CIF signal with a raw data rate of 37.3 mbps can be compressed to about 128–384 kbps, with reasonable quality, while a QCIF signal with a raw data rate of 9.3 mbps can be compressed to 64–128 kbps. A later standard, H.263, can achieve better quality than H.261, at the same bit rate. For example, it is possible to compress a QCIF image to about 20 kbps, while maintaining a quality similar to or better than H.261 at 64 kbps. This enables videophone communications over a 28.8 kbps modem line.

In parallel with the effort of ITU-T, the ISO-MPEG also defined a series of digital video formats. The SIF (Source Intermediate Format) is essentially a quarter the size of the active area in the BT.601 signal, and is about the same as CIF. This format is targeted for video applications requiring medium quality, such as video games and CD movies. As with BT.601, there are two SIF formats: one with a frame rate of 30 Hz and a line number of 240, and another with a frame rate of 25 and line number of 288; both have 352 pixels/line. There is also a corresponding set of SIF-I formats, which are 2:1 interlaced. The MPEG-1 algorithm can compress a typical SIF video signal with a raw data rate of 30 mbps to about 1.1 mbps, with a quality similar to the resolution seen on a VHS VCR, which is lower than that of broadcast television. The rate of 1.1 mbps enables the playback of digital movies on CD-ROMs, which have an access rate of 1.5 mbps. Distribution of MPEG-1 movies on video CDs (VCD) marked the entrance of digital video into the consumer market in the early 1990s. MPEG-2–based DVDs, which started in the mid-'90s, opened the era of high-quality digital video entertainment. MPEG-2 technology is also the cornerstone of the next generation of TV systems, which will be fully digital, employing digital compression and transmission technology. Table 1.3 lists the details of the video formats discussed, along with their main applications, compression methods, and compressed bit rates. (More on compression standards will be presented in Chapter 13.)

The BT.601 format is the standard picture format for digital TV (DTV). To further enhance video quality, several HDTV formats have also been standardized by the Society of Motion Picture and Television Engineers (SMPTE), which are also listed in Table 1.3. A distinct feature of HDTV is its wider aspect ratio, 16:9 as opposed to 4:3 in SDTV. The picture resolution is doubled to tripled in both horizontal and vertical dimensions. Furthermore, progressive scan is used to reduce interlacing artifacts. A high profile has been developed in the MPEG-2 video compression standard for compressing HDTV video. Typically it can reduce the data rate to about 20 mbps, while retaining the very high quality required. This video bit rate is chosen so that the combined bit stream with audio, when transmitted using digital modulation techniques, can still fit into a 6 MHz terrestrial channel, which is the assigned channel bandwidth for HDTV broadcast in the United States. For more information on DTV including HDTV, the reader is referred to Whitaker [14].

TABLE 1.4   DIGITAL VIDEO TAPE FORMATS

Tape format	Video format	Source rate	Compressed rate	Compression method	Intended application
Uncompressed formats					
SMPTE D1	BT.601 4:2:2	216 mbps	N/A	N/A	Professional
SMPTE D2	BT.601 composite	114 mbps	N/A	N/A	Professional
SMPTE D3	BT.601 composite	114 mbps	N/A	N/A	Professional/ consumer
SMPTE D5	BT.601 4:2:2 (10 bit)	270 mbps	N/A	N/A	Professional
Compressed formats					
Digital Betacam	BT.601 4:2:2	166 mbps	80 mbps	Frame DCT	Professional
Betacam SX	BT.601 4:2:2	166 mbps	18 mbps	MPEG-2 (I and B mode only)	Consumer
DVCPRO50	BT.601 4:2:2	166 mbps	50 mbps	frame/field DCT	Professional
DVCPRO25 (DV)	BT.601 4:1:1	124 mbps	25 mbps	frame/field DCT	Consumer

1.5.4 Digital Video Recording

To store video in digital formats, various digital video tape recorder (DVTR) formats have been developed, which differ in video format handled and technology for error-correction-coding and storage density. Table 1.4 lists some standard and proprietary tape formats. The D1–D5 formats store video in its raw, uncompressed formats, whereas others precompress the video. Only a conservative amount of compression is employed, so as not to degrade the video quality beyond that acceptable for the intended application. A good review of DVTRs can be found in [12]. Extensive coverage of the underlying physics of magnetic recording and operation of DVTRs can be found in [13].

In addition to magnetic tape recorders, VCD and DVD are two video storage devices using optical disks. By incorporating MPEG-1 and MPEG-2 compression technologies, they can store SIF and BT.601 videos, respectively, with sufficient quality. At present, VCD and DVD are read-only, so that they are mainly used for distribution of prerecorded video, as opposed to tools for recording video by consumers.

Hard-disk–based video recording systems, such as [11] and [10], are also on the horizon. These systems enable consumers to record up to 30 hours of live TV programs onto hard disks in MPEG-2 compressed formats, which can be viewed later with the usual VCR features such as fast-forward, slow motion, and so on. They also allow instant pause of a live program that is being watched, by storing the live video from the time of the pause onto the disk. As the price for hard disks drops continuously, hard-disk–based DVTRs may eventually overtake tape-based systems, which are slower and have less storage capacity.

1.5.5 Video Quality Measure

To conduct video processing, it is necessary to define an objective criterion that can measure the difference between the original and the processed signal. This is especially important, for example, in video coding applications where one must measure the distortion caused by compression. Ideally, such a measure should correlate well with the perceived difference between two video sequences. Finding such a measure, however, turns out to be an extremely difficult task. Although various quality measures have been proposed, those that correlate well with visual perception are quite complicated to compute. Most video processing systems of today are designed to minimize the *mean squared error* (MSE) between two video sequences ψ_1 and ψ_2, which is defined as

$$\text{MSE} = \sigma_e^2 = \frac{1}{N} \sum_k \sum_{m,n} (\psi_1(m, n, k) - \psi_2(m, n, k))^2, \qquad (1.5.5)$$

where N is the total number of pixels in either sequence. For a color video, the MSE is computed separately for each color component.

Instead of the MSE, the *peak signal-to-noise ratio* (PSNR) in decibel (dB) is more often used as a quality measure in video coding. The PSNR is defined as

$$\text{PSNR} = 10 \log_{10} \frac{\psi_{\max}^2}{\sigma_e^2} \qquad (1.5.6)$$

where $\psi_{\max}$ is the peak (maximum) intensity value of the video signal. For the most common eight bit/color video, $\psi_{\max} = 255$. Note that for a fixed peak value, the PSNR is completely determined by the MSE. The PSNR is more commonly used than the MSE because people tend to associate the quality of an image with a certain range of PSNR. As a rule of thumb, for the luminance component, a PSNR higher than 40 dB typically indicates an excellent image (i.e., being very close to the original), between 30–40 dB usually means a good image (i.e., the distortion is visible but acceptable); between 20–30 dB is quite poor; and, finally, a PSNR lower than 20 dB is unacceptable.

It is worth noting that to compute the PSNR between two sequences, it is *incorrect* to calculate the PSNR between every two corresponding frames and then take the average of the values obtained over individual frames. Rather, one should compute the MSE between corresponding frames, average the resulting MSE values over all frames, and finally convert the MSE value to PSNR.

A measure that is sometimes used in place of the MSE, mainly for reduced computation, is the *mean absolute difference* (MAD), defined as

$$\text{MAD} = \frac{1}{N} \sum_k \sum_{m,n} |\psi_1(m, n, k) - \psi_2(m, n, k)|. \qquad (1.5.7)$$

For example, for motion estimation, the MAD is usually used to find the best matching block in another frame for a given block in a current frame.

It is well known that the MSE and PSNR do not correlate very well with perceptual distortion between images. But these measures have been used almost exclusively as objective distortion measures in image and video coding, motion-compensated prediction, and image restoration, partly because of their mathematical tractability, and partly because of the lack of better alternatives. Designing objective distortion measures that are easy to compute and yet correlate well with perceptual distortion is still an open research issue. In this book, we will mostly use MSE or PSNR as the distortion measure.

1.6 SUMMARY

Color Generation, Perception, and Specification (Section 1.1)

- The color of a light depends on its spectral content. Any color can be created by mixing three primary colors. The most common primary set includes red, green, and blue colors.
- The human eye perceives color by means of receptors (cones) in the retina that are tuned to red, green, and blue wavelengths. Color sentation can be described by three attributes: luminance (brightness), hue (color tone), and saturation (color purity). The human eye is most sensitive to luminance, then to hue, and finally to saturation.
- A color can be specified by three numbers: either those corresponding to the contributions of the three primary colors (i.e., tristimulus values), or a luminance value and two chrominance values.

Analog Video (Section 1.3)

- Analog video signals used in broadcast TV, video camcorders, video display, and so on, are stored in a raster scan format. The visual quality and bandwidth of a raster scan depends on its frame rate and line number.
- Interlaced scan is a mechanism to trade off vertical resolution for enhanced temporal resolution. It also leads to interlacing artifacts, however.

Analog Color TV Systems (Section 1.4)

- There are three analog color TV systems in use worldwide: NTSC, PAL, and SECAM. They all use 2:1 interlace, but differ in frame rate, line number, color coordinate, and luminance and chrominance multiplexing.
- In color TV systems, the luminance and two chrominance components, as well as the associated audio signal, are multiplexed into a composite signal, using modulation (frequency shifting) techniques. The multiplexing methods are designed so that the color TV system is downward compatible with the monochrome TV system. Furthermore, the modulation frequencies for individual components are chosen to minimize the interference among them.

Digital Video (Section 1.5)

- BT.601 is a digital video format, resulting from sampling analog color TV signals. The sampling rate is chosen so that the horizontal and vertical sampling rates are similar, and the data rates for NTSC and PAL/SECAM systems are the same.

- The chrominance components can be sampled at a lower rate than the luminance component. There are different color subsampling formats defined in BT.601.
- Compression is necessary to reduce the raw data rate of a digital video signal, and thus reduce the storage and transmission cost. Different video compression standards have been developed for video intended for different applications.

1.7 PROBLEMS

1.1 Describe the mechanism by which human beings perceive color.

1.2 What is the perceived color if you have a light that has approximately the same energy at frequencies corresponding to red, green, and blue, and zero energy at other frequencies? What about red and green frequencies only?

1.3 What is the perceived color if you mix red, green, and blue dyes in equal proportion? What about red and green dyes only?

1.4 For the following colors in the RGB coordinate, determine their values in the YIQ and YUV coordinates, respectively.
(a) (1, 1, 1)   (b) (0, 1, 0)   (c) (1, 1, 0)   (d) (0, 1, 1)

1.5 For the following colors in the digital RGB coordinate, determine their values in the YCbCr coordinate.
(a) (255, 255, 255)   (b) (0, 255, 0)   (c) (255, 255, 0)   (d) (0, 255, 255)

1.6 In Section 1.5.2, we say that the maximum value of C_r corresponds to red, whereas the minimum value yields cyan. Similarly, the maximum and minimum values of C_b correspond to blue and yellow, respectively. Verify these statements using the YCbCr-to-RGB coordinate transformation.

1.7 In Figure 1.4, we show the spectrum of a typical raster signal. Why is the spectrum of the video signal nearly periodic? What does the width of harmonic lobes depend on?

1.8 What are the pros and cons of progressive versus interlaced scans? For the same line number per frame, what is the relation between the maximum temporal frequency that a progressive raster can have and that of an interlaced raster that divides each frame into two fields? What about the relation between the maximum vertical frequencies?

1.9 In Section 1.4.3, we estimate the bandwidth of the NTSC signal based on its scan parameters. Following the same approach, estimate the bandwidth of the PAL and SECAM signals.

1.10 Describe the process for forming a composite color video signal. How should you select the color and audio subcarrier frequencies?

1.11 What are the pros and cons of using component versus composite formats?

1.12 Project: Use an oscilloscope to (a) draw the waveform, and (b) measure the spectrum of a composite video signal output from a TV set or a camcorder.

1.13 Project: Digitize a composite video signal using an A/D converter, and use MATLAB to determine the spectrum. Also, perform filtering to separate the luminance, chrominance, and audio signals.

1.8 BIBLIOGRAPHY

[1] Blinn, J. F. NTSC: nice technology, super color. *IEEE Computer Graphics and Applications Magazine* (March 1993):17–23.

[2] Boynton, R. M. *Human Color Vision*. New York: Holt, Rinhart, and Winston, 1979.

[3] Gonzalez, R. G., and R. E. Woods. *Digital Image Processing*. 2nd ed. Reading, MA: Addison–Wesley, 1992.

[4] Grob, B., and C. E. Herndon. *Basic Television and Video Systems*. 6th ed. New York: McGraw Hill, 1999.

[5] Hashimoto, Y., M. Yamamoto, and T. Asaida. Cameras and display systems. *IEEE* (July 1995), 83(7):1032–43.

[6] Haskell, B. G., A. Puri, and A. N. Netravali. *Digital Video: An Introduction to MPEG-2*. New York: Chapman & Hall, 1997.

[7] ITU-R. BT.601-5: Studio encoding parameters of digital television for standard 4:3 and wide-screen 16:9 aspect ratios, 1998. (Formerly CCIR601).

[8] Netravali, A. N., and B. G. Haskell. *Digital Pictures—Representation, Compression and Standards*. 2nd ed. New York: Plenum Press, 1995.

[9] Pritchard, D. H. US color television fundamentals. *IEEE Trans. Consum. Electron.* (1977), CE-23(4):467–78.

[10] ReplayTV, Inc.: http://www.replaytv.com

[11] TiVo, Inc.: http://www.tivo.com

[12] Umemoto, M., Y. Eto, and T. Fukinuki. Digital video recording. *IEEE* (July 1995), 83(7):1044–54.

[13] Watkinson, J. *The Art of Digital Video*. 2nd ed. Oxford: Focal Press, 1994.

[14] Whitaker, J. C. *DTV Handbook: The Revolution in Digital Video*. 3rd ed. New York: McGraw Hill, 2001.

[15] Whitaker J. C. and K. B. Benson. ed. *Standard Handbook of Video and Television Engineering*. 3rd ed. New York: McGraw-Hill, 2000.

[16] Wyszecki, G., and W. S. Stiles. *Color Science*. New York: John Wiley, 1967.

[17] Young, T. On the theory of light and colors. *Philosophical Transactions of the Royal Society of London* (1802), 92:20–71.

2

Fourier Analysis of Video Signals and Frequency Response of the Human Visual System

Fourier analysis is an important tool for signal analysis. We assume that the reader is familiar with Fourier transforms for one- and two-dimensional spaces as well as signal processing tools using such transforms. In this chapter, we first extend these results to K dimensions (K-D), where K can be any positive integer. We then focus on their applications for video signals, which are three-dimensional. We will explore the meaning of spatial and temporal frequencies, and their interrelationship. Finally, we discuss visual sensitivity to different frequency components.

2.1 MULTIDIMENSIONAL CONTINUOUS-SPACE SIGNALS AND SYSTEMS

Most of the theorems and techniques for multidimensional signals and systems are direct extensions of those developed for 1-D and 2-D signals and systems. In this section, we introduce some important concepts and theorems for signal analysis in the K-D real space, $\mathcal{R}^K = \{[x_1, x_2, \ldots, x_K]^T | x_k \in \mathcal{R}, k \in \mathcal{K}\}$, where $\mathcal{R}$ is the set of real numbers, and $\mathcal{K} = \{1, 2, \ldots, K\}$. We start by defining K-D signals, common operations between K-D signals, and special K-D signals. We then define the Fourier transform representation of K-D signals. Finally, we define K-D systems and properties of the linear and shift-invariant systems. This presentation is intentionally kept brief. We also intentionally leave out discussion of the convergence conditions of various integral formulations. For a more substantial treatment of the subject, the reader is referred to [2].

Definition 2.1. A K-D continuous space signal $\psi(\mathbf{x})$ is a function of a K-D continuous variable $\mathbf{x} = [x_1, x_2, \ldots, x_K] \in \mathcal{R}^K$. The function can take on any values, real or complex, and is called a real function if it takes only real values.

Signals can be combined according to vector space operations of addition and scaling. Another important operation between K-D functions is convolution, as defined in the following.

Definition 2.2.   The convolution of two signals, $\psi(\mathbf{x})$ and $h(\mathbf{x})$, both defined over $\mathcal{R}^K$, is defined as

$$\psi(\mathbf{x}) * h(\mathbf{x}) = \int_{\mathcal{R}^K} \psi(\mathbf{x} - \mathbf{y}) h(\mathbf{y}) \, d\mathbf{y}. \tag{2.1.1}$$

Among several special functions, the delta function (also known as the impulse function) plays an important role in characterizing signals and linear systems.

Definition 2.3.   The delta function in $\mathcal{R}^K$ satisfies:

$$\delta(\mathbf{x}) = \begin{cases} \infty, & \mathbf{x} = \mathbf{0}, \\ 0, & \text{otherwise}, \end{cases} \quad \text{and} \quad \int_{\mathcal{R}^K} \delta(\mathbf{x}) \, d\mathbf{x} = 1. \tag{2.1.2}$$

Properties of the Delta Function.   The following are several important identities associated with the delta function, where $\psi(\cdot)$ is an arbitrary K-D function:

$$\psi(\mathbf{x}) * \delta(\mathbf{x} - \mathbf{x}_0) = \psi(\mathbf{x} - \mathbf{x}_0), \tag{2.1.3}$$

$$\int_{\mathcal{R}^K} \psi(\mathbf{x})\delta(\mathbf{x} - \mathbf{x}_0) \, d\mathbf{x} = \psi(\mathbf{x}_0), \tag{2.1.4}$$

$$\int_{\mathcal{R}^K} \exp\left(j 2\pi \mathbf{f}_0^T \mathbf{x}\right) d\mathbf{x} = \delta(\mathbf{f}_0). \tag{2.1.5}$$

Definition 2.4 (Continuous-Space Fourier Transform).   The continuous-space Fourier transform (CSFT)* of a signal $\psi(\mathbf{x})$ is defined as,[†]

$$\Psi_c(\mathbf{f}) = \int_{\mathcal{R}^K} \psi(\mathbf{x}) \exp(-j 2\pi \mathbf{f}^T \mathbf{x}) \, d\mathbf{x}, \tag{2.1.6}$$

where $\mathbf{f} = [f_1, f_2, \ldots, f_K]^T \in \mathcal{R}^K$ represents the continuous domain frequency variable.

Theorem 2.1 (Inverse Continuous-Space Fourier Transform).   A function can be obtained from its CSFT by

$$\psi(\mathbf{x}) = \int_{\mathcal{R}^K} \Psi_c(\mathbf{f}) \exp(j 2\pi \mathbf{f}^T \mathbf{x}) \, d\mathbf{f}. \tag{2.1.7}$$

∎

---

*In the 1-D case, the CSFT is usually called the *continuous time Fourier transform* or simply Fourier transform, because the 1-D signal is typically a function of time.

[†]In this book, we define CSFT in terms of the frequency variable $\mathbf{f}$, rather than the radian frequency $\Omega = 2\pi\mathbf{f}$. With our definition, the normalization by 2π is not needed, as each transform basis function in terms of $\mathbf{f}$ has a unit norm.

To prove the preceding theorem, substituting Equation (2.1.6) into the right side of Equation (2.1.7), we obtain

$$\int_{\mathcal{R}^K} \Psi_c(\mathbf{f}) \exp(j2\pi\mathbf{f}^T\mathbf{x}) \, d\mathbf{f} = \int_{\mathcal{R}^K} \psi(\mathbf{y}) \int_{\mathcal{R}^K} \exp(j2\pi\mathbf{f}^T(\mathbf{x}-\mathbf{y})) \, d\mathbf{f} \, d\mathbf{y}$$

$$= \int_{\mathcal{R}^K} \psi(\mathbf{y})\delta(\mathbf{x}-\mathbf{y}) \, d\mathbf{y} = \psi(\mathbf{x}).$$

The second equality comes from Equation (2.1.5), and the third follows from Equation (2.1.4).

The inverse CSFT shows that any signal $\psi(\mathbf{x})$ can be expressed as a linear combination of complex exponential functions with different frequencies. The CSFT at a particular frequency represents the contribution of the corresponding complex exponential basis function.*

Most of the properties of the 1-D Fourier transform can be extended to the K-D case, including linearity, translation, convolution, and so on. Here we list only the convolution theorem, which is very important for signal analysis.

Theorem 2.2 (Convolution Theorem).   The convolution of two K-D signals in the spatial domain is equivalent to the product of the two signals in the frequency domain; that is,

$$\phi(\mathbf{x}) = \psi(\mathbf{x}) * h(\mathbf{x}) \longleftrightarrow \Phi_c(\mathbf{f}) = \Psi_c(\mathbf{f}) H_c(\mathbf{f}). \qquad (2.1.8)$$

Conversely, the product of two signals in the spatial domain corresponds to the convolution in the Fourier domain; that is,

$$\phi(\mathbf{x}) = \psi(\mathbf{x})h(\mathbf{x}) \longleftrightarrow \Phi_c(\mathbf{f}) = \Psi_c(\mathbf{f}) * H_c(\mathbf{f}). \qquad (2.1.9)$$

■

Definition 2.5.   A system with K-D input and output signals is in general described by:

$$\phi(\mathbf{x}) = T\{\psi(\mathbf{x})\}, \quad \mathbf{x} \in \mathcal{R}^K. \qquad (2.1.10)$$

Let $\phi_1(\mathbf{x})$ and $\phi_2(\mathbf{x})$ represent the output signals of a system corresponding to input signals $\psi_1(\mathbf{x})$ and $\psi_2(\mathbf{x})$, respectively. The system is linear if for any $\alpha_1, \alpha_2 \in \mathcal{R}$,

$$T\{\alpha_1\psi_1(\mathbf{x}) + \alpha_2\psi_2(\mathbf{x})\} = \alpha_1\phi_1(\mathbf{x}) + \alpha_2\phi_2(\mathbf{x}). \qquad (2.1.11)$$

---

*The Fourier integral defined in Equation (2.1.6) does not converge for all functions in $\mathcal{R}^K$. But if we allow the Fourier transform to contain delta functions, then Fourier analysis can be applied to a wide variety of functions. In this book, we choose not to discuss the various convergence conditions of the Fourier transform for both the continuous- and discrete-time cases. More rigorous definitions of Fourier transforms can be found in [9, 2].

Furthermore, the system is shift-invariant if

$$T\{\psi(\mathbf{x} + \mathbf{x}_0)\} = \phi(\mathbf{x} + \mathbf{x}_0). \tag{2.1.12}$$

The system is called *linear and shift-invariant* (LSI)* if it is both linear and shift-invariant. The output signal of an LSI system to an impulse function $\delta(\mathbf{x})$, $T\{\delta(\mathbf{x})\}$, is known as the *impulse response* of the system, usually denoted by $h(\mathbf{x})$. Its CSFT is called the *frequency response* of the system, usually denoted by $H(\mathbf{f})$.

Theorem 2.3.   In an LSI system with input and output signals in $\mathcal{R}^K$, the output signal $\phi(\mathbf{x})$ corresponding to any input signal $\psi(\mathbf{x})$ can be described as a convolution of $\psi(\mathbf{x})$ with the system impulse response function $h(\mathbf{x})$; that is,

$$\phi(\mathbf{x}) = \psi(\mathbf{x}) * h(\mathbf{x}). \tag{2.1.13}$$

In the frequency domain, they are related by

$$\Phi_c(\mathbf{f}) = \Psi_c(\mathbf{f}) H_c(\mathbf{f}). \tag{2.1.14}$$

∎

We wish to emphasize that the preceding result is solely the consequence of the linearity and shift-invariance of the system. Obviously, the frequency response $H(\mathbf{f})$ of the system describes how the input signal will be altered by the system at different frequencies. *An LSI system is completely characterized by its impulse response $h(\mathbf{x})$ or the frequency response $H(\mathbf{f})$.*

2.2 MULTIDIMENSIONAL DISCRETE-SPACE SIGNALS AND SYSTEMS

The previous section considers signals defined over a K-D continuous space, $\mathcal{R}^K$. There are also applications where the signal is discrete by nature, that is, the signal is defined only over a set of discrete points. We can always index each discrete point by an integer vector, so that the signal is defined over a K-D integer space, $\mathcal{Z}^K = \{[n_1, n_2, \ldots, n_K]^T | n_k \in \mathcal{Z}, k \in \mathcal{K}\}$, where $\mathcal{Z}$ is the set of all integers. In general, we call such signals K-D discrete signals or sequences. A special case of the discrete signal is the sampled version of a continuous signal; we discuss the sampling process and properties of sampled signals in the next chapter. In this section, we formally define multidimensional discrete signals and systems and their characterization by the discrete-space Fourier transform.

Definition 2.6.   A K-D discrete space signal $\psi(\mathbf{n})$ is a function with a K-D discrete variable $\mathbf{n} = [n_1, n_2, \ldots, n_K] \in \mathcal{Z}^K$. The function can take on any values, real or complex, and is called a real function if it takes only real values.

---

*In the I-D case, LSI is called LTI, meaning linear time-shift invariant, when the I-D signal is a function of time.

Definition 2.7.   The discrete convolution of two discrete-space signals $\psi(\mathbf{n})$ and $h(\mathbf{n})$ is defined as

$$\psi(\mathbf{n}) * h(\mathbf{n}) = \sum_{\mathbf{m} \in \mathcal{R}^K} \psi(\mathbf{n} - \mathbf{m}) h(\mathbf{m}). \qquad (2.2.1)$$

Definition 2.8.   The discrete delta function $\delta(\mathbf{n})$ is defined as

$$\delta(\mathbf{n}) = \begin{cases} 1, & \mathbf{n} = \mathbf{0}, \\ 0, & \text{otherwise.} \end{cases} \qquad (2.2.2)$$

As with the continuous delta function, there are special properties associated with the discrete delta function. These can be obtained by replacing the integral with sum in Equations (2.1.3–5).

Definition 2.9 (Discrete-Space Fourier Transform).   The discrete-space Fourier transform (DSFT)* of a K-D discrete-space signal $\psi(\mathbf{n})$ is defined as

$$\Psi_d(\mathbf{f}) = \sum_{\mathbf{n} \in \mathcal{Z}^K} \psi(\mathbf{n}) \exp(-j2\pi \mathbf{f}^T \mathbf{n}) \qquad (2.2.3)$$

where $\mathbf{f} \in \mathcal{R}^K$ represents the K-D discrete frequency variable.

Notice that because of the complex exponential term in the right-hand side of Equation (2.2.3), the DSFT is periodic in each dimension, with a period of 1. The fundamental period in the K-D space is the unit hypercube, denoted by $\mathcal{I}^K = \{\mathbf{f} | f_k \in (-\frac{1}{2}, \frac{1}{2}), k \in \mathcal{K}\}$. This fundamental period repeats at all integer points. For this reason, one only need specify the DSFT of a signal over the unit hypercube, $\mathcal{I}^K$. We will formally define the periodicity of a K-D signal using the concept of lattice in the next chapter (Section 3.1).

Theorem 2.4 (Inverse Discrete-Space Fourier Transform).   A discrete-space signal can be reconstructed from its DSFT using:

$$\psi(\mathbf{n}) = \int_{\mathcal{I}^K} \Psi_d(\mathbf{f}) \exp(j2\pi \mathbf{f}^T \mathbf{n}) \, d\mathbf{f}. \qquad (2.2.4)$$

∎

The preceding equality is easy to prove by recognizing that

$$\int_{\mathcal{I}^K} \exp(j2\pi \mathbf{f}^T \mathbf{n}) \, d\mathbf{f} = \delta(\mathbf{n}). \qquad (2.2.5)$$

Most of the properties of the CSFT can be carried over to the DSFT. In particular, the convolution theorem still applies; that is,

$$\psi(\mathbf{n}) * h(\mathbf{n}) \longleftrightarrow \Psi_d(\mathbf{f}) H_d(\mathbf{f}). \qquad (2.2.6)$$

---

*In the 1-D case, DSFT is better known as *discrete time Fourier transform* or DTFT, because the 1-D signal is typically a function of discrete time indices.

Similarly to the continuous-space case, we can define a discrete-space system with discrete input and output signals. If the system is linear and shift-invariant, then it can be completely characterized by the impulse response, $h(\mathbf{n}) = \mathcal{T}\{\delta(\mathbf{n})\}$, the system output to a discrete delta function. The output to any input signal can be described by the convolution of the input signal with the impulse response.

2.3 FREQUENCY DOMAIN CHARACTERIZATION OF VIDEO SIGNALS

A video is a 3-D signal, having two spatial dimensions (horizontal and vertical) and one temporal dimension. One can apply the CSFT and DSFT described in the previous section to analog and digital video signals, respectively, in a straightforward manner with $K = 3$. One can also apply the various concepts and properties presented for multidimensional linear systems to a video processing system. Instead of using the general notation of (x_1, x_2, x_3) to describe a point in the 3-D signal domain, we will use (x, y, t), with x and y indicating the horizontal and vertical positions and t the temporal dimensions. Similarly, we will use (f_x, f_y, f_t) to represent the frequencies associated with these coordinates. It is natural to wonder what these frequencies mean in the physical world, and how the visual system perceives them. In this section, we attempt to answer the first question. In the next section, we describe the frequency response of the human visual system.

2.3.1 Spatial and Temporal Frequencies

Spatial Frequency   The 2-D spatial frequency is a measure of how fast image intensity or color changes in the 2-D image plane. One can measure the spatial frequency along different directions. The spatial frequency in a given direction is measured in terms of cycles per unit length in that direction, which could be 1 meter, the picture-height of a TV monitor, and so on. The spatial variation of a 2-D pattern can be completely characterized by the frequencies in two orthogonal directions. One can project the frequency in any other direction onto these two. Usually, we characterize the spatial frequency of a 2-D image signal by a pair (f_x, f_y) representing the horizontal and vertical frequencies, respectively. For example, a sinusoidal pattern described by $\psi(x, y) = \sin(10\pi x)$ has a frequency of $(5, 0)$, in that it changes five cycles per unit length in the horizontal direction, and it stays constant in the vertical direction. On the other hand, an image $\psi(x, y) = \sin(10\pi x + 20\pi y)$ has a frequency of $(5, 10)$, because there are five and ten cycles per unit length in the horizontal and vertical directions, respectively. This image can also be characterized by a frequency of $\sqrt{f_x^2 + f_y^2} \approx 11$ cycles per unit length along the direction $\theta = \arctan(f_y/f_x) \approx 64°$. These two pure sinusoidal patterns are illustrated in Figure 2.1.

The preceding examples are pure sinusoidal patterns that have single frequencies. Using the Fourier transform, an arbitrary signal can be decomposed into many sinusoidal patterns, as described by the inverse CSFT in Equation (2.1.7).

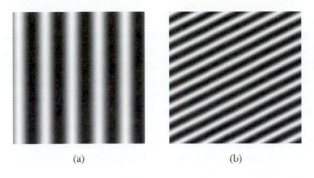

(a)   (b)

Figure 2.1   Two-dimensional sinusoidal signals: (a) $(f_x, f_y) = (5, 0)$; (b) $(f_x, f_y) = (5, 10)$. The horizontal and vertical units are the width and height of the image, respectively. Therefore, $f_x = 5$ means that there are five cycles along each row.

Angular Frequency   In the preceding, we defined the spatial frequency f_s along a particular direction in terms of a given unit length. Such a measure is however not very useful, as the perceived speed of spatial variation increases with the viewing distance. A more useful measure of the spatial frequency is in terms of cycles per degree of the viewing angle. As illustrated in Figure 2.2, if a picture has a height of h and is viewed at a distance of d, the vertical span of the viewing angle θ can be approximated, if $h/2 \ll d$, by

$$\theta = 2 \arctan \left(\frac{h}{2d} \right) \text{ (radian)} \approx \frac{h}{d} \text{(radian)} = \frac{180}{\pi} \frac{h}{d} \text{(degree)}. \tag{2.3.1}$$

If there are f_s cycles per picture-height, then the number of cycles per viewing degree is

$$f_\theta = f_s / \theta = \frac{\pi}{180} \frac{d}{h} f_s \text{ (cycle/degree)}. \tag{2.3.2}$$

We call f_θ the *angular frequency,* which has a unit of cycle/degree or cpd. Equation (2.3.2) shows the relation between the spatial and angular frequencies. For the same picture, f_θ increases as the viewing distance increases. On the other hand, for a fixed viewing distance, a larger screen size leads to lower angular frequency. These results match well with our intuition: the same picture appears to change more rapidly when viewed farther away, and it changes more slowly if viewed from a larger screen. Although the preceding relation is determined by evaluating the number of cycles vertically, angular frequency can be defined along the horizontal as well as any other

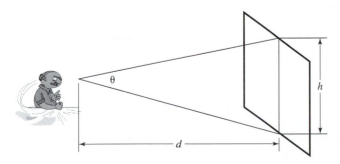

Figure 2.2   Relation of viewing angle and distance; see Equation (2.3.2).

direction. Note that *the angular frequency is not a characterization of the signal alone. It depends on both the spatial frequency in the signal and the viewing conditions.*

Temporal frequency   With a 1-D temporal function, the temporal frequency is very clear: the number of cycles per second. With a video signal that consists of varying 2-D frames, the temporal frequency is 2-D–position dependent. For a fixed 2-D position (x, y), its temporal frequency is defined as the number of cycles per second, usually denoted by Hz. The maximum temporal frequency of a video signal refers to the maximum of the temporal frequencies at all points in the imaged area.

2.3.2 Temporal Frequencies Caused by Linear Motion

The temporal frequency of a video signal depends on the rate at which the imaged scene varies, which may be due to camera motion, object motion, or both. There are usually many objects in a scene, which may be undergoing different motions. Therefore, it is not easy to relate the motion and the temporal frequency directly. Here, we consider the temporal frequency associated with an object that is undergoing a linear motion (i.e., with a constant velocity). The analysis also applies when the entire scene is undergoing a global linear motion; for example, due to camera translation. As we will see, *the temporal frequency depends not only on the motion, but also on the spatial frequency of the object.*

Let the image pattern of the object at time zero be described by $\psi_0(x, y)$ and its velocities in the horizontal and vertical directions by v_x and v_y. If the scene is under homogeneous ambient illumination, so that the same object point will have the same image intensity at different times, then the image pattern at time t will be

$$\psi(x, y, t) = \psi_0(x - v_x t, y - v_y t), \tag{2.3.3}$$

because the point (x, y) at time t corresponds to the point $(x - v_x t, y - v_y t)$ at time zero (see Figure 2.3). This is known as the *constant intensity assumption.* Conditions under which this assumption is valid will be explained further in Section 5.2

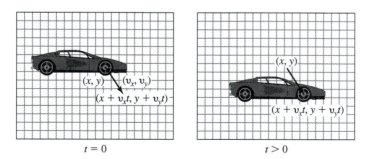

$t = 0$ $\qquad\qquad\qquad$ $t > 0$

Figure 2.3   Illustration of the constant intensity assumption under motion. Every point (x, y) at $t = 0$ is shifted by $(v_x t, v_y t)$ to $(x + v_x t, y + v_y t)$ at time t, without change in color or intensity. Alternatively, a point (x, y) at time t corresponds to a point $(x - v_x t, y - v_y t)$ at time zero.

(Equation (5.2.11)). Performing the CSFT for the preceding signal, we obtain

$$\Psi(f_x, f_y, f_t) = \int\int\int \psi(x, y, t) \exp(-j2\pi(f_x x + f_y y + f_t t)) \, dx \, dy \, dt$$

$$= \int\int \psi_0(x - v_x t, y - v_y t) \exp(-j2\pi(f_x(x - v_x t)$$

$$+ f_y(y - v_y t))) \, dx \, dy \cdot \int \exp(-j2\pi(f_t + f_x v_x + f_y v_y)t) \, dt$$

$$= \Psi_0(f_x, f_y) \int \exp(-j2\pi(f_t + f_x v_x + f_y v_y)t) \, dt$$

$$= \Psi_0(f_x, f_y)\delta(f_t + f_x v_x + f_y v_y),$$

where $\Psi_0(f_x, f_y)$ represents the 2-D CSFT of $\psi_0(x, y)$. The last equality is based on Equation (2.1.5).

This result shows that $\Psi(f_x, f_y, f_t)$ is nonzero only on the plane defined by

$$f_t + f_x v_x + f_y v_y = 0. \tag{2.3.4}$$

This means that a spatial pattern characterized by (f_x, f_y) in the object will lead to a temporal frequency

$$f_t = -f_x v_x - f_y v_y, \tag{2.3.5}$$

if the object is moving at a velocity of (v_x, v_y).* Figure 2.4(a) illustrates the relation between f_x, f_y and f_t for a given velocity. We can see that the nonzero region of the frequency spectrum in the 3-D frequency space is a plane with a normal vector defined by $(v_x, v_y, 1)$. If the spatial signal has a finite bandwidth, $(f_{x,max}, f_{y,max})$, then, the temporal bandwidth is $f_{t,max} = v_x f_{x,max} + v_y f_{y,max}$. In Figure 2.4(a), $f_{x,max} = f_{y,max} = 5$; therefore, $f_{t,max} = 35$ for $(v_x, v_y) = (3, 4)$.

From Equation (2.3.5), *the temporal frequency depends not only on the velocity but also on the spatial frequency.* In fact, it is the projection of the velocity vector onto the spatial frequency vector, as shown in Figure 2.4(b). The following observations are immediate:

- When $f_x = f_y = 0$, $f_t = 0$ regardless of the values of v_x, v_y. This means that if the object has a completely flat pattern (i.e., with a homogeneous brightness or color), then no temporal changes can be observed no matter how fast the object moves along the image plane.
- The temporal frequency $f_t = 0$ if the direction of motion, (v_x, v_y), and spatial frequency direction, (f_x, f_y), are orthogonal. The spatial frequency direction (f_x, f_y) is the direction along which the speed of change is highest. Its orthogonal direction is the one along which there is no spatial change. This result indicates

---

*Note that with a real signal, the CSFT is symmetric, so that for every frequency component at (f_x, f_y), there is also a component at $(-f_x, -f_y)$ with the same magnitude. The corresponding temporal frequency caused by this other component is $f_x v_x + f_y v_y$.

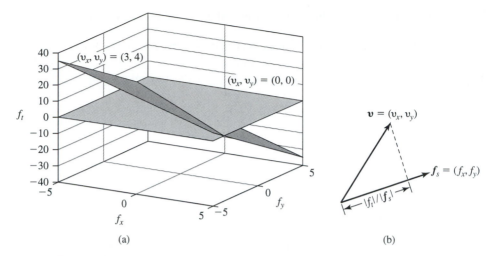

Figure 2.4   Relation between spatial and temporal frequencies under linear motions. (a) The spatiotemporal frequency plane in the (f_x, f_y, f_t) space, corresponding to two different velocity vectors; (b) the temporal frequencies is equal to the projection of the velocity onto the spatial gradient.

that an object moving in the direction in which object pattern does not change will not produce any temporal variations. The temporal frequency is greatest when the object moves in the direction in which the spatial change is greatest.

Consider an infinite plane with a vertical sinusoidal bar pattern, as shown in Figure 2.1(a). The spatial frequency direction is horizontal in this case. If the plane moves vertically from top to bottom (orthogonal to the spatial frequency direction), then the eye will not perceive any change, no matter how fast the plane moves. Once its motion is slightly tilted from the vertical direction, the eye will start to perceive temporal changes. The perceived change is most rapid when the plane moves horizontally from left to right (along the spatial frequency direction).

The preceding analysis was conducted for an object undergoing a constant speed of motion. For a video containing a more complex scene, with multiple objects moving in different ways, we can divide the imaged area into small regions so that each region can be considered as undergoing a uniform motion with a constant speed. The local spatial frequency and speed at this region determine the local temporal frequency associated with the region.

2.4 FREQUENCY RESPONSE OF THE HUMAN VISUAL SYSTEM

Any video system is ultimately targeted for human viewers. Therefore, it is extremely important to understand how the human visual system perceives a video signal. In Section 1.1, we described the color perception mechanism; in this section, we focus on the perception of spatial and temporal changes in image luminance. Similar results follow for the perception of chrominance variation, being proportionally lower than for the luminance. As will be shown, the sensitivity of the HVS to a visual pattern depends on the pattern's spatial and temporal frequency content. The visual sensitivity is highest at some intermediate spatial and temporal frequencies. It then falls off quickly and diminishes at some cut-off frequencies. Spatial or temporal changes above these frequencies are invisible to the human eye. Knowledge of the visual frequency response is very important in designing a video system. For example, the temporal and spatial cut-off frequencies form the basis for determining the frame and line rates in a video capture or display system. In the following, we first describe the spatial and temporal frequency responses of the HVS. We then introduce the joint spatiotemporal frequency response. Finally, we describe how to translate the visual frequency response when the eye is tracking a moving object.

2.4.1 Temporal Frequency Response and Flicker Perception

The temporal frequency response of the HVS refers to the visual sensitivity to a temporally varying pattern at different frequencies. Many experiments have been conducted to determine the temporal frequency response of the HVS. It has been found that the temporal response of an observer depends on many factors, including viewing distance, display brightness, and ambient lighting. Figure 2.5 shows the results obtained by an experiment conducted by Kelly [5]. In this experiment, the viewer was presented a flat screen whose brightness varied sinusoidally in the form of

$$\psi(t) = B(1 + m\cos 2\pi f t). \qquad (2.4.1)$$

For a fixed mean brightness level B and frequency f, the modulation level m was varied and the viewer was asked to identify the lowest modulation level m_{min} at which the temporal variation of the screen brightness (i.e., flicker) became just noticeable. Obviously, the inverse of m_{min} represents how sensitive the viewer is to the temporal changes at the given frequency f. Therefore, $1/m_{min}$, which is also known as *contrast sensitivity*, has been used to describe the visual sensitivity or response. The contrast sensitivity as a function of frequency is referred to as the *modulation transfer function* (MTF).

It can be seen that the temporal response of the HVS is similar to that of a bandpass filter, with peaks at some intermediate frequencies and the response falling off quickly afterwards, up to about 4.5 times the peak frequency. The peak increases with the mean brightness of the image. For example, at a mean brightness of 0.65 trolands,[*] the peak response occurs at about 5 Hz, and the cut-off frequency, where the response essentially diminishes, is at about 20–25 Hz. On the other hand, at a mean brightness of 850 trolands, the response is highest at about 15–20 Hz and diminishes at about 75 Hz.

One reason that the eye has reduced sensitivity at higher temporal frequencies is because the eye can retain the sensation of an image for a short time interval even when the actual image has been removed. This phenomenon is known as the *persistence*

---

[*]A troland is the unit used to describe the intensity of light entering the retina.

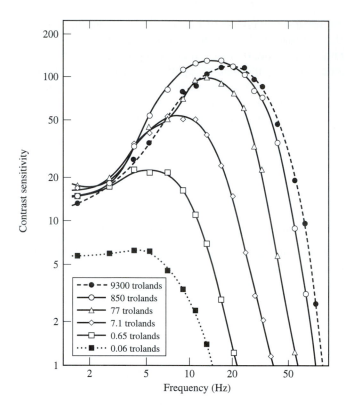

Figure 2.5   The temporal frequency response of the HVS obtained by a visual experiment. Different curves represent the responses obtained with different mean brightness levels, *B*, measured in trolands. The horizontal axis represents the flicker frequency *f*, measured in Hz. Reprinted from D. H. Kelly, Visual responses to time-dependent stimuli. I. Amplitude sensitivity measurements, *J. Opt. Soc. Am.* (1961) 51:422–29, by permission of the Optical Society of America.

*of vision.** On one hand, this causes temporal blurring of the observed pattern, if a pattern changes at a rate faster than the refresh rate of the HVS. The visual response falls off very quickly beyond this frequency. On the other hand, this vision persistence allows the display of a video signal as a consecutive sequence of frames. As long as the frame interval is shorter than the vision persistence period, then the eye perceives a continuously varying image. When the frame interval is longer than this period, the eye will observe *frame flicker,* which refers to the perception of discontinuous frames.

---

*This is caused by the *temporal summation* mechanism of the HVS, which integrates the incoming light. *Bloch's law* says that the integration or persistence period is inversely proportional to the intensity of the light [10]. The brighter the light source, the shorter the integration period. This gives the eye higher temporal sensitivity when the display is brighter, which is consistent with the results shown in Figure 2.5.

The lowest frame rate at which the eye does not perceive flicker is known as the *critical flicker frequency*. By definition, this is equivalent to the temporal cut-off frequency discussed previously. The frame rate used by a video capture or display system should exceed the critical flicker frequency to avoid the perception of flicker.

From Figure 2.5, the critical flicker frequency ranges from 20 to 80 Hz, depending on the mean brightness of the display. The brighter the display, the higher the critical frequency. In a movie theater, the mean brightness is very low; on the other hand, a CRT display is much brighter, close to 9600 trolands [8]. This is why the motion picture system can use a lower frame rate than the TV system. Presently, the motion picture industry uses 24 frames/second, whereas the TV industry uses 50 (in PAL and SECAM systems) and 60 fields/second (in NTSC). We can see that these frame and field rates are close to the cut-off frequencies associated with the mean brightness levels in their intended applications. Computer displays use a much higher rate, 72 fps, because a computer user sits much closer to the screen than does a TV viewer, and at a shorter distance the visual threshold is higher. More discussion of how to determine the spatial-temporal sampling rates of a video system is given in Section 3.3.1.

2.4.2 Spatial Frequency Response

The spatial frequency response of the HVS refers to the visual sensitivity to a stationary spatial pattern with different spatial frequencies. Assuming the visual sensitivity is isotropic with respect to the direction of spatial variation, then the spatial frequency response can be evaluated with respect to an arbitrary spatial direction; usually the horizontal or vertical direction is used.* For the response to be independent of the viewing distance, the spatial frequency response is normally expressed as a function of the angular frequency. Many experiments have been done to evaluate the spatial frequency response of the HVS. It has been found that the spatial frequency response of the HVS is also similar to that of a band-pass filter, with a peak response at about 2–5 cycle/degree (cpd), and diminishing at about 30 cpd. The result of a study conducted by Kelly [6] is shown in Figure 2.6. This result was obtained by viewing a vertical bar pattern that changed sinusoidally in the horizontal direction; that is,

$$\psi(x, y, t) = B(1 + m \cos 2\pi f x). \tag{2.4.2}$$

The pattern was displayed on a CRT monitor with a very high refresh rate (1000 fps), so that it could be considered as temporally stationary. For each given spatial frequency f, the modulation m was changed, and the viewer was asked to identify the minimum modulation level, m_{min}, at which the spatial change became just noticeable. The vertical axis in Figure 2.6 represents $1/m_{min}$. Three curves are shown, which were obtained using different stabilization settings to remove the effect of eye movements.

---

*In fact, visual sensitivity is higher to changes in the horizontal and vertical directions than in the other directions.

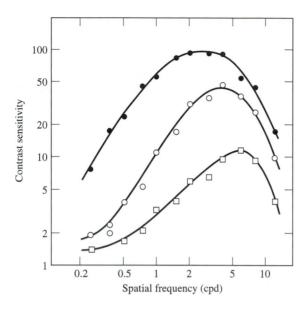

Figure 2.6 The spatial frequency response of the HVS, obtained by a visual experiment. The three curves result from different stabilization settings used to remove the effect of saccadic eye movements. Filled circles were obtained under normal, unstablized conditions; open squares, with optimal gain setting for stabilization; open circles, with the gain changed about 5 percent. Reprinted from D. H. Kelly, Motion and vision. I. Stabilized images of stationary gratings, *J. Opt. Soc. Am.* (1979), 69:1266–74, by permission of the Optical Society of America.

The eye often jumps from one fixation position to another very rapidly when observing a scene. This is known as *saccadic eye movement* [4]. It has been found that saccadic eye movement has the effect of enhancing contrast sensitivity, but reduces the frequency at which peak responses occur. For example, in Figure 2.6, the viewer's sensitivity is about ten fold higher with normal eye movement than without (i.e., with stabilization). On the other hand, the peak response occurs at about 2 cpd with normal eye movement, but the peak is shifted to about 4 cpd with complete removal of eye movement.

The band-pass character of the spatial frequency response can be attributed to the band-pass filtering operation effectively conducted by the HVS. It has been found that the impulse response of the HVS has a shape similar to the Laplacian of a Gaussian (the shape of a Mexican sombrero), with positive weightings to nearby visual cells but negative weightings to farther cells. The positive portion provides smoothing, whereas the negative portion effectively enhances sharp edges. The negative weighting effect is caused by lateral inhibition in the visual cortex [1].

In Section 3.3.1, we will show how the spatial frequency response of the HVS is taken into consideration in determining the horizontal and vertical sampling rate in a video system.

2.4.3 Spatiotemporal Frequency Response

The spatial frequency response presented previously is evaluated at a zero temporal frequency, and vice versa. Here we discuss the visual response when both the spatial and temporal frequencies are nonzero. It has been found that, at higher temporal frequencies,

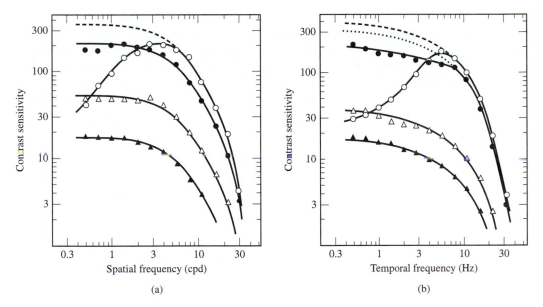

Figure 2.7   Spatiotemporal frequency response of the HVS. (a) Spatial frequency responses for different temporal frequencies of 1 Hz (open circles), 6 Hz (filled circles), 16 Hz (open triangles), and 22 Hz (filled triangles). (b) Temporal frequency responses for different spatial frequencies of 0.5 cpd (open circles), 4 cpd (filled circles), 16 cpd (open triangles), and 22 cpd (filled triangles). Reprinted from J. G. Robson, Spatial and temporal contrast sensitivity functions of the visual systems, *J. Opt. Soc. Am.* (1966), 56:1141–42, by permission of the Optical Society of America.

both the peak and cut-off frequencies in the spatial frequency response shift downwards. A similar trend happens with the temporal frequency response.

Figure 2.7 shows experimental results by Robson [12]. The test pattern in this experiment was described by

$$\psi(x, y, t) = B(1 + m\cos(2\pi f_x x) \cos(2\pi f_t t)). \tag{2.4.3}$$

For a fixed set of f_x and f_t, the modulation m was varied, and the viewer was asked to identify the minimum modulation $m_{\min}$ at which both spatial and temporal variation were just noticeable. The vertical axis in Figure 2.7 represents $1/m_{\min}$. We can see that, at nearly zero temporal (spatial) frequencies, the spatial (temporal) frequency response has band-pass characteristics. This is consistent with the results shown previously. But at higher temporal (spatial) frequencies, the spatial (temporal) response becomes more like a low-pass filter, with the peak response decreasing when the temporal (spatial) frequency increases. This reveals that when an image pattern moves very fast, the eye will not be able to differentiate the very high spatial frequency in this pattern. The eye can resolve a much higher spatial frequency when the pattern is stationary. Similarly, at higher spatial frequencies, the temporal response becomes low-pass, with the transition frequency shifting downwards at higher spatial frequencies.

One implication of this reciprocal relation between the spatial and temporal responses for the design of a video system is that one can trade off the spatial resolution with temporal resolution, and vice versa. This property has been exploited judiciously in the interlaced scan mechanism used in analog TV systems. With interlaced scan, in order to render very rapidly changing scenes with a limited frame rate, a frame is split into two fields, each with half the number of lines in a frame. Because the HVS has a reduced spatial frequency response when the temporal frequency is high, the eye will not be able to differentiate the very high spatial frequencies in a rapidly changing scene, even if it is rendered at a high spatial resolution. Therefore, using a lower spatial resolution is justified. On the other hand, when the imaged scene is stationary, the lines in the two separate fields combine to provide a high spatial resolution, to accommodate the higher spatial frequency resolving power of the HVS.

The reciprocity between the spatial and temporal responses also explains why we perceive *line flicker* when we watch TV from a close distance. In this case, the lines are separated further and the angular frequency becomes low. At low angular frequencies, the human eye is more temporally acute and can therefore perceive flickers more easily.

Robson's experiment was conducted under normal saccadic eye movement. Kelly later conducted an experiment in which the effect of saccadic motion was removed by stablization techniques [7]. For testing the temporal response, he used a traveling wave, instead of a flickering wave. The trends in the resulting spatiotemporal frequency responses were similar, but the temporal response caused by motion was about twice that by flickering. This means that the eye is more sensitive to temporal variations caused by motion than by flickering.

2.4.4 Smooth Pursuit Eye Movement

Although the experimental results described in Section 2.4.1 indicate that the HVS cannot resolve temporal frequencies greater than 80 Hz, in reality the human eye can see clearly a fast-moving object when the eye is tracking the object—for example, when a baseball batter tracks a fastball. This is because the relative motion of the object becomes smaller when the eye is tracking the object. If the tracking were perfect, the object would appear still. The phenomenon that the eyes automatically move to track observed objects is known as *smooth pursuit eye movement* [11, 4].

To account for the eyes' tracking effect, we need to translate the spatiotemporal frequency at the display screen to the retina coordinate when the eye is moving [3]. Assuming that the eye is moving with a speed of v_x and v_y, then the image observed at the retina coordinate, $\tilde{\psi}(x, y, t)$, is related to the image at the display screen, $\psi(x, y, t)$, by

$$\tilde{\psi}(x, y, t) = \psi(x + v_x t, y + v_y t, t). \tag{2.4.4}$$

Taking the CSFT of both sides yields

$$\tilde{\Psi}(f_x, f_y, f_t) = \Psi(f_x, f_y, f_t - v_x f_x - v_y f_y). \tag{2.4.5}$$

This relation means that the observed frequencies in the retina, $(\tilde{f}_x, \tilde{f}_y, \tilde{f}_t)$, are related to the frequencies in the display screen, (f_x, f_y, f_t), by

$$\tilde{f}_x = f_x, \quad \tilde{f}_y = f_y, \quad \tilde{f}_t = f_t + v_x f_x + v_y f_y. \tag{2.4.6}$$

When the temporal frequency of the displayed object is zero, the perceived temporal frequency increases with the eye motion speed as well as with the spatial frequencies of the observed object. When the displayed object is moving with a constant speed, recall that the resulting temporal frequency is described by Equation (2.3.5). Therefore, the effective temporal frequency at the retina is zero, when the eye movement matches that of the object movement. In general, by tracking the object motion, the effective temporal frequency at the retina is reduced.

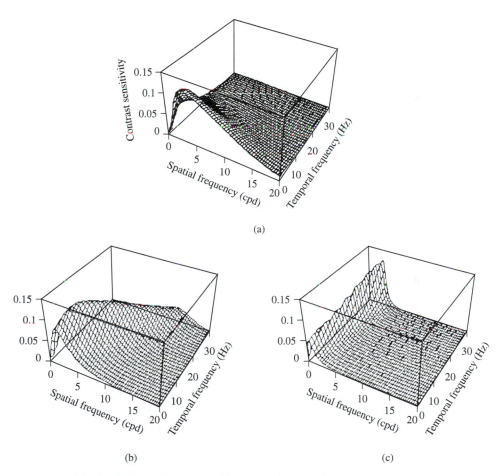

Figure 2.8   Spatiotemporal response of the HVS under smooth pursuit eye movements: (a) without smooth pursuit eye movement; (b) with eye velocity of 2 deg/s; (c) with eye velocity of 10 deg/s. Reprinted from Girod, B. "Motion compensation: visual aspects, accuracy, and fundamental limits." In Sezan, M. I., and R. L. Lagendijk, eds., *Motion Analysis and Image Sequence Processing,* Boston: Kluwer Academic Publishers, 1993, 126–52, by permission of Kluwer Academic Publishers.

Using the coordinate transformation of Equation (2.4.6), Girod redrew the visual frequency response measured by Robson. Figure 2.8 shows the visual frequency response without smooth pursuit eye movement, and the response under two different eye-movement speeds. In this figure, the contrast sensitivity is measured differently than those in Figures 2.5–7, and therefore has a different scale. Clearly, smooth pursuit eye movement has the effect of extending the nonzero region of the visual response to a larger temporal frequency range. In fact, temporal frequencies up to 1000 Hz can be perceived [3, 13].

Smooth pursuit eye movement has important consequences for the design of video display and processing systems, as suggested by Girod [3]. In order to avoid visible blur of fast-moving objects, the display system must be able to display temporal frequencies up to 1000 Hz. Likewise, any type of temporal filtering must be carried out with motion compensation (i.e., filtering along the motion trajectory) to avoid visible blur of the picture content.

2.5 SUMMARY

Fourier Analysis of Multidimensional Signals and Systems (Sections 2.1–2)

- A K-D continuous (discrete) signal can be represented by a continuous-(discrete-) space Fourier transform (Equations (2.1.6–7) for CSFT or Equations (2.2.3–4) for DSFT). The existence of inverse transform means that the signal can be decomposed into infinitely many complex sinusoidal functions with different frequencies.

- An LSI system can be completely characterized by its impulse response, or equivalently, its frequency response, the Fourier transform of the impulse response. (Theorem 2.3)

- The input and output signals of an LSI system are related by a convolution with the impulse response in the spatial domain, or a product with the frequency response in the transform domain, as given in Equation (2.1.13) or (2.1.14), respectively.

Spatial and Temporal Frequencies in Video Signals (Section 2.3)

- The 2-D spatial frequency of a video signal can be specified along any two orthogonal dimensions, usually the horizontal and vertical directions.

- The temporal frequency of a moving object depends on its velocity as well as its spatial frequency. More specifically, the temporal frequency at any image point is proportional to the projection of the velocity vector onto the spatial frequency direction at that point on the object, as given by Equation (2.3.5) and Figure 2.4.

Spatial and Temporal Frequency Responses of the HVS (Section 2.4)

- The human eye is most sensitive to certain intermediate spatial and temporal frequency components. Under normal viewing conditions, the eye is most sensitive to spatial frequencies between 2–5 cpd, and temporal frequencies of 5–20 Hz.

- Beyond certain cut-off frequencies, the eye cannot perceive spatiotemporal varia-tions. These occur at about 30 cpd in spatial frequency, and between 20–80 Hz in temporal frequency. These visual thresholds set the guidelines for designing video processing and display systems, which need only accommodate frequencies up to twice the visual thresholds.
- The eye will not perceive individual frames as long as the frame rate (field rate for interlaced display) is beyond the critical flicker frequency, which is below 80 Hz for most people.
- The visual frequency responses depend on the mean brightness of the display. In general, a brighter display can make the eye more sensitive to high-frequency components.
- When the eye is tracking a moving object, the perceived temporal frequency at the retina is reduced. If the tracking is perfect, than the temporal frequency at the retina plane is zero.
- The cut-off frequencies of the HVS are driving factors for determining the spatial and temporal sampling resolutions in video capture, processing, and display. (This is further discussed in Section 3.3.1).

2.6 PROBLEMS

2.1 The impulse response of a camera can usually be modeled by

$$h(x, y, t) = \begin{cases} \frac{1}{T_x T_y T_e} & |x| < \frac{T_x}{2}, |y| < \frac{T_y}{2}, t \in (0, T_e), \\ 0 & \text{otherwise}, \end{cases} \tag{2.6.1}$$

where T_x, T_y are the horizontal and vertical size of the camera aperture and T_e is the exposure time. Find the CSFT of $h(x, y, t)$. Plot the magnitude response. Comment on the effect of the parameters T_x, T_y, T_e on the frequency response of the camera.

2.2 Suppose the camera described in Problem 2.1 is looking at a scene that consists of a cube with width B on each side moving in parallel with the camera imaging plane. The projected image on the camera plane can be described by

$$\psi(x, y, t) = \begin{cases} 1 & -B/2 + v_x t < x < B/2 + v_x t, -B/2 < y < B/2, \\ 0 & \text{otherwise} \end{cases}$$

$$\tag{2.6.2}$$

Derive the signal captured by the camera. Assume $B \gg T_x$, $B \gg T_y$.

2.3 What are the considerations you would use to determine the frame rate and line number when designing a video capture or display system? Are the parameters chosen for NTSC color TV broadcasting appropriate?

2.4 Why does a computer monitor use a higher temporal refresh rate and line number than does a typical TV monitor?

2.5 Consider a horizontal bar pattern on a TV screen with 100 cycle/picture-height. If the picture-height is 1 meter, and the viewer sits at 3 meters from the screen, what is the equivalent angular frequency in cpd? What if the viewer sits at 1 meter or

5 meters distance? In either case, would the viewer be able to perceive the vertical variation properly?

2.6 Consider an object that has a flat, homogeneously textured surface with maximum spatial frequency of $(f_x, f_y) = (3, 4)$ cycle/meter, and that is moving at a constant speed of $(v_x, v_y) = (1, 1)$ meter/second. What is the temporal frequency of the object surface at any point? What are the results for the following speeds (in meter/second): $(4, -3)$, $(4, 0)$, $(0, 1)$?

2.7 (Continued from Problem 2.6.) Suppose that the eye tracks the moving object at a speed that is equal to the object speed. What are the perceived temporal frequencies at the retina for the different speeds? What if the eye moves at a fixed speed of $(2, 2)$ meter/second?

2.7 BIBLIOGRAPHY

[1] Cornsweet, T. N. *Visual Perception*. New York: Academic Press, 1970.

[2] Dudgeon, D. E., and R. M. Mersereau. *Multidimensional Digital Signal Processing*. Englewood Cliffs, NJ: Prentice Hall, 1984.

[3] Girod, B. Motion compensation: visual aspects, accuracy, and fundamental limits. In Sezan, M. I., and R. L. Lagendijk, eds., *Motion Analysis and Image Sequence Processing,* Boston: Kluwer Academic Publishers, 1993, 126–52.

[4] Hallett, P. Eye movements. In Boff, K., L. Kaufman, and J. P. Thomas, eds., *Handbook of Perception and Human Performance: Sensory Processes and Perception*. New York: Wiley-Interscience, 1986.

[5] Kelly, D. H. Visual responses to time-dependent stimuli. I. Amplitude sensitivity measurements. *J. Opt. Soc. Am.* (1961), 51:422–29.

[6] Kelly, D. H. Motion and vision. I. Stabilized images of stationary gratings. *J. Opt. Soc. Am.* (1979), 69:1266–74.

[7] Kelly, D. H. Motion and vision. II. Stabilized spatio-temporal threshold surface. *J. Opt. Soc. Am.* (1979), 69:1340–49.

[8] Netravali, A. N., and B. G. Haskell. *Digital Pictures—Representation, Compression and Standards*. 2nd ed. New York: Plenum Press, 1995.

[9] Oppenheim, A. V., and R. W. Schafer. *Discrete-Time Signal Processing*. Englewood Cliffs: Prentice Hall, 1989.

[10] Owen, W. G. Spatio-temporal integration in the human peripheral retina. *Vision Research* (1972), 12:1011–26.

[11] Robinson, D. A. The mechanics of human smooth pursuit eye movement. *J. Physiology* (Oct. 1965), 180(3):569–91.

[12] Robson, J. G. Spatial and temporal contrast sensitivity functions of the visual system. *J. Opt. Soc. Am.* (Aug. 1966), 56:1141–42.

[13] Tonge, G. J. Time-sampled motion portrayal. In *Second International Conf. Image Processing and Its Applications,* 1986, London.

3

Video Sampling

The very first step of any digital video processing task is the conversion of an intrinsically continuous video signal to a digital one. The digitization process consists of two steps: sampling and quantization. These could be implemented by a digital camera, which directly digitizes the video of a continuous physical scene, or by digitizing an analog signal produced by an analog camera. We also frequently need to convert a digital video signal from one format (in terms of spatial and temporal resolution) to another—for example, converting a video recording in the PAL format to the NTSC format. In this chapter we consider the sampling problem. The sampling rate conversion problem is discussed in Chapter 4; the quantization problem is discussed in Chapter 8, together with other means for signal compression.

When designing a video sampling system, three questions must be resolved: (1) What are the necessary sampling frequencies in the spatial and temporal directions? (2) Given an overall sampling rate (i.e., product of the horizontal, vertical, and temporal sampling rates), how do we sample in the 3-D space to obtain the best representation? and (3) How can we avoid the sampling artifact known as aliasing, given that we can only use a finite sampling rate? In this chapter, we start by describing the general theory for sampling a multidimensional signal, which relates the frequency spectrum of a signal and its necessary sampling pattern (Sections 3.1–2). We then focus on the sampling of video signals (Section 3.3). We first describe the factors that determine the necessary sampling rates for video signals, and trade-offs that are often made in practical systems. We then discuss 2-D sampling schemes and compare progressive and interlaced scans. We also consider how to sample a color video and, as an example, revisit the BT.601 format. Then we introduce several 3-D sampling lattices and compare their efficiency. Finally, we describe how common video cameras and displays accomplish prefiltering and interpolation functions implicitly, and discuss some practical limitations (Section 3.4).

3.1 BASICS OF THE LATTICE THEORY

In this and the next section, we extend the well-known sampling theorem for 1-D and 2-D signals to a general multidimensional space. Recall that, for 1-D signals, samples are usually taken at a regular spacing. With 2-D signals, samples are normally taken on a rectangular grid. In fact, one can also take samples on a nonrectangular grid, as long as the grid has a structure that allows the specification of the grid points using integer vectors. Mathematically, this type of grid is known as a lattice. In this section, we introduce the concept of a lattice in a K-D space. In the next section, we describe how to sample a continuous signal using a lattice.

The theory of sampling multidimensional signals on a lattice was first presented by Petersen and Middleton [7]. An excellent review of the related theory and an extension to sampling using the union of selected cosets of a sublattice in a lattice was given by Dubois [4]. In this chapter, we introduce only those concepts and properties that are essential for us to analyze the sampling process in a multidimensional space. For a comprehensive treatment of the lattice theory, the reader is referred to [2]. The definitions and theorems introduced here are based primarily on [4].

Definition 3.1. A lattice, Λ, in the real K-D space, $\mathcal{R}^K$, is the set of all possible vectors that can be represented as integer-weighted combinations of a set of K linearly independent basis vectors, $\mathbf{v}_k \in \mathcal{R}^K$, $k \in \mathcal{K} = \{1, 2, \ldots, K\}$. That is,

$$\Lambda = \left\{ \mathbf{x} \in \mathcal{R}^K \,\middle|\, \mathbf{x} = \sum_{k=1}^{K} n_k \mathbf{v}_k, \quad \forall n_k \in \mathcal{Z} \right\}. \tag{3.1.1}$$

The matrix $[\mathbf{V}] = [\mathbf{v}_1, \mathbf{v}_2, \ldots, \mathbf{v}_K]$ is called the generating matrix.[*]

Example 3.1

Consider two lattices in $\mathcal{R}^2$ with the following generating matrices:

$$[\mathbf{V}_1] = \begin{bmatrix} 1 & 0 \\ 0 & 1 \end{bmatrix}; \quad [\mathbf{V}_2] = \begin{bmatrix} \sqrt{3}/2 & 0 \\ 1/2 & 1 \end{bmatrix}. \tag{3.1.2}$$

To sketch the actual lattice from a given generating matrix, we first draw the two points corresponding to the basis vectors, and then determine points corresponding to typical integer combinations of the basis vectors; for example, $\mathbf{v}_1 + \mathbf{v}_2$, $\mathbf{v}_1 - \mathbf{v}_2$, and so on. Based on these points, we can usually extrapolate all the other possible points by visual inspection. Using this procedure, we draw the lattices determined by the two given generating matrices in Figure 3.1(a) and (b). As expected, the first lattice is a simple rectangular grid, because $[\mathbf{V}_1]$ is a diagonal matrix. The second lattice is the so-called hexagonal lattice. Although

---

[*]It is also referred to as the *sampling matrix*.

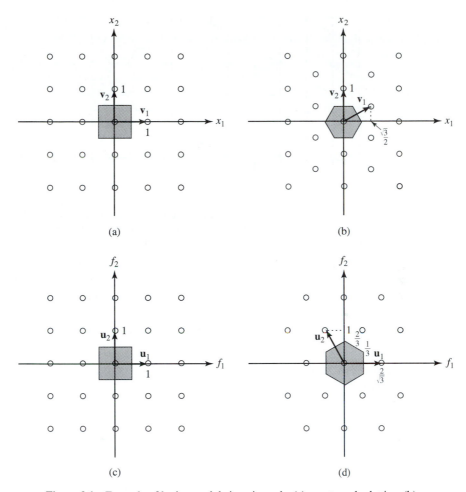

Figure 3.1   Example of lattices and their reciprocals: (a) a rectangular lattice; (b) a hexagonal lattice; (c) the reciprocal of the rectangular lattice; (d) the reciprocal of the hexagonal lattice. The shaded area in each lattice indicates the Voronoi cell of that lattice.

a set of basis vectors or a generating matrix completely defines a lattice, the basis or generating matrix associated with a lattice is nonunique. In fact, one can easily find more than one basis or generating matrix that can generate the same lattice. For example, one can verify that for the lattice in Figure 3.1(b), another set of basis vectors is $\mathbf{v}_1 = [\sqrt{3}/2, 1/2]^T$, $\mathbf{v}_2 = [\sqrt{3}, 0]^T$.

For a given generating matrix, any point in the lattice can be indexed by the integer coefficients, n_k, $k \in \mathcal{K}$, associated with it. Using matrix notation, we can represent any point in the lattice by an integer vector $\mathbf{n} = [n_1, n_2, \ldots, n_K]^T \in \mathcal{Z}^K$. The actual position

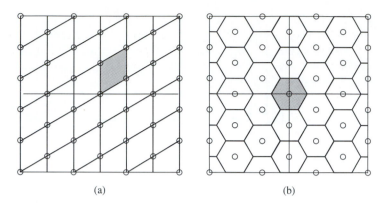

Figure 3.2   Unit cells associated with the lattice shown in Figure 3.1(b): (a) the
fundamental parallelepiped; (b) the Voronoi cell. A unit cell and its translations to all
lattice points form a partition of the underlying continuous space.

of this point is $\mathbf{x} = [\mathbf{V}]\mathbf{n}$. *The fact that any point in a lattice can be indexed by an integer
vector makes the lattice an elegant tool for sampling a continuous signal with a regular
geometry.*

In the following, we introduce some important properties of the lattice.

Theorem 3.1.   Given a lattice Λ, one can find a unit cell $\mathcal{U}(\Lambda)$ such that its
translations to all lattice points form a partition (i.e., nonoverlapping covering) of the
entire space $\mathcal{R}^K$; that is,

$$\bigcup_{\mathbf{x} \in \Lambda} (\mathcal{U} + \mathbf{x}) = \mathcal{R}^K \quad \text{and} \quad (\mathcal{U} + \mathbf{x}) \cap (\mathcal{U} + \mathbf{y}) = \emptyset, \quad \text{if } \mathbf{x} \neq \mathbf{y}, \qquad (3.1.3)$$

where $\mathcal{U} + \mathbf{x} = \{\mathbf{p} + \mathbf{x} \mid \mathbf{p} \in \mathcal{U}\}$ denotes the translation of $\mathcal{U}$ by $\mathbf{x}$, and $\emptyset$ represents the
empty set.   ∎

Theorem 3.1 tells us that the space $\mathcal{R}^K$ can be represented as a *tiling* by a chosen
unit cell and its translations, as illustrated in Figure 3.2. This representation is useful
when we consider the quantization of the space $\mathcal{R}^K$. The two partitions shown in
Figure 3.1 are two ways of uniformly quantizing the space $\mathcal{R}^2$.

The unit cell associated with a lattice is nonunique. In fact, if $\mathcal{U}$ is a unit cell of
Λ, then $\mathcal{U} + \mathbf{x}, \forall \mathbf{x} \in \Lambda$ is also a unit cell. Among several interesting unit cells, the
fundamental parallelepiped and the Voronoi cell, introduced in the following definitions,
are most useful.

Definition 3.2.   The fundamental parallelepiped of a lattice with basis vectors
$\mathbf{v_k}, \mathbf{k} \in \mathcal{K}$, is the set defined by

$$\mathcal{P}(\Lambda) = \left\{ \mathbf{x} \in \mathcal{R}^K \,\middle|\, \mathbf{x} = \sum_k \alpha_k \mathbf{v}_k, \quad \forall\, 0 \leq \alpha_k \leq 1 \right\}. \qquad (3.1.4)$$

In words, this is the polygon enclosed by the vectors corresponding to the basis vectors.

The fundamental parallelepiped of the lattice given in Figure 3.1(b) is shown in Figure 3.2(a). As shown in the figure, the fundamental parallelepiped and its translations to all the lattice points form a partition of the space $\mathcal{R}^K$, and therefore the fundamental parallelepiped is a unit cell. Obviously, there are many fundamental parallelepipeds associated with a lattice, because of the nonuniqueness of the generating basis.

Definition 3.3.   The Voronoi cell of a lattice is the set of points that are closer to the origin than any other points in the lattice. That is:

$$\mathcal{V}(\Lambda) = \{\mathbf{x} \in \mathcal{R}^K \,|\, d(\mathbf{x}, \mathbf{0}) \leq d(\mathbf{x}, \mathbf{p}), \quad \forall \mathbf{p} \in \Lambda\}. \tag{3.1.5}$$

The Voronoi cell of the lattice given in Figure 3.1(b) is shown in Figure 3.2(b). As with the fundamental parallelepiped, the Voronoi cell and its translations to all the lattice points also form a partition of the space $\mathcal{R}^K$. Therefore, the Voronoi cell is also a unit cell. We will see that the Voronoi cell is very useful for analyzing the sampling process. In Figure 3.1, the shaded region in each lattice indicates its Voronoi cell. As expected, the Voronoi cell of a rectangular lattice is simply a rectangle. The Voronoi cell of the second lattice is a hexagon with six sides of equal length. This is why the lattice is called the hexagonal lattice.

In the 2-D case, the Voronoi cell of a lattice can be determined by first drawing a straight line between the origin and each one of the closest nonzero lattice points, and then drawing a perpendicular line that is half way between the two points. This line is the equidistant line between the origin and this lattice point. The polygon formed by all such equidistant lines surrounding the origin is then the Voronoi cell, as illustrated in Figure 3.3. In the 3-D case, this procedure can be extended by replacing the equidistant lines with equidistant planes.

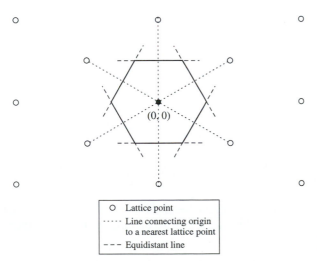

○	Lattice point
·····	Line connecting origin to a nearest lattice point
− − −	Equidistant line

Figure 3.3 Determining the Voronoi cell by drawing equidistant lines.

Volume of the Unit Cell and Sampling Density   Note that although the unit cell associated with a lattice is not unique, the volume of the unit cell is unique. This is because the same number of unit cells are required to cover a finite subspace of $\mathcal{R}^K$, regardless of the shape of the unit cell. From basic algebraic theory, the volume of a parallelepiped formed by column vectors in a matrix $[\mathbf{V}]$ is equal to the absolute value of the determinant of the matrix. Therefore, the volume of the fundamental parallelepiped and henceforth any unit cell is $|\det [\mathbf{V}]|$. Obviously, the smaller the unit cell, the more lattice points exist in a given volume. Therefore, the inverse of the volume of the unit cell measures the *sampling density,* which we denote by

$$d(\Lambda) = \frac{1}{|\det [\mathbf{V}]|}. \tag{3.1.6}$$

This variable describes how many lattice points exist in a unit volume in $\mathcal{R}^K$. For the two lattices given in Figure 3.1(a) and (b), it is easy to verify that their sampling densities are $d_1 = 1$ and $d_2 = 2/\sqrt{3}$, respectively.

Definition 3.4.   Given a lattice with a generating matrix V, its reciprocal lattice Λ^* is defined as a lattice with a generating matrix

$$[\mathbf{U}] = ([\mathbf{V}]^T)^{-1} \quad \text{or} \quad [\mathbf{V}]^T[\mathbf{U}] = \mathbf{I}. \tag{3.1.7}$$

By definition, if $\mathbf{x} = [\mathbf{V}]\mathbf{m} \in \Lambda, \mathbf{y} = [\mathbf{U}]\mathbf{n} \in \Lambda^*$, then $\mathbf{x}^T\mathbf{y} = \mathbf{m}^T[\mathbf{V}]^T[\mathbf{U}]\mathbf{n} = \mathbf{m}^T\mathbf{n} \in \mathcal{Z}$. That is, the inner product of any two points, one from each lattice, is an integer. The relation in Equation (3.1.7) implies that the basis vectors in Λ and Λ^* are orthonormal to each other, that is, $\mathbf{v}_k^T\mathbf{u}_l = \delta_{kl}, \forall k, l \in \mathcal{K}$, where $\delta_{kl} = 1$, if $k = l$, and $\delta_{kl} = 0$, otherwise. Because $\det [\mathbf{V}] = 1/\det [\mathbf{U}]$, the sampling densities of the two lattices are inversely related by $d(\Lambda^*) = \frac{1}{d(\Lambda)}$. *This means that the denser Λ is, the sparser Λ^* is, and vice versa.*

Example 3.2

For the two lattices given in Example 3.1 and shown in Figure 3.1(a) and (b), the generating matrices of their reciprocals are

$$[\mathbf{U}_1] = \begin{bmatrix} 1 & 0 \\ 0 & 1 \end{bmatrix}; \qquad [\mathbf{U}_2] = \begin{bmatrix} 2/\sqrt{3} & -1/\sqrt{3} \\ 0 & 1 \end{bmatrix}. \tag{3.1.8}$$

Based on the basis vectors in $[\mathbf{U}_i]$, we can determine all the points in the reciprocal lattices, using the same procedure as for determining the original lattices. The results are shown in Figure 3.1(c) and (d). The Voronoi cells of the reciprocal lattices are also indicated in these two figures. We see that with the rectangular lattice, the Voronoi cells in both the original and the reciprocal lattices are rectangles, whereas with the hexagonal lattice, the original and reciprocal Voronoi cells are both hexagons, but with different orientations. We also see that Λ_2 is denser than Λ_1, but Λ_2^* is sparser than Λ_1^*.

Theorem 3.2.   Let Λ_1 and Λ_2 represent two lattices with generating matrices $[\mathbf{V}_1]$ and $[\mathbf{V}_2]$, respectively. If $[\mathbf{V}_1]^{-1}[\mathbf{V}_2]$ is a matrix of rational numbers, then the intersection

$$\Lambda_1 \cap \Lambda_2 = \{\mathbf{x} \in \mathcal{R}^K \,|\, \mathbf{x} \in \Lambda_1 \quad \text{and} \quad \mathbf{x} \in \Lambda_2\}$$

is also a lattice, and it is the largest lattice that is contained in both Λ_1 and Λ_2. Under the same condition, the sum

$$\Lambda_1 + \Lambda_2 = \{\mathbf{x} + \mathbf{y} \mid \forall \mathbf{x} \in \Lambda_1, \quad \forall \mathbf{y} \in \Lambda_2\}$$

is also a lattice and it is the smallest lattice that contains both Λ_1 and Λ_2. Furthermore, $(\Lambda_1 + \Lambda_2)^* = \Lambda_1^* \cap \Lambda_2^*$. ∎

      Theorem 3.3.   If $\Lambda_1 \subset \Lambda_2$, then $\Lambda_1^* \supset \Lambda_2^*$. Also, if $\Lambda_1 \subset \Lambda_2$, $\mathcal{V}(\Lambda_1) \supset \mathcal{V}(\Lambda_2)$. ∎

The preceding results can be proven by making use of the fact that a lattice must contain points that can be represented as $[\mathbf{V}]\mathbf{n}$, where $\mathbf{n} \in \mathcal{Z}^K$ and $[\mathbf{V}]$ is a nonsingular matrix. The usefulness of these theorems will become clear in Section 4.1, when we discussion sampling rate conversion.

      Characterization of Periodicity Using the Lattice Concept   We are all familiar with the concept of periodicity in 1-D. We say that a function is periodic with a period of T if $\psi(x) = \psi(x + nT)$, $\forall n \in \mathcal{Z}$. One important application of the lattice concept is for describing the periodicity of a multidimensional function, as defined in the following.

      Definition 3.5.   A function is periodic with a nonsingular periodicity matrix $[\mathbf{V}]$ if $\psi(\mathbf{x}) = \psi(\mathbf{x} + [\mathbf{V}]\mathbf{n})$, for any $\mathbf{n} \in \mathcal{Z}^K$.

      Obviously, the set of all repetition centers, $[\mathbf{V}]\mathbf{n}$, $\mathbf{n} \in \mathcal{Z}^K$, forms a lattice with a generating matrix $[\mathbf{V}]$. For this reason, when a lattice is used to describe the periodicity of a function, it is also referred to as a *periodicity lattice*. Recall that the entire space $\mathcal{R}^K$ can be partitioned into infinitely many shifted versions of a unit cell of the lattice. Therefore, one can think of a periodic function as one that repeats a basic pattern over a chosen unit cell at all other translated unit cells. We call the Voronoi cell of the lattice the *fundamental period* of this function. For a 1-D function with a period of T, the fundamental period is the interval centered at the origin, $(-T/2, T/2)$. A periodic function essentially consists of its fundamental period and the shifted versions of this period over all lattice points. Obviously, one need only specify a periodic function over its fundamental period.

3.2 SAMPLING OVER LATTICES

The lattice structure provides an elegant tool for sampling a continuous signal with a uniform but not necessarily hypercubic grid. It enables one to index all samples with integer vectors, which simplifies the description of the resulting sampled signal. Sampling over a lattice is a generalization of uniform sampling over a rectangular grid in 2-D or a hypercubic grid in K-D. As will be shown, with this sampling structure, all the theorems known for 1-D and 2-D uniform sampling still apply. In particular, a generalized Nyquist sampling theory exists, which governs the necessary density and structure of the sampling lattice for a given signal spectrum.

3.2.1 Sampling Process and Sampled-Space Fourier Transform

Definition 3.6.   Given a continuous signal $\psi_c(\mathbf{x}), \mathbf{x} \in \mathcal{R}^K$, a sampled signal over a lattice Λ with a generating matrix $[\mathbf{V}]$ is defined as:

$$\psi_s(\mathbf{n}) = \psi_c([\mathbf{V}]\mathbf{n}), \quad \mathbf{n} \in \mathcal{Z}^K. \tag{3.2.1}$$

An alternative way of defining the sampled signal is by considering it as a continuous-space signal with impulses at the sampled points; that is,

$$\psi_s(\mathbf{x}) = \sum_{\mathbf{n} \in \mathcal{Z}^K} \psi_c(\mathbf{x})\delta(\mathbf{x} - [\mathbf{V}]\mathbf{n}), \quad \mathbf{x} \in \mathcal{R}^K. \tag{3.2.2}$$

In Section 2.2, we defined the discrete-space Fourier transform. Similarly, we can define the Fourier transform of a sampled signal over a lattice [5, 4]. Here we call it the *sampled-space Fourier transform* (SSFT). Compared to the DSFT, which is defined in terms of the abstract indices of sampled points, the definition of the SSFT keeps the underlying sampling structure of a discrete signal in the picture, and enables one to make connections with actual physical dimensions.

Definition 3.7.   The sampled-space Fourier transform (SSFT) of a signal sampled over a lattice Λ with a generating matrix $[\mathbf{V}]$ is defined as:

$$\Psi_s(\mathbf{f}) = \sum_{\mathbf{n} \in \mathcal{Z}^K} \psi_s(\mathbf{n}) \exp(-j2\pi\mathbf{f}^T[\mathbf{V}]\mathbf{n}) \tag{3.2.3}$$

Using the definition in Equation (3.2.2), it can be shown that the CSFT of the sampled signal is equal to the SSFT. Thus, the definition of the SSFT is consistent with the CSFT. Also, the SSFT reduces to the DSFT when the lattice is a hypercube, that is, when $[\mathbf{V}]$ is a K-D identity matrix.

Notice that because $\exp(j2\pi\mathbf{f}^T\mathbf{x}) = 1$, if $\mathbf{f}^T\mathbf{x} = n \in \mathcal{Z}$, we have

$$\Psi_s(\mathbf{f} + [\mathbf{U}]\mathbf{m}) = \Psi_s(\mathbf{f}), \quad \text{with } [\mathbf{U}] = ([\mathbf{V}]^T)^{-1}.$$

This means that the SSFT is periodic with a periodicity matrix $[\mathbf{U}]$, and that the spectral repetition centers fall on the reciprocal of the sampling lattice, Λ^*. The fundamental period is the Voronoi cell of the reciprocal lattice, $\mathcal{V}(\Lambda^*)$. For this reason, one need only specify the SSFT of a sampled signal over $\mathcal{V}(\Lambda^*)$. This basic pattern repeats at all points in the reciprocal lattice (this is illustrated and described in detail in Section 3.2.2).

Theorem 3.4 (Inverse Sampled-Space Fourier Transform).   A sampled signal over a lattice Λ with generating matrix $[\mathbf{V}]$ can be obtained from its SSFT by

$$\psi_s(\mathbf{n}) = \frac{1}{d(\Lambda)} \int_{\mathcal{V}(\Lambda^*)} \Psi_s(\mathbf{f}) \exp(j2\pi\mathbf{f}^T[\mathbf{V}]\mathbf{n}) \, d\mathbf{f}, \quad \text{for } \mathbf{n} \in \mathcal{Z}^K \tag{3.2.4}$$

∎

The validity of the inverse relation can be proven by substituting $\Psi_s(\mathbf{f})$ in Equation (3.2.4) with the forward transform in Equation (3.2.3), and making use of the equality:

$$\int_{\mathcal{V}(\Lambda^*)} \exp(j2\pi\mathbf{x}^T\mathbf{f})\,d\mathbf{f} = \begin{cases} \text{volume}\{\mathcal{V}(\Lambda^*)\} = d(\Lambda), & \text{for } \mathbf{x} = \mathbf{0}, \\ 0, & \text{otherwise.} \end{cases} \quad (3.2.5)$$

The inverse relation in Equation (3.2.4) reveals that one can represent a sampled signal as an infinite sum of complex exponential functions with frequencies defined in the Voronoi cell of the reciprocal lattice.

All the properties of the CSFT can be carried over to the SSFT. Here we only state the convolution theorem. The proof is straightforward and is left as an exercise (Problem 3.4).

Definition 3.8.   The convolution of two sampled signals over the same lattice Λ is defined as:

$$\psi_s(\mathbf{n}) * h_s(\mathbf{n}) = \sum_{\mathbf{m}\in\mathcal{Z}^K} \psi_s(\mathbf{n}-\mathbf{m})h_s(\mathbf{m}) = \sum_{\mathbf{x}_m\in\Lambda} \psi_c([\mathbf{V}](n-m))h_c([\mathbf{V}]m), \quad (3.2.6)$$

where $\psi_c(\mathbf{x})$ and $h_c(\mathbf{x})$ are the continuous versions of $\psi_s(\mathbf{n})$ and $h_s(\mathbf{n})$, respectively.

Theorem 3.5 (Convolution Theorem over the Sampled Space).   As with the CSFT and DSFT, the convolution of two signals in a sampled space is equivalent to the product of their SSFTs; that is,

$$\psi_s(\mathbf{n}) * h_s(\mathbf{n}) \longleftrightarrow \Psi_s(\mathbf{f})H_s(\mathbf{f}). \quad (3.2.7)$$

∎

Characterization of Sampled Linear Shift-Invariant Systems   Given a linear shift-invariant system over the continuous space $\mathcal{R}^K$ with an impulse response $h_c(\mathbf{x})$, if one samples the input and output signals over a lattice Λ, it can be shown (Problem 3.5) that the sampled input signal $\psi_s(\mathbf{n}) = \psi_c([\mathbf{V}]\mathbf{n})$ and output signal $\phi_s(\mathbf{n}) = \phi_c([\mathbf{V}]\mathbf{n})$ are related by the previously defined convolution in Λ, and that the filter will be exactly the sampled impulse response $h_s(\mathbf{n}) = h_c([\mathbf{V}]\mathbf{n})$. Therefore, the sampled system is completely characterized by the sampled impulse response $h_s(\mathbf{n})$.

3.2.2 The Generalized Nyquist Sampling Theorem

Theorem 3.6 (Generalized Nyquist Sampling Theorem).   If a continuous signal $\psi_c(\mathbf{x})$, $\mathbf{x} \in \mathcal{R}^K$, is sampled over a lattice Λ with a generating matrix $[\mathbf{V}]$, then the SSFT of the sampled signal, $\psi_s(\mathbf{n}) = \psi_c([\mathbf{V}]\mathbf{n})$, $\mathbf{n} \in \mathcal{Z}^K$, is the summation of the CSFT of the original continuous signal and its translated versions at all points of the reciprocal lattice Λ^*, scaled by the density of the lattice. That is,

$$\Psi_s(\mathbf{f}) = d(\Lambda) \sum_{\mathbf{m}\in\mathcal{Z}^K} \Psi_c(\mathbf{f} - [\mathbf{U}]\mathbf{m}). \quad (3.2.8)$$

It is possible to recover the original continuous signal from the sampled signal perfectly if and only if the nonzero region (known as the *support region*) of the CSFT of the original signal is limited within the Voronoi cell of the reciprocal lattice, that is,

$$\Psi_c(\mathbf{f}) = 0, \quad \text{for } \mathbf{f} \notin \mathcal{V}(\Lambda^*). \tag{3.2.9}$$

Furthermore, the perfect reconstruction can be accomplished by filtering the sampled signal using a reconstruction filter with the following frequency response (in terms of the CSFT):

$$H_r(\mathbf{f}) = \begin{cases} \dfrac{1}{d(\Lambda)}, & \mathbf{f} \in \mathcal{V}(\Lambda^*), \\ 0, & \text{otherwise.} \end{cases} \tag{3.2.10}$$

∎

One can easily recognize that the Theorem 3.6 is a generalization of the 1-D Nyquist sampling theorem. In that special case, $[\mathbf{V}] = [\Delta]$, $[\mathbf{U}] = [1/\Delta]$, and $d(\Lambda) = 1/\Delta$, where Δ is the sampling interval. As with 1-D and 2-D signals, we refer to the repeated spectra as *alias components*. The theorem tells us that *if Λ is the sampling lattice in the spatial domain, then Λ^* encompasses centers of the alias components in the frequency domain.* If the continuous signal has a support region larger than $\mathcal{V}(\Lambda^*)$, then adjacent alias components would overlap so that the original high-frequency components would fold-over to a lower-frequency range. This effect is known as *aliasing*. As we have learned, the denser Λ is, the sparser Λ^* is. Therefore, using a denser sampling grid will further separate the alias components in the frequency domain, thus reducing the aliasing artifact in the sampled signal.

For a fixed sampling lattice, in order to avoid aliasing, the original continuous signal should be band-limited using the following prefilter:

$$H_p(\mathbf{f}) = \begin{cases} 1, & \mathbf{f} \in \mathcal{V}(\Lambda^*), \\ 0, & \text{otherwise.} \end{cases} \tag{3.2.11}$$

We see that both the prefilter and the reconstruction filter should be, in the ideal case, low-pass filters with support regions defined by the Voronoi cell of the reciprocal lattice, $\mathcal{V}(\Lambda^*)$.

To prove Theorem 3.6, we can make use of the inverse CSFT. From Equation (2.1.7),

$$\psi_s(\mathbf{n}) = \psi_c([\mathbf{V}]\mathbf{n}) = \int_{\mathcal{R}^K} \Psi_c(\mathbf{f}) \exp(j2\pi \mathbf{f}^T [\mathbf{V}]\mathbf{n}) \, d\mathbf{f}$$

$$= \sum_{\mathbf{m} \in \mathcal{Z}^K} \int_{\mathcal{V}(\Lambda^*)} \Psi_c(\mathbf{f} - [\mathbf{U}]\mathbf{m}) \exp(j2\pi (\mathbf{f} - [\mathbf{U}]\mathbf{m})^T [\mathbf{V}]\mathbf{n}) \, d\mathbf{f}$$

$$= \sum_{\mathbf{m} \in \mathcal{Z}^K} \int_{\mathcal{V}(\Lambda^*)} \Psi_c(\mathbf{f} - [\mathbf{U}]\mathbf{m}) \exp(j2\pi \mathbf{f}^T [\mathbf{V}]\mathbf{n}) \, d\mathbf{f}$$

$$= \int_{\mathcal{V}(\Lambda^*)} \sum_{\mathbf{m} \in \mathcal{Z}^K} \Psi_c(\mathbf{f} - [\mathbf{U}]\mathbf{m}) \exp(j2\pi \mathbf{f}^T [\mathbf{V}]\mathbf{n}) \, d\mathbf{f}$$

Comparing the last equality with Equation (3.2.4), the result in Equation (3.2.8) follows.

Example 3.3

   Consider a 2-D signal with a spectrum as shown in Figure 3.4(a), which has a circular support region with radius $r = 1/\sqrt{3}$. If we use the lattice presented in Figure 3.1(a) to sample this signal, the spectrum of the sampled signals will be as shown in Figure 3.4(b), which is obtained by repeating the original spectra over points in the reciprocal lattice, shown previously in Figure 3.1(c). Obviously, because the signal support region is larger than the Voronoi cell of the reciprocal lattice, there is overlap between the alias components in the sampled spectrum. To avoid the overlap, we can prefilter the continuous signal so that its spectrum is nonzero only in the Voronoi cell of the reciprocal lattice, as illustrated in Figure 3.4(c). If we now sample this signal with the same lattice, all the alias components will cover the entire spectrum exactly, without overlapping, as shown in Figure 3.4(d). The reconstructed signal using the ideal reconstruction filter will be exactly the same as the prefiltered signal, given in Figure 3.4(c). Obviously, this signal has lost some frequency components compared to the original signal. If the application demands retention of the full bandwidth of the original signal, then a sampling lattice with a reciprocal Voronoi cell equal to or larger than the support region of the signal is needed. For example, if we use the hexagonal lattice in Figure 3.1(b), then the signal spectrum fits entirely within the Voronoi cell of its reciprocal lattice, as illustrated in Figure 3.4(e), and the spectrum of the sampled signal will be as shown in Figure 3.4(f), with no aliasing. In this case, the reconstructed signal will be perfect. Rather than using the hexagonal lattice given in Figure 3.1(b), one can still use a rectangular lattice, with a smaller sampling interval $\Delta_1 = \Delta_2 = \sqrt{3}/2 < 1$. This will lead to a higher sampling density. Recall that the sampling densities of the first and second lattices are 1 and $2/\sqrt{3}$, respectively, as derived previously, whereas the preceding scaled rectangular lattice has a sampling density of $4/3$, higher than that of the hexagonal lattice. From this example, we can see that a hexagonal lattice is more efficient than a rectangular one, because it requires a lower sampling density to obtain alias-free sampling.

3.2.3 Sampling Efficiency

As illustrated by Example 3.3, to avoid aliasing, the sampling lattice must be designed so that the Voronoi cell of its reciprocal lattice completely covers the signal spectrum. Obviously, one can always design a very dense lattice to satisfy this requirement; however, it will require a very high sampling density. To minimize the sampling density, one should design the lattice so that its reciprocal Voronoi cell covers the signal spectrum as tightly as possible. The solution will depend on the signal spectrum, which in general could have an arbitrarily shaped support region.

   Fortunately, most real-world signals are symmetric in frequency contents in different directions (after properly scaling the frequency axes), so that their spectrum support regions can be approximated well by a sphere. Therefore, to compare the sampling efficiencies of different lattices, we can evaluate the sampling densities required for a signal that has a spherical support with radius 1. To avoid aliasing, the Voronoi cell of the reciprocal of the sampling lattice, $\mathcal{V}(\Lambda^*)$, must enclose the sphere. The tighter the fit of $\mathcal{V}(\Lambda^*)$ to the sphere, the lower the required sampling density. This leads to

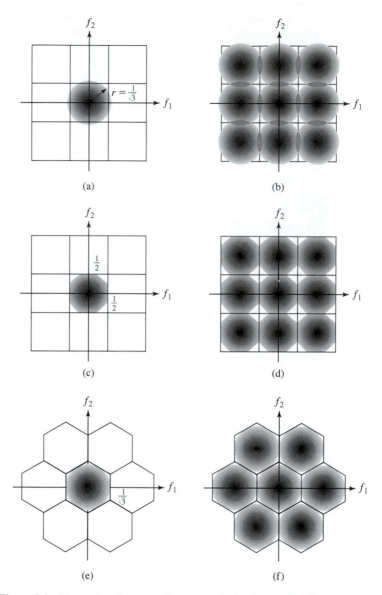

Figure 3.4   Illustration of the sampling process in the spectral domain: (a) the original signal; (b) the sampled signal using the rectangular lattice shown in Figure 3.1(c); (c) the prefiltered signal; (d) the sampled signal after prefiltering; (e) the original signal spectrum overlaid on the hexagonal partition given in Figure 3.1(d); (f) the sampled signal using the hexagonal lattice. In this figure, the darkest shade represents the highest magnitude of the spectrum.

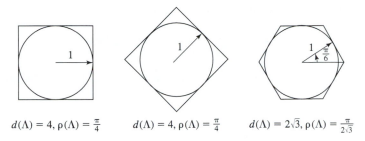

$$d(\Lambda) = 4,\; \rho(\Lambda) = \tfrac{\pi}{4} \qquad d(\Lambda) = 4,\; \rho(\Lambda) = \tfrac{\pi}{4} \qquad d(\Lambda) = 2\sqrt{3},\; \rho(\Lambda) = \tfrac{\pi}{2\sqrt{3}}$$

Figure 3.5   Illustration of the sphere-covering capabilities of 2-D rectangular, diamond, and hexagonal lattices.

the definition of the *sampling efficiency** by

$$\rho(\Lambda) = \frac{\text{volume(unit sphere)}}{\text{volume}(\mathcal{V}(\Lambda^*))} = \frac{\text{volume(unit sphere)}}{d(\Lambda)}. \tag{3.2.12}$$

The closer the above value is to one, the more efficient the lattice is.

Recall that the volume of a unit sphere is π in the 2-D case, and $\tfrac{4}{3}\pi$ in the 3-D case. On the other hand, the minimum cube covering the sphere has a volume of 4 in 2-D, and 8 in 3-D. Therefore, the efficiency of the cubic lattice is $\pi/4$ in 2-D and $\pi/6$ in 3-D. Thus, the cubic lattice is not very efficient for sampling a signal with spherical support.

Figure 3.5 shows the efficiency in covering a sphere of several 2-D lattices. It can be seen that a hexagonal lattice covers the sphere most tightly. For this reason, the sampling efficiency can also be defined relative to the density of the hexagonal lattice; then the efficiency of the hexagonal lattice becomes one.

3.2.4 Implementation of the Prefilter and Reconstruction Filter

The ideal prefilter and reconstruction filter were described in the frequency domain in the preceding discussion. One can obtain the equivalent filters in the spatial domain by taking the inverse CSFT. Although the prefiltering operation must be implemented entirely over the continuous domain, the reconstruction operation can be simplified, because the input signal exists only over sampled points. To derive the equivalent interpolation operation in the spatial domain, let us think of the sampled signal as a continuous signal with impulses at the sampled points, as described by Equation (3.2.2).

---

*An alternative definition of the sampling efficiency [4] is obtained by comparing the volume of $\mathcal{V}(\Lambda^*)$ to that of the unit cube covering the sphere; that is,

$$\rho(\Lambda) = \frac{\text{volume}(\mathcal{V}(\Lambda^*))}{\text{volume(unit cube)}}.$$

With this definition, a smaller value represents a higher efficiency.

The filtering operation can then be simplified to:

$$\psi_r(\mathbf{x}) = \int_{\mathbf{y}\in\mathcal{R}^K} h_r(\mathbf{x}-\mathbf{y})\psi_s(\mathbf{y})\,d\mathbf{y}$$

$$= \sum_{\mathbf{n}\in\mathcal{Z}^K} \psi_s(\mathbf{n}) \int_{\mathbf{y}\in\mathcal{R}^K} h_r(\mathbf{x}-\mathbf{y})\delta(\mathbf{y}-[\mathbf{V}]\mathbf{n})\,d\mathbf{y}$$

$$= \sum_{\mathbf{n}\in\mathcal{Z}^K} \psi_s(\mathbf{n})h_r(\mathbf{x}-[\mathbf{V}]\mathbf{n}). \tag{3.2.13}$$

So the interpolated value at $\mathbf{x}$ is a weighted sum of all sampled values, with the weight for sample $\mathbf{n}$ being $h_r(\mathbf{x}-[\mathbf{V}]\mathbf{n})$. This is the K-D version of the well-known interpolation formula in 1-D.

Note that the ideal prefilter and interpolation filter will have infinite spatial extent because of the sharp transition band in the desired frequency response. In practice, only finite-extent filters are implementable. We discuss the filter design problem in Section 4.1.4.

3.2.5 Relation between Fourier Transforms over Continuous, Discrete, and Sampled Spaces

Note that the sampled signal $\psi_s(\mathbf{n})$ is a special case of the discrete-space signal, considered in Section 2.2. Thus, one can apply the DSFT defined in Equation (2.2.3) to a sampled signal over an arbitrary lattice. Compared to the SSFT, the DSFT does not take into account the actual sampling geometry. By definition, we have

$$\Psi_s(\mathbf{f}) = \sum_{\mathbf{n}\in\mathcal{Z}^K} \psi_s(\mathbf{n})\exp(-j2\pi\mathbf{f}^T[\mathbf{V}]\mathbf{n}), \tag{3.2.14}$$

$$\Psi_d(\mathbf{f}) = \sum_{\mathbf{n}\in\mathcal{Z}^K} \psi_s(\mathbf{n})\exp(-j2\pi\mathbf{f}^T\mathbf{n}). \tag{3.2.15}$$

Comparing these equations, we see that

$$\Psi_d(\mathbf{f}) = \Psi_s([\mathbf{U}]\mathbf{f}), \quad \mathbf{f}\in\mathcal{I}^K, \quad \text{and} \quad \Psi_s(\mathbf{f}) = \Psi_d([\mathbf{V}]^T\mathbf{f}), \quad \mathbf{f}\in\mathcal{V}(\Lambda^*). \tag{3.2.16}$$

This tells us that the DSFT in the unit hypercube $\mathcal{I}^K$ is the image of the SSFT over $\mathcal{V}(\Lambda^*)$ warped to $\mathcal{I}^K$, or the SSFT in $\mathcal{V}(\Lambda^*)$ is the image of the DSFT over $\mathcal{I}^K$ warped to $\mathcal{V}(\Lambda^*)$.* The discrete frequency variable $\mathbf{f}_d$ is related to the continuous frequency variable $\mathbf{f}_c$ by:

$$\mathbf{f}_c = [\mathbf{U}]\mathbf{f}_d = ([\mathbf{V}]^T)^{-1}\mathbf{f}_d, \quad \text{and} \quad \mathbf{f}_d = [\mathbf{V}]^T\mathbf{f}_c = [\mathbf{U}]^{-1}\mathbf{f}_c. \tag{3.2.17}$$

Recalling the relation between $\Psi_s(\mathbf{f})$ and $\Psi_c(\mathbf{f})$ in Equation (3.2.8), we have:

$$\Psi_d(\mathbf{f}) = d(\Lambda) \sum_{\mathbf{m}\in\mathcal{Z}^K} \Psi_c([\mathbf{U}]\mathbf{f}-[\mathbf{U}]\mathbf{m}). \tag{3.2.18}$$

---

*Note that the spatial region corresponding to $[\mathbf{U}]\mathbf{f}$ for $\mathbf{f}\in\mathcal{I}^K$ does not correspond to $\mathcal{V}(\Lambda^*)$ in general. But because of the periodicity of both $\Psi_s(\mathbf{f})$ and $\Psi_d(\mathbf{f})$, the function $\Psi_d(\mathbf{f})$ over $\mathbf{f}\in\mathcal{I}^K$ comes from different parts of $\mathcal{V}(\Lambda^*)$. This is further considered in Problem 3.6.

If $\psi_c(\mathbf{x})$ is band-limited within $\mathcal{V}(\Lambda^*)$, then

$$\Psi_d(\mathbf{f}) = d(\Lambda)\Psi_c([\mathbf{U}]\mathbf{f}), \quad \mathbf{f} \in \mathcal{I}^K. \tag{3.2.19}$$

Obviously, $\Psi_d(\mathbf{f})$ is periodic with repetition centers at all integer points, which correspond to the points in Λ^* in the continuous frequency domain.

Remember that in the 1-D case, if we sample a continuous signal with a sampling frequency f_s, the discrete frequency is related to the continuous frequency by $f_d = f_c/f_s$, and the basic period $(-1/2, 1/2)$ in the discrete frequency domain corresponds to the basic period $(-f_s/2, f_s/2)$ in the continuous frequency domain. This is a special case of the situation described here for the K-D signal, with $[\mathbf{V}] = [1/f_s]$, $[\mathbf{U}] = [f_s]$.

3.3 SAMPLING OF VIDEO SIGNALS

The preceding sections presented the sampling theory for general K-D signals. In the remaining sections of this chapter, we focus on 3-D video signals. The two spatial dimensions and the temporal dimension are asymmetric, in that they have different characteristics and that the visual sensitivities to the spatial and temporal frequencies are different. This asymmetry has led to the development of several interesting techniques for video sampling.

Although a video signal is continuously varying in space and time, cameras of today cannot capture the entire signal continuously in all three dimensions. Most motion picture cameras sample a scene in the temporal direction, and store a sequence of analog image frames on film. On the other hand, most TV cameras capture a video sequence by sampling it in temporal and vertical directions. The resulting signal is stored in a 1-D raster scan, which is a concatenation of the color variations along successive horizontal scan lines. To obtain a full digital video, one can sample in 2-D the analog frames resulting from a motion picture camera, sample in 1-D along the raster scan resulting from a TV camera, or acquire discrete video frames directly using a digital camera by sampling a scene in 3-D. Different sampling schemes correspond to different sampling lattices. In designing a video sampling system, two questions to be answered are: (1) What are the necessary sampling rates for video? and (2) Which sampling lattice is the most efficient under a given total sampling rate? In this section, we first describe the factors that determine the necessary sampling rates for video. We then discuss 2-D sampling schemes, and compare progressive and interlaced scans. Next, we introduce several 3-D sampling lattices and compare their efficiency. Finally, we illustrate the effect of spatial and temporal aliasing by presenting a special example.

3.3.1 Required Sampling Rates

The fundamental question one must answer when designing a video sampling system, in either 1-D, 2-D, or 3-D, is what the spatial and temporal sampling resolutions should be. The answer is governed by several factors: the frequency content of the underlying signal; the visual thresholds in terms of the spatial and temporal cut-off frequencies; the capture and display device characteristics; and, finally, the affordable processing, storage, and transmission cost. Based on the sampling theorem, if one uses a cubic lattice the sampling rate in each dimension should be at least twice the highest frequency along that direction. On the other hand, from Section 2.4, we know that the human eye cannot distinguish spatial and temporal variations beyond certain high frequencies. Given that the maximum frequency in the signal could vary significantly, the visual cut-off frequencies, which are the highest spatial and temporal frequencies that can be observed by the HVS, should be the driving factor in determining the sampling rates for video; there is no need to accommodate frequency components beyond these values. It is fair to assume that the technology can always be advanced to accommodate the human demand.

As described in Section 2.4, visual sensitivity depends on the mean brightness of the display. For TV signals, which are very bright, the visual thresholds call for a frame rate of greater than 70 Hz, and a spatial resolution of at least 30 cpd. At a normal viewing distance of three times the screen height, a spatial frequency of 25 cpd translates to about 500 line/frame (lpf). To sample each line, the horizontal sampling interval should approximately match the vertical interval, so that the resulting pixels are square (i.e., have a PAR of 1). This leads to about 670 pixel/line for a display with 500 lines and an IAR of 4:3.

These sampling rates, required by the visual cut-off frequencies, were beyond what practical technologies could offer at the time analog TV systems were developed. In order to reduce the data rate and consequently the cost for video capture, transmission, and display, interlaced scan was developed, which trades off vertical resolution for increased temporal resolution, for a given total data rate (product of the frame and line rates). For example, as explained in Section 1.4, in the NTSC TV system, 60 fields are captured per second, but each field contains only half the desired number of lines (240 active line/field). The total data rate is the same as that with a progressive scan using 30 fps and 480 lpf. By interlacing the lines scanned in each field, it can produce the same quality as a progressive scan using 60 fps and 480 lpf, if the underlying scene is stationary. But for a high-motion scene with special patterns (vertical line patterns), it can lead to the notorious "interlacing artifact." The next generation of digital TV systems can operate in either the 30 fps and 480 lpf mode, or the 60 field/s and 240 line/field mode. The interlaced format is retained mainly for compatibility with the analog TV system. The HDTV systems further enhance the visual impact by employing a wider screen with an IAR of 16:9, and a sampling resolution of 60 fps and 720 lpf. Again, for compatibility purpose, an interlaced format with 60 field/s and 540 line/field can also be used (see Table 1.3).

For motion pictures, because of the reduced visual sensitivity in a movie theater where the ambient brightness is kept low, a frame rate of 24 fps (progressive) is usually sufficient. As described in Section 2.4.1, although the original image is captured at 24 fps, when it is played back, a blade is placed before the projection lens that rotates three times per frame, so that the effective playback rate is 72 fps. This suppresses the flicker artifacts that might otherwise be experienced by some acute viewers. Most older motion pictures were made using a screen IAR of 4:3; for more dramatic visual impact, newer motion pictures are usually made with an IAR of up to 2:1.

For computer displays, much higher temporal and spatial sampling rates are needed. For example, an SVGA display has a frame rate of 72 fps (progressive) and a spatial resolution of 1024×720 pixels. This is to accommodate the very close viewing distance (normally between one and two times the picture-height) and the high-frequency content of the displayed material (line graphics and text).

3.3.2 Sampling Video in Two Dimensions: Progressive versus Interlaced Scans

We described the mechanism for raster scan in Section 1.3, including both progressive and interlaced scans. It should be clear that a video raster is actually a version of a 3-D video signal sampled in the temporal and vertical directions. As already hinted, the motivation for using the interlaced scan is to trade off vertical resolution for an enhanced temporal resolution, given the total number of lines that can be recorded within a given time. In this section, we attempt to gain more insight into the artifacts associated with these two sampling schemes by analyzing their aliasing patterns in the spectral domain.

To ease the discussion, we ignore the horizontal direction, and consider the video signal as a 2-D signal in the space spanned by the temporal and vertical directions. Let Δ_t represent the field interval and Δ_y the line interval. Then the sampling lattices employed by progressive and interlaced scans are as shown in Figure 3.6(a) and (b), respectively. In the figure, we also indicate the basis vectors for generating each lattice. From these basis vectors, we derive the following generating matrices for the original and reciprocal lattices:

$$\text{Progressive Scan:} \quad [\mathbf{V}_1] = \begin{bmatrix} 2\Delta_t & 0 \\ 0 & \Delta_y \end{bmatrix}, \quad [\mathbf{U}_1] = \begin{bmatrix} 1/2\Delta_t & 0 \\ 0 & 1/\Delta_y \end{bmatrix}, \quad (3.3.1)$$

$$\text{Interlaced Scan:} \quad [\mathbf{V}_2] = \begin{bmatrix} 2\Delta_t & \Delta_t \\ 0 & \Delta_y \end{bmatrix}, \quad [\mathbf{U}_2] = \begin{bmatrix} 1/2\Delta_t & 0 \\ -1/2\Delta_y & 1/\Delta_y \end{bmatrix}. \quad (3.3.2)$$

Based on the matrices $[\mathbf{U}_1]$ and $[\mathbf{U}_2]$, we draw the reciprocal lattices in Figure 3.6(c) and (d), in which we have highlighted the three points (filled circles) that are nearest to the origin in the first quadrant of the frequency plane. These are the centers of the alias components that are closest to the original signal spectrum, and are the primary causes for perceivable distortion.

Note that when drawing the lattices, we have scaled the spatial and temporal dimensions so that a spatial frequency equal to the vertical sampling rate $f_{s,y} = 1/\Delta_y$ is assigned the same length as a temporal frequency equal to the field rate $f_{s,t} = 1/\Delta_t$. Similarly, the spatial interval Δ_y and the temporal interval Δ_t are given the same length. Ideally, we would like to equate the temporal and spatial frequencies based on the visual sensitivity, by assigning the same length to the spatial and temporal cut-off frequencies. Because the spatial and temporal sampling rates are usually chosen based on the respective cut-off frequencies, equating the sampling frequencies in different directions is in general appropriate.

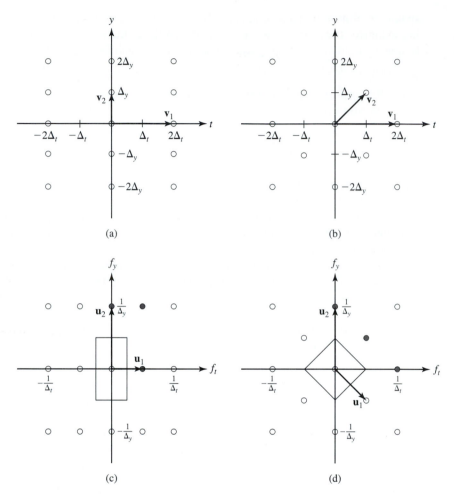

Figure 3.6 Comparison of progressive and interlaced scans: (a) sampling lattice for progressive scan; (b) sampling lattice for interlaced scan; (c) reciprocal lattice for progressive scan; (d) reciprocal lattice for interlaced scan. Filled circles in (c) and (d) indicate the nearest aliasing components.

Comparing the original and reciprocal lattices of these two scans, we arrive at the following observations:

1. They have the same sampling density, that is, $d(\Lambda_1) = d(\Lambda_2) = 1/2\Delta_t\Delta_y$.
2. They have the same nearest aliases, at $1/\Delta_y$ along the vertical frequency axis. This means that in the absence of motion, the two sampling lattices have the same vertical resolving power. This is because although there are only half the number of lines in each field, when the imaged scene is stationary, the lines sampled at two separate field times appear as if they were sampled at the same time. When

there is motion in the scene, however, the vertical resolution of the interlaced scan is lower than that of the progressive scan.

3. They have different nearest aliases along the temporal frequency axis. For the progressive scan, the first alias occurs at $1/2\Delta_t$, whereas for the interlaced scan, this occurs at $1/\Delta_t$. Recall that a temporal frequency component that is larger than half the nearest temporal alias will cause a flicker artifact. We can see that with the interlaced scan flickering is less likely to happen when the object has a flat or slowly varying vertical pattern.

4. They have different *mixed aliases*. The mixed alias is defined as the nearest off-axis alias component. A frequency component that is close to the mixed alias causes *interline flicker* and *line crawl*. For the progressive scan, the mixed alias occurs at $(1/2\Delta_t, 1/\Delta_y)$, whereas for the interlaced scan, this occurs at $(1/2\Delta_t, 1/2\Delta_y)$. Because the interlaced scan has a mixed alias that is closer to the origin, interline flicker and line crawl are more visible in interlaced scans—these are the notorious interlacing artifacts.

5. For a signal with isotropic spectral support, the interlaced scan is more efficient. Specifically, the maximum radius of the signal spectrum that can be represented without aliasing is equal to $1/4\Delta_t$ with the progressive scan, whereas this is increased to $1/2\sqrt{2}\Delta_t$ with the interlaced scan. In interpreting this result, it is important to remember that this is based on our way of equating the spatial and temporal frequencies.

Note that the preceding comparisons are between interlaced and progressive scans that have the same total sampling rate. In this case, the interlaced scan has some advantages over the progressive scan, as enumerated. To improve the quality further, the interlaced scan can be deinterlaced, to yield a progressive scan that has twice the total sampling rate. (We consider the deinterlacing problem in Section 4.2.1.)

An interesting analysis of the spectrum of an interlaced frame under constant linear motion is given by Beuker and Shah [1]. Based on this analysis, they consider techniques for deinterlacing and other rate conversion problems.

3.3.3 Sampling a Raster Scan: BT.601 Format Revisited

As described in the beginning of this chapter, one way to obtain a fully digital video signal is to sample a raster scan. Recall that a raster is a 1-D signal consisting of successive horizontal scan lines in successive frames (or fields, for an interlaced raster). Therefore, the sampling interval along the scan line directly determines the horizontal sampling interval. To determine the sampling interval, several factors must be taken into account. First, the resulting horizontal sampling spacing should match the vertical spacing between scan lines, so that the sampling frequencies in the horizontal and vertical directions are similar. Second, the resulting samples in the 3-D space should follow a desired sampling lattice. For example, if the samples on each frame or field are to form a rectangular grid, the sampling interval should be an integer divisor of the line interval. These criteria were used in designing the BT.601 digital video format described in Section 1.5.2. By selecting the sampling interval according to Equation (1.5.2), the samples on successive horizontal scan lines are vertically aligned, with the horizontal and vertical sampling spacings being approximately equal.

The preceding discussion assumes that the underlying video raster has a single luminance component. For a color video raster, with one luminance and two chrominance components, a straightforward approach is to use the same sampling frequency for all components. This will lead to the BT.601 4:4:4 format, previously shown in Figure 1.8(a). However, it is well known that the HVS is more sensitive to the luminance information than to the chrominance. Therefore, it is wise to sample the chrominance components at resolutions lower than that for the luminance. Ideally, sampling frequencies in all three dimensions should be reduced equally. However, because a raster is a stack of horizontal scan lines, reducing the sampling frequency for the raster only lowers the horizontal sampling rate; the vertical and temporal sampling rates are unaffected. For example, if the sampling rate for the chrominance components, $f_{s,c}$, is half of that for the luminance component, f_s, the temporal and vertical sampling intervals are the same for the chrominance component and the luminance component, but the horizontal sampling interval for the chrominance is twice that for the luminance. In fact, this is the color subsampling format used in the BT.601 4:2:2 format shown in Figure 1.8(b). If we down-sample the chrominance component by a factor of 4 along the raster scan, we will arrive at the BT.601 4:1:1 format of Figure 1.8(c). Given a 4:2:2 signal, we can further down-sample the vertical dimension by a factor of 2, which will lead to the BT.601 4:2:0 format, given in Figure 1.8(d). Note that the 4:2:0 format cannot be obtained directly from the raster signal by uniform sampling. Conversion between BT.601 4:2:2 and 4:2:0 signals is considered in Chapter 4 (Problems 4.6–7).

The 4:1:1 and 4:2:0 formats have the same overall sampling rates (samples/s), but the 4:2:0 format can represent a video signal more accurately because it captures horizontal and vertical details in the chrominance components with the same resolution, half of that for the luminance component. On the other hand, with the 4:1:1 format, the vertical detail is represented with the same resolution as the luminance component, whereas the horizontal is represented with only a quarter of that resolution.

The preceding discussion assumes that the three color components in the raster signal are separated. To sample a composite color video, one must first separate the individual color components, and then perform sampling as discussed.

3.3.4 Sampling Video in Three Dimensions

In the sampling schemes described in the previous section, the horizontal samples are aligned vertically in all fields. Such a sampling scheme is by no means optimal. One can also take samples in an interlaced or more complicated pattern over the x-y plane. More generally, one can directly sample the 3-D space using a desired lattice. It is difficult to make a camera that can implement a complicated 3-D sampling structure. However, one can first acquire samples using a dense but simply structured lattice and then down-convert it to the desired lattice. In this section, we compare the efficiency of several lattices that have been used for video sampling. The materials presented here are from

[4]. To ease the presentation, we will assume that the frequency axes, f_x, f_y, f_t, are scaled by the maximum frequencies of the signal, $1/f_{x,max}$, $1/f_{y,max}$, $1/f_{t,max}$, respectively. The horizontal, vertical, and temporal axes, x, y, t, are correspondingly scaled by $f_{x,max}$, $f_{y,max}$, $f_{t,max}$. The following discussion is in terms of the scaled unitless variables. To evaluate the sampling efficiency of each lattice, we further assume that the support region of the signal is a sphere with radius 1, in the scaled frequency space.

First, let us consider the sampling of a progressively scanned raster with frame interval Δ_t and line interval Δ_y. If the samples are aligned vertically with a horizontal interval of Δ_x, then the equivalent 3-D sampling lattice is simply cubic or orthorhombic (ORT), as given in Figure 3.7(a). Obviously, to avoid aliasing, we must have $\Delta_x = \Delta_y = \Delta_t = \frac{1}{2}$, with a sampling density $d(\text{ORT}) = 8$.

Next, let us consider the sampling of an interlaced raster with a field interval of $\Delta_t/2$, and line interval within each field of $2\Delta_y$. If the samples in different fields are aligned, with horizontal interval Δ_x, the equivalent 3-D sampling lattice is illustrated in Figure 3.7(b), and is referred to as ALI. It can be shown that to avoid aliasing, we must have $\Delta_x = \Delta_y = \frac{1}{2}$, $\Delta_t = \frac{1}{\sqrt{3}}$, which leads to a sampling density of $d(\text{ALI}) = 4\sqrt{3}$. This is more efficient than the ORT lattice.

To improve the sampling efficiency, we can shift the samples in the even and odd lines by half, as shown in Figure 3.7(c). This result is referred to as a body-centered orthorhombic lattice (BCO). It can be shown that the required sampling intervals for avoiding aliasing are $\Delta_x = \Delta_t = \frac{1}{\sqrt{2}}$, $\Delta_y = \frac{1}{2\sqrt{2}}$, with a sampling density of $4\sqrt{2}$. Therefore, this lattice is more efficient than the ALI.

A variation of the BCO sampling lattice, known as the face-centered orthorhombic (FCO), is shown in Figure 3.7(d). With this sampling structure, each field contains all the lines, but the samples in the same field are interleaved in the even and odd lines. The sampling efficiency of this lattice is the same as that of the BCO, but because samples in the even and odd fields are more evenly blended, it usually leads to a better visual quality.

The minimum sampling density and associated sampling intervals for these lattices are summarized in Table 3.1.

TABLE 3.1   ALIAS-FREE SAMPLING LATTICES FOR SIGNALS WITH A UNIT-SPHERE SPECTRUM.

Lattice	Δ_x	Δ_y	Δ_t	Density	Efficiency
ORT	1/2	1/2	1/2	8	$\pi/6 = 0.524$
ALI	1/2	1/2	$1/\sqrt{3}$	$4\sqrt{3}$	$\pi/3\sqrt{3} = 0.605$
BCO	$1/\sqrt{2}$	$1/2\sqrt{2}$	$1/\sqrt{2}$	$4\sqrt{2}$	$\pi/3\sqrt{2} = 0.740$
FCO	1/2	1/2	$1/\sqrt{2}$	$4\sqrt{2}$	$\pi/3\sqrt{2} = 0.740$

3.3.5 Spatial and Temporal Aliasing

In Section 2.3.2, we described the relation between the temporal frequency, spatial frequency, and velocity of an object undergoing a linear motion. In general, the temporal frequency of an object is higher if it has a higher velocity (unless it is moving in a

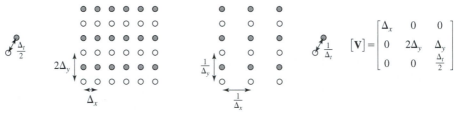

(a) Orthorhombic lattice (ORT)

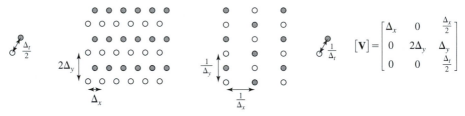

(b) Lattice obtained by vertically aligned sampling of 2:1 interlaced signal (ALI)

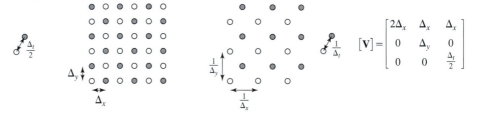

(c) Body-centered orthorhombic lattice (BCO)

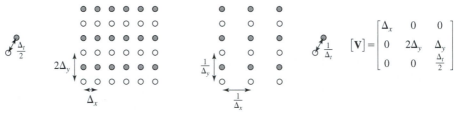

(d) Face-centered orthorhombic lattice (FCO)

Figure 3.7   Several 3-D video sampling lattices and their reciprocals. In each part of the figure, the sampling lattice is on the left, and the reciprocal lattice is on the right. The matrix $[\mathbf{V}]$ indicates the generating matrix for the sampling lattice. The open and filled circles represent lattice points on two different time (left) or temporal frequency (right) planes, with a distance indicated by the arrows shown on the left and right, respectively. For example, for (b), on the left, the open circles represent the samples taken at time $t = 0$, and the filled circles at time $t = \Delta_t/2$. On the right, the open circles represent the aliasing components at temporal frequency $f_t = 0$, and the filled circles at $f_t = 1/\Delta_t$. For (a), the filled circles fall directly behind the open circles and are therefore not shown.

direction orthogonal to the spatial frequency direction). Here, we illustrate the effect of temporal aliasing by again considering an object with linear motion.

Example 3.4

Consider a scene with a pure vertical sinusoidal color pattern with a horizontal frequency of 1 cycle/cm (see Figure 2.1(a)). Suppose that the pattern is moving horizontally with a velocity $v_x = 3$ cm/s. If we sample it using a rate of $f_{s,x} = f_{s,y} = 3$ samples/cm, $f_{s,t} = 3$ fps, what will be the apparent motion and sinusoidal frequency?

Because the signal has a spatial frequency of 1 cycle/cm and is moving at 3 cm/s, it advances 3 cps or 1 cycle/frame at a temporal sampling rate of 3 fps. Thus the captured frames at two succeeding frames are exactly the same, as illustrated in Figure 3.8(a). Because the spatial sampling rate is high enough ($f_{s,x} > 2 f_{x,0}$), there is no spatial aliasing.

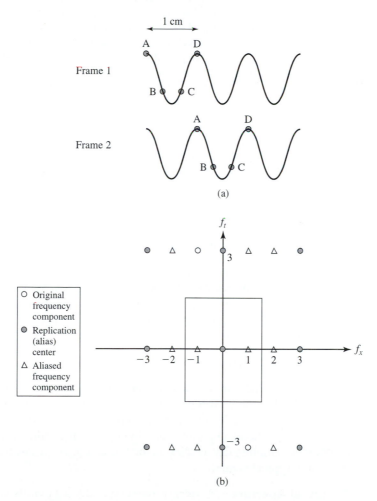

Figure 3.8   Sampling a sinusoidal pattern (see Example 3.4).

Therefore, the sampled signal will appear as a stationary sinusoidal with the same pattern as the continuous signal.

Now, let us explain this phenomenon using the lattice sampling theorem. Obviously, this signal has a pair of spatial frequency components at $(f_{x,0}, f_{y,0}) = \{(1, 0), (-1, 0)\}$. Based on Equation (2.3.4), the temporal frequency caused by the motion is $f_{t,0} = -(f_{x,0}v_x + f_{y,0}v_y) = \{-3, 3\}$ (cps). Therefore, the spectrum of the signal has a single pair of impulses at frequencies $(f_{x,0}, f_{y,0}, f_{t,0}) = \{(1, 0, -3), (-1, 0, 3)\}$, as illustrated in Figure 3.8(b). (In this figure, we have only drawn the $f_x - f_t$ plane at $f_y = 0$, as the signal is zero in all other planes.) After sampling, the signal spectrum will replicate at all multiples of the sampling frequencies, as shown in the figure. We can see that because $f_{s,t} < 2f_{t,0}$, a pair of alias components fall in the Voronoi region of the reciprocal lattice, which in this case is simply a rectangular region bounded by $\pm f_{s,x}/2$ and $\pm f_{s,t}/2$. In fact, the original signal component falls out of this region. If an ideal low-pass filter is used to recover the continuous signal from the sampled signal, then the recovered signal would be a pure sinusoid with a spatial frequency of $(1, 0)$ and a temporal frequency of 0. This corresponds to a stationary vertical sinusoid with horizontal frequency of 1 cycle/cm. This is the signal that will be perceived by the eye, which functions as a low-pass filter.

In order to retain the true motion in the sampled signal, the temporal sampling rate should be at least $2f_{t,0} = 6$ fps. At this rate, each frame is the previous frame moved by half a cycle.

Example 3.4 presents an extreme case of the temporal aliasing effect: a moving object becomes stationary. More generally, a rapidly moving object may appear to move at a slower speed when the frame rate is not high enough. Depending on the spatial pattern of the moving object, the directions of some line patterns may also be altered. This is why a moving stagecoach wheel in a Western sometimes appears to turn backwards and move slower.

3.4 FILTERING OPERATIONS IN CAMERAS AND DISPLAY DEVICES

The previous sections presented the sampling theory for general K-D signals and video signals. We have derived, for several typical video formats, the required prefilters and reconstruction filters. In this section, we discuss how practical cameras and display devices accomplish these tasks in a crude way, and how the HVS partially accomplishes the required interpolation task.

3.4.1 Camera Apertures

Consider a camera that samples a continuously varying scene at regular intervals $\Delta_x, \Delta_y, \Delta_t$ in the horizontal, vertical, and temporal directions, respectively. This corresponds to using a simple cubic lattice. The sampling frequencies are $f_{s,x} = 1/\Delta_x$, $f_{s,y} = 1/\Delta_y$, $f_{s,t} = 1/\Delta_t$. The ideal prefilter should be a low-pass filter with cut-off frequencies at half of the sampling frequencies. In the following, we discuss the actual prefilters implemented in typical cameras.

Temporal Aperture   A video camera typically accomplishes a certain degree of prefiltering in the capturing process. First, the intensity values read out at any frame instant are not the sensed values at that time; rather, they are the averages of the sensed signal over a certain time interval, Δ_e, known as the exposure time. Therefore, the camera is applying a prefilter in the temporal domain with an impulse response of the form:

$$h_{p,t}(t) = \begin{cases} \frac{1}{\Delta_e} & t \in (0, \Delta_e), \\ 0 & \text{others.} \end{cases} \tag{3.4.1}$$

The frequency response of this filter is:

$$H_{p,t}(f_t) = \exp(-j\pi f_t \Delta_e) \frac{\sin(\pi f_t \Delta_e)}{\pi f_t \Delta_e}, \tag{3.4.2}$$

We can see that it reaches zero at $f_t = 1/\Delta_e$. Recall that $1/\Delta_t$ is the temporal sampling rate, and that the ideal prefilter for this task is a low-pass filter with cut-off frequency at half of the sampling rate. By choosing $\Delta_e \geq \Delta_t$, the camera can suppress temporal alias components near the sampling rate. But too large a Δ_e will blur the signal. In practice, the effect of blurring is sometimes more noticeable than aliasing. Therefore, the exposure time Δ_e must be chosen to reach a proper trade-off between aliasing and blurring effects.

Spatial Aperture   In addition to temporal integration, the camera also performs spatial integration. The value read out at any pixel (a position on a scan line in a tube-based camera or a sensor in a CCD camera) is not the optical signal at that point alone, but rather a weighted integration of the signals in a small window surrounding it, called the *aperture*. The shape of the aperture and the weighting values constitute the camera's *spatial aperture function*. This aperture function serves as the spatial prefilter, and its Fourier transform is known as the *modulation transfer function* (MTF) of the camera. With most cameras, the spatial aperture function can be approximated by a circularly symmetric Gaussian function,

$$h_{p,x,y}(x, y) = \frac{1}{\sqrt{2\pi}\sigma} \exp(-(x^2 + y^2)/2\sigma^2). \tag{3.4.3}$$

The spectrum of this function is also Gaussian:

$$H_{p,x,y}(f_x, f_y) = \exp\left(-(f_x^2 + f_y^2)/2\beta^2\right), \quad \text{with } \beta = \frac{1}{2\sigma\pi}. \tag{3.4.4}$$

The value of σ or β depends on the aperture size and shape, which are usually designed so that the frequency response is 0.5 at half of the vertical and horizontal sampling rates. Assuming $f_{s,x} = f_{s,y} = f_{s,x,y}$, we see that this requires $\beta = f_{s,x,y}/2\sqrt{\ln 2}$. For the MTF of several camera tubes, see [6].

Combined Aperture   The overall camera aperture function or prefilter is

$$h_p(x, y, t) = h_{p,t}(t)h_{p,x,y}(x, y), \tag{3.4.5}$$

with a frequency response of

$$H_p(f_x, f_y, f_t) = H_{p,t}(f_t)H_{p,x,y}(f_x, f_y). \tag{3.4.6}$$

The impulse response of a camera with $\Delta_e = \Delta_t = 1/60$ second and $f_{s,y} = 480$ (line/picture-height) is shown in Figure 3.9(a); its frequency response is given in Figure 3.9(b). Only the $f_y - f_t$ plane at $f_x = 0$ is shown. Obviously, it is far from the ideal half-band low-pass filter, which should have a square pass-band defined by $|f_y| \leq f_{s,y}/2, |f_t| \leq f_{s,t}/2$. On one hand, it attenuates the frequency components inside the desired pass-band (the Voronoi cell), thereby reducing the signal resolution unnecessarily; on the other hand, it does not completely remove the frequency components in the desired stop-band, which will cause aliasing in the sampled signal. It has been found that viewers are more annoyed by loss of resolution than aliasing artifacts.

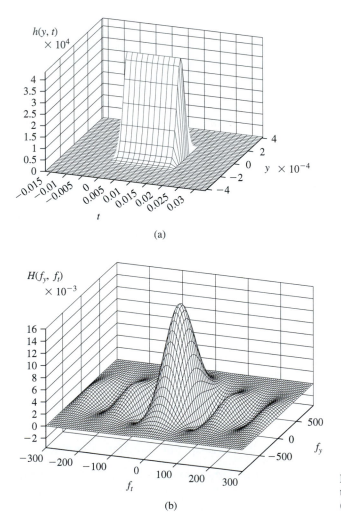

(a)

(b)

Figure 3.9 The aperture function of a typical camera: (a) impulse response, (b) frequency response.

This is partly because aliasing artifacts only cause noticeable visual artifacts when the image contains high-frequency periodic patterns that are close to the lowest aliasing frequency, which are rare in natural scene imagery. For this reason, preservation of the signal in the pass-band is more important than suppression of the signal outside the pass-band.

In order to accomplish more precise prefiltering, one can use digital filters. This would involve three steps: (1) sample the signal at a rate higher than the desired sampling rate, (2) use a digital filter to suppress the frequencies outside the desired pass-band, and (3) down-convert this digital signal to the desired sampling rate. Although a filter with a very sharp transition gives better results in terms of mean squared error, such a filter will cause ringing at sharp edges, known as the Gibbs effect. Furthermore, to realize a sharp transition, a very high-order filter is needed, which may not be feasible in video applications.

3.4.2 Display Apertures

As described in Section 1.2, in a CRT monitor, an electronic gun emits an electronic beam across the screen line by line, striking phosphors with intensities proportional to the intensity of the video signal at corresponding locations. To display a color image, three beams are emitted by three separate guns, striking red, green, and blue phosphors with the desired intensity combination at each location. The beam thickness essentially determines the vertical filtering: a very thin beam will make the image look sharper but will also cause the perception of scan lines if the viewer sits too close to the screen; on the other hand, a thicker beam will blur the image. Normally, to minimize the loss of spatial resolution, thin beams are used so that very little vertical filtering is exercised by the display device. Temporal filtering is determined by the phosphors. The P22 phosphors used in color TVs decay to less than 10 percent of peak responses in 10 μs to 1 ms [3, 4], much shorter than the field time (16.7 ms). Thus, virtually no temporal filtering is performed.

Fortunately, the HVS has a low-pass or band-pass characteristic, depending on the regime of the temporal and spatial frequencies in the imagery (this was described in Section 2.4). Therefore, the eye performs to some degree the required interpolation task. For improved performance, one can use a digital filter to up-convert the sampled signal to a higher resolution, which is then fed to a high-resolution display system.

It has been found that the combination of camera and display apertures in most current television systems results in a maximum vertical resolution of only about 0.7 of the theoretical limit (which is half of the line rate). In other words, the maximum vertical frequency that can be rendered properly by a TV system with a line number of N_y is about $0.7N_y/2$. This factor is referred to as the *Kell factor* [4], denoted by K. For a discussion of how the parameters of an imaging system can influence the Kell factor, the reader is referred to [8]. By using digital filters for prefiltering and interpolation, as described, Kell factors much closer to unity may be possible. In Section 1.4.3, we saw how the Kell factor affects the bandwidth of a TV signal.

3.5 SUMMARY

Sampling of K-D Signals over Arbitrary Lattices (Section 3.2)

- A K-D continuous-space signal can be sampled over a lattice. One can define a sampled-space Fourier transform (SSFT), which satisfies similar properties as the DSFT.

- An SSFT is periodic with alias centers located at points of the reciprocal of the sampling lattice.

- The original signal can be reconstructed perfectly from the sampled signal if the spectrum (i.e., CSFT) of the original signal is nonzero only within the Voronoi cell of the reciprocal lattice.

- The sampling lattice should be designed such that the support region of the signal falls within the Voronoi cell of the reciprocal lattice. Otherwise, aliasing would occur. Alternatively, one can design a prefilter to limit the support region of the signal.

- A sampling lattice is more efficient if the Voronoi cell of its reciprocal lattice covers the support region of the signal more tightly. For example, for a signal with a circular spectrum support, a rectangular sampling lattice is less efficient than a hexagonal lattice.

Sampling of Video Signals (Section 3.3)

- The spatial and temporal sampling frequencies required for a video signal depend on visual thresholds (i.e., the cut-off frequencies of the spatiotemporal frequency response of the HVS) in addition to the spectrum content of the signal.

- Interlaced scanning is a way to trade off vertical resolution for enhanced temporal resolution. An interlaced scan can be considered to use a nonrectangular sampling lattice on the vertical-temporal plane.

- More generally, one can use interleaved scans in both horizontal and vertical dimensions. These correspond to noncubic lattices in 3-D. Compared to cubic lattices, they can reduce aliasing artifacts under a given total sampling rate.

3.6 PROBLEMS

3.1 Suppose that the basis vectors of two sampling lattices are:

$$\text{Lattice A: } \mathbf{v}_1 = [\sqrt{3}, 1]^T \quad \text{and} \quad \mathbf{v}_2 = [0, 2]^T;$$
$$\text{Lattice B: } \mathbf{v}_1 = [2\sqrt{2}, 0]^T \quad \text{and} \quad \mathbf{v}_2 = [\sqrt{2}, \sqrt{2}]^T;$$

For each of these sampling lattices:

(a) Sketch the basis vectors and the sample points. Illustrate the Voronoi unit cell. Illustrate how the entire spatial domain is tiled by the shifted versions of the unit cell. Determine the sampling density.

(b) Determine the basis vectors of the reciprocal lattice. Repeat part (a) for this lattice.

(c) For a signal with a circular spectrum (i.e., the support region is circle), which lattice is better? Construct a spectrum (sketch its support region) that will cause aliasing by one lattice but not the other.

(d) The basis vectors for a given lattice are not unique. Can you find another set of basis vectors for these lattices?

3.2 Prove Theorem 3.2.

3.3 Prove Theorem 3.3.

3.4 Prove Theorem 3.5.

3.5 Consider a linear shift-invariant system over the continuous space $\mathcal{R}^K$ with an impulse response $h_c(\mathbf{x})$. Show that if one samples the input and output signals over a lattice Λ, the sampled input signal $\psi_s(\mathbf{n}) = \psi_c([\mathbf{V}]\mathbf{n})$ and output signal $\phi_s(\mathbf{n}) = \phi_c([\mathbf{V}]\mathbf{n})$ are related by a sampled space convolution in Λ, and the filter will be exactly the sampled impulse response $h_s(\mathbf{n}) = h_c([\mathbf{V}]\mathbf{n})$.

3.6 In Section 3.2.5, we showed that the DSFT in the unit hypercube $\mathcal{I}^K$ is the warped image of the SSFT in a region warped from $\mathcal{I}$ using the transformation $[\mathbf{U}]\mathbf{f}$. Illustrate the warping process, for the case where the sampling lattice is hexagonal. More specifically, consider a signal with a spherical support. Start with a signal sampled over the hexagonal lattice, and draw both SSFT and DSFT.

3.7 Consider a 2-D signal defined by

$$\psi(x, y) = \frac{1}{\sqrt{2\pi}} e^{-\frac{x^2+y^2}{2}}. \tag{3.6.1}$$

(a) Determine its CSFT.

(b) Suppose we want to sample this signal using a hexagonal lattice Λ given by

$$[\mathbf{V}] = \alpha \begin{bmatrix} \sqrt{3}/2 & 0 \\ 1/2 & 1 \end{bmatrix} \tag{3.6.2}$$

Choose an appropriate scaling factor α so that the aliasing effect is not severe. For example, choose α so that $\Psi(f_x, f_y) = 0.1$ at the border of the Voronoi cell of Λ^*.

(c) Determine the sampled signal and its spectrum (SSFT).

(d) Draw the original continuous signal, the sampled signal, and the continuous and sampled spectrums using MATLAB. Do you see the repeated peaks in the SSFT?

(e) Determine and draw the DSFT of the sampled signal, and compare it with the SSFT. Are the DSFT and SSFT related by the warping relation described by Equation (3.2.16)?

3.8 For the 3-D sampling lattices shown in Figure 3.7, prove that, in order to avoid aliasing, the sampling intervals should be chosen as in Table 3.1 if the support region of the signal is a unit sphere.

3.9 Consider a sinusoidal bar pattern described by

$$\psi(x, y) = \sin(4\pi(x - y)).$$

Assume that the unit for x and y directions is the meter (m). Suppose that this pattern is moving at a speed of (v_x, v_y) m/s. Determine the necessary sampling rates in horizontal, vertical, and temporal directions, for the following velocities:

(a) $(v_x, v_y) = (1, 1)$

(b) $(v_x, v_y) = (-1, 1)$

(c) $(v_x, v_y) = (2, 1)$

3.10 (Continued from Problem 3.9). Suppose that the sinusoidal bar is moving at a speed of $(v_x, v_y) = (3, 0)$. If the signal is sampled at $f_{s,x} = f_{s,y} = 6$ sample/m, and $f_{s,t} = 6$ sample/s, what are the apparent spatial frequency, the orientation, and the speed of the sampled bar pattern?

3.11 Consider a scene under constant linear motion, such that

$$\psi(x, y, t) = \psi(x - v_x t, y - v_y t, 0) = \psi_0(x - v_x t, y - v_y t). \qquad (3.6.3)$$

If this continuous imagery is sampled using an interlaced lattice to form even and odd fields, how is the DSFT of the even field at time t related to the CSFT of $\psi_0(x, y, t)$? What about the DSFT of the odd field and that of the interlaced frame? (*Hint:* Refer to [1].)

3.12 In Section 3.4.1, we considered a camera with a spatial aperture function that has a Gaussian shape. Show that indeed its spectrum is as given by Equation (3.4.4). What should the value of β be for the frequency response to be 0.5 at $f_x = f_{s,x}/2$, $f_y = f_{s,y}/2$?

3.13 For some practical cameras, their spatial aperture functions can be better modeled by a box function:

$$h_{p,x,y}(x, y) = \begin{cases} \frac{1}{T_x T_y} & |x| < T_x/2, |y| < T_y/2, \\ 0 & \text{otherwise.} \end{cases} \qquad (3.6.4)$$

Assume that the temporal impulse response is the same as that given in Equation (3.4.1). Find the CSFT of the overall aperture function of the camera, $h_p(x, y, t)$. Plot the magnitude response, assuming that $T_x = \Delta_x = 1/720$ picture-width, $T_y = \Delta_y = 1/480$ picture-height, $\Delta_e = \Delta_t = 1/60$ s. Compare the spectrum with that of the camera with a Gaussian aperture (Figure 3.9). Which performs prefiltering more effectively?

3.7 BIBLIOGRAPHY

[1] Beuker, R. A., and I. A. Shah. Analysis of interlaced video signals and its applications. *IEEE Trans. Image Process.* (Sept. 1994), 3(5):501–12.

[2] Cassels, J. W. S. *An Introduction to the Geometry of Numbers*. Berlin: Springer-Verlag, 1959.

[3] Diakides, N. A. Phosphor screens. In D. G. Fink, ed., *Electronics Engineers' Handbook*. New York: McGraw Hill, 1975, 11-33–39.

[4] Dubois, E. The sampling and reconstruction of time-varying imagery with application in video systems. *IEEE* (1985), 73:502–22.

[5] Dudgeon, D. E., and R. M. Mersereau. *Multidimensional Digital Signal Processing*. Englewood Cliffs, NJ: Prentice Hall, 1984.

[6] Miller, L. D. A new method of specifying the resolving power of television camera tubes using the RCA P-300 test chart. *Journal of Society of Motion Picture Television Engineering,* (Apr. 1980), 89:249–56.

[7] Petersen, D. P., and D. Middleton. Sampling and reconstruction of wave-number-limited functions in *N*-dimensional Euclidean spaces. *Information Control* (1962), 5:279–323.

[8] Tonge, G. J. The television scanning process. *Journal of Society of Motion Picture Television Engineers*, (July 1984), 93:657–66.

4

Video Sampling Rate Conversion

In Chapter 3, we considered sampling, which is required for converting an intrinsically continuous video signal to a digital one. Once in the digital domain, we often need to convert a digital video signal from one format (in terms of spatial and temporal resolution) to another—for example, converting from the PAL format to the NTSC format, from interlaced to progressive scan, from an HDTV format signal to an SDTV format, and so on. In this chapter, we consider the sampling rate conversion problem. Section 4.1 describes the general theory governing the conversion of multidimensional signals sampled on different lattices. Section 4.2 focuses on the sampling rate conversion problem of 3-D video signals and, as examples, discusses some of the aforementioned video conversion problems.

4.1 CONVERSION OF SIGNALS SAMPLED ON DIFFERENT LATTICES

Given a sampled signal defined over one lattice Λ_1, it is often desired to generate a signal defined over another lattice Λ_2. This constitutes the sampling rate conversion problem. The solution depends on the relation between the two lattices. For example, if $\Lambda_1 \subset \Lambda_2$, that is, every point in Λ_1 is also in Λ_2, then the problem is an *up-conversion* (or interpolation) problem. We can first fill all points that are in Λ_2 but not in Λ_1 by zeros (i.e., zero-padding), and then estimate the values of these points, which can be accomplished by an interpolation filter over Λ_2. (We discuss the desired form of this interpolation filter in Section 4.1.1.)

On the other hand, if $\Lambda_1 \supset \Lambda_2$, this is a *down-conversion* (or decimation) problem. We could simply take those samples from Λ_1 that are also in Λ_2. However, in order to avoid aliasing in the down-sampled signal, we need to prefilter the signal to limit

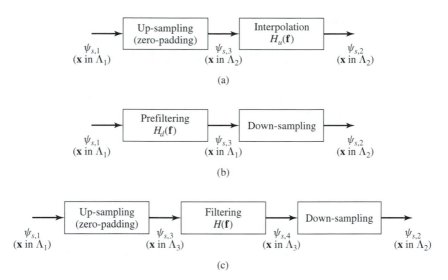

Figure 4.1   Sampling rate conversion: (a) up-conversion, $\Lambda_1 \subset \Lambda_2$;
(b) down-conversion, $\Lambda_2 \subset \Lambda_1$; and (c) arbitrary rate conversion, $\Lambda_3 = \Lambda_1 + \Lambda_2$.

its bandwidth to the Voronoi cell of Λ_2^*. The process for up- and down-conversion is illustrated in Figure 4.1(a) and (b).

In general, if Λ_1 and Λ_2 are not contained in each other, we need to find another lattice Λ_3 that contains both Λ_1 and Λ_2. We can first up-sample the signal from Λ_1 to Λ_3, and then down-sample the signal from Λ_3 to Λ_2. This process is illustrated in Figure 4.1(c). There the intermediate filter in Λ_3 fulfills two purposes: first, it interpolates the missing samples in Λ_1; second, it limits the spectrum of the signal in Λ_3 to the Voronoi cell of Λ_2^*.

In the following subsections, we discuss the preceding scenarios separately. For notation simplicity, we will use $\mathcal{V}_i^*$ to represent $\mathcal{V}(\Lambda_i^*)$, the Voronoi cell of Λ_i^*.

4.1.1 Up-Conversion

As previously mentioned, if $\Lambda_1 \subset \Lambda_2$, then we can first transfer all the samples in Λ_1 to Λ_2 with the missing samples filled by zeros. This *up-sampling* process can be described by:

$$\psi_{s,3}(\mathbf{x}) = \mathcal{U}\{\psi_{s,1}\}(\mathbf{x}) = \begin{cases} \psi_{s,1}(\mathbf{x}), & \mathbf{x} \in \Lambda_1, \\ 0, & \mathbf{x} \in \Lambda_2 \backslash \Lambda_1, \end{cases} \qquad (4.1.1)$$

where $\Lambda_2 \backslash \Lambda_1$ represents the set of points in Λ_2 but not in Λ_1. This signal is now defined over the lattice Λ_2.

To fill the zero-padded samples, we need to apply an interpolation filter that is defined over Λ_2. To determine the appropriate interpolation filter, let us return to the original continuous signal $\psi_c(\mathbf{x})$ defined over the continuous space $\mathcal{R}^K$. If this signal is directly sampled over Λ_2 with a generating matrix $[\mathbf{V}_2]$, then the SSFT of the sampled signal would be

$$\Psi_{s,2}(\mathbf{f}) = d(\Lambda_2) \sum_{\mathbf{m} \in \mathcal{Z}^K} \Psi_c(\mathbf{f} - [\mathbf{U}_2]\mathbf{m}), \tag{4.1.2}$$

with $[\mathbf{U}_2] = ([\mathbf{V}_2]^T)^{-1}$. On the other hand, the SSFT of the signal sampled over Λ_1 with a generating matrix $[\mathbf{V}_1]$ is:

$$\Psi_{s,1}(\mathbf{f}) = d(\Lambda_1) \sum_{\mathbf{m} \in \mathcal{Z}^K} \Psi_c(\mathbf{f} - [\mathbf{U}_1]\mathbf{m}), \tag{4.1.3}$$

with $[\mathbf{U}_1] = ([\mathbf{V}_1]^T)^{-1}$. If Λ_1 satisfies the alias-free condition in Equation (3.2.9), then there will be no overlapping between the alias components in Λ_1^*.

Because $\Lambda_1 \subset \Lambda_2$, according to Theorem 3.3, $\Lambda_1^* \supset \Lambda_2^*$; and consequently, $\mathcal{V}_1^* \subset \mathcal{V}_2^*$. To derive $\Psi_{s,2}(\mathbf{f})$ from $\Psi_{s,1}(\mathbf{f})$, we need only remove from $\mathcal{V}_2^*$ the portion of $\Psi_{s,1}(\mathbf{f})$ that is not in the original spectrum. Given that the original spectrum is not known exactly, the best effort is to assume that it is the same as the portion of $\Psi_{s,1}(\mathbf{f})$ in its Voronoi cell, $\mathcal{V}_1^*$. Therefore, the frequency response of the interpolation filter should be, in the ideal case:

$$H_u(\mathbf{f}) = \begin{cases} d(\Lambda_2)/d(\Lambda_1), & \mathbf{f} \in \mathcal{V}_1^*, \\ 0, & \mathbf{f} \in \mathcal{V}_2^* \backslash \mathcal{V}_1^*. \end{cases} \tag{4.1.4}$$

In this equation, $H_u(\mathbf{f})$ represents the SSFT of a sampled-space filter defined over Λ_2.

Example 4.1

Consider the up-conversion from the sampling lattice Λ_1 to Λ_2 in Figure 4.2. Their respective generating matrices are

$$[\mathbf{V}_1] = \begin{bmatrix} 2 & 1 \\ 0 & 1 \end{bmatrix}; \qquad [\mathbf{V}_2] = \begin{bmatrix} 1 & 0 \\ 0 & 1 \end{bmatrix}. \tag{4.1.5}$$

Using Equation (3.1.7), the generating matrices for the reciprocal lattices are

$$[\mathbf{U}_1] = \begin{bmatrix} 1/2 & 0 \\ -1/2 & 1 \end{bmatrix}; \qquad [\mathbf{U}_2] = \begin{bmatrix} 1 & 0 \\ 0 & 1 \end{bmatrix}. \tag{4.1.6}$$

(The reciprocal lattices and their Voronoi cells are also shown in the figure.) Assuming that the spectrum of the original continuous signal has a circular support, we also draw in the figure the replicated spectra corresponding to different sampling patterns. We can see that, indeed, $\mathcal{V}_2^* \supset \mathcal{V}_1^*$. The ideal interpolation filter is

$$H_u(\mathbf{f}) = \begin{cases} d(\Lambda_2)/d(\Lambda_1) = 2, & \mathbf{f} \in \mathcal{V}_1^*, \\ 0, & \mathbf{f} \in \mathcal{V}_2^* \backslash \mathcal{V}_1^*. \end{cases} \tag{4.1.7}$$

as illustrated on lattice Λ_2^* in Figure 4.2. The filter serves to eliminate alias components in Λ_1^* that should not appear in Λ_2^*.

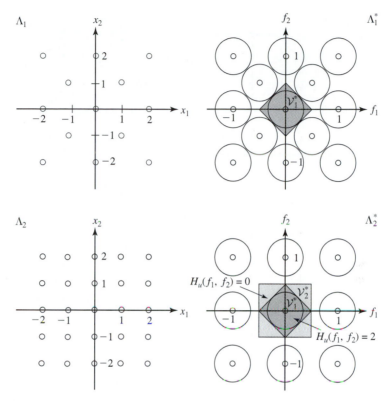

Figure 4.2   An example of up-conversion, from lattice Λ_1 to lattice Λ_2. The ideal interpolation filter is illustrated in Λ_2^* (see Example 4.1).

4.1.2 Down-Conversion

If $\Lambda_1 \supset \Lambda_2$, then we can obtain $\psi_{s,2}$ easily by retaining the samples in Λ_1 that are also in Λ_2, and discarding all other samples. This down-sampling process is described by:

$$\psi_{s,2}(\mathbf{x}) = \mathcal{D}\{\psi_{s,1}\}(\mathbf{x}) = \psi_{s,1}(\mathbf{x}), \quad \text{for } \mathbf{x} \in \Lambda_2. \tag{4.1.8}$$

The signal so obtained would be exactly the same as the one resulting from sampling the original continuous signal over Λ_2 if the same prefilter were used. Let us assume that the support region of the original continuous signal, or the one after prefiltering when obtaining $\psi_{s,1}$, is equal to or smaller than $\mathcal{V}_1^*$, so that there is no aliasing in $\psi_{s,1}$. Then, in general, there will be aliasing in $\psi_{s,2}$, as $\mathcal{V}_2^*$ is smaller. To avoid aliasing, one must prefilter $\psi_{s,1}$ to remove the frequency content that is in $\mathcal{V}_1^*$ but not in $\mathcal{V}_2^*$. Therefore, the appropriate prefilter over Λ_1 is, in terms of the SSFT over Λ_1:

$$H_d(\mathbf{f}) = \begin{cases} d(\Lambda_2)/d(\Lambda_1), & \mathbf{f} \in \mathcal{V}(\Lambda_2^*), \\ 0, & \mathbf{f} \in \mathcal{V}_1^* \backslash \mathcal{V}_2^*. \end{cases} \tag{4.1.9}$$

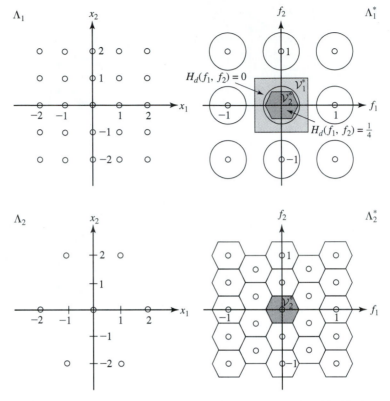

Figure 4.3 An example of down-conversion, from lattice Λ_1 to Λ_2. The ideal prefilter is illustrated in $\mathcal{V}_1^*$ in the figure (see Example 4.2).

Example 4.2

Consider the down-conversion example given in Figure 4.3. The associated generating matrices are:

$$[\mathbf{V}_1] = \begin{bmatrix} 1 & 0 \\ 0 & 1 \end{bmatrix}; \qquad [\mathbf{V}_2] = \begin{bmatrix} 2 & 1 \\ 0 & 2 \end{bmatrix}; \tag{4.1.10}$$

$$[\mathbf{U}_1] = \begin{bmatrix} 1 & 0 \\ 0 & 1 \end{bmatrix}; \qquad [\mathbf{U}_2] = \begin{bmatrix} 1/2 & 0 \\ -1/4 & 1/2 \end{bmatrix}. \tag{4.1.11}$$

The ideal prefilter is

$$H_d(\mathbf{f}) = \begin{cases} 1/4, & \mathbf{f} \in \mathcal{V}_2^*, \\ 0, & \mathbf{f} \in \mathcal{V}_1^* \backslash \mathcal{V}_2^*, \end{cases} \tag{4.1.12}$$

which is illustrated in Λ_1^* in the figure. Without prefiltering, the original circular spectrum would have caused aliasing in Λ_2^*. By applying the prefilter in Λ_1^*, only the portion of the spectrum that is contained in V_2^* is retained, so that no aliasing will occur. Note that if there exists aliasing in $\psi_{s,1}$, it will be retained in $\psi_{s,2}$. On the other hand, if the original signal has a support region that is within V_2^*, then there is no need for using the prefilter, when obtaining the original sampled signal over Λ_1, and when down-sampling $\psi_{s,1}$ to obtain $\psi_{s,2}$.

4.1.3 Conversion between Arbitrary Lattices

As explained in the beginning of this section, in the general case when Λ_1 and Λ_2 are not subsets of one another, we need to introduce another lattice Λ_3 so that both Λ_1 and Λ_2 are subsets of Λ_3. Obviously, to minimize the sampling density in Λ_3, we would like to find the smallest lattice that will satisfy this criterion. Once Λ_3 is determined, the conversion from Λ_1 to Λ_2 can be accomplished by concatenating the up-conversion and down-conversion steps discussed in the previous two subsections. The interpolation filter and the prefilter in these two steps are, respectively,

$$H_u(\mathbf{f}) = \begin{cases} d(\Lambda_3)/d(\Lambda_1), & \mathbf{f} \in V_1^*, \\ 0, & \mathbf{f} \in V_3^* \backslash V_1^*, \end{cases} \tag{4.1.13}$$

$$H_d(\mathbf{f}) = \begin{cases} d(\Lambda_2)/d(\Lambda_3), & \mathbf{f} \in V_2^*, \\ 0, & \mathbf{f} \in V_3^* \backslash V_2^*. \end{cases} \tag{4.1.14}$$

Obviously, these two filters, both defined over Λ_3, can be merged into one filter, with the following frequency response in terms of SSFT:

$$H(\mathbf{f}) = \begin{cases} d(\Lambda_2)/d(\Lambda_1), & \mathbf{f} \in (V_1^* \cap V_2^*), \\ 0, & \mathbf{f} \in V_3^* \backslash (V_1^* \cap V_2^*). \end{cases} \tag{4.1.15}$$

The preceding discussion assumed that Λ_3 can somehow be determined. Based on Theorem 3.2, Λ_3 can be determined by the sum of Λ_1 and Λ_2, that is, $\Lambda_3 = \Lambda_1 + \Lambda_2$, if $[\mathbf{V}_1]^{-1}[\mathbf{V}_2]$ contains only rational numbers. Recall that in the 1-D case, to enable conversion between two sampling rates, a necessary condition is that the ratio between the two sampling frequencies, $f_{s,1}$ and $f_{s,2}$ (or, equivalently, the two sampling intervals, Δ_1 and Δ_2), must be a rational number; that is, $r = f_{s,2}/f_{s,1} = \Delta_1/\Delta_2 = p/q$, where p and q are both integers. In this case, the sampling rate conversion by a factor of r can be accomplished by first up-sampling by a factor of p and then down-sampling by a factor of q. Obviously, this is a special case of the more general procedure described here. Specifically, these two steps correspond to the up-conversion from Λ_1 to Λ_3, and down-conversion from Λ_3 to Λ_2, respectively. When the condition of Theorem 3.2 is not satisfied, one can first determine the union of Λ_1 and Λ_2, which may not be a legitimate lattice. One then must extend this set to a lattice by adding additional points, usually by inspection.

Example 4.3

Consider the conversion between two sampling lattices Λ_1 and Λ_2 with generating matrices:

$$[\mathbf{V}_1] = \begin{bmatrix} 2 & 1 \\ 0 & 1 \end{bmatrix}, \qquad [\mathbf{V}_2] = \begin{bmatrix} 2 & 1 \\ 0 & 2 \end{bmatrix}. \tag{4.1.16}$$

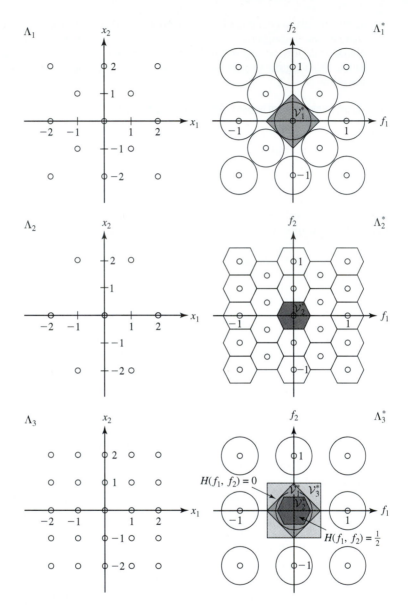

Figure 4.4 Conversion between two noninclusive lattices, Λ_1 and Λ_2. Λ_3 is the intermediate lattice that includes both Λ_1 and Λ_2. The ideal filter is illustrated in Λ_3^* (see Example 4.3).

The generating matrices of their corresponding reciprocal lattices are:

$$[\mathbf{U}_1] = \begin{bmatrix} 1/2 & 0 \\ -1/2 & 1 \end{bmatrix}, \qquad [\mathbf{U}_2] = \begin{bmatrix} 1/2 & 0 \\ -1/4 & 1/2 \end{bmatrix}. \qquad (4.1.17)$$

First, we sketch the input and output sampling lattices Λ_i and reciprocal lattices Λ_i^*, $i = 1, 2$, based on the generating matrices, as illustrated in Figure 4.4. In this case, Λ_1 and Λ_2 are not subsets of one another. Therefore, we need to determine a lattice Λ_3 that is the smallest lattice that contains both Λ_1 and Λ_2. Because $[\mathbf{V}_1]^{-1}[\mathbf{V}_2]$ includes only rational numbers, $\Lambda_3 = \Lambda_1 + \Lambda_2$. In general, it is difficult to determine the generating matrix for Λ_3 analytically. However, by visual inspection, one can see that Λ_3 in this case must be the square lattice sketched in the figure. For this lattice, one obvious generating matrix is the identity matrix, that is,

$$[\mathbf{V}_3] = \begin{bmatrix} 1 & 0 \\ 0 & 1 \end{bmatrix}, \tag{4.1.18}$$

The reciprocal lattice is also a square lattice with the generating matrix being the identity matrix, that is, $[\mathbf{U}_3] = ([\mathbf{V}_3]^T)^{-1} = [\mathbf{V}_3]$.

From the sketched reciprocal lattices, Λ_i^*, $i = 1, 2, 3$, we can determine their Voronoi cells $\mathcal{V}_i^*$ individually, which are shown in Figure 4.4. In this case, $\mathcal{V}_1^* \supset \mathcal{V}_2^*$, so that $\mathcal{V}_1^* \cap \mathcal{V}_2^* = \mathcal{V}_2^*$. Therefore, the intermediate filter should be:

$$H(\mathbf{f}) = \begin{cases} \dfrac{d(\Lambda_2)}{d(\Lambda_1)} = \dfrac{1}{2}, & \mathbf{f} \in \mathcal{V}_2^*, \\ 0, & \mathbf{f} \in \mathcal{V}_3^* \backslash \mathcal{V}_2^*, \end{cases}$$

as illustrated on top of Λ_3^* in the figure.

4.1.4 Filter Implementation and Design, and Other Interpolation Approaches

In the preceding discussion, we have specified the desired filter only in terms of its SSFT. The equivalent filter in the spatial domain $h(\mathbf{x})$, $\mathbf{x} \in \Lambda_3$ can be determined by the inverse SSFT. Theoretically, to implement the filtering operation, we need to first up-sample the input signal from Λ_1 to Λ_3, filter it in Λ_3, and then down-sample it from Λ_3 to Λ_2, as shown in Figure 4.1(c). The filtering operation in Λ_3 is:

$$\psi_{s,4}(\mathbf{x}) = \psi_{s,3}(\mathbf{x}) * h(\mathbf{x}) = \sum_{\mathbf{y} \in \Lambda_3} \mathcal{U}\{\psi_{s,1}\}(\mathbf{y})h(\mathbf{x} - \mathbf{y}), \quad \mathbf{x} \in \Lambda_3. \tag{4.1.19}$$

In reality, the filtering output need not be calculated for all the samples in Λ_3, as only the samples in Λ_2 are desired in the output. In addition, because the up-sampled input signals are nonzero only in Λ_1, the sum in Equation (4.1.19) involves only input samples in Λ_1. Overall, the output signal over Λ_2 can be directly obtained from the input over Λ_1 by

$$\psi_{s,2}(\mathbf{x}) = \mathcal{D}\{\psi_{s,4}\}(\mathbf{x}) = \sum_{\mathbf{y} \in \Lambda_1} \psi_{s,1}(\mathbf{y})h(\mathbf{x} - \mathbf{y}), \quad \mathbf{x} \in \Lambda_2. \tag{4.1.20}$$

This means that a sample $\mathbf{x} \in \Lambda_2$ is estimated by a weighted average of the samples $\mathbf{y} \in \Lambda_1$, with weights equal to $h(\mathbf{x} - \mathbf{y})$.

In previous discussions, we have simply given the desired frequency responses of the prefilter and reconstruction filter, without considering how to design the filters. From the specified frequency response $H(\mathbf{f})$, $\mathbf{f} \in \mathcal{V}_3^*$, in Equation (4.1.15), we can perform the inverse SSFT to obtain the ideal filter $h(\mathbf{x})$, $\mathbf{x} \in \Lambda_3$. However, this filter will have an infinite extent, because of the very sharp transition in $H(\mathbf{f})$. If we simply truncate the filter, it will cause the notorious ringing artifacts in the filtered signal. Therefore, we must carefully modify the desired filter response by applying an appropriate windowing function, so that the corresponding filter has a finite extent. We can also use various optimization techniques to directly design finite extent filters so that their frequency responses are as close to the desired responses as possible. Filter design techniques are beyond the scope of this book; interested readers are referred to [4], which has good coverage of multidimensional filter design. Most of the digital filter design techniques are developed assuming that the frequency response is specified in terms of the discrete frequency. To solve the problem here, we must warp the desired frequency response in terms of the continuous frequency to the discrete frequency, as specified in Equation (3.2.16). Based on this warped frequency response, we can apply the digital filter design technique to derive the desired filter over the discrete space. We can implement the convolution operation in the abstract discrete space, or in the sampled space directly.

As will be seen, for video signal sampling and sampling rate conversion, simple filters that are far from the ideal frequency response are often used, in order to reduce the computational complexity. Example filters for several important conversion problems are described in Section 4.2.

Up to now, we have derived the desired interpolation filters based on the frequency domain consideration. One can also develop interpolation methods based on sample domain requirements. For the interpolation of a continuous signal from samples, the problem is to construct a continuous surface that passes through the given sample values. For the up-conversion task, the problem is to estimate missing samples from surrounding known samples. The general approach is to use a polynomial function with a finite order to approximate the continuous signal over each small region, and determine the polynomial coefficients based on the specified sample values in that region, using a least-squares fitting procedure. Once the polynomial is determined, one can either reconstruct the entire continuous surface, or derive the missing samples. To maintain the continuity of the surfaces derived for adjacent regions, one can use spline functions that are special classes of the piecewise polynomials (the most popular being the cubic-spline approach that has been described for 2-D signal interpolation [6]).

4.2 SAMPLING RATE CONVERSION OF VIDEO SIGNALS

Sampling rate conversion is often required in video systems. For example, display of a PAL signal on an NTSC TV system requires the conversion of an interlaced signal with sampling rate of $f_{s,t} = 50$ field/s, $f_{s,y} = 625$ line/picture-height into an interlaced signal with $f_{s,t} = 60$ field/s, $f_{s,y} = 525$ line/picture-height. To convert a motion picture on film for NTSC TV broadcasting and display requires the conversion from $f_{s,t} = 24$ fps to $f_{s,t} = 60$ field/s. Another interesting problem is to convert an interlaced raster into a progressive raster, known as *deinterlacing,* which is required to display made-for-TV material on a computer screen, which uses progressive display. Some advanced TV systems already have a built-in deinterlacing function, so that interlaced material is displayed in a progressive mode for improved visual quality.

As described in Section 4.1, the general procedure for solving the sampling rate conversion problem is as follows: (1) determine the equivalent sampling lattices of the input and output signals, and an intermediate lattice that covers the samples in both input and output signals; (2) based on the Voronoi cells of the three lattices, determine the desired filter frequency response; (3) design a filter that approximates the desired response. In reality, because of the very high data rate of the video signal, one cannot afford to use a filter with many coefficients, especially in the temporal direction. To reduce the computational complexity, one can also decompose a spatiotemporal conversion problem into spatial conversion followed by temporal conversion, or vice versa.

In the following, we describe the solutions for deinterlacing and for conversion between PAL and NTSC in more detail. We consider operations in the vertical and temporal directions only, as the sampling rate conversion in the horizontal direction can be done in a separate step with conventional 1-D techniques. Haskell, Puri, and Netravali [13] has an extensive discussion of various video conversion problems and their practical solutions; some of these problems are considered in the problem section at the end of this chapter.

4.2.1 Deinterlacing

The problem of deinterlacing is to fill in the skipped lines in each field, as illustrated in Figure 4.5. Consider an interlaced NTSC signal, with field rate $f_{s,t} = 60$ fields/s, and line rate $f_{s,y} = 525$ lines/picture-height. The sampling intervals are $\Delta_t = 1/60$ s, $\Delta_y = 1/525$ picture-height. The corresponding sampling lattice and its reciprocal are illustrated in Figure 4.6(a). The sampling lattice for the deinterlaced signal is a square

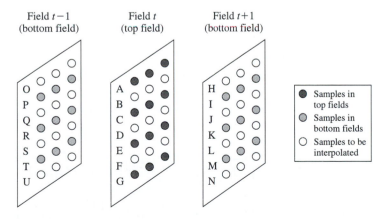

Figure 4.5   The deinterlacing process; fields t and $t + 1$ form one interlaced frame.

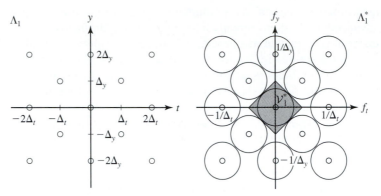

(a) Sampling lattice and reciprocal corresponding to interlaced scan

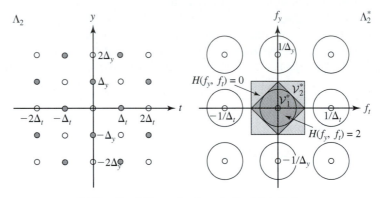

(b) Sampling lattice and reciprocal after deinterlacing

Figure 4.6   The sampling lattices and their reciprocal lattices before and after deinterlacing. Open circles represent original samples, filled circles the samples to be interpolated.

lattice, shown in Figure 4.6(b). As before, we have scaled the vertical and temporal axes so that Δ_y and Δ_t are represented with the same length. Similarly, $f_{s,y}$ and $f_{s,t}$ are assigned the same length in the frequency domain.

Obviously, the generating matrices for these two lattices and their reciprocal lattices are

$$[\mathbf{V}_1] = \begin{bmatrix} 2\Delta_t & \Delta_t \\ 0 & \Delta_y \end{bmatrix}, \qquad [\mathbf{U}_1] = \begin{bmatrix} 1/2\Delta_t & 0 \\ -1/2\Delta_y & 1/\Delta_y \end{bmatrix}, \qquad (4.2.1)$$

$$[\mathbf{V}_2] = \begin{bmatrix} \Delta_t & 0 \\ 0 & \Delta_y \end{bmatrix}, \qquad [\mathbf{U}_2] = \begin{bmatrix} 1/\Delta_t & 0 \\ 0 & 1/\Delta_y \end{bmatrix}. \qquad (4.2.2)$$

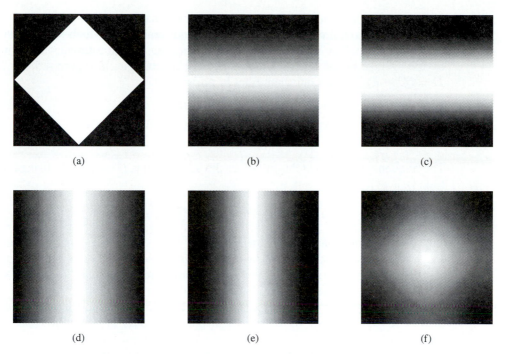

Figure 4.7 Frequency responses of several deinterlacing filters: (a) the ideal filter (Equation 4.2.3); (b) line averaging (Equation 4.2.5); (c) four-point-vertical interpolation (Equation 4.2.7); (d) field merging (Equation 4.2.9); (e) field averaging (Equation 4.2.11); (f) line-field averaging (Equation 4.2.13). In each figure, the vertical axis represents the vertical frequency, and the horizontal axis the temporal frequency. The frequency range is $|f_y| \leq f_{s,y}/2, |f_t| \leq f_{s,t}/2$.

In this case, $\Lambda_2 \supset \Lambda_1$, and the problem is an up-conversion problem, and the ideal interpolation filter is:

$$H(f_y, f_t) = \begin{cases} \dfrac{d(\Lambda_2)}{d(\Lambda_1)} = 2, & (f_y, f_t) \in \mathcal{V}_1^*, \\ 0, & (f_y, f_t) \in \mathcal{V}_2^* \backslash \mathcal{V}_1^*. \end{cases} \qquad (4.2.3)$$

The Voronoi cells $\mathcal{V}_1^*$ and $\mathcal{V}_2^*$ are also illustrated in Figure 4.6. The magnitude of the frequency response of this filter is shown in Figure 4.7(a). We can see that this is a low-pass filter with a diamond-shaped pass-band. Notice that this filter is not separable (that is, cannot be decomposed into a product of a temporal filter and a vertical filter), so one must use 2-D filter design technique. To simplify the design problem, one can first design a separable low-pass filter with a square-shaped pass-band, and then rotate the resulting filter.

In practice, many simpler filters have been proposed for deinterlacing. One option is to use vertical interpolation within the same field, which is a 1-D factor-of-two up-conversion problem. The ideal vertical filter is a half-band low-pass filter [7]; however, this filter requires infinite length and is not realizable. Much shorter filters are used in practice. The simplest one is *line averaging,* which estimates a missing line by the average of the line above and below. For example, in Figure 4.5, for field t, $D = (C + E)/2$. The equivalent filter in Λ_2 is

$$h(y, t) = \begin{cases} 1, & (y, t) = (0, 0), \\ 1/2, & (y, t) = (\Delta_y, 0), (-\Delta_y, 0), \\ 0, & \text{others.} \end{cases} \tag{4.2.4}$$

The frequency response of this filter (i.e., the SSFT of $h(y, t)$ over Λ_2) is, using Equation (3.2.3),

$$H(f_y, f_t) = 1 + \frac{1}{2}(e^{j2\pi \Delta_y f_y} + e^{-j2\pi \Delta_y f_y}) = 1 + \cos(2\pi \Delta_y f_y). \tag{4.2.5}$$

This frequency response is illustrated in Figure 4.7(b). We can see that it deviates from the desired response shown in (a) significantly. Because no temporal filtering is applied, it has an all-pass characteristic along the temporal frequency axis. Along the vertical frequency axis, it is far from the ideal half-band low-pass filter.

To improve the performance, longer vertical interpolation filters that have frequency responses closer to the ideal half-band low-pass filter can be used. For example, in Figure 4.5, a satisfactory interpolation method for lines in field t is $D = (A + 7C + 7E + G)/16$. The equivalent interpolation filter over the deinterlaced grid (the vertical axis in Λ_2) is

$$h = [1, 0, 7, 16, 7, 0, 1]/16. \tag{4.2.6}$$

Following the same approach as before, we can easily show that the frequency response of this filter is

$$H(f_y, f_t) = 1 + \frac{7}{8} \cos(2\pi \Delta_y f_y) - \frac{1}{8} \cos(6\pi \Delta_y f_y), \tag{4.2.7}$$

which is shown in Figure 4.7(c).

The previous two approaches use vertical interpolation only. An alternative is to use temporal interpolation. Notice that for every missing line in a field, there is a corresponding line in the other field of the same frame (see Figure 4.5). A simple temporal interpolation scheme is to copy this corresponding line. That is, $D = K$, $J = C$. This method is known as *field merging,* because each deinterlaced frame is obtained by merging two fields. The equivalent filter used for deinterlacing the top field (assuming it appears before the bottom field) is:

$$h(y, t) = \begin{cases} 1, & (y, t) = (0, 0), (0, -\Delta_t), \\ 0, & \text{others.} \end{cases} \tag{4.2.8}$$

The filter for the bottom field is a time-reversed version of this filter. The frequency response of the filter in Equation (4.2.8) is

$$H(f_y, f_t) = 1 + e^{-j2\pi \Delta_t f_t}. \tag{4.2.9}$$

Because the filter is asymmetric in time, it has a complex frequency response. In addition, the temporal interpolation directions are opposite for the two fields in a frame. (These may yield visual artifacts for certain special patterns.) The magnitude response is illustrated in Figure 4.7(d), which again deviates from the desired response significantly. Because filtering is applied only in the temporal direction, it is all-pass in the vertical direction.

To improve the performance, a symmetric filter can be used—for example, averaging the corresponding lines in the previous and following fields to obtain a missing line in the current field. For example, for field t, $D = (K + R)/2$. We refer to this method as *field averaging*. The equivalent interpolation filter is:

$$h(y, t) = \begin{cases} 1, & (y, t) = (0, 0), \\ 1/2, & (y, t) = (0, \Delta_t), (0, -\Delta_t), \\ 0, & \text{others}, \end{cases} \tag{4.2.10}$$

or

$$H(f_y, f_t) = 1 + \cos(2\pi \Delta_t f_t). \tag{4.2.11}$$

The frequency response is shown in Figure 4.7(e). Although this filter has a better frequency response than the previous one, it involves three fields for interpolating any field. This requires the storage of two frames, which represents a nontrivial increase in memory and delay requirements compared to the field merging approach.

To achieve a compromise between spatial and temporal artifacts, a better approach is to use both vertical and temporal interpolation. For example, one can interpolate a missing pixel by taking the average of the pixels above and below in the same field, and the pixels in the previous and following fields. In Figure 4.5, this corresponds to $D = (C + E + K + R)/4$. This method is referred to as *line and field averaging*. The equivalent filter on the deinterlaced lattice is:

$$h(y, t) = \begin{cases} 1, & (y, t) = (0, 0), \\ 1/4, & (y, t) = (\Delta_y, 0), (-\Delta_y, 0), (0, \Delta_t), (0, -\Delta_t), \\ 0, & \text{others}. \end{cases} \tag{4.2.12}$$

The frequency response is

$$H(f_y, f_t) = 1 + \frac{1}{2}(\cos(2\pi f_y \Delta_y) + \cos(2\pi f_t \Delta_t)), \tag{4.2.13}$$

which is illustrated in Figure 4.7(f). We can see that it is closer to the desired response than all previous filters. As with field averaging, this method requires the storage of two frames. An approach that requires storage of only one frame is discussed in Problem 4.1.

When the imaged scene is stationary between adjacent fields, the missing even lines in an odd field should be exactly the same as the corresponding even lines in the previous and following even field. Therefore, temporal interpolation will yield perfect estimates. On the other hand, when there is motion in the scene, the corresponding lines in adjacent fields may not correspond to the same object location and temporal interpolation will yield unacceptable artifacts. Imagine a vertical grating pattern moving horizontally so that the grating patterns captured in two adjacent fields are shifted by half of the grating interval. Field averaging will yield serrated vertical gratings. The methods that use both spatial and temporal averaging will produce less severe artifacts, but the resulting signal may still be unacceptable for certain applications. For better quality, motion-compensated interpolation should be used, which is discussed in Section 4.2.3.

4.2.2 Conversion between PAL and NTSC Signals

In this section, we consider the conversion from PAL to NTSC format signals (both are interlaced). The conversion from NTSC to PAL will follow the same principle, and is considered in Problem 4.4. Recall that the line and field rates of the PAL signal are $f_{y,1} = 625$ lines/picture-height and $f_{t,1} = 50$ fields/s, and those for the NTSC signal are $f_{y,2} = 525$ lines/picture-height and $f_{t,2} = 60$ fields/s. The sampling lattices for the PAL (Λ_1) and NTSC (Λ_2) formats are sketched in Figure 4.8(a) and (b), respectively, where $\Delta_t = 1/60$ s, $\Delta_y = 1/525$ picture-height. Based on the points on each sampling lattice, one can easily determine its generating matrix, and consequently determine the [**V**] and [**U**] matrices. From the [**U**] matrix, one can then determine the reciprocal lattice and its Voronoi cell. All these are given in Figure 4.8. In drawing the lattices, we have scaled the temporal and vertical axes by equating Δ_t and Δ_y in the spatial-temporal domain, and equating $1/\Delta_t$ and $1/\Delta_y$ in the frequency domain.

The conversion between PAL and NTSC is difficult because the vertical and temporal sampling rates of these two lattices are not integer multiples of each other. First, we derive the ideal solution using the technique described in Section 4.1. Towards this purpose, we need to determine a lattice Λ_3 that contains both Λ_1 and Λ_2. By observing that the minimal common multiple of $f_{y,1} = 625$ and $f_{y,2} = 525$ is $f_{y,3} = 13,125$, and that for $f_{t,1} = 50$ and $f_{t,2} = 60$ is $f_{t,3} = 300$, we can see that a rectangular lattice with line rate of $f_{y,3}$ and frame rate of $f_{t,3}$ is the desired intermediate lattice, which is the lattice Λ_3 in Figure 4.8(c). From the figure, it is easy to see that the desired frequency response of the filter over Λ_3 is

$$H(f_y, f_t) = \begin{cases} \dfrac{d(\Lambda_2)}{d(\Lambda_1)} = 126/125, & (f_y, f_t) \in \mathcal{V}_1^* \cap \mathcal{V}_2^*, \\ 0, & (f_y, f_t) \in \mathcal{V}_3^* \backslash (\mathcal{V}_1^* \cap \mathcal{V}_2^*). \end{cases} \tag{4.2.14}$$

This ideal conversion filter is illustrated on Λ_3^* in Figure 4.8(c) (and will also be seen in Figure 4.11(a)). Obviously, because of the very complicated shape of $\mathcal{V}_1^* \cap \mathcal{V}_2^*$, this filter is not easy to design.

The preceding solution calls for directly converting a signal from Λ_1 to Λ_3 and then back to Λ_2, as shown in Figure 4.9(a). In practice, the problem is more often solved in four steps: (1) deinterlacing each field in the PAL signal so that each resulting frame contains 625 lines, (2) line rate down-conversion from 625 to 525 lines within each deinterlaced frame, (3) frame rate up-conversion from 50 to 60 Hz, and finally (4) splitting each frame to two interlacing fields. This sequential implementation is illustrated in Figure 4.9(b).

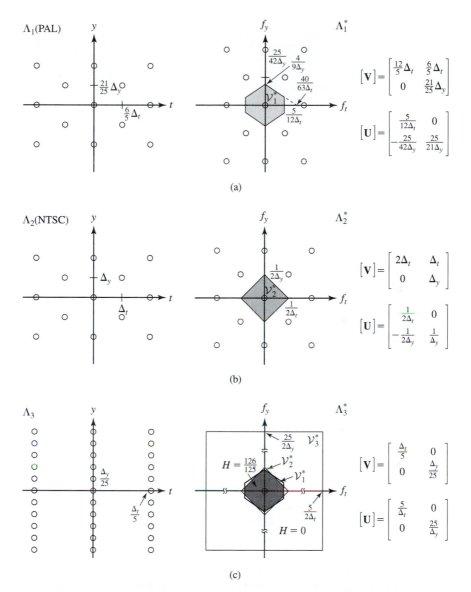

Figure 4.8   The sampling lattices involved in PAL to NTSC conversion. Λ_1 and Λ_2 are sampling lattices for PAL and NTSC formats, and Λ_3 is the intermediate lattice including both Λ_1 and Λ_2.

The general deinterlacing problem has been discussed in Section 4.2.1, and the solution for the current specific set-up is illustrated in Figure 4.10(a). For the vertical conversion from 625 to 525 lines, ideally one should first convert the line rate to 13,125 lines, filtering the signal in that high resolution, and then down-sample to 525 lines,

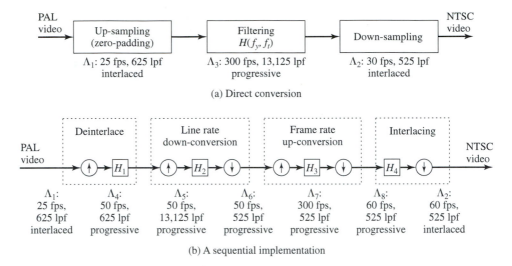

Figure 4.9   Two different methods for converting PAL to NTSC formats.

as indicated in Figure 4.9(b). The ideal solution is illustrated in Figure 4.10(a–c). For the temporal conversion from 50 to 60 fps, one must first convert the frame rate to 300 fps, filtering the signal in that high resolution, and then down-sample to 50 fps, as indicated in Figure 4.9(b). The ideal solution is illustrated in Figure 4.10(c–e). Finally, to create an interlaced signal from the progressive frames, one can simply skip every other line in each frame. But to avoid aliasing, a prefilter is needed, and the solution is given in Figure 4.10(e). The equivalent overall filter response is shown in Figure 4.10(f). Compared to the ideal filter shown in Figure 4.8(a), this filter has the same pass-band, but a different stop-band. The frequency response corresponding to the concatenation of H_2 and H_3 is shown in Figure 4.11(b). Although the direct conversion and the sequential method lead to the same frequency response, it can be shown that the sequential operation can reduce the computation requirement (see Problem 4.3).

Note that to convert 625 lines to 525 lines is equivalent to converting every 25 lines to 21 lines [13]. Instead of using the ideal interpolation filter in the 13,125 line resolution, a simple approach is to find, for each of the 21 lines to be determined, its two nearest neighbors in the given 25 lines, and use linear interpolation to determine the sample values, as shown in Figure 4.12. To determine the interpolation coefficients associated with the two known lines, one should determine their distances to the line to be interpolated. If the distance to the left known line is d_l, and that to the right is d_r, the coefficients should be $h_l = d_r/(d_l + d_r)$ and $h_r = d_l/(d_l + d_r)$, respectively. Let $\Delta_{y,3}$ represent the spacing between every two lines in the 13,125-line grid, and $\Delta_{y,1}$ and $\Delta_{y,2}$ the spacings in the 625- and 525-line grids, respectively. Obviously $\Delta_{y,1} = 21\Delta_{y,3}, \Delta_{y,2} = 25\Delta_{y,3}$. For line l in the 525-line grid, if $k = l * \Delta_{y,2}/\Delta_{y,1} = \frac{25}{21}l$ is an integer, then it can be directly copied from line k in the 625 line-grid. This happens when $l = 0, 21, 42, \ldots$. For the in-between lines, the indices for the left and right

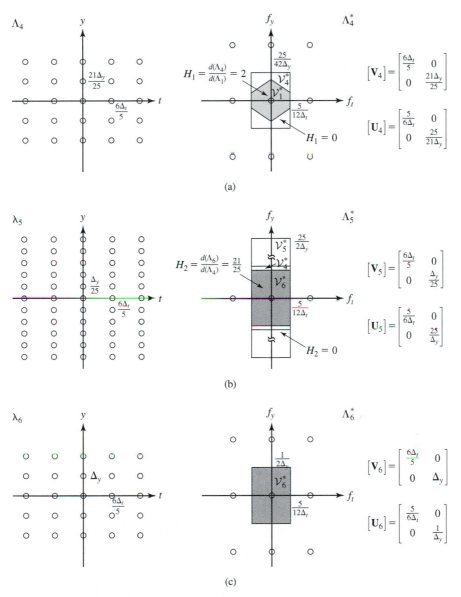

Figure 4.10   Ideal operations involved in the sequential implementation of PAL to NTSC conversion. (*Continued on next page*)

adjacent lines in the 625-line grid are $k = \lfloor l * \Delta_{y,2}/\Delta_{y,1} \rfloor$ and $k+1$, respectively. The corresponding distances are $d_l = l\Delta_{y,2} - k\Delta_{y,1}$, and $d_r = (k+1)\Delta_{y,1} - l\Delta_{y,2}$. Therefore, the interpolation coefficients are $h_l = k+1-l*25/21$ and $h_r = l*25/21-k$. The weight for the left known line for each interpolated line is indicated in Figure 4.12.

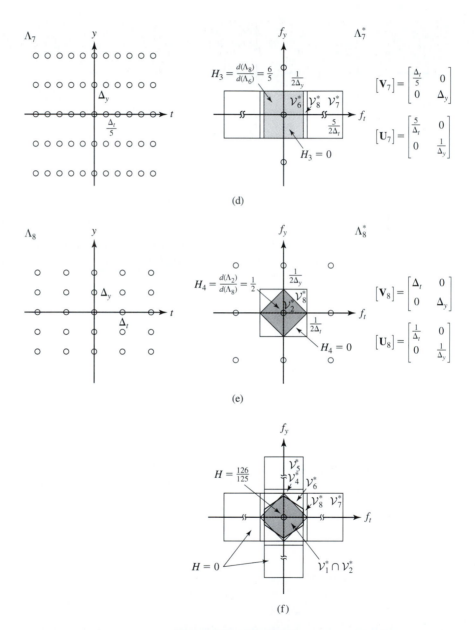

Figure 4.10 (*Continued*)

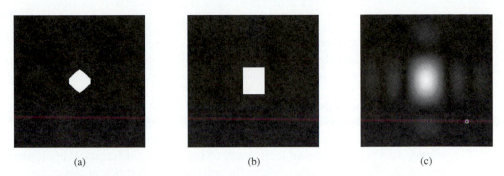

(a)   (b)   (c)

Figure 4.11   Frequency responses of PAL to NTSC conversion filters: (a) ideal filter; (b) composite filter corresponding to concatenation of H_2 and H_3 in Figure 4.9, when H_2 and H_3 are both ideal filters; (c) composite filter when line rate down-conversion (H_2) and frame rate up-conversion (H_3) are implemented by Equations (4.2.15–16). Part (a) considers deinterlacing and interlacing filters, whereas (b) and (c) do not. In each part, the vertical axis represents f_y, and the horizontal axis f_t. Only the frequency region defined by $|f_y| \leq 2.5/\Delta_y$, $|f_t| \leq 2.5/\Delta_t$ is drawn, although the filter response is defined over $|f_y| \leq 12.5/\Delta_y$, $|f_t| \leq 2.5/\Delta_t$.

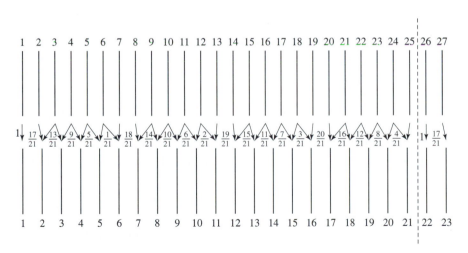

Figure 4.12   Accomplishing 625 lines to 525 lines down-conversion by converting every 25 lines to 21 lines and using the two nearest known lines (in the upper row) for interpolating each new line (in the lower row) [5]. For each line in the bottom row, the interpolation coefficient associated with the left neighboring line in the upper row is given in the figure. The interpolation coefficient for the right neighboring line is one minus the left coefficient.

The equivalent vertical filter in Λ_3 is

$$h_v(y) = \begin{cases} (1 - |n|/21), & y = n\Delta_{y,3}, |n| = 0, 1, \ldots, 20, \\ 0 & \text{otherwise.} \end{cases} \qquad (4.2.15)$$

This is the linear interpolation filter for interpolating from 625 lines to 13,125 lines, or inserting 21 lines between every two lines. But at most one of the lines is retained in the 525-line grid, using two out of 21 possible interpolation coefficients given in Equation (4.2.15).

To implement the temporal conversion from 50 Hz to 60 Hz, instead of first up-converting the frame rate from 50 to 300 fps and then down-sampling to 60 fps, a much simpler approach is to convert every five frames to six frames as illustrated in Figure 4.13 [13]. As shown there, each of the six output frames is interpolated from two of the five input frames, where the interpolation coefficient associated with each input frame is inversely proportional to the distance between the output frame and the input frame. Following the procedure described previously for the line rate conversion, one can determine the interpolation coefficients, which are also shown in Figure 4.13. The equivalent temporal filter over the 300-frame grid is

$$h_t(t_k) = \begin{cases} (1 - |k|/6), & t_k = k\Delta_{t,3}, |k| = 0, 1, \ldots, 5, \\ 0 & \text{otherwise.} \end{cases} \qquad (4.2.16)$$

The frequency response of the combined spatiotemporal filter is illustrated in Figure 4.11(c). Compared to the desired response shown in Figure 4.11(b), we see that linear interpolation leads to blurred transitions and many ripples in the stop-band.

4.2.3 Motion-Adaptive Interpolation

From the examples given in the previous two subsections, we can see that there are advantages and disadvantages associated with using interpolation in the spatial and temporal directions, respectively. In regions that are stationary, temporal interpolation will yield an accurate result, which will indeed increase the resolution of a given frame. On the other hand, in regions undergoing rapid temporal changes between successive fields or frames, pixels with the same spatial indices in two separate fields or frames

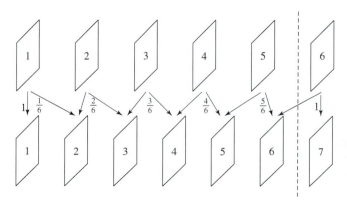

Figure 4.13 Accomplishing 50 to 60 frames up-conversion by converting every five frames into six, and using the two nearest known frames for interpolating each new frame [13]. For each frame in the bottom row, the interpolation coefficient associated with the left neighboring frame in the upper row is given in the figure. The interpolation coefficient for the right neighboring frame is one minus the left coefficient.

may correspond to different object regions in the scene. In this case, temporal interpolation yields erroneous results, and spatial interpolation alone may be better. To overcome such problems, motion-adaptive interpolation filters can be used. With such a filter, one switches between vertical and temporal interpolation at each pixel, depending on the output of a motion detector. If motion is detected, vertical interpolation only is used; otherwise, temporal interpolation only is used. Motion detection at each pixel can be based on the temporal variation in a small neighborhood surrounding the pixel. A key to the success of this approach is the accuracy of the motion-detection process.

Rather than using hard switching between temporal and spatial interpolation, one can also use a weighted average of both, where the weights can be determined by measures of spatial and temporal variation in a small neighborhood surrounding the pixel being interpolated. Let σ_s^2 and σ_t^2 represent the spatial and temporal variation measures, respectively, and $\hat{\psi}_s$ and $\hat{\psi}_t$ the interpolated values using spatial and temporal interpolations, respectively. The final result will be:

$$\hat{\psi} = w_t \hat{\psi}_t + w_s \hat{\psi}_s, \quad \text{with } w_t = \frac{\sigma_s^2}{\sigma_s^2 + \sigma_t^2}, \quad w_s = \frac{\sigma_t^2}{\sigma_s^2 + \sigma_t^2}. \quad (4.2.17)$$

To further improve the performance in regions undergoing motion, motion-compensated interpolation has also been investigated. With this approach, if a pixel is deemed to fall in a motion region, its corresponding position in adjacent fields is determined and used to provide the temporal interpolation result, that is, $\hat{\psi}_t$ in Equation (4.2.17). The most challenging problem is motion estimation—finding corresponding points in adjacent fields or frames. We discuss motion estimation at length in Chapter 6. For a more in-depth study of motion-compensated sampling rate conversion, the reader is referred to [2, 3]. Another alternative is to adapt the sampling grid based on the spatial frequency content [1].

4.3 SUMMARY

Ideal Solutions for Sampling Rate Conversion (Section 4.1)

- The ideal solutions for different conversion problems are illustrated in Figure 4.1. The ideal filters are given in Equations (4.1.4, 4.1.9, 4.1.15). The up-sampling problem is a special case of the general conversion problem with $\Lambda_3 = \Lambda_2$; the down-sampling problem is a special case with $\Lambda_3 = \Lambda_1$.

- In the general rate conversion problem, the filtering operation need not be implemented in Λ_3. To interpolate each sample in Λ_2, filtering need only be applied to samples in Λ_1, as indicated by Equation (4.1.20).

Practical Solutions for Video Sampling Rate Conversion (Section 4.2)

- Common video rate conversion problems include: deinterlacing; NTSC to/from PAL or, equivalently, BT.601 525/60 to/from 625/50; SDTV to/from HDTV; BT.601 4:2:2 to/from 4:2:0; film to/from NTSC or PAL.

- Very short temporal and spatial filters are typically used in practice, to reduce the storage and computation requirements.
- It is important to determine the ideal solution and use it as a guideline for evaluating practical schemes.
- Motion-adaptive operations are important for temporal-direction filtering. The performance depends on the accuracy of the motion detection and estimation.

4.4 PROBLEMS

4.1 In Section 4.2.1, we discussed several practical filters for deinterlacing. We see that the line-field averaging method is closest to the ideal filter; however, it requires storage of two frames. To minimize the frame storage, a single-frame interpolation filter has been proposed [13]. It uses vertical interpolation within the same field and temporal interpolation from the opposite field in the same frame. Referring to Figure 4.5, line D in field t is estimated by $D = (C + E)/2 + K/4 - (I + M)/8$. Determine the equivalent filter in the deinterlaced lattice and its frequency response. Draw the magnitude response (using MATLAB) and compare it to those shown in Figure 4.7. Does this filter have a real frequency response? If not, why not?

4.2 Consider the following deinterlacing method:

$$\hat{f}(t, m) = \frac{1}{2} f(t - 1, m) + \frac{9}{32}[f(t, m - 1) + f(t, m + 1)]$$
$$- \frac{1}{32}[f(t, m - 3) + f(t, m + 3)]$$

Here, $f(t, m)$ represents the image value at field t and line m. For field t, we assume that the lines $m + 2k$, $k = 0, 1, \ldots$ are missing. Find the equivalent interpolation filter for the given operation, and calculate its frequency response. Use MATLAB to draw the frequency response, and compare it to the ideal interpolation filter for deinterlacing.

4.3 Compare the computation requirements for the direct and sequential methods for PAL to NTSC conversion, as discussed in Section 4.2.2. Assume that filtering over a domain with a total number of N samples takes $\alpha N \log_2 N$ operations. Compare the total numbers of operations for the two methods. You need only count the operations involved in filtering; the operations involved in up- and down-conversion are trivial and can be neglected.

4.4 Consider the two methods shown in Figure P4.4 for conversion of NTSC (525 lpf, 30 fps, interlaced) to PAL (625 lpf, 25 fps, interlaced).

    (a) For each method, draw the sampling lattice and its reciprocal lattice at each intermediate step, and determine the desired frequency responses of the filters involved. For simplicity, do this in the temporal-vertical plane only.

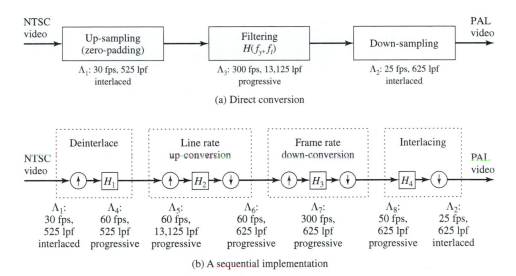

Figure P4.4   Two different methods for converting NTSC to PAL formats.

When sketching the lattice and Voronoi regions in the frequency domain, assume that the temporal and spatial frequency axes have been scaled so that $\frac{1}{\Delta t}$ cps $= \frac{1}{\Delta y}$ cycles/picture-height, for $\Delta t = \frac{1}{60}$ s, $\Delta y = \frac{1}{525}$ picture-height.

(b) The sequential implementation can be combined so that it can be accomplished by a single operation, as with the direct method. Find the response of the equivalent overall filter and compare it to that of the direct method.

(c) Compare the total numbers of operations for the two methods, using the same assumptions as in Problem 4.3.

(d) Based on your solutions for (b) and (c), comment on the advantages and disadvantages of the two methods.

4.5 Consider the problem of converting a film captured at 24 fps, progressive, to an NTSC TV format with 30 fps, interlaced.

(a) Draw the lattices corresponding to these two sampling formats. Determine the intermediate lattice for the conversion, and the ideal filter over this lattice.

(b) In practice, a simple 3:2 pull-down operation as illustrated in Figure P4.5 is often used. We can see that half of the frames are converted into three fields each, whereas the other half are converted into two fields each. For simplicity, no filtering is used. Can you find the equivalent filter in the intermediate lattice determined in (a)?

(c) Can you come up with other simple approaches that are better than (b)?

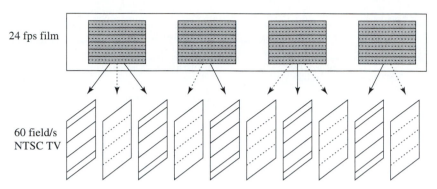

24 fps film

60 field/s
NTSC TV

Figure P4.5   The 3:2 pull-down operation used for converting 24 fps film to 60 field/s NTSC TV [5].

4.6 Consider the conversion of a BT.601 4:2:2 signal to 4:2:0. (Remember that the Y component is the same in these two formats, but the Cr or Cb component has half the number of the lines in the 4:2:0 format as in the 4:2:2.) Because both formats are interlaced, this is not a simple 2:1 down-sampling problem. The conversion problem for the chrominance component is illustrated in Figure P4.6.

   (a) Illustrate the sampling lattices for both BT.601 4:2:2 and 4:2:0 formats, and determine the intermediate lattice and the ideal interpolation filter.

   (b) For simplicity, one can restrict filtering within the same field; for example, samples in field 1 (field 2) are down-sampled by filtering in field 1 (field 2) only. In fact, this is the technique recommended in the MPEG-2 standard for converting the 4:2:2 format to 4:2:0 [13, Section 5.4.1]. In this case, field 1 must be down-sampled by a factor of two. The recommended prefiltering operation is

$$G' = (-29A + 88E + 138G + 88I - 29M)/138.$$

   For field 2, interpolation by a factor of two must be applied to obtain the desired samples. The recommended interpolation operation is

$$S = (D + 7F + 7H + J)/16.$$

   For this resampling scheme, determine the equivalent vertical filter for both field 1 and field 2, as well as the overall vertical-temporal filter. Find their frequency responses. Compare the result with the ideal filter of (a).

4.7 Problem 4.6 considered the conversion from 4:2:2 to 4:2:0 format. Here we consider the conversion from 4:2:0 to 4:2:2 format.

   (a) Draw the intermediate lattice and determine the ideal interpolation filter over this lattice.

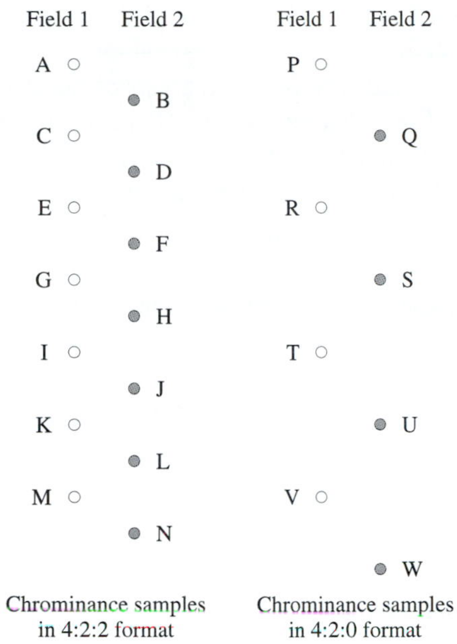

Field 1   Field 2   Field 1   Field 2

A ○   P ○
            ● B
C ○   ● Q
            ● D
E ○   R ○
            ● F
G ○   ● S
            ● H
I ○   T ○
            ● J
K ○   ● U
            ● L
M ○   V ○
            ● N
                                         ● W

Chrominance samples   Chrominance samples
in 4:2:2 format   in 4:2:0 format

Figure P4.6   Relation between chrominance samples in BT.601 4:2:2 and 4:2:0 formats.

(b) Determine the equivalent filter and its frequency response for the following practical scheme [13]: For field 1, interpolation by a factor of two is needed, and the missing samples are obtained by an average of two neighboring samples,

$$G = (R + T)/2.$$

For field 2, interpolation by a factor of four is needed, followed with downsampling by a factor of two. The end-operations are

$$D = (3Q + S)/4, \qquad F = (Q + 3S)/4.$$

Determine the equivalent vertical filters for both field 1 and field 2, as well as the overall vertical-temporal filter, in the interpolated grid. Find their frequency responses. Compare with the result in (a).

4.5 BIBLIOGRAPHY

[1] Belfor, R. A. F., M. P. A. Hesp, and R. L. Lagendijk. Spatially adaptive subsampling of image sequences. *IEEE Trans. Image Process.* (Sept. 1994), 3(5):492–500.

[2] Belfor, R. A. F., R. L. Lagendijk, and J. Biemond. Subsampling of digital image sequences using motion information. In M. I. Sezan and R. L. Lagendijk, eds., *Motion*

Analysis and Image Sequence Processing. Boston: Kluwer Academic Publishers, 1993, 189–223.

[3] Beuker, R. A., and I. A. Shah. Analysis of interlaced video signals and its applications. *IEEE Trans. Image Process.* (Sept. 1994), 3(5):501–12.

[4] Dudgeon, D. E., and R. M. Mersereau. *Multidimensional Digital Signal Processing*. Englewood Cliffs, NJ: Prentice Hall, 1984.

[5] Haskell, B. G., A. Puri, and A. N. Netravali. *Digital Video: An Introduction to MPEG-2*. New York: Chapman & Hall, 1997.

[6] Hou, H. S., and H. C. Andrews. Cubic splines for image interpolation and digital filtering. *IEEE Trans. Acoust., Speech, Signal Process.* (Dec. 1978), ASSP-26: 508–17.

[7] Mitra, S. K. *Digital Signal Processing—A Computer-Based Approach*. New York: McGraw Hill, 1998.

5

Video Modeling

One of the tasks of digital video processing is to describe the change between consecutive images of a video sequence in terms of object motion and other effects, such as illumination changes and camera motion. In order to relate changes in the real world to changes in a video sequence, we need parametric models that describe the real world and the image generation process. The most important models are *scene, object, camera,* and *illumination* models. These models describe the assumptions that we make about the real world. Depending on the selected model, we are able to describe the real world with more or less detail and precision. Using image analysis tools, we estimate the parameters of the parametric models from the video sequence of images of the real world. Using the parametric models and their estimated parameters, we can reconstruct a model world that is an approximation of the real world. Table 5.1 shows the terms that we use to name the parametric models, their corresponding real-world entities, and the entities reconstructed according to the parametric model and presented in the model world. We distinguish items of the real world from items in the model world using the qualifiers *real* and *model*. When the context is clear, we may omit the qualifier.

In the following sections, we describe the camera model (Section 5.1), the illumination model (Section 5.2), the object model (Section 5.3), and the scene model (Section 5.4). Object models in most current video processing applications assume objects with 2-D motion. We present the relevant 2-D motion models in Section 5.5.

Depending on the model used, the estimation of the model parameters can be difficult. In Chapters 6 and 7, we discuss the estimation of model parameters for 2-D and 3-D motions, respectively.

TABLE 5.1 NAMING CONVENTIONS FOR REAL-WORLD AND MODEL ENTITIES

Real world	Parametric model	Model world
real scene	scene model	model scene
real object	object model	model object
real texture	texture model	model texture
real shape	shape model	model shape
real motion	motion model	model motion
real camera	camera model	model camera
real image	image model	model image
real illumination	illumination model	model illumination

5.1 CAMERA MODEL

The camera model describes the projection of real objects in the real scene onto the *imaging plane* of the real camera. The imaging plane is also referred to as the *camera target*. The image on the imaging plane is then converted into a digital image, as described in Chapter 1.

5.1.1 Pinhole Model

A widely used approximation of the projection of real objects onto a real camera target is the *pinhole camera* model, illustrated in Figure 5.1. In this figure, F represents the *focal length* of the camera, and $\mathbf{C}$ the *focal center* (also known as the camera center). The projected position $\mathbf{x}$ of a 3-D point $\mathbf{X}$ is the intersection of the line connecting $\mathbf{X}$ and $\mathbf{C}$ with the imaging plane. Because the imaging plane is behind the focal center, the imaged position is reversed from its true 3-D position. Most camera models used in video processing place the imaging plane on the same side of the focal center as the objects (Figure 5.2(a)), to avoid dealing with reversed positions in the imaging plane.

Without loss of generality, we assume (see Figure 5.2(a)) that the origin of the 3-D coordinate system (also referred to as the *world* coordinate system) is located at the focal center; that its XY-plane is parallel to the imaging plane; and that the world coordinate (X, Y, Z) follows the right-hand system, with the positive direction of the Z-axis being in the imaging direction. Furthermore, we assume that the imaging plane uses the same distance unit as the 3-D coordinate. From the similar triangles illustrated in Figure 5.2(a), we have

$$\frac{x}{F} = \frac{X}{Z}, \qquad \frac{y}{F} = \frac{Y}{Z}, \tag{5.1.1}$$

or

$$x = F\frac{X}{Z}, \qquad y = F\frac{Y}{Z}. \tag{5.1.2}$$

This relation is known as *perspective projection*. A notable character of perspective projection is that the image of an object is smaller if it is farther away from the camera. Mathematically, this is described by the inverse relation between the projected x and y values and the depth value Z.

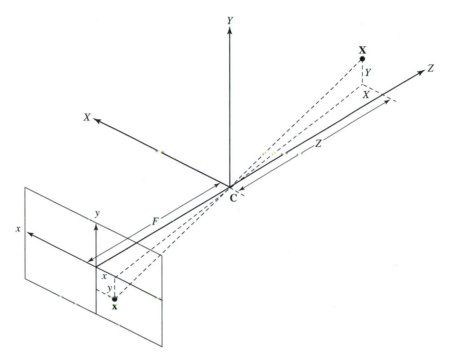

Figure 5.1   Perspective projection by a pinhole camera.

When the imaged object is very far from the camera plane, perspective projection can be approximated by *orthographic projection,* which is also known as *parallel projection* (Figure 5.2(b)):

$$x = X, \qquad y = Y. \tag{5.1.3}$$

Obviously, this relation is much simpler and can greatly simplify transformation between 3-D and 2-D. In practice, the condition that the object must be very far away from the camera can be relaxed. As long as the relative depth variation of the object surface is negligible compared to the distance of the object from the camera, this approximation can be used reliably.

Note that both perspective and orthographic projections are many-to-one mappings. In fact, all the 3-D points on a ray connecting the focal center **C** and the image point **x** would have **x** as its projection. We name this ray the *line of sight*. On the imaging plane we see only the first object point that this ray encounters. The many-to-one mapping of the projection used in camera models makes the estimation of the 3-D structure and motion of objects based on their 2-D images a very difficult and challenging problem.

The pinhole camera model with its perspective projection is only an approximation of most real cameras. It does not consider the misalignment of the camera axis and the image center, the low-pass filter effect of the finite-size aperture of a real lens, the finite exposure time, and other distortions of the lens.

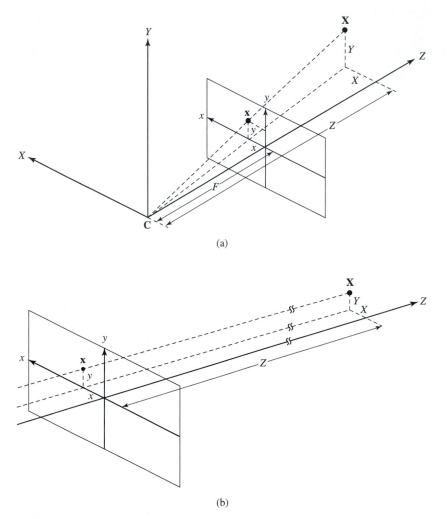

Figure 5.2   Camera model: (a) using perspective projection as in a pinhole camera, (b) using parallel projection as an approximation of a pinhole camera. In both cases, perfect alignment of the *XY*-plane of the world coordinate system with the image plane is assumed.

5.1.2 CAHV Model

For some applications, we like to describe a camera such that we can accommodate camera motion. Furthermore, we would like to calibrate the camera to compensate for geometrical differences between the pinhole camera model and the real camera. The *CAHV camera model* [16] describes perspective projection for a pinhole camera model using four vectors:

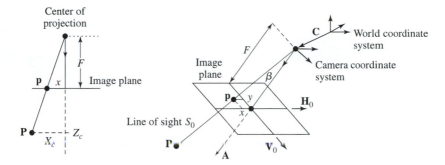

Figure 5.3 Perspective projection of a point **P** in space onto a point **p** in the imaging plane using the CAHV camera model. From W. Niem, Robust and fast modelling of 3-D natural objects from multiple views, *SPIE Image and Video Processing II,* Vol. 2182 (1994). Reprinted by permission of SPIE.

C: Vector to the camera center,
A: Unit vector in the direction of the optical axis,
$\mathbf{H}_0$: Unit vector in the direction of the horizontal axis of the imaging plane,
$\mathbf{V}_0$: Unit vector in the direction of the vertical axis of the imaging plane.

This geometry is depicted in Figure 5.3. Adapting the perspective projection of Equation (5.1.2) by projecting the vector $(\mathbf{P}-\mathbf{C})$ onto the camera axis **A** and the imaging plane axis $\mathbf{H}_0$ and $\mathbf{V}_0$, a point **P** is projected onto the image point **p** according to

$$\mathbf{p} = \begin{pmatrix} x \\ y \end{pmatrix} = \frac{F}{\mathbf{A}^{\mathbf{T}} \cdot (\mathbf{P} - \mathbf{C})} \cdot \begin{pmatrix} \mathbf{H}_0^T \cdot (\mathbf{P} - \mathbf{C}) \\ \mathbf{V}_0^T \cdot (\mathbf{P} - \mathbf{C}) \end{pmatrix}. \tag{5.1.4}$$

In reality, the imaging plane coordinate and the world coordinate are not always in perfect alignment as assumed by the pinhole camera model according to Equation (5.1.2). The CAHV camera model can characterize a practical camera system by its extrinsic parameters **C** and **A** and its intrinsic parameters $\mathbf{H}_0$, $\mathbf{V}_0$, and F. These parameters enable us to describe an imaging plane that is off the camera axis, as well as certain distortions introduced by the optical system. Additional intrinsic camera parameters describe radial distortions of the lens and the modulation transfer function (MTF) of the lens. The MTF can be modeled by a low-pass filter. Camera calibration algorithms are used to estimate these parameters; interested readers are referred to [2]. Precise knowledge of camera parameters is especially useful when estimating 3-D shapes and motion from a video sequence. Recently, algorithms were described that do not require a calibrated camera in order to extract 3-D shape and motion from an image sequence [10].

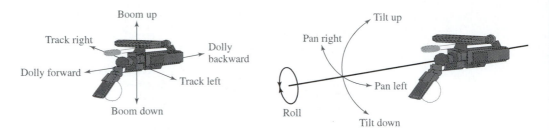

Figure 5.4   Typical types of camera motion. [Courtesy of MPEG-7].

5.1.3 Camera Motions

Figure 5.4 shows typical camera motions. Translations of the camera along the horizontal ($\mathbf{H}_0$) and vertical ($\mathbf{V}_0$) axes of the imaging plane are called *track* and *boom*, respectively. Translation in the direction of the optical camera axis $\mathbf{A}$ is called *dolly*. *Pan* refers to the turning around the vertical axis ($\mathbf{V}_0$), whereas *tilt* refers to the turning around the horizontal axis ($\mathbf{H}_0$). The camera rotation around the optical axis ($\mathbf{A}$) is called *roll*. In addition to the motion types, a camera can also change its focal length, which is called *zoom*. As mentioned, the CAHV camera model is well suited to express these various camera motions.

5.2 ILLUMINATION MODEL

As described in Section 1.1.1, there are two types of light sources: illuminating and reflecting. An illumination model describes how the light incident on an object influences the reflected light distribution, which is what we see. In this section, we describe several such models. Illumination models can be divided into *spectral* and *geometric* models. Spectral illumination models are used if we want to model changes of color resulting from several colored light sources, indirect illumination with objects of different colors, or colored reflecting surfaces. Geometric models describe the amplitude and directional distribution of incident light. Spectral properties were discussed in Section 1.1.1; geometric models are discussed in this section.

Geometric illumination models are available for ambient and for point light sources. For each type of light source, we can also distinguish between *local* and *global* illumination models. A local illumination model assumes that the illumination is independent of object location and other objects. Global illumination models also consider effects between objects, such as shadows and light reflected from object surfaces.

5.2.1 Diffuse and Specular Reflection

We can see only objects that reflect light, and their perceived color depends on the range of wavelengths reflected. In general, the reflection can be decomposed into two components: the *diffuse reflection,* which has equal energy distribution in all directions, and the *specular reflection,* which is strongest in the mirror direction of the incident light. (For a good treatment of both diffuse and specular reflection, see [3].) Surfaces that exhibit only diffuse reflection are known as *Lambertian surfaces,* more commonly described as dull or matte—wood surfaces and cement walls belong to this category. Due to diffuse reflection, we can perceive the color of an object. Specular reflection can be observed with shiny surfaces and mirrors. Specular reflection does not show the color of the object but the color of the incident light; therefore, we cannot actually perceive the color of an object with a surface that shows only specular reflection. Except for mirrors, surfaces usually have diffuse as well as specular reflections. Only the diffuse reflection determines the color of the object surface. Looking at a human eye in daylight, we usually see the eye itself except for a white spot which is caused by specular reflection of the sunlight. From experience, we know that this spot moves on the surface if we change our position relative to the eye.

5.2.2 Radiance Distribution under Differing Illumination and Reflection Conditions

In video processing, the illumination model is mainly used to describe the temporal changes in the video sequence caused by the changing illumination of the real world. The illumination of a background may change because of an object that moves together with its cast shadow. Since the object surface reflects light, this reflecting source changes the overall illumination of the scene.

When discussing the interaction of a light source with an object surface, there are three types of energy involved. First, *incident flux* refers to the rate at which energy is emitted from the light source, measured in watts (W). Second, *incident irradiance* is the incident flux per unit surface area on the object, with a unit of W/m^2. (Note that the irradiance at an object point depends on the angle between the incident light and the surface normal at that point.) Finally, *reflected radiance* measures the light energy reflected from an object surface.

The distribution of the reflected radiance C depends on the distribution of incident irradiance E and the object surface *reflectance function r* at this point. The most general relation can be described by

$$C(\mathbf{L}, \mathbf{V}, \mathbf{N}, \mathbf{X}, t, \lambda) = r(\mathbf{L}, \mathbf{V}, \mathbf{N}, \mathbf{X}, t, \lambda) \cdot E(\mathbf{L}, \mathbf{N}, \mathbf{X}, t, \lambda), \qquad (5.2.1)$$

where $\mathbf{X}$ is the location on the object surface, $\mathbf{N}$ is the surface normal vector at the location $\mathbf{X}$, $\mathbf{L}$ is the illumination direction, $\mathbf{V}$ is the viewing direction connecting $\mathbf{X}$ to the focal point of the camera, and λ is the wavelength of light (Figure 5.5) [12]. Obviously, $\mathbf{L}$, $\mathbf{V}$, and $\mathbf{N}$ are functions of $\mathbf{X}$ and t. The reflectance function r is defined as the ratio between the reflected light intensity (i.e., the flux) and the incident light intensity. This scalar function r is also known as the *diffuse-reflection coefficient,* or *reflection coefficient* for short. The reflectance function depends on the wavelength of the incident light, the surface geometry, and material properties. When the object is moving, the reflection coefficient changes in time at the same location. Note that $r(\mathbf{L}, \mathbf{V}, \mathbf{N}, \mathbf{X}, t, \lambda)$ is defined only for those $\mathbf{X}$ that belong to a surface at time t.

In the following paragraphs, we introduce several simplifying assumptions in order to learn more about the reflected radiance (Equation 5.2.1). We start by assuming

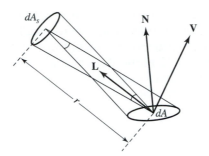

Figure 5.5 Surface patch dA with normal vector **N** illuminated from direction **L** by a point light source with an infinitesimally small area dA_s. The patch is viewed from direction **V**.

opaque object surfaces and temporally invariant illumination (as well as viewing) direction. In this case, Equation (5.2.1) simplifies to

$$C(\mathbf{N}, \mathbf{X}, t, \lambda) = r(\mathbf{N}, \mathbf{X}, t, \lambda) \cdot E(\mathbf{N}, \mathbf{X}, t, \lambda). \tag{5.2.2}$$

Note that although the illumination and viewing directions (**V** and **L**) are fixed, the incident irradiance is still time varying because the object is moving.

An *ambient source* radiates the same amount of energy in every direction at any point, hence it illuminates objects without casting shadows. When the incident light is such an ambient source and the object surface is diffuse reflecting, the reflected radiance intensity distribution is

$$C(\mathbf{X}, t, \lambda) = r(\mathbf{X}, t, \lambda) \cdot E_a(t, \lambda), \tag{5.2.3}$$

where $E_a(t, \lambda)$ represents the intensity of the ambient light at time t. Since the light source is ambient, E_a does not depend on the surface location **X** or surface normal **N**. Since the surface is diffuse reflecting, the reflectance function r does not depend on the surface normal **N**. This ambient light source model is a local illumination model, because we can no longer model global effects like shadows. In rooms, illuminated white walls can often be modeled as ambient light sources. Outdoors, the sun, when covered by clouds, provides ambient illumination.

Now we discuss the reflected radiance due to a *point source*. We assume that the light source is far away from the scene, such that the position of an object has no influence on the incident light. With a local illumination model and a diffuse-reflecting surface, the reflected radiance at any object surface point **X** depends on the angle between the incident light direction **L** and the surface normal **N** at this point, denoted by θ. Obviously, $\mathbf{L}^T\mathbf{N} = \cos(\theta)$ holds. These quantities are illustrated in Figure 5.6.

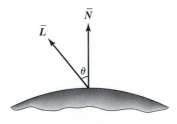

Figure 5.6 Diffuse reflection of a point light source is constant in all directions. The intensity depends on the angle θ between the surface normal **N** and the direction **L** to the light source.

Let $E_p(t, \lambda)$ represents the maximum irradiance from the light source, which is the irradiance intensity when the light is perpendicular to the surface. Then the irradiance with the light at an arbitrary angle θ is $E_p(t, \lambda) \cos \theta$. Starting with Equation (5.2.2), the reflected radiance intensity at $\mathbf{X}$ simplifies [3] to

$$C(\mathbf{X}, t, \lambda) = r(\mathbf{X}, t, \lambda) \cdot E_p(t, \lambda) \cdot \max(0, \mathbf{L}^T \mathbf{N}). \qquad (5.2.4)$$

The max operator in Equation (5.2.4) prevents negative reflected intensities from those parts of the object that do not receive light from the point light source. Typical point light sources are spotlights and the sun. When both ambient and point light sources are present, the total reflected radiance at any point is the superposition of the reflected radiance from each light source according to Equations (5.2.3–4).

    If a point light source is far away from the object surface, we can approximate the incident light as a parallel light. This approximation is valid for daylight and sometimes even indoors. In this case, the maximum incident light irradiance E_p is no longer dependent on the surface point $\mathbf{X}$, but on the surface normal $\mathbf{N}$. Assuming that the scene is illuminated by one stationary, distant point light source and an ambient light, both invariant in time and space, the description of the incident irradiance simplifies to:

$$E(\mathbf{N}, \lambda) = E_a(\lambda) + E_p(\lambda) \cdot \max(0, \mathbf{L}^T \cdot \mathbf{N}). \qquad (5.2.5)$$

This is the shading model of Phong used in early computer graphics [11]. Assuming that the object has a homogeneous surface, that is $r(\mathbf{X}, \lambda) = r(\lambda)$, the corresponding reflected radiance becomes

$$C(\mathbf{N}, \lambda) = r(\lambda) \cdot E(\mathbf{N}, \lambda). \qquad (5.2.6)$$

5.2.3 Changes in the Image Function Due to Object Motion

Now, we investigate how the illumination model can be used to describe illumination effects in the image plane. Assuming white illumination, and that the surface normals $\mathbf{N}$ (and therefore the object shape) are known, the illumination model in Equation (5.2.5) can be used to determine the illumination changes in an image sequence due to object motion. Three parameters must be estimated: the ratio between ambient and point light irradiance E_a / E_p and the two angles describing the direction $\mathbf{L}$ of the point light source.

    The model given by Equation (5.2.5) has been implemented in an object-based analysis-synthesis coder by Stauder [13]. In the 2-D image plane, he assumes that the luminance ψ at pixel $\mathbf{x}$ and time t is proportional to the reflected radiance at the 3-D point $\mathbf{X}$ corresponding to $\mathbf{x}$; that is,

$$\psi(\mathbf{x}, t) = k \cdot C(\mathbf{X}) = k \cdot r(\mathbf{X}) \cdot E(\mathbf{N}), \qquad (5.2.7)$$

where k is a constant and $\mathbf{N}$ denotes the normal direction corresponding to $\mathbf{X}$. Thus, the luminance intensity of a point moving from $\mathbf{x}_k$ to $\mathbf{x}_{k+1}$ (with corresponding surface normals $\mathbf{N}_k$ and $\mathbf{N}_{k+1}$), from time t to $t + d_t$, changes according to

$$\psi(\mathbf{x}_{k+1}, t + d_t) = \psi(\mathbf{x}_k, t) \cdot \frac{E(\mathbf{N}_{k+1})}{E(\mathbf{N}_k)}. \qquad (5.2.8)$$

Another simple illumination model assumes that the image signal $\psi(\mathbf{x})$ depends on the illumination $E(\mathbf{x})$ and the reflectance function $r(\mathbf{x})$, both defined in the image plane. The illumination $E(\mathbf{x})$ in general depends on both ambient and point light sources. Assuming ambient illumination, diffuse-reflecting surfaces, and a camera model with orthographic projection, the image signal is given by the reflection model

$$\psi(\mathbf{x}, t) = k_b \cdot E(\mathbf{x}) \cdot r(\mathbf{x}, t), \qquad (5.2.9)$$

where k_b is a constant. In the image plane, the signal $\psi(\mathbf{x})$ of a point moving from $\mathbf{x}_k$ to $\mathbf{x}_{k+1}$ changes according to

$$\psi(\mathbf{x}_{k+1}, t + d_t) = k_b \cdot E(\mathbf{x}_{k+1}) \cdot r(\mathbf{x}_k, t), \qquad (5.2.10)$$

because $r(x_{k+1}, t+d_t) = r(x_k, t)$ holds; that is, the reflectance function moves with the object. Equation (5.2.10) assumes that the illumination of the scene does not change due to the motion; that is, that the object does not cast shadows in the scene. This has proven to be useful in 2-D motion estimation using block matching, 3-D motion estimation, and segmentation of moving objects and their shadows [1, 4, 9, 14].

The simplest and yet most widely used illumination model simply assumes that $E(\mathbf{x})$ is a constant. In this case, the luminance of a moving point does not change, and Equation (5.2.10) simplifies to

$$\psi(\mathbf{x}_{k+1}, t + d_t) = \psi(\mathbf{x}_k, t). \qquad (5.2.11)$$

This is sometimes referred to as the *constant intensity assumption*. It is valid for spatially and temporally invariant ambient illumination sources and diffuse-reflecting surfaces. Under these conditions, there are no moving shadows and no reflections due to glossy surfaces. Unless otherwise indicated, the algorithms discussed in this book make use of this assumption.

In the preceding equations, we have assumed that the incident light intensity is time invariant. However, the reflected radiance and hence the luminance image is in general space and time varying, as long as the object surface is not completely flat and stationary. The preceding discussion can be extended to the chrominance components of a video image by assuming a colored incident light. Therefore, ψ in Equations (5.2.7–11) can represent either a luminance or a color function.

5.3 OBJECT MODEL

The object model describes assumptions about real objects. Objects are entities in a scene that we consider to be separate items. The interpretation of what an object might be can change, depending on the application. For traffic surveillance, we might consider each car to be an object; whereas for video coding, we might define an object according to its motion, thus segmenting the car into the rotating wheels and the translationally moving body. An object is described by *shape, motion,* and *texture* models [7].

The texture model describes the surface properties of an object. Here, we assume that the texture of an object m is described by the color parameters S_m. These color parameters contain the luminance as well as the chrominance reflectance. Since we usually assume constant intensity according to Equation (5.2.11), these color parameters correspond to the image signal $\psi(\mathbf{x})$.

While all object models described in this book use the same texture model, they employ different motion and shape models. Motion and shape models are usually derived from an approximated 3-D description of objects in order to simplify the estimation of the model parameters. In this section, we start with the most general shape and motion models. Simplification of these models will be introduced as needed (see discussions in Section 5.5 and Chapter 7).

5.3.1 Shape Model

The shape of a 3-D object is described by the 3-D space that it occupies. If we quantize the 3-D space into *voxels,* small cubes of edge length w, we can determine for each voxel whether it belongs to the object or not. This is illustrated in Figure 5.7, where w denotes the length of each side of the cube. Object shapes can be convex or concave, can have holes. We assume that an object is topologically connected; that is, that we can draw a path from any point in the object to any other point without leaving the object. The projection of an object onto an image plane might result in topologically disconnected regions, due to covered object parts or parts of the object being outside of the image. Since we are not much interested in the interior of an object, a voxel representation is inefficient in terms of memory. Therefore, we usually describe the shape of an object by its surface. Often, a mesh of triangles commonly referred to as a *wireframe* is used. The mesh of triangles is put up by vertices referred to as *control points* $\mathbf{P}_C^{(i)}$. In addition to these control points, we need a list that defines which control points define triangles (Figure 5.8). Assuming that the list of control points is numbered as in Figure 5.8(b), we can use these numbers as indices for the control points. This simplifies the definition of surface patches or face to that of giving a list of the indices that define the patches. This list of indices is called an *indexed faceset* list (Figure 5.8(c)). Sometimes, faces are defined by more than three points.

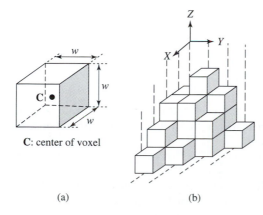

(a)   (b)

Figure 5.7   Voxels: (a) volume element, (b) shape representation with voxels in a 3-D lattice. From W. Niem, Robust and fast modelling of 3-D natural objects from multiple views, *SPIE Image and Video Processing II,* Vol. 2182 (1994). Reprinted by permission of SPIE.

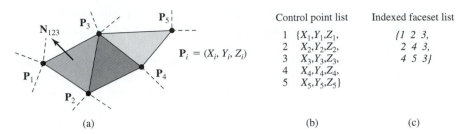

Figure 5.8   Wireframe for object shape description: (a) mesh of triangles representing the object surface, (b) representation of the mesh of triangles as a list of control points, and (c) a list of triangles that references the list of control points. From W. Niem, Robust and fast modelling of 3-D natural objects from multiple views, *SPIE Image and Video Processing II*, Vol. 2182 (1994). Reprinted by permission of SPIE.

The control points of a wireframe are located on the object surface. The number and location of control points are determined by the object shape, as well as the accuracy with which the wireframe model is to describe the object shape. As long as control points cannot be moved independently, the object is *rigid* and cannot change its shape. Otherwise, the object is *flexible,* and it can change its shape.

5.3.2 Motion Model

As just described, an object can be rigid or flexible. The motion of a rigid object can be described in terms of a translation vector $\mathbf{T} = (T_x, T_y, T_z)^T$ and a rotation matrix $[\mathbf{R}]$. Let us first describe how a point of an object moves in 3-D space using the motion parameters $\mathbf{T}$ and $[\mathbf{R}]$. The translation vector $\mathbf{T}$ describes a displacement of a point from $\mathbf{X}$ to $\mathbf{X}'$ by T_x, T_y, T_z in the directions of the coordinate axes X, Y, Z, respectively:

$$\mathbf{X}' = \mathbf{X} + \mathbf{T}. \tag{5.3.1}$$

If an object is translated, Equation (5.3.1) holds for all object points.

If an object rotates around the origin of the 3-D space, we describe the motion of its points with the rotation matrix $[\mathbf{R}]$:

$$[\mathbf{R}] = [\mathbf{R}_z] \cdot [\mathbf{R}_y] \cdot [\mathbf{R}_x]. \tag{5.3.2}$$

This rotation matrix rotates a point around the axes $X, Y,$ and Z in that order. It is computed from the rotation matrices that rotate a point around just one axis. These individual rotation matrices are

$$[\mathbf{R}_x] = \begin{bmatrix} 1 & 0 & 0 \\ 0 & \cos\theta_x & -\sin\theta_x \\ 0 & \sin\theta_x & \cos\theta_x \end{bmatrix}, \tag{5.3.3}$$

$$[\mathbf{R}_y] = \begin{bmatrix} \cos\theta_y & 0 & \sin\theta_y \\ 0 & 1 & 0 \\ -\sin\theta_y & 0 & \cos\theta_y \end{bmatrix}, \tag{5.3.4}$$

and

$$[\mathbf{R}_z] = \begin{bmatrix} \cos\theta_z & -\sin\theta_z & 0 \\ \sin\theta_z & \cos\theta_z & 0 \\ 0 & 0 & 1 \end{bmatrix}, \tag{5.3.5}$$

where θ_x, θ_y, and θ_z are the rotation angles with respect to each axis. Finally, we have $[\mathbf{R}]$ as a function of the rotation angles θ_x, θ_y, and θ_z

$$[\mathbf{R}] = \begin{bmatrix} \cos\theta_y \cos\theta_z & \sin\theta_x \sin\theta_y \cos\theta_z - \cos\theta_x \sin\theta_z & \cos\theta_x \sin\theta_y \cos\theta_z + \sin\theta_x \sin\theta_z \\ \cos\theta_y \sin\theta_z & \sin\theta_x \sin\theta_y \sin\theta_z + \cos\theta_x \cos\theta_z & \cos\theta_x \sin\theta_y \sin\theta_z - \sin\theta_x \cos\theta_z \\ -\sin\theta_y & \sin\theta_x \cos\theta_y & \cos\theta_x \cos\theta_y \end{bmatrix}. \tag{5.3.6}$$

The rotation matrix $[\mathbf{R}]$ is an orthonormal matrix; that is, it satisfies

$$[\mathbf{R}]^T = [\mathbf{R}]^{-1}, \tag{5.3.7}$$

$$\det[\mathbf{R}] = \pm 1. \tag{5.3.8}$$

Now, we can express the motion of a point $\mathbf{X}$ on the object surface from $\mathbf{X}$ to $\mathbf{X}'$ as

$$\mathbf{X}' = [\mathbf{R}] \cdot \mathbf{X} + \mathbf{T}. \tag{5.3.9}$$

For many motion estimation algorithms, the nonlinear rotation matrix according to Equation (5.3.2) must be linearized with respect to the rotation angles. Assuming that rotation angles are small, such that the approximations $cos(\alpha) \approx 1$ and $sin(\alpha) \approx \alpha$ are valid, Equation (5.3.6) simplifies to

$$[\mathbf{R}] \approx [\mathbf{R}'] = \begin{bmatrix} 1 & -\theta_z & \theta_y \\ \theta_z & 1 & -\theta_x \\ -\theta_y & \theta_x & 1 \end{bmatrix}. \tag{5.3.10}$$

Equation (5.3.9) rotates the point $\mathbf{X}$ on the object surface around the center of the world coordinate system. In the case that the object is far away from the center of the coordinate system and the object is rotating around its own center only, a large translation vector is required for its motion representation. This can be avoided by defining a local coordinate system for each object, and by defining rotation and translation of the object with respect to its own center $\mathbf{C} = (C_x, C_y, C_z)^T$, which would also be the center of the local coordinate system (see Figure 5.9):

$$\mathbf{X}' = [\mathbf{R}] \cdot (\mathbf{X} - \mathbf{C}) + \mathbf{C} + \mathbf{T}. \tag{5.3.11}$$

Assuming that $\mathbf{C} \neq 0$, the rotation matrices in Equations (5.3.9) and (5.3.11) are identical, whereas the translation vectors differ for identical physical motion.

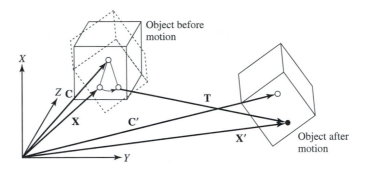

Figure 5.9   World coordinate system, showing the motion of a point **X**.

An object is rigid if all its points move with the same 3-D motion parameters. As shown in the preceding, the motion of a rigid object can be described by the parameters $A^{(m)} = (T_x^{(m)}, T_y^{(m)}, T_z^{(m)}, \theta_x^{(m)}, \theta_y^{(m)}, \theta_z^{(m)})$. A rigid object does not change its shape when it is moved with the parameter set $A^{(m)}$.

Not all real objects are rigid. One way to describe a flexible object is by decomposing it into two or more rigid *components*. Each component has its own set of motion parameters according to Equation (5.3.11). Since each component is defined by its control points, the components are linked by triangles interconnecting control points in different components. Through these triangles, components are flexibly connected. Figure 5.10 shows a scene with the objects "Background" and "Claire." The model object "Claire" consists of two components "Head" and "Shoulder." Alternatively, flexible objects can be described by superposing small local motions onto the rigid motion parameters $A^{(m)}$. If we have a sailing ship with a flag, we might describe the motion of the entire ship using a set of rigid parameters $A^{(m)}$, and use local motion

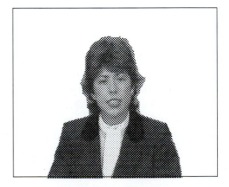

Figure 5.10   A model object consisting of two components. The shape is represented by a wireframe; the surface texture is extracted from an image.

to model the waving of its flag in the wind. Since flexible objects change their shape due to local motion, it is not obvious whether this flexibility should be described by shape or motion parameters.

5.4 SCENE MODEL

Having discussed how to model the illumination source, objects, and cameras, we are now ready to consider the modeling of an imaged scene. The scene model describes the world with illumination sources, objects, and cameras. Depending on the models used for different components, we arrive at different scene models. Figure 5.11 shows a complex model scene that is created using a *3-D scene model*. It is obtained using a camera model assuming perspective projection. This model allows for describing object motion and object occlusion, as well as cast shadows. Each dashed line describes a *line*

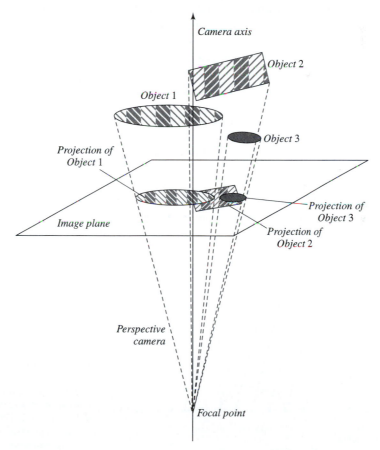

Figure 5.11   A complex 3-D scene with several objects, cameras, and diffuse illumination (not shown).

of sight according to the camera model. A point on the object surface is projected onto the image plane at the point where the line of sight intersects the image plane. The projections onto the image plane change as the objects move. Objects 1 and 3 occlude part of Object 2. All objects occlude part of the *image background* which is defined as the image that we would have if there were no moving objects present in the scene. (Due to the perspective projection of the camera model, Object 3 extends to image areas not covered by Object 2. The scene in Figure 5.11 is illuminated by diffuse illumination only; therefore, we do not see any shadows.) If the objects are in motion, we distinguish between four image areas: *static background, moving object, uncovered background* (or region), and *newly covered background* (or region). Newly covered regions are also known as occluded regions or background.

Figure 5.12 shows two image frames with a moving object. Comparing frames k and $k + 1$, we can distinguish the *changed* and the *unchanged regions*. The unchanged regions show the static background in both images. The moving object is part of the changed regions in images k and $k + 1$. In frame k, the changed region is defined as the area of the moving object, and the *background to be covered* in frame $k + 1$ due to the object motion. In frame $k + 1$, the changed region is defined as the area of the moving object and the *uncovered background* that was not visible in frame k.

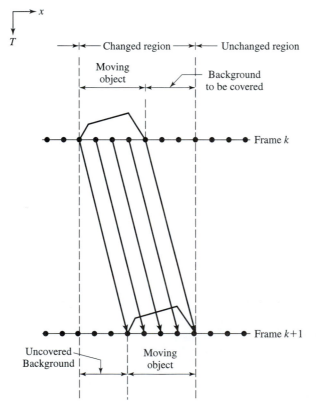

Figure 5.12 The separation of changed areas into moving objects, uncovered background, and background to be covered (covered background). Reprinted from *Signal Processing,* Vol. 15, No. 3, M. Hötter and R. Thoma, Image segmentation based on object oriented mapping parameter estimation, pp. 315–34, copyright 1988, with permission from Elsevier Science.

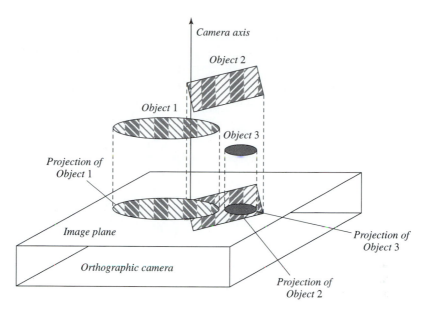

Figure 5.13   A 2.5-D scene with several objects, a camera, and ambient illumination (not shown).

In Figure 5.13, we see a scene with layered 2-D objects. The camera model uses orthographic projection instead of perspective projection. Therefore depth has no influence on the image of this scene. We can describe how objects are layered on top of each other when viewed from the camera. This scene model does not allow the description of effects due to 3-D motion, since we do not explicitly use the third dimension of space, and is commonly referred to as a *2.5-D scene model*. The MPEG-4 video coding standard supports layering of video objects and hence supports this scene model (MPEG-4 is treated in Chapter 13).

Objects 1, 2, and 3 in Figure 5.13 are at the same 3-D locations as in Figure 5.11. Due to the use of parallel projection, the lines of sight are now parallel. Hence, Object 3 no longer extends beyond Object 2 in the image plane.

Figure 5.14 shows the simplest scene model that is useful for image processing. It assumes that all objects are flat and lay on the same image plane. 2-D objects are limited to motion in a 2-D plane. The concept of object layering is not supported. This simple *2-D scene model* is used in the video coding standards H.261, H.263, MPEG-1, and MPEG-2 (treated in Chapter 13).

The selection of a specific scene model limits the choices for illumination, camera, and object models. If we choose a 2-D scene model, it makes no sense to pick an object model that explicitly describes 3-D shapes. For some applications however, it may be useful to choose a 3-D scene model and yet an object model that only allows for 2-D shapes.

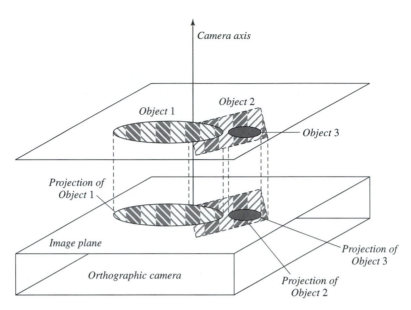

Figure 5.14   A 2-D scene with several objects, a camera, and ambient illumination, (not shown).

5.5 TWO-DIMENSIONAL MOTION MODELS

5.5.1 Definition and Notation

When the camera or the object in the scene is moving, the image of the same 3-D object point will also change. Figure 5.15 illustrates the projection of a moving object onto an image plane, and the relation between 3-D and 2-D motions. When an object point is moved from $\mathbf{X} = [X, Y, Z]^T$ at time t_1 to $\mathbf{X}' = [X', Y', Z']^T = [X + D_X, Y + D_Y, Z + D_Z]^T$ at time $t_2 = t_1 + d_t$, its projected image is changed from $\mathbf{x} = [x, y]^T$ to $\mathbf{x}' = [x', y']^T = [x + d_x, y + d_y]^T$. We call the 3-D displacement, $\mathbf{D}(\mathbf{X}; t_1, t_2) = \mathbf{X}' - \mathbf{X} = [D_X, D_Y, D_Z]^T$, the *3-D motion vector* at $\mathbf{X}$, and the 2-D displacement, $\mathbf{d}(\mathbf{x}; t_1, t_2) = \mathbf{x}' - \mathbf{x} = [d_x, d_y]^T$, the *2-D motion vector* at $\mathbf{x}$. Note that in general the motion vectors (MVs) are position-dependent. As a function of all image positions $\mathbf{x}$ at time t_1, $\mathbf{d}(\mathbf{x}; t_1, t_2)$ represents a *2-D motion field* from t_1 to t_2. When the underlying t_1 and t_2 are clear, we simply write $\mathbf{d}(\mathbf{x})$.

Instead of describing the displacement between corresponding points, it is sometimes more convenient to specify, for each point $\mathbf{x}$ at time t_1, its corresponding position at t_2, $\mathbf{w}(\mathbf{x}; t_1, t_2) = \mathbf{x}'$. We call $\mathbf{w}(\mathbf{x}; t_1, t_2)$ or $\mathbf{w}(\mathbf{x})$ for short, the *mapping function*. Obviously, it is uniquely related to the motion field by $\mathbf{w}(\mathbf{x}) = \mathbf{x} + \mathbf{d}(\mathbf{x})$.

In this book, we deal only with digital video signals that have finite and discrete image domains, described by a truncated lattice, Λ. The notation $\mathbf{x} = [x, y]^T \in \Lambda$ represents a pixel index. We further assume that the time interval $d_t = t_2 - t_1$ is equal to either the temporal sampling interval (i.e., the frame interval) or an integer multiple

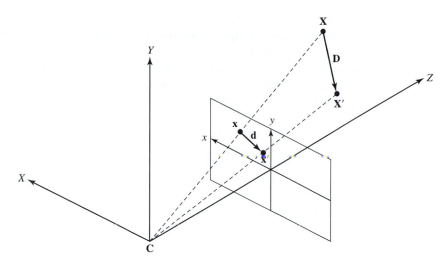

Figure 5.15   Projection of a moving object: relation between 3-D and 2-D motion vectors.

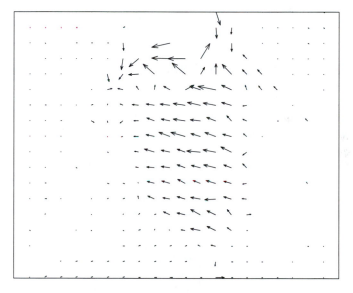

Figure 5.16   A typical 2-D motion field.

of this interval. The motion field for a given time interval is a finite set of 2-D vectors arranged in a 2-D array, in the same order as the pixels. Such a *discrete motion field* is often depicted by a vector graph, as shown in Figure 5.16, where the direction and magnitude of each arrow represents the direction and magnitude of the MV at the pixel where the arrow originates.

Instead of using the actual displacements over a given time interval, one can use the velocity vectors to characterize the motion. The velocity vectors are known as *flow vectors*, defined as $\mathbf{v} = \frac{\partial \mathbf{d}}{\partial t} = \left[\frac{\partial d_x}{\partial t}, \frac{\partial d_x}{\partial t} \right]^T$. If the time interval d_t is sufficiently small, the motion within this time interval can be assumed to be constant, that is, $\mathbf{v} = \mathbf{d}/d_t$. When dealing with digital video, we often consider $d_t = 1$, so that the flow vectors are synonymous with the motion vectors. As with the definition of the motion field, one can define the *flow field* over the entire image domain, as $\mathbf{v}(\mathbf{x}; t_1, t_2), \mathbf{x} \in \Lambda$.

In a real video sequence, motion vectors are not always defined at every point. As described in Section 5.4, there may be overlapping objects in a scene, one behind the other. The most common example (see Figure 5.12) is a moving object in front of a static background. Motion vectors are not defined over uncovered background. A complete definition of the 2-D motion field at any given time should consist of a masking image and a motion field image. For example, in Figure 5.12, if the motion is defined from frame $k + 1$ to frame k, then the pixels in the uncovered background in frame $k + 1$ should be labeled as uncovered.

5.5.2 Two-Dimensional Motion Models Corresponding to Typical Camera Motions

In this section, we describe 2-D motion models that correspond to some of the camera motions depicted in Figure 5.4. These models are derived by assuming the perspective imaging geometry shown in Figure 5.2(a). That is, the image coordinates (x, y) are related to the world coordinates (X, Y, Z) by the perspective mapping given in Equation (5.1.2).

To derive the 2-D motion resulting from a camera motion, we first determine the new 3-D coordinate of an object point with respect to the new camera coordinate, and then determine its image position by using Equation (5.1.2).

Camera Translation within the Image Plane (Track and Boom)   Recall that track and boom refer to the translations of the optical center along the X- and Y-axes, respectively (see Figure 5.4(a)). Let T_x and T_y denote the actual translations. Then the 3-D position of any point (X, Y, Z) in the old camera coordinate will change to

$$\begin{bmatrix} X' \\ Y' \\ Z' \end{bmatrix} = \begin{bmatrix} X \\ Y \\ Z \end{bmatrix} + \begin{bmatrix} T_x \\ T_y \\ 0 \end{bmatrix}. \tag{5.5.1}$$

Using (5.1.2), the imaged position (x, y) changes to

$$\begin{bmatrix} x' \\ y' \end{bmatrix} = \begin{bmatrix} x \\ y \end{bmatrix} + \begin{bmatrix} FT_x/Z \\ FT_y/Z \end{bmatrix}.$$

Therefore, the displacement of a pixel (x, y) depends on the Z coordinate of its corresponding 3-D point. However, when the relative variation of the depth of the imaged object is small compared to the average distance $\bar{Z}$ between the object and the camera center, one can approximate the motion field by a constant vector; that is,

$$\begin{bmatrix} d_x(x, y) \\ d_y(x, y) \end{bmatrix} = \begin{bmatrix} t_x \\ t_y \end{bmatrix},$$

where $t_x = FT_x/\bar{Z}, t_y = FT_y/\bar{Z}$. Note that a translation along the Z-axis (i.e., dolly) will lead to a substantially more complex 2-D motion field, as will become clear.

Camera Pan and Tilt   Pan and tilt refer to the rotation of the camera around the Y- and X-axes, respectively (see Figure 5.4(b)). Let θ_y and θ_x denote the corresponding rotation angles. The new camera coordinate is related to the old by

$$\mathbf{X}' = [\mathbf{R}_x][\mathbf{R}_y]\mathbf{X} \tag{5.5.2}$$

where $[\mathbf{R}_x]$ and $[\mathbf{R}_y]$ are as defined previously in Equations (5.3.3–4). When the rotation angles are small, we have

$$[\mathbf{R}_x][\mathbf{R}_y] = \begin{bmatrix} 1 & 0 & \theta_y \\ 0 & 1 & -\theta_x \\ -\theta_y & \theta_x & 1 \end{bmatrix}. \tag{5.5.3}$$

If $Y\theta_x \ll Z$ and $X\theta_y \ll Z$, then $Z' \approx Z$. Using Equation (5.1.2), we arrive at

$$\begin{bmatrix} x' \\ y' \end{bmatrix} = \begin{bmatrix} x \\ y \end{bmatrix} + \begin{bmatrix} \theta_y F \\ -\theta_x F \end{bmatrix}. \tag{5.5.4}$$

or

$$\begin{bmatrix} d_x(x, y) \\ d_y(x, y) \end{bmatrix} = \begin{bmatrix} \theta_y F \\ -\theta_x F \end{bmatrix}. \tag{5.5.5}$$

The preceding result implies that when the pan and tilt angles are small, and when the imaged object is far from the camera, so that its extensions in the X and Y directions are significantly smaller than its distance from the camera, the image of the object is displaced uniformly.

Camera Zoom   Let F and F' represent the focal lengths before and after zooming. Using Equation (5.1.2), it is easily shown that

$$\begin{bmatrix} x' \\ y' \end{bmatrix} = \begin{bmatrix} \rho x \\ \rho y \end{bmatrix}, \tag{5.5.6}$$

where $\rho = F'/F$ is the zoom factor. The 2-D motion field is

$$\begin{bmatrix} d_x(x, y) \\ d_y(x, y) \end{bmatrix} = \begin{bmatrix} (1 - \rho)x \\ (1 - \rho)y \end{bmatrix}. \tag{5.5.7}$$

A typical motion field corresponding to zoom is illustrated in Figure 5.17(a).

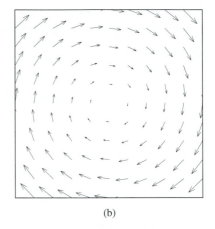

(a)   (b)

Figure 5.17   Motion fields corresponding to (a) camera zoom, and (b) camera rotation.

Camera Roll   Recall that roll refers to the rotation around the Z-axis (see Figure 5.4(b)). Because such a rotation does not change the Z coordinate, the net effect after perspective projection is that the image coordinate undergoes a rotation; that is,

$$\begin{bmatrix} x' \\ y' \end{bmatrix} = \begin{bmatrix} \cos\theta_z & -\sin\theta_z \\ \sin\theta_z & \cos\theta_z \end{bmatrix} \begin{bmatrix} x \\ y \end{bmatrix} \approx \begin{bmatrix} 1 & -\theta_z \\ \theta_z & 1 \end{bmatrix} \begin{bmatrix} x \\ y \end{bmatrix}, \tag{5.5.8}$$

where the approximation is valid when θ_z is small. The corresponding motion field is

$$\begin{bmatrix} d_x(x, y) \\ d_y(x, y) \end{bmatrix} = \begin{bmatrix} -\theta_z y \\ \theta_z x \end{bmatrix}. \tag{5.5.9}$$

Figure 5.17(b) illustrates a typical motion field due to rotation.

Four-Parameter Model   Consider a camera that goes through translation, pan, tilt, zoom, and rotation in sequence. Using the approximations presented in the preceding, the imaged positions of the same 3-D point before and after the camera motions are related by

$$\begin{bmatrix} x' \\ y' \end{bmatrix} = \rho \begin{bmatrix} \cos\theta_z & -\sin\theta_z \\ \sin\theta_z & \cos\theta_z \end{bmatrix} \begin{bmatrix} x + \theta_y F + t_x \\ y - \theta_x F + t_y \end{bmatrix} \tag{5.5.10}$$

$$= \begin{bmatrix} c_1 & -c_2 \\ c_2 & c_1 \end{bmatrix} \begin{bmatrix} x \\ y \end{bmatrix} + \begin{bmatrix} c_3 \\ c_4 \end{bmatrix}. \tag{5.5.11}$$

This mapping function has four parameters, and is a special case of the affine mapping, which generally has six parameters. This specialized form is known as the *geometric mapping,* which can characterize any combination of scaling, rotation, and translation in 2-D. If the order of pan, zoom, and rotation is changed, this form would still be valid, albeit the relations between the parameters c_k and the camera motion parameters $(\rho, \theta_x, \theta_y, \theta_z, t_x, t_y)$ would change.

5.5.3 Two-Dimensional Motion Corresponding to Three-Dimensional Rigid Motion

The previous 2-D motion models do not cover all possible 3-D motions of the camera. Specifically, they does not consider camera translation in the Z direction, nor rotation around an arbitrary axis. An arbitrary camera motion in 3-D can be decomposed into a 3-D rotation and a 3-D translation. Similarly, an imaged object may undergo a rigid motion (rotation plus translation) while the camera is held steady. In general, both the camera and object can undergo a rigid motion, and it is hard to separate the motions of the camera and object. The overall effect, however, is equivalent to assuming that one item (object or camera) is stationary and the other item is undergoing a rigid motion. In this section, we consider the 2-D motion induced by such an arbitrary 3-D rigid motion. Without loss of generality, we will assume that the camera is stationary and the object is undergoing a rigid motion.

General Case   As described in Section 5.3, the 3-D positions of any object point before and after a rigid motion are related by

$$\begin{bmatrix} X' \\ Y' \\ Z' \end{bmatrix} = \begin{bmatrix} r_1 & r_2 & r_3 \\ r_4 & r_5 & r_6 \\ r_7 & r_8 & r_9 \end{bmatrix} \begin{bmatrix} X \\ Y \\ Z \end{bmatrix} + \begin{bmatrix} T_x \\ T_y \\ T_z \end{bmatrix}. \tag{5.5.12}$$

Although the rotation matrix has nine entries, it is completely determined by three rotation angles (see Equation (5.3.6)). Together with the three translation parameters, there are six parameters.

Substituting Equation (5.1.2) into Equation (5.5.12), after some manipulation, one can obtain the relation between the image coordinates before and after the motion:

$$\begin{aligned} x' &= F\frac{(r_1 x + r_2 y + r_3 F)Z + T_x F}{(r_7 x + r_8 y + r_9 F)Z + T_z F}, \\ y' &= F\frac{(r_4 x + r_5 y + r_6 F)Z + T_y F}{(r_7 x + r_8 y + r_9 F)Z + T_z F}. \end{aligned} \tag{5.5.13}$$

Note that if we scale the translational parameters T_x, T_y, T_z and the depth Z by the same factor, the correspondence between (x, y) and (x', y') will not change. This indicates that based on the correspondences of imaged coordinates, the parameters T_x, T_y, T_z, are unique only up to a scaling factor. Alternatively, we say that only the direction but not the length of the translation vector can be determined. In general, when the object surface (i.e., the Z value corresponding to every (x, y)) is spatially varying, the mapping function changes from point to point, and hence cannot be characterized by a small number of parameters.

Projective Mapping   When there is no translational motion in the Z direction, or when the imaged object has a planar surface,* it is easy to show (Problem (5.4)

---

*This means that the 3-D coordinates of points on the object satisfy $aX + bY + cZ = 1$ for some constants a, b, c.

that Equation (5.5.13) can be simplified into the following eight-parameter *projective mapping:*[*]

$$x' = \frac{a_0 + a_1 x + a_2 y}{1 + c_1 x + c_2 y}, \quad y' = \frac{b_0 + b_1 x + b_2 y}{1 + c_1 x + c_2 y}. \tag{5.5.14}$$

There are eight free parameters in the projective mapping, which are dependent on the five motion parameters (recall that the three translation parameters are unique only up to a scale and therefore only two of them are free) and three plane parameters of the object surface.

The projective mapping is an important relation in the study of motion between two video frames, or equivalently the registration of two video frames. It models *exactly* the relation between images of the same object after camera or object motion, if either of the following is true: (1) the object has a planar surface (in this case, the camera or object or both can undergo arbitrary 3-D rotation and translation); or (2) neither the camera nor the object undergoes translation along the imaging axis (i.e., the Z-axis[†]; in this case, the object geometry can be arbitrary). In practice, when the imaged scene does not have a planar surface, we can divide the imaged domain into small elements so that the 3-D patch corresponding to each 2-D element is flat, and make use of the projective model over each patch. Note that the projective mapping includes as special cases pure translation and affine mapping.

When the imaging geometry can be approximated by an orthographic projection (see Equation (5.1.3)), it is easy to show that a planar patch undergoing 3-D rigid motion can be described by an affine function (see Problem 5.2).

The projective mapping can characterize the two unique phenomena associated with perspective imaging [6], namely: (1) the chirping effect, which refers to the increase in the perceived spatial frequency as the distance from the camera increases; and (2) the converging effect (also known as the keystone effect), which refers to the fact that two parallel lines appear to move closer at a distance and eventually converge to a single point. These effects are illustrated in Figure 5.18.

Another important property of projective mapping is that the concatenation of two projective mappings is another projective mapping [6]. This has a beneficial implication for estimation of a projective mapping between two images: one can estimate the parameters in several iterations; as long as each step leads to a projective mapping, the overall mapping is guaranteed to be projective.

Projective mapping acting on a square block is illustrated in Figure 5.19(d). The projective function can map between two arbitrary quadrilaterals. Instead of using the parameters a_l, b_l, and c_l, one can use the displacement $\mathbf{d_k} = (x'_k - x_k, y'_k - y_k), k = 1, 2, 3, 4$, between two quadrangles at the four corners to parameterize a projective mapping. Problem 5.9 considers how to determine the projective mapping parameters from $\mathbf{x}_k$ and $\mathbf{d}_k$.

---

[*]Such a mapping is sometimes referred to as *perspective mapping*. Here, to avoid confusion with the relation between 3-D and 2-D coordinates in a perspective imaging system, we use the term "projective."

[†]A motion that does not include translation in the imaging direction is sometimes referred as *zero-parallax* motion.

Non-chirping models   Chirping models

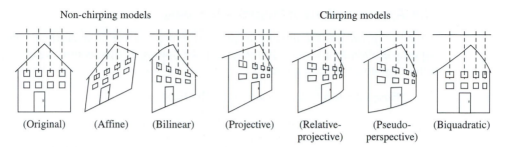

(Original)   (Affine)   (Bilinear)   (Projective)   (Relative-   (Pseudo-   (Biquadratic)
                                                        projective)   perspective)

Figure 5.18   Pictorial effects of different mapping functions. Reprinted from
S. Mann and R. W. Picard, Video or bits of the projective group: A simple approach
to featureless estimation of parameters, *IEEE Trans. Image Process.* (Sept. 1997),
6:1281–95. Copyright 1997 IEEE.

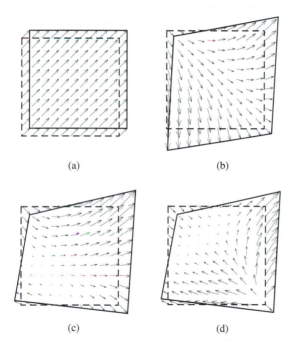

(a)   (b)

(c)   (d)

Figure 5.19   Illustration of basic motion
models: (a) translational, (b) affine,
(c) bilinear, (d) projective.

The motion field corresponding to a projective mapping is

$$d_x(x, y) = \frac{a_0 + a_1 x + a_2 y}{1 + c_1 x + c_2 y} - x, \qquad d_y(x, y) = \frac{b_0 + b_1 x + b_2 y}{1 + c_1 x + c_2 y} - y. \quad (5.5.15)$$

We should point out that the motion field itself cannot be described by a projective
function. In [6], the mapping corresponding to a motion field that is approximated by a
projective mapping is called *relative projective*. Its effect is shown in Figure 5.18. We
can see that it can approximate a projective mapping relatively well.

5.5.4 Approximations of Projective Mapping

In practice, in order to circumvent the problem caused by its rational form, the projective mapping is often approximated by polynomial mapping. In this section, we discuss some of the common approximations.

Affine Motion   *Affine motion* has the form

$$\begin{bmatrix} d_x(x, y) \\ d_y(x, y) \end{bmatrix} = \begin{bmatrix} a_0 + a_1 x + a_2 y \\ b_0 + b_1 x + b_2 y \end{bmatrix}. \tag{5.5.16}$$

It has six parameters, $a_k, b_k, k = 0, 1, 2$. As we have shown, the projected 2-D motion of most camera motions can be described by an affine function. Furthermore, a planar patch undergoing arbitrary 3-D rigid motion can be described by an affine function under orthographic projection. The affine model, however, cannot capture either the chirping or converging effect of the projective mapping, as illustrated in Figure 5.18.

Affine motion can be visualized as the deformation of one triangle into another, by moving the corners of the triangle. The affine parameters are completely determined by the MVs of the three corners. Therefore, an affine motion can also be parameterized by these MVs. In the example shown in Figure 5.19(b), one can specify the displacements at three corners only, and the displacement at the fourth corner will be determined by the MVs at the other three corners.

Bilinear Motion   *Bilinear motion* has the form of

$$\begin{bmatrix} d_x(x, y) \\ d_y(x, y) \end{bmatrix} = \begin{bmatrix} a_0 + a_1 x + a_2 y + a_3 xy \\ b_0 + b_1 x + b_2 y + b_3 xy \end{bmatrix}. \tag{5.5.17}$$

It has eight parameters, $a_k, b_k, k = 0, 1, 2, 3$. From Figure 5.18, we see that bilinear mapping can capture the converging effect of projective mapping, but not the chirping effect.

Bilinear motion can be visualized as the warping of a square into a quadrangle, as illustrated in Figure 5.19(c). The eight parameters are completely determined by the MVs of the four corners of the original quadrilateral. Note that, unlike projective mapping, the bilinear function cannot map between two arbitrary quadrangles, even though such a mapping has only eight degrees of freedom. This is because the bilinear function maps a straight line to a curved line, unless the original line is horizontal or vertical. In general, the bilinear function maps a quadrilateral to a curvilinear quadrilateral [15].

Other Polynomial Models   In general, any motion function can be approximated by a polynomial mapping of the form

$$\begin{bmatrix} d_x(x, y) \\ d_y(x, y) \end{bmatrix} = \sum_{\substack{0 \le i, j \le N_1 \\ i+j \le N_2}} \begin{bmatrix} a_{i,j} \\ b_{i,j} \end{bmatrix} x^i y^j. \tag{5.5.18}$$

The translation, affine, and bilinear motions are special cases of this model with $N_1 = N_2 = 0$ for translation, $N_1 = N_2 = 1$ for affine, and $N_1 = 1, N_2 = 2$ for bilinear. The case corresponding to $N_1 = 2$, $N_2 = 2$ is

$$\begin{bmatrix} d_x(x, y) \\ d_y(x, y) \end{bmatrix} = \begin{bmatrix} a_0 + a_1 x + a_2 y + a_3 x^2 + a_4 y^2 + a_5 xy \\ b_0 + b_1 x + b_2 y + b_3 x^2 + b_4 y^2 + b_5 xy \end{bmatrix}. \qquad (5.5.19)$$

This is known as *biquadratic mapping,* which has twelve parameters. As shown in Figure 5.18, the biquadratic mapping can produce the chirping but not the converging effect.

In [6], Mann and Picard compared the advantages and disadvantages of various polynomial approximations for projective mapping. They found that the affine function, although having fewer parameters, is more susceptible to noise because it lacks the correct degrees of freedom needed to properly track the actual image motion. On the other hand, the bilinear function, although possessing the correct number of parameters, fails to capture the chirping effect. A better eight-parameter model is the following *pseudo-perspective mapping:*

$$\begin{bmatrix} d_x(x, y) \\ d_y(x, y) \end{bmatrix} = \begin{bmatrix} a_0 + a_1 x + a_2 y + a_3 xy + b_3 x^2 \\ b_0 + b_1 x + b_2 y + b_3 xy + a_3 y^2 \end{bmatrix}. \qquad (5.5.20)$$

From Figure 5.18, we see that this mapping can produce both the chirping and converging effects, and is the best approximation of projective mapping using low-order polynomials.

5.6 SUMMARY

In this chapter, we discussed parametric models that allow us to describe how an image sequence is generated. Depending on the application, we can use models that describe the real world with an appropriate amount of detail.

- **Camera Model (Section 5.1)**   The camera model describes the projection of the 3-D world on to the image plane by means of the real camera. Depending on our application, we can use camera models of different complexity. If the objects in the 3-D world are far away from the camera we might use a simple camera model with orthographic projection. A pinhole camera model uses a perspective projection, and enables us to describe the change of object size in an image sequence as an object changes its distance from the camera.

- **Illumination Model (Section 5.2)**   In order to see an object, we need to illuminate the observed scene. Describing illumination and the reflection of light on object surfaces usually requires complex models. However, we assume the most simple illumination model for many video processing applications: ambient light and diffuse reflecting surfaces. Under this model, the surface reflectance of an object does not change as the object moves. This model is sometimes referred to as the constant intensity assumption. (The motion estimation algorithms presented in Chapters 6 and 7 make use of this assumption extensively.) Note that, with this model, we cannot describe shadows or glossy reflections.

- **Object Model (Section 5.3)**   Objects are described by shape, motion, and texture. The 3-D shape of an object can be described by its volume, using cubes of unit length named voxels; or by its surface, using a wireframe. We express 3-D object motion by means of a 3-D translation vector and a 3×3 rotation matrix that is computed from the rotation angles around the three coordinate axes. Simple models can be derived by simplifying the 3-D object shape and motion models.

- **Scene Model (Section 5.4)**   The scene model describes how the moving objects and the camera of a 3-D scene are positioned with respect to each other. In video coding, we often use a 2-D scene model that assumes 2-D objects moving parallel to the imaging plane of the camera. The slightly more complex 2.5-D scene model allows us to handle object occlusion. Finally, the 3-D scene model can realistically describe the real world.

- **2-D Motion Model (Section 5.5)**   Object or camera motions in 3-D lead to 2-D motions. The 2-D motion model depends not only on the 3-D motion model, but also the illumination and camera models. The most important 2-D motion model is the projective mapping, which is valid over small 2-D regions, when the object or camera undergoes rigid motion, the camera can be modeled by perspective projection, and the constant intensity assumption applies. In practice, this mapping is often approximated by affine or bilinear mappings.

5.7 PROBLEMS

5.1 Assume a pinhole camera with focal length $F = 9$ mm, a target of $1'' \times 1.33''$, an image resolution of 352×288 pels, and an object point with a distance of $Z = 2$ m from the camera center. Determine the projection of this point into the image as a function of its (X, Y) position in 3-D space. How much does the point have to move in the direction of the Z-axis in order to move its image point by 1 pel? What is the answer if we assume a camera model with orthographic projection?

5.2 Show that, under orthographic projection, the projected 2-D motion of a planar patch undergoing translation, rotation, and scaling (because of camera zoom) can be described by an affine function.

5.3 Show that, under perspective projection, the projected 2-D motion of a planar patch undergoing rigid motion can be described by a projective mapping.

5.4 Show that Equation (5.5.13) can be simplified into the projective mapping given in Equation 5.5.14, when the imaged object has a planar surface. That is, the 3-D coordinates of points on the object satisfy $aX + bY + cZ = 1$ for some constants a, b, c.

5.5 Consider a triangle whose original corner positions are at $\mathbf{x}_k$, $k = 1, 2, 3$. Suppose that each corner is moved by $\mathbf{d}_k$. Determine the affine parameters that can realize such a mapping.

5.6 Suppose that you are given the motion vectors $\mathbf{d}_k$ at $K > 3$ points $\mathbf{x}_k$ in an image, and that you want to use the affine mapping to approximate the overall motion. How do you determine the affine parameters?
(*Hint:* use a least-squares fitting method.)

5.7 Repeat Problem 5.5 for the bilinear mapping between two quadrangles. Assume that displacements at $K = 4$ points are available.

5.8 Repeat Problem 5.6 for the bilinear mapping, assuming $K > 4$.

5.9 Repeat Problem 5.5 for the projective mapping between two quadrangles. Assume that displacements at $K = 4$ points are available.
(*Hint:* you can still set up a linear equation in terms of the parameters a_k, b_k, c_k. The constants in the left-hand side of the equation will contain both $\mathbf{x}_k$ and $\mathbf{x}'_k = \mathbf{x}_k + \mathbf{d}_k$.)

5.10 Repeat Problem 5.6 for the projective mapping, assuming $K > 4$.

5.8 BIBLIOGRAPHY

[1] Bozdagi, G., A. M. Tekalp, and L. Onural. 3-D motion estimation and wireframe adaptation including photometric effects for model-based coding of facial image sequences. *IEEE Trans. Circuits Syst. for Video Technology* (June 1994), 4(3): 246–56.

[2] Faugeras, O. *Three-Dimensional Computer Vision—A Geometric Viewpoint.* Cambridge, MA: MIT Press, 1993.

[3] Foley, J., A. van Dam, S. K. Feiner, J. F. Hughes, and R. L. Phillips. *Introduction to Computer Graphics.* Reading, MA: Addison-Wesley, 1994.

[4] Gilge, M., Motion estimation by scene adaptive block matching (SABM) and illumination correction. In *SPIE Image Processing Algorithms and Techniques* (Feb. 1990), 1244:355–66.

[5] Hötter, M., and R. Thoma. Image segmentation based on object oriented mapping parameter estimation. *Signal Processing* (Oct. 1988), 15(3):315–34.

[6] Mann, S., and R. W. Picard. Video orbits of the projective group: A simple approach to featureless estimation of parameters. *IEEE Trans. Image Process.* (Sept. 1997), 6:1281–95.

[7] Musmann, H. G., M. Hötter, and J. Ostermann. Object oriented analysis-synthesis coding of moving images. *Signal Processing: Image Commun.* (Oct. 1989), 1:119–38.

[8] Niem, W. Robust and fast modelling of 3-D natural objects from multiple views. *SPIE Image and Video Processing II* (1994), 2182:388–97.

[9] Ostermann, J. Segmentation of image areas changed due to object motion considering shadows. In Y. Wang, et al., eds., *Multimedia Communications and Video Coding.* New York: Plenum Press, 1996.

[10] Pollefeys, M., R. Koch, and L. Van Gool. Self-calibration and metric reconstruction in spite of varying and unknown internal camera parameters. In *Int. Conf. Computer Vision (ICCV'98),* 90–95.

[11] Rogers, D. F. *Procedural Elements for Computer Graphics.* Singapore: McGraw-Hill, 1985.

[12] Stauder, J. An illumination estimation method for 3-D-object-based analysis-synthesis coding. In *COST 211 European Workshop,* 4.5.1–6. University of Hannover, Germany, December 1993.

[13] Stauder, J. Estimation of point light source parameters for object-based coding. *Signal Processing: Image Communications* (1995), 46(7):355–79.

[14] Stauder, J., R. Mech, and J. Ostermann. Detection of moving cast shadows for object segmentation. *IEEE Transactions on Multimedia* (March 1999), 1(1):65–76.

[15] Wolberg, G. *Digital Image Warping.* Los Alamitos, CA: IEEE Computer Society Press, 1990.

[16] Yakimovsky, Y., and R. Cunningham. A system for extracting three-dimensional measurements from a stereo pair of TV cameras. *Computer Graphics and Image Processing* (April 1978), 7(2):195–210.

6

Two-Dimensional Motion Estimation

Motion estimation is an important part of any video processing system. In this chapter, we are concerned with the estimation of 2-D motion; in Chapter 7, we discuss estimation of actual 3-D motion. As will be seen, 2-D motion estimation is often a preprocessing step required for 3-D structure and motion estimation. Also, 2-D motion estimation itself has a wide range of applications, including video compression, sampling rate conversion, filtering, and so on. Depending on the intended applications for the resulting 2-D motion vectors, motion estimation methods can be very different. For example, for computer vision applications, where the 2-D motion vectors will be used to deduce 3-D structure and motion parameters, a sparse set of 2-D motion vectors at critical feature points may be sufficient. (The motion vectors must be physically correct for them to be useful.) On the other hand, for video compression applications, the estimated motion vectors are used to produce a motion-compensated prediction of a frame to be coded from a previously coded reference frame. The ultimate goal is to minimize the total bits used for coding the motion vectors and the prediction errors. There is a trade-off that one can make between the accuracy of the estimated motion, and the number of bits used to specify the motion. Sometimes, even when the estimated motion is not an accurate representation of the true physical motion, it can still lead to good temporal prediction and, in that regard, be considered a good estimate. In this chapter, we focus on the type of motion estimation algorithms targeted for motion-compensated processing (prediction, filtering, interpolation, etc.). For additional readings on this topic, the reader is referred to the review papers by Musmann, Pirsch, and Grallert [28] and Stiller and Konrad [38]. For a good treatment of motion estimation methods for computer vision applications, please see the article by Aggarwal and Nandhahumar [1].

All the motion estimation algorithms are based on temporal changes in image intensities (more generally color). In fact, the observed 2-D motions based on intensity changes may not be the same as the actual 2-D motions. To be more precise, the velocity of observed or apparent 2-D motion vectors is referred to as *optical flow*. Optical flow can be caused not only by object motions, but also camera movements or illumination condition changes. In this chapter, we start by defining optical flow. We then derive the optical flow equation, which imposes a constraint between image gradients and flow vectors. This is a fundamental equality that many motion estimation algorithms are based on. Next, we present the general methodologies for 2-D motion estimation. As will be seen, a motion estimation problem is usually converted to an optimization problem, and involves three key components: parameterization of the motion field, formulation of the optimization criterion, and searching for the optimal parameters. Finally, we present motion estimation algorithms that have been developed based on different parameterizations of the motion field and different estimation criteria. Unless specified otherwise, the word "motion" refers in this chapter to 2-D motion.

6.1 OPTICAL FLOW

6.1.1 Two-Dimensional Motion versus Optical Flow

The human eye perceives motion by identifying corresponding points at different times. The correspondence is usually determined by assuming that the color or brightness of a point does not change after the motion. It is interesting to note that observed 2-D motion can be different from the actual projected 2-D motion under certain circumstances. Figure 6.1 illustrates two special cases. In the first example, a sphere with a uniform flat surface is rotating under a constant ambient light. Because every point on the sphere reflects the same color, the eye cannot observe any change in the color pattern of the imaged sphere and thus considers the sphere as being stationary. In the second example, the sphere is stationary, but is illuminated by a point light source that is rotating around the sphere. The motion of the light source causes the movement of the reflecting light

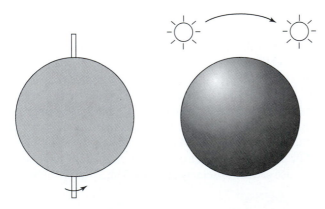

Figure 6.1 The optical flow is not always the same as the true motion field. In (a), a sphere is rotating under a constant ambient illumination, but the observed image does not change. In (b), a point light source is rotating around a stationary sphere, causing the highlight point on the sphere to rotate. Adapted from B. K. P. Horn, *Robot Vision,* MIT Press, 1986. Copyright 1986 MIT Press.

spot on the sphere, which in turn can make the eye believe the sphere is rotating. Observed or apparent 2-D motion is referred to as optical flow in computer vision literature. The preceding examples reveal that the optical flow may not be the same as the true 2-D motion. When only image color information is available, the best one can hope to estimate accurately is the optical flow. However, in the remaining part of this chapter, we will use the term "2-D motion" or simply "motion" to describe optical flow. The reader should bear in mind that sometimes it may be different from the true 2-D motion.

6.1.2 Optical Flow Equation and Ambiguity in Motion Estimation

Consider a video sequence whose luminance variation is represented by $\psi(x, y, t)$.* Suppose an imaged point (x, y) at time t is moved to $(x + d_x, y + d_y)$ at time $t + d_t$. Under the constant intensity assumption introduced in Section 5.2.3 (Equation (5.2.11)), the images of the same object point at different times have the same luminance value. Therefore,

$$\psi(x + d_x, y + d_y, t + d_t) = \psi(x, y, t). \qquad (6.1.1)$$

Using Taylor's expansion, when d_x, d_y, d_t are small, we have

$$\psi(x + d_x, y + d_y, t + d_t) = \psi(x, y, t) + \frac{\partial \psi}{\partial x} d_x + \frac{\partial \psi}{\partial y} d_y + \frac{\partial \psi}{\partial t} d_t. \qquad (6.1.2)$$

Combining Equations (6.1.1) and (6.1.2) yields

$$\frac{\partial \psi}{\partial x} d_x + \frac{\partial \psi}{\partial y} d_y + \frac{\partial \psi}{\partial t} d_t = 0. \qquad (6.1.3)$$

Equation (6.1.3) is written in terms of the motion vector (d_x, d_y). Dividing both sides by d_t yields

$$\frac{\partial \psi}{\partial x} v_x + \frac{\partial \psi}{\partial y} v_y + \frac{\partial \psi}{\partial t} = 0 \quad \text{or} \quad \nabla \psi^T \mathbf{v} + \frac{\partial \psi}{\partial t} = 0. \qquad (6.1.4)$$

where (v_x, v_y) represents the *velocity vector* (also called *flow vector*), $\nabla \psi = [\frac{\partial \psi}{\partial x}, \frac{\partial \psi}{\partial y}]^T$ is the spatial gradient vector of $\psi(x, y, t)$. In arriving at Equation (6.1.4), commonly known as the *optical flow equation*, we have assumed that d_t is small, so that $v_x = d_x/d_t$, $v_y = d_y/d_t$.† The conditions for this relation to hold are the same as for the constant intensity assumption, and were discussed in Section 5.2.3.

---

*In this text, we consider motion estimation based solely on the luminance intensity information, although the same methodology can be applied to the full color information.

†Another way to derive the optical flow equation is by representing the constant intensity assumption as $d\psi(x, y, t)/dt = 0$. Expanding $d\psi(x, y, t)/dt$ in terms of the partials will lead to the same equation.

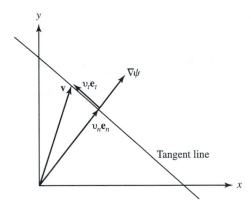

Figure 6.2 Decomposition of motion **v** into normal ($v_n\mathbf{e}_n$) and tangent ($v_t\mathbf{e}_t$) components. Given $\nabla\psi$ and $\frac{\partial\psi}{\partial t}$, any MV on the tangent line satisfies the optical flow equation.

As shown in Figure 6.2, the flow vector **v** at any point **x** can be decomposed into two orthogonal components as

$$\mathbf{v} = v_n\mathbf{e}_n + v_t\mathbf{e}_t, \tag{6.1.5}$$

where $\mathbf{e}_n$ is the direction vector of the image gradient $\nabla\psi$, to be called the normal direction, and $\mathbf{e}_t$ is orthogonal to $\mathbf{e}_n$, to be called the tangent direction. The optical flow equation in Equation (6.1.4) can be written as

$$v_n\|\nabla\psi\| + \frac{\partial\psi}{\partial t} = 0, \tag{6.1.6}$$

where $\|\nabla\psi\|$ is the magnitude of the gradient vector. Three consequences from Equation (6.1.4) or (6.1.6) are:

1. At any pixel **x**, one cannot determine the velocity vector **v** based on $\nabla\psi$ and $\partial\psi/\partial t$ alone. There is only one equation for two unknowns (v_x and v_y, or v_n and v_t). In fact, the underdetermined component is v_t. To solve both unknowns, one must impose additional constraints. The most common constraint is that the flow vectors should vary smoothly spatially, so that one can make use of the intensity variation over a small neighborhood surrounding **x** to estimate the motion at **x**.

2. Given $\nabla\psi$ and $\partial\psi/\partial t$, the projection of the velocity vector along the normal direction is fixed, with $v_n = -(\partial\psi/\partial t)/\|\nabla\psi\|$, whereas the projection onto the tangent direction, v_t, is undetermined. Any value of v_t would satisfy the optical flow equation. In Figure 6.2, this means that any point on the tangent line will satisfy the optical flow equation. This ambiguity in estimating the motion vector is known as the *aperture problem*. The word "aperture" here refers to the small window over which to apply the constant intensity assumption. The motion can be estimated uniquely only if the aperture contains at least two different gradient directions, as illustrated in Figure 6.3.

3. In regions with constant brightness, such that $\|\nabla\psi\|=0$, the flow vector is indeterminate. This is because there are no perceived brightness changes when the

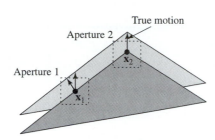

Figure 6.3 The aperture problem in motion estimation: To estimate the motion at x_1 using aperture 1, it is impossible to determine whether the motion is upward or perpendicular to the edge, because there is only one spatial gradient direction in this aperture. On the other hand, the motion at x_2 can be determined accurately, because the image has gradients in two different directions in aperture 2.

underlying surface has a flat pattern. The estimation of motion is reliable only in regions with brightness variation; in other words, regions with edges or nonflat textures.

The preceding observations are consistent with the relation between spatial and temporal frequencies discussed in Section 2.3.2. There, we showed that the temporal frequency of a moving object is zero if the spatial frequency is zero, or if the motion direction is orthogonal to the spatial frequency. When the temporal frequency is zero, no changes can be observed in image patterns, and consequently, motion is indeterminate.

As will be seen in the following sections, the optical flow equation (or, equivalently, the constant intensity assumption) plays a key role in all motion estimation algorithms.

6.2 GENERAL METHODOLOGIES

In this chapter, we consider the estimation of motion between two given frames, $\psi(x, y, t_1)$ and $\psi(x, y, t_2)$. Recall from Section 5.5.1 that the MV at x between time t_1 and t_2 is defined as the displacement of this point from t_1 to t_2. We will call the frame at time t_1 the *anchor frame,* and the frame at t_2 the *target frame.** Depending on the intended application, the anchor frame can be either before or after the target frame in time. As illustrated in Figure 6.4, the problem is referred to as *forward motion estimation,* when $t_1 < t_2$, and as *backward motion estimation,* when $t_1 > t_2$. For notational convenience, we will use $\psi_1(x)$ and $\psi_2(x)$ to denote the anchor and target frames, respectively. In general, we can represent the motion field as $d(x; a)$, where $a = [a_1, a_2, \ldots, a_L]^T$ is a vector containing all the motion parameters. Similarly, the mapping function can be denoted by $w(x; a) = x + d(x; a)$. The motion estimation problem is equivalent to estimating the motion parameter vector a. Methods that have been developed can be categorized into two groups: *feature-based* and *intensity-based.* In the feature-based approach, correspondences between pairs of selected feature points in two video frames are first established. The motion model parameters are then obtained by a least-squares fitting of the established correspondences into the chosen motion model. This approach is applicable only to parametric motion models, and can

---

*In video coding, a current frame $\psi(x, y, t_1)$ is predicted from a previously coded frame $\psi(x, y, t_2)$, where $t_1 > t_2$ in forward motion compensation, and $t_1 < t_2$ in backward motion compensation. In video coding literature, $\psi(x, y, t_2)$ is called the reference frame.

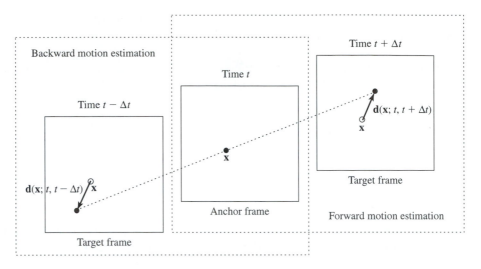

Figure 6.4 Forward and backward motion estimation.

be quite effective in, say, determining global motions. The intensity-based approach applies the constant intensity assumption or the optical flow equation at every pixel, and requires the estimated motion to satisfy this constraint as closely as possible. This approach is more appropriate when the underlying motion cannot be characterized by a simple model, and an estimate of a pixelwise or blockwise motion field is desired.

In this chapter, we consider only intensity-based approaches, which are more widely used in applications requiring motion-compensated prediction and filtering. In general, the intensity-based motion estimation problem can be converted into an optimization problem, in which three key questions must be answered: (1) How do we parameterize the underlying motion field? (2) What criterion do we use to estimate the parameters? and (3) How do we search for the optimal parameters? In this section, we first describe several ways to represent a motion field. Then we introduce different types of estimation criteria. Finally, we present search strategies commonly used for motion estimation. Specific motion estimation schemes using different motion representations and estimation criteria will be introduced in subsequent sections.

6.2.1 Motion Representation

A key problem in motion estimation is how to parameterize the motion field. As shown in Section 5.5, the 2-D motion field resulting from a camera or object motion can usually be described by a few parameters. However, there are typically multiple objects in the imaged scene that move differently. Therefore, a global parameterized model is usually inadequate. The most direct and unconstrained approach is to specify the motion vector at every pixel. This is the so-called *pixel-based representation.* Such a representation is universally applicable, but it requires the estimation of a large number of unknowns (twice the number of pixels!) and the solution can often be physically incorrect unless a proper physical constraint is imposed during the estimation step. On the other hand, if only the camera is moving, or the imaged scene contains a single moving object with a planar surface, one can use a *global motion representation* to characterize the entire motion field. In general, for scenes containing multiple moving objects, it is more appropriate to divide an image frame into multiple regions, so that the motion within each region can be characterized well by a parameterized model. This is known as *region-based motion representation,** which consists of a region segmentation map and several sets of motion parameters, one for each region. The difficulty with such an approach is that one does not know in advance which pixels have similar motions. Therefore, segmentation and estimation must be accomplished iteratively, which requires intensive computations that may not be feasible in practice.

One way to reduce the complexity associated with region-based motion representation is by using a fixed partition of the image domain into many small blocks. As long as each block is small enough, the motion variation within each block can be characterized well by a simple model, and the motion parameters for each block can be estimated independently. This brings us to the popular *block-based representation*. The simplest version models the motion in each block by a constant translation, so that the estimation problem becomes that of finding one MV for each block. This method provides a good compromise between accuracy and complexity, and has found great success in practical video coding systems. One main problem with the block-based approach is that it does not impose any constraint on the motion transition between adjacent blocks. The resulting motion is often discontinuous across block boundaries, even when the true motion field is changing smoothly from block to block. One way to overcome this problem is by using a *mesh-based representation,* in which the underlying image frame is partitioned into nonoverlapping polygonal elements. The motion field over the entire frame is described by the MVs at the nodes (corners of polygonal elements) only, and the MVs at the interior points of an element are interpolated from the nodal MVs. This representation induces a motion field that is continuous everywhere. It is more appropriate than the block-based representation over interior regions of an object, which usually undergo a continuous motion, but it fails to capture motion discontinuities at object boundaries. Adaptive schemes that allow discontinuities when necessary are needed for more accurate motion estimation. Figure 6.5 illustrates the effects of several of the motion representations just described for a head-and-shoulder scene. In the next few sections, we will introduce motion estimation methods using different motion representations.

6.2.2 Motion Estimation Criteria

For a chosen motion model, the problem is how to estimate the model parameters. In this section, we describe several different criteria for estimating motion parameters.

Criterion Based on the Displaced Frame Difference   The most popular criterion for motion estimation is the sum of the differences between the luminance

---

*This is sometimes called *object-based motion representation* [39]. Here we use the term "region-based" to acknowledge the fact that we are only considering 2-D motions, and that a region with a coherent 2-D motion may not always correspond to a physical object.

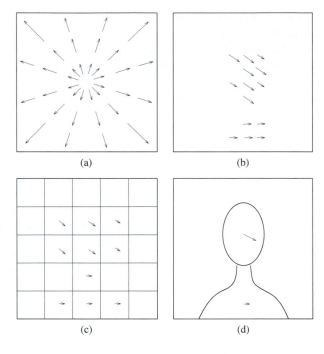

(a)   (b)

(c)   (d)

Figure 6.5 Different motion representations: (a) global, (b) pixel-based, (c) block-based, and (d) region-based. Reprinted from C. Stiller and J. Konrad, Estimating motion in image sequences, *IEEE Signal Processing Magazine* (July 1999), 16:70–91. Copyright 1999 IEEE.

values of every pair of corresponding points between the anchor frame ψ_1 and the target frame ψ_2. Recall that $\mathbf{x}$ in ψ_1 is moved to $\mathbf{w}(\mathbf{x}; \mathbf{a})$ in ψ_2. Therefore, the objective function can be written as,

$$E_{\mathrm{DFD}}(\mathbf{a}) = \sum_{\mathbf{x} \in \Lambda} |\psi_2(\mathbf{w}(\mathbf{x}; \mathbf{a})) - \psi_1(\mathbf{x})|^p, \tag{6.2.1}$$

where Λ is the set of all pixels in ψ_1, and p is a positive number. When $p = 1$, the above error is called the *mean absolute difference* (MAD), and when $p = 2$, the *mean squared error* (MSE). The error image, $e(\mathbf{x}; \mathbf{a}) = \psi_2(\mathbf{w}(\mathbf{x}; \mathbf{a})) - \psi_1(\mathbf{x})$, is usually called the *displaced frame difference (DFD) image,* and the preceding measure the *DFD error.*

The necessary condition for minimizing E_{DFD} is that its gradient vanishes. In the case of $p = 2$, this gradient is

$$\frac{\partial E_{\mathrm{DFD}}}{\partial \mathbf{a}} = 2 \sum_{\mathbf{x} \in \Lambda} (\psi_2(\mathbf{w}(\mathbf{x}; \mathbf{a})) - \psi_1(\mathbf{x})) \frac{\partial \mathbf{d}(\mathbf{x})}{\partial \mathbf{a}} \nabla \psi_2(\mathbf{w}(\mathbf{x}; \mathbf{a})) \tag{6.2.2}$$

where

$$\frac{\partial \mathbf{d}}{\partial \mathbf{a}} = \begin{bmatrix} \dfrac{\partial d_x}{\partial a_1} & \dfrac{\partial d_x}{\partial a_2} & \cdots & \dfrac{\partial d_x}{\partial a_L} \\ \dfrac{\partial d_y}{\partial a_1} & \dfrac{\partial d_y}{\partial a_2} & \cdots & \dfrac{\partial d_y}{\partial a_L} \end{bmatrix}^T .$$

Criterion Based on the Optical Flow Equation   Instead of minimizing the DFD error, another approach is to solve the system of equations that is established based on the optical flow (OF) constraint given in Equation (6.1.3). Let $\psi_1(x, y) = \psi(x, y, t)$, $\psi_2(x, y) = \psi(x, y, t+d_t)$. If d_t is small, we can assume that $(\partial\psi/\partial t)\, dt = \psi_2(\mathbf{x}) - \psi_1(\mathbf{x})$. Then, Equation (6.1.3) can be written as

$$\frac{\partial\psi_1}{\partial x}d_x + \frac{\partial\psi_1}{\partial y}d_y + (\psi_2 - \psi_1) = 0 \quad \text{or} \quad \nabla\psi_1^T\mathbf{d} + (\psi_2 - \psi_1) = 0. \qquad (6.2.3)$$

This discrete version of the optical flow equation is more often used for motion estimation in digital video. Solving the preceding equations for all $\mathbf{x}$ can be turned into a minimization problem with the following objective function:

$$E_{\text{OF}}(\mathbf{a}) = \sum_{\mathbf{x}\in\Lambda} |(\nabla\psi_1(\mathbf{x}))^T\mathbf{d}(\mathbf{x}; \mathbf{a}) + \psi_2(\mathbf{x}) - \psi_1(\mathbf{x})|^p. \qquad (6.2.4)$$

The gradient of E_{OF} is, when $p = 2$,

$$\frac{\partial E_{\text{OF}}}{\partial\mathbf{a}} = 2\sum_{\mathbf{x}\in\Lambda} \left((\nabla\psi_1(\mathbf{x}))^T\mathbf{d}(\mathbf{x}; \mathbf{a}) + \psi_2(\mathbf{x}) - \psi_1(\mathbf{x})\right)\frac{\partial\mathbf{d}(\mathbf{x})}{\partial\mathbf{a}}\nabla\psi_1(\mathbf{x}). \qquad (6.2.5)$$

If the motion field is constant over a small region Λ'—that is, if $\mathbf{d}(\mathbf{x}; \mathbf{a}) = \mathbf{d}_0$, $\mathbf{x} \in \Lambda'$—then Equation (6.2.5) becomes

$$\frac{\partial E_{\text{OF}}}{\partial\mathbf{d}_0} = \sum_{\mathbf{x}\in\Lambda'} \left((\nabla\psi_1(\mathbf{x}))^T\mathbf{d}_0 + \psi_2(\mathbf{x}) - \psi_1(\mathbf{x})\right)\nabla\psi_1(\mathbf{x}). \qquad (6.2.6)$$

Setting the preceding gradient to zero yields the least-squares solution for $\mathbf{d}_0$:

$$\mathbf{d}_0 = \left(\sum_{\mathbf{x}\in\Lambda'} \nabla\psi_1(\mathbf{x})(\nabla\psi_1(\mathbf{x}))^T\right)^{-1}\left(\sum_{\mathbf{x}\in\Lambda'} (\psi_1(\mathbf{x}) - \psi_2(\mathbf{x}))\nabla\psi_1(\mathbf{x})\right). \qquad (6.2.7)$$

When the motion is not a constant, but can be related to the model parameters linearly, one can still derive a similar least-squares solution (see Problem 6.6).

An advantage of the preceding method is that the minimizing function is a quadratic function of the MVs, when $p = 2$. If the motion parameters are linearly related to the MVs, then the function has a unique minimum and can be solved easily. This is not true with the DFD error given in Equation (6.2.1). However, the optical flow equation is valid only when the motion is small, or when an initial motion estimate $\tilde{\mathbf{d}}(\mathbf{x})$ that is close to the true motion can be found and one can pre-update $\psi_2(\mathbf{x})$ to $\psi_2(\mathbf{x} + \tilde{\mathbf{d}}(\mathbf{x}))$. When this is not the case, it is better to use the DFD error criterion, and find the minimal solution using the gradient descent or exhaustive search method.

Regularization   Minimizing the DFD error or solving the optical flow equation does not always give a physically meaningful motion estimate. This is partially because the constant intensity assumption is not always correct. The imaged intensity of the same object point may vary after an object motion because of the various reflectance and shadowing effects. A second reason is that, in a region with flat texture, many different motion estimates can satisfy the constant intensity assumption or the optical flow equation. Finally, if the motion parameters are the MVs at every pixel, the optical flow equation does not constrain the motion vector completely. These factors can make motion estimation an ill-posed problem.

To obtain a physically meaningful solution, one must impose additional constraints to regularize the problem. One common regularization approach is to add a penalty term to the error function in (6.2.1) or (6.2.4), which should force the resulting motion estimate to bear the characteristics of common motion fields. One well-known property of a typical motion field is that it usually varies smoothly from pixel to pixel, except at object boundaries. To enforce smoothness, one can use a penalty term that measures the differences between the MVs of adjacent pixels, that is,

$$E_s(\mathbf{a}) = \sum_{\mathbf{x} \in \Lambda} \sum_{\mathbf{y} \in \mathcal{N}_\mathbf{x}} \|\mathbf{d}(\mathbf{x}; \mathbf{a}) - \mathbf{d}(\mathbf{y}; \mathbf{a})\|^2, \qquad (6.2.8)$$

where $\|\cdot\|$ represents the two-norm and $\mathcal{N}_\mathbf{x}$ represents the set of pixels adjacent to $\mathbf{x}$. Either the four- or eight-connectivity neighborhood can be used.*

The overall minimization criterion can be written as

$$E = E_{\mathrm{DFD}} + w_s E_s. \qquad (6.2.9)$$

The weighting coefficient w_s should be chosen based on the importance of motion smoothness relative to the prediction error. To avoid over-blurring, one should reduce the weighting at object boundaries. This, however, requires accurate detection of object boundaries, which is not a trivial task.

Bayesian Criterion   The Bayesian estimator is based on a probabilistic formulation of the motion estimation problem, pioneered by Konrad and Dubois [22, 38]. Under this formulation, given an anchor frame ψ_1, the image function at the target frame ψ_2 is considered a realization of a random field Ψ, and the motion field $\mathbf{d}$ is a realization of another random field $\mathcal{D}$. The a posteriori probability distribution of the motion field $\mathcal{D}$, given a realization of Ψ and ψ_1, can be written, using the Bayes rule, as

$$P(\mathcal{D} = \mathbf{d} \mid \Psi = \psi_2; \psi_1) = \frac{P(\Psi = \psi_2 \mid \mathcal{D} = \mathbf{d}; \psi_1) P(\mathcal{D} = \mathbf{d}; \psi_1)}{P(\Psi = \psi_2; \psi_1)}. \qquad (6.2.10)$$

(In this notation, the semicolon indicates that subsequent variables are deterministic parameters.) An estimator based on the Bayesian criterion attempts to maximize the a posteriori probability. But for a given ψ_1 and ψ_2, maximizing this probability is equivalent to maximizing the numerator only. Therefore, the maximum a posteriori (MAP) estimate of $\mathbf{d}$ is

$$\mathbf{d}_{\mathrm{MAP}} = \mathrm{argmax}_\mathbf{d}\{P(\Psi = \psi_2 \mid \mathcal{D} = \mathbf{d}; \psi_1) P(\mathcal{D} = \mathbf{d}; \psi_1)\}. \qquad (6.2.11)$$

The first probability denotes the likelihood of an image frame given the motion field and the anchor frame. Let $\mathcal{E}$ represent the random field corresponding to the DFD

---

*The four-connectivity neighborhood of $\mathbf{x}$ contains the four adjacent pixels to the east, west, north, and south of $\mathbf{x}$. The eight-connectivity neighborhood includes the additional four pixels in the northeast, northwest, southeast and southwest directions.

image $e(\mathbf{x}) = \psi_2(\mathbf{x} + \mathbf{d}) - \psi_1(\mathbf{x})$ for a given $\mathbf{d}$ and ψ_1; then

$$P(\Psi = \psi_2 \mid \mathcal{D} = \mathbf{d}; \psi_1) = P(\mathcal{E} = e),$$

and Equation (6.2.11) becomes

$$\mathbf{d}_{\mathrm{MAP}} = \mathrm{argmax}_{\mathbf{d}} \{ P(\mathcal{E} = e) P(\mathcal{D} = \mathbf{d}; \psi_1) \}$$

$$= \mathrm{argmin}_{\mathbf{d}} \{ -\log P(\mathcal{E} = e) - \log P(\mathcal{D} = \mathbf{d}; \psi_1) \}. \qquad (6.2.12)$$

Source coding theory which will be discussed in Chapter 8, tells us that the minimum coding length for a source $\mathcal{X}$ is its entropy, $-\log P(\mathcal{X} = x)$. We see that the MAP estimate is equivalent to minimizing the sum of the coding lengths for the DFD image e and the motion field $\mathbf{d}$. As will be shown in Chapter 9, this is precisely what a video coder using motion-compensated prediction must code. Therefore, the MAP estimate for $\mathbf{d}$ is equivalent to a *minimum description length* (MDL) estimate [34]. Because the purpose of motion estimation in video coding is to minimize the bit rate, the MAP criterion is a better choice than minimizing the prediction error.

The most common model for the DFD image is a zero-mean independently identically distributed (i.i.d.) Gaussian field, with distribution

$$P(\mathcal{E} = e) = (2\pi\sigma^2)^{-|\Lambda|/2} \exp \frac{-\sum_{\mathbf{x} \in \Lambda} e^2(\mathbf{x})}{2\sigma^2}, \qquad (6.2.13)$$

where $|\Lambda|$ denotes the size of Λ (i.e., the number of pixels in Λ). With this model, minimizing the first term in Equation (6.2.12) is equivalent to minimizing the previously defined DFD error (when $p = 2$).

For the motion field $\mathcal{D}$, a common model is a Gibbs/Markov random field [11]. Such a model is defined by a neighborhood structure called a *clique*. Let $\mathcal{C}$ represent the set of cliques; the model assumes that

$$P(\mathcal{D} = \mathbf{d}) = \frac{1}{Z} \exp \left(-\sum_{c \in \mathcal{C}} V_c(\mathbf{d}) \right), \qquad (6.2.14)$$

where Z is a normalization factor. The function $V_c(\mathbf{d})$ is called the potential function, which is usually defined to measure the difference between pixels in the same clique:

$$V_c(\mathbf{d}) = \sum_{(\mathbf{x}, \mathbf{y}) \in c} |\mathbf{d}(\mathbf{x}) - \mathbf{d}(\mathbf{y})|^2. \qquad (6.2.15)$$

Under this model, minimizing the second term in Equation (6.2.12) is equivalent to minimizing the smoothing function in Equation (6.2.8). Therefore, the MAP estimate is equivalent to the DFD-based estimator with an appropriate smoothness constraint.

6.2.3 Optimization Methods

The error functions presented in Section 6.2.2 can be minimized using various optimization methods. Here we consider only exhaustive search and gradient-based search methods. Usually, for the exhaustive search, the MAD is used for reasons of computational simplicity, whereas for the gradient-based search, the MSE is used for its mathematical tractability.

Obviously, the advantage of the exhaustive search method is that it guarantees reaching the global minimum. However, such a search is feasible only if the number of unknown parameters is small, and each parameter takes only a finite set of discrete values. To reduce the search time, various fast algorithms can be developed, which achieve suboptimal solutions.

The most common gradient descent methods include the *steepest gradient descent* and the *Newton-Ralphson method* (a brief review of these methods is provided in Appendix B). A gradient-based method can handle unknown parameters in a high-dimensional continuous space. However, it can only guarantee convergence to a local minimum. The error functions introduced in the previous section are in general not convex, and can have many local minima that are far from the global minimum. Therefore, it is important to obtain a good initial solution through the use of a priori knowledge, or by adding a penalty term to make the error function convex.

With the gradient-based method, one must calculate the spatiotemporal gradients of the underlying signal. Appendix A reviews methods for computing first- and second-order gradients from digital sampled images. Note that the methods used for calculating the gradient functions can have profound impact on the accuracy and robustness of the associated motion estimation methods, as shown by Barron, Fleet, and Beauchemin [4]. Using a Gaussian prefilter followed by a central difference generally leads to significantly better results than the simple two-point difference approximation.

One important search strategy is to use a multiresolution representation of the motion field and conduct the search in a hierarchical manner. The basic idea is to first search the motion parameters in a coarse resolution, propagate this solution into a finer resolution, and then refine the solution in the finer resolution. This can combat both the slowness of exhaustive search methods and the nonoptimality of gradient-based methods. We will present the multiresolution method in more detail in Section 6.9.

6.3 PIXEL-BASED MOTION ESTIMATION

In pixel-based motion estimation, one tries to estimate an MV for every pixel. Obviously, this problem is ill-defined. If one uses the constant intensity assumption, for every pixel in the anchor frame there will be many pixels in the target frame that have exactly the same image intensity. If one uses the optical flow equation instead, the problem is again indeterminate, because there is only one equation for two unknowns. To circumvent this problem, there are in general four approaches: First, one can use regularization techniques to enforce smoothness constraints on the motion field, so that the MV of a new pixel is constrained by those found for surrounding pixels. Second, one can assume that the MVs in a neighborhood surrounding each pixel are the same, and apply the constant intensity assumption or the optical flow equation over the entire neighborhood. Third, one can make use of additional invariance constraints; in addition to intensity invariance, which leads to the optical flow equation, one can assume that the intensity gradient is invariant under motion, as proposed in [29, 26, 15]. Finally, one can also make use of the relation between the phase functions of the frame before and after motion [9]. In [4], Barron, Fleet, and Beauchemin evaluated various methods for optical flow computation, by testing these algorithms on both synthetic and real world images. In this section, we describe the first two approaches only. We also introduce the pel-recursive type of algorithms that are developed for video compression applications.

6.3.1 Regularization Using the Motion Smoothness Constraint

Horn and Schunck [16] proposed to estimate motion vectors by minimizing the following objective function, which is a combination of the flow-based criterion and a motion smoothness criterion:

$$E(\mathbf{v}(\mathbf{x})) = \sum_{\mathbf{x} \in \Lambda} \left(\frac{\partial \psi}{\partial x} v_x + \frac{\partial \psi}{\partial y} v_y + \frac{\partial \psi}{\partial t} \right)^2 + w_s (\|\nabla v_x\|^2 + \|\nabla v_y\|^2). \qquad (6.3.1)$$

In their original algorithm, the spatial gradients of v_x and v_y are approximated by $\nabla v_x = [v_x(x, y) - v_x(x - 1, y), v_x(x, y) - v_x(x, y - 1)]^T$, $\nabla v_y = [v_y(x, y) - v_y(x - 1, y), v_y(x, y) - v_y(x, y - 1)]^T$. The minimization of this error function is accomplished by a gradient-based method known as the Gauss-Siedel method.

Nagle and Enkelmann conducted a comprehensive evaluation of the effect of smoothness constraints on motion estimation [30]. In order to avoid over-smoothing of the motion field, Nagel suggested an *oriented-smoothness* constraint in which smoothness is imposed along the object boundaries, but not across the boundaries [29]. This has resulted in significant improvement in motion estimation accuracy [4].

6.3.2 Using a Multipoint Neighborhood

In this approach, when estimating the MV at a pixel $\mathbf{x}_n$, we assume that the MVs of all the pixels in a neighborhood $\mathcal{B}(\mathbf{x}_n)$ surrounding $\mathbf{x}_n$ are the same, being $\mathbf{d}_n$. To determine $\mathbf{d}_n$, one can either minimize the prediction error over $\mathcal{B}(\mathbf{x}_n)$, or solve the optical flow equation using a least-squares method. Here we present the first approach. To estimate $\mathbf{d}_n$ we minimize the DFD error over $\mathcal{B}(\mathbf{x}_n)$:

$$E(\mathbf{d}_n) = \frac{1}{2} \sum_{\mathbf{x} \in \mathcal{B}(\mathbf{x}_n)} w(\mathbf{x})(\psi_2(\mathbf{x} + \mathbf{d}_n) - \psi_1(\mathbf{x}))^2, \qquad (6.3.2)$$

where the $w(\mathbf{x})$ are the weights assigned to pixel $\mathbf{x}$. (Usually, the weight decreases as the distance from $\mathbf{x}$ to $\mathbf{x}_n$ increases.)

The gradient with respect to $\mathbf{d}_n$ is

$$\mathbf{g}(d_n) = \frac{\partial E}{\partial \mathbf{d}_n} = \sum_{\mathbf{x} \in \mathcal{B}(\mathbf{x}_n)} w(\mathbf{x}) e(\mathbf{x}, \mathbf{d}_n) \left. \frac{\partial \psi_2}{\partial \mathbf{x}} \right|_{\mathbf{x} + \mathbf{d}_n}, \qquad (6.3.3)$$

where $e(\mathbf{x}, \mathbf{d}_n) = \psi_2(\mathbf{x} + \mathbf{d}_n) - \psi_1(\mathbf{x})$ is the DFD at $\mathbf{x}$ with the estimate $\mathbf{d}_n$. Let $\mathbf{d}_n^{(l)}$ represent the estimate at the lth iteration; then the first-order gradient descent method would yield the following update algorithm,

$$\mathbf{d}_n^{(l+1)} = \mathbf{d}_n^{(l)} - \alpha \mathbf{g}(\mathbf{d}_n^{(l)}). \qquad (6.3.4)$$

From Equation (6.3.3), the update at each iteration depends on the sum of the image gradients at various pixels, scaled by the weighted DFD values at those pixels.

One can also derive an iterative algorithm using the Newton-Ralphson method. From Equation (6.3.3), the Hessian matrix is

$$[\mathbf{H}(\mathbf{d}_n)] = \frac{\partial^2 E}{\partial \mathbf{d}_n^2} = \sum_{\mathbf{x}\in\mathcal{B}(\mathbf{x}_n)} w(\mathbf{x}) \left. \frac{\partial \psi_2}{\partial \mathbf{x}} \left(\frac{\partial \psi_2}{\partial \mathbf{x}}\right)^T \right|_{\mathbf{x}+\mathbf{d}_n} + w(\mathbf{x})e(\mathbf{x}, \mathbf{d}_n) \left. \frac{\partial^2 \psi_2}{\partial \mathbf{x}^2} \right|_{\mathbf{x}+\mathbf{d}_n}$$

$$\approx \sum_{\mathbf{x}\in\mathcal{B}(\mathbf{x}_n)} w(\mathbf{x}) \left. \frac{\partial \psi_2}{\partial \mathbf{x}} \left(\frac{\partial \psi_2}{\partial \mathbf{x}}\right)^T \right|_{\mathbf{x}+\mathbf{d}_n} .$$

The Newton-Ralphson update algorithm (see Appendix B) is then:

$$\mathbf{d}_n^{(l+1)} = \mathbf{d}_n^{(l)} - \alpha \left[\mathbf{H}(\mathbf{d}_n^{(l)})\right]^{-1} \mathbf{g}(\mathbf{d}_n^{(l)}). \qquad (6.3.5)$$

This algorithm converges faster than the first-order gradient descent method, but it requires more computation in each iteration.

Instead of using gradient-based update algorithms, one can use exhaustive search to find the $\mathbf{d}_n$ that yields the minimal error within a defined search range. This will lead to the exhaustive block-matching algorithm, to be presented in Section 6.4.1. The difference is that the neighborhood used here is a sliding window, and an MV is determined for each pixel by minimizing the error in its neighborhood; the neighborhood in general does not have to be a rectangular block.

6.3.3 Pel-Recursive Methods

In a video coder using motion-compensated prediction, one must specify both the MVs and the DFD image. With a pixel-based motion representation, one would need to specify an MV for each pixel, which is very costly. In pel-recursive motion estimation methods, which are developed for video coding applications, the MVs are obtained recursively. Specifically, the MV at a current pixel is updated from those of its neighboring pixels that have been coded before. The decoder can derive the same MV following the same update rule, so that the MVs do not need to be coded. A variety of such algorithms have been developed, where the update rules all follow some type of gradient descent method [31].

Although pel-recursive methods are very simple, their motion estimation accuracy is quite poor. As a result, the prediction error is still large and requires a significant number of bits to code. These algorithms have been used in earlier generations of video codecs because of their simplicity. Today's codecs use more sophisticated motion estimation algorithms, which can provide a better trade-off between the bits used for specifying MVs and the DFD image; the most popular is the block-matching algorithm discussed in the next section.

6.4 BLOCK-MATCHING ALGORITHM

As we have seen, a problem with pixel-based motion estimation is that one must impose smoothness constraints to regularize the problem. One way of imposing smoothness constraints on the estimated motion field is to divide the image domain into nonoverlapping small regions, called blocks, and assume that the motion within each block can be characterized by a simple parametric model—for example, constant, affine, or bilinear. If the block is sufficiently small, then this model can be quite accurate. In this section, we describe motion estimation algorithms developed using this block-based motion representation. Let $\mathcal{B}_m$ represent the mth image block, M the number of blocks, and $\mathcal{M} = \{1, 2, \ldots, M\}$; the partition into blocks should satisfy

$$\bigcup_{m \in \mathcal{M}} \mathcal{B}_m = \Lambda, \quad \text{and} \quad \mathcal{B}_m \bigcap \mathcal{B}_n = \emptyset, m \neq n.$$

Theoretically, a block can have any polygonal shape. In practice, however, the square shape is used almost exclusively. (The triangular shape has also been used, and it is more appropriate when the motion in each block is described by an affine model.)

In the simplest case, the motion in each block is assumed to be constant, that is, the entire block undergoes a translation. This is called the *blockwise translational model*. In this section, we consider this simple case only, where the motion estimation problem is to find a single MV for each block. This type of algorithm is collectively referred as the *block-matching algorithm* (BMA). In the next section, we will consider the more general case where the motion in each block is characterized by a more complex model.

6.4.1 The Exhaustive Block-Matching Algorithm

Given an image block in the anchor frame $\mathcal{B}_m$, the motion estimation problem at hand is to determine a matching block $\mathcal{B}'_m$ in the target frame such that the error between these two blocks is minimized. The displacement vector $\mathbf{d}_m$ between the spatial positions of these two blocks (in terms of the center or a selected corner) is the MV of this block. Under the blockwise translational model, $\mathbf{w}(\mathbf{x}; \mathbf{a}) = \mathbf{x} + \mathbf{d}_m, \mathbf{x} \in \mathcal{B}_m$, the error in Equation (6.2.1) can be written as:

$$E(\mathbf{d}_m, \forall m \in \mathcal{M}) = \sum_{m \in \mathcal{M}} \sum_{\mathbf{x} \in \mathcal{B}_m} |\psi_2(\mathbf{x} + \mathbf{d}_m) - \psi_1(\mathbf{x})|^p . \tag{6.4.1}$$

Because the estimated MV for a block affects the prediction error in that block only, one can estimate the MV for each block individually, by minimizing the prediction error accumulated over each block, which is:

$$E_m(\mathbf{d}_m) = \sum_{\mathbf{x} \in \mathcal{B}_m} |\psi_2(\mathbf{x} + \mathbf{d}_m) - \psi_1(\mathbf{x})|^p , \tag{6.4.2}$$

One way to determine the $\mathbf{d}_m$ that minimizes this error is by using exhaustive search, and this method is called the *exhaustive block-matching algorithm* (EBMA). As illustrated in Figure 6.6, the EBMA determines the optimal $\mathbf{d}_m$ for a given block $\mathcal{B}_m$ in the anchor frame by comparing it with all candidate blocks $\mathcal{B}'_m$ in the target frame within a predefined search region and finding the one with the minimum error. The displacement between the two blocks is the estimated MV.

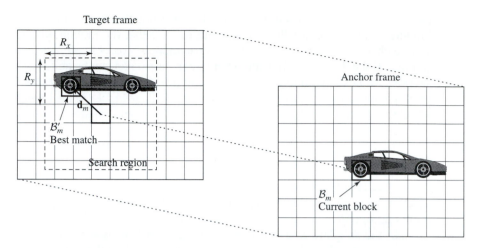

Figure 6.6   The search procedure of the exhaustive block-matching algorithm.

To reduce the computational load, the MAD error ($p = 1$) is often used. The search region is usually symmetric with respect to the current block, up to R_x pixels to the left and right, and up to R_y pixels above and below, as illustrated in the figure. If it is known that the dynamic range of the motion is the same in both horizontal and vertical directions, then $R_x = R_y = R$. The estimation accuracy is determined by the search stepsize, which is the distance between two nearby candidate blocks in the horizontal or vertical direction. Normally, the same stepsize is used along the two directions. In the simplest case, the stepsize is one pixel, which is known as *integer-pel accuracy search.*

Let the block size be $N \times N$ pixels, and the search range be $\pm R$ pixels in both horizontal and vertical directions (see Figure 6.6). With a stepsize of one pixel, the total number of candidate matching blocks for each block in the anchor frame is $(2R + 1)^2$. Let an operation be defined as consisting of one subtraction, one absolute value computation, and one addition. The number of operations for calculating the MAD for each candidate estimate is N^2. The number of operations for estimating the MV for one block is then $(2R + 1)^2 N^2$. For an image of size $M \times M$, there are $(M/N)^2$ blocks (assuming that M is a multiple of N). The total number of operations for a complete frame is then $M^2(2R + 1)^2$. *It is interesting to note that the overall computational load is independent of the block size N.*

As an example, consider the case in which $M = 512$, $N = 16$, and $R = 16$: the total operation count per frame is 2.85×10^8. For a video sequence with a frame rate of 30 fps, the number of operations required per second is 8.55×10^9, an astronomical number! This example shows that EBMA requires intense computation, which poses a challenge to applications requiring software-only implementation. Because of this problem, various fast algorithms have been developed, which trade off estimation accuracy for reduced computation. Some fast algorithms are presented in Section 6.4.3. One advantage of EBMA is that it can be implemented in hardware using simple and modular design, and speed-up can be achieved by using multiple modules in parallel.

There have been many research efforts on efficient realization of the EBMA using VLSI chips, which sometimes involve slight modifications of the algorithm to trade off accuracy for reductions in computation, memory space, or memory access. For a good review of VLSI architecture for implementing EBMA and other fast algorithms for block matching, see [21, 32, 14].

6.4.2 Fractional Accuracy Search

As already hinted, the stepsize for searching the corresponding block in the BMA need not be an integer. For more accurate motion representation, fractional-pel accuracy is needed. A problem with using fractional stepsizes is that there may not be corresponding sample points in the target frame for certain sample points in the anchor frame—these samples must be interpolated from the available sample points. Bilinear interpolation is commonly used for this purpose. In general, to realize a stepsize of $1/K$ pixel, the target frame must first be interpolated by a factor of K. An example of $K = 2$ is shown in Figure 6.7, which is known as *half-pel accuracy search*. It has been shown that half-pel accuracy search can provide a significant improvement in estimation accuracy over integer-pel accuracy search, especially for low-resolution video.

A question that naturally arises is what the appropriate search stepsize is for motion estimation. Obviously, it depends on the intended application of the estimated motion vectors. For video coding, where the estimated motion is used to predict a current frame (the anchor frame) from a previously coded frame (the target frame), it is the prediction error image (i.e., the DFD image) that should be minimized. A statistical analysis of the relation between the prediction error and search precision has been considered by Girod [12] and will be presented in Chapter 9.

Obviously, with a fractional-pel stepsize, the complexity of the EBMA is further increased. For example, with half-pel search, the number of search points is quadrupled over that using integer-pel accuracy. The overall complexity is more than quadrupled, considering the extra computation required for interpolating the target frame.

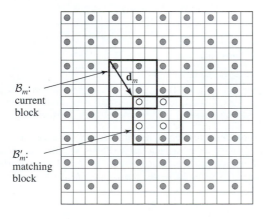

$\mathcal{B}_m$:
current
block

$\mathcal{B}'_m$:
matching
block

Figure 6.7 Half-pel accuracy block matching. Filled circles are samples existing in the original target frame, open circles are samples to be interpolated for calculating the matching error, for a candidate MV $\mathbf{d}_m = (-1, -1.5)$. Instead of calculating these samples on demand for each candidate MV, a better approach is to preinterpolate the entire target frame.

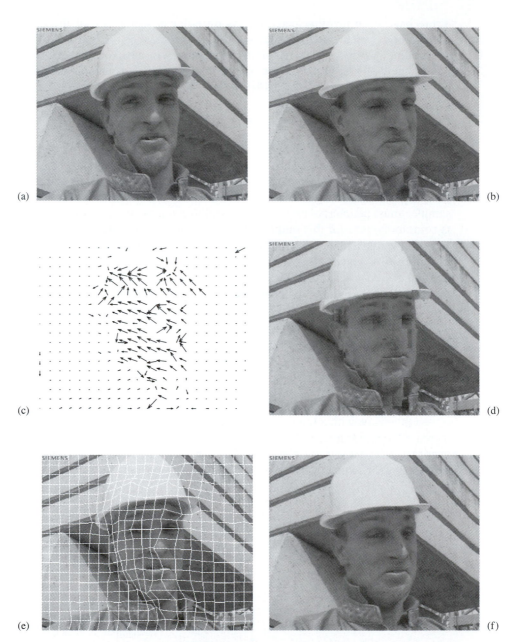

(a)

(b)

(c)

(d)

(e)

(f)

Figure 6.8   Example motion estimation results: (a) the target frame; (b) the anchor frame; (c–d) motion field and predicted image for the anchor frame (PSNR = 29.86 dB) obtained by half-pel accuracy EBMA; (e–f) motion field (represented by the deformed mesh overlaid on the target frame) and predicted image (PSNR = 29.72 dB) obtained by the mesh-based motion estimation scheme in [43].

Example 6.1

Figure 6.8(c) shows the motion field estimated by a half-pel EBMA algorithm for two frames given in Figure 6.8(a–b). Figure 6.8(d) shows the predicted anchor frame based on the estimated motion. This is obtained by replacing each block in the anchor frame by its best matching block in the target frame. The image size is 352×288 and the block size is 16×16. We can see that a majority of blocks are predicted accurately; however, there are blocks that are not well predicted. Some of these blocks undergo nontranslational motions, such as those blocks covering the eyes and the mouth. Other blocks contain both the foreground object and the background and only the foreground object is moving. There are also blocks where the image intensity change is due to the change in the reflection patterns when the head turns. The motion variation over these blocks cannot be approximated well by a constant MV, and the EBMA algorithm simply identifies a block in the target frame that has the smallest absolute error from a given block in the anchor frame. Furthermore, the predicted image is discontinuous along certain block boundaries, which is the notorious *blocking artifact* common with the EBMA algorithm. Such artifacts are due to the inherent limitation of the blockwise translational motion model, and the fact that the MV for a block is determined independent of the MVs of its adjacent blocks.

The accuracy between a predicted image and the original is usually measured by PSNR, as defined previously in Equation (1.5.6). The PSNR of the image predicted by the half-pel EBMA is 29.86 dB. With the integer-pel EBMA, the resulting predicted image is visually very similar, although the PSNR is slightly lower.

6.4.3 Fast Algorithms

As we have shown, the EBMA requires a very large amount of computation. To speed up the search, various fast algorithms for block matching have been developed. The key to reducing the computation is reducing the number of search candidates. As previously described, for a search range of $\pm R$ and a stepsize of 1 pixel, the total number of candidates is $(2R + 1)^2$ with EBMA. Various fast algorithms differ in the ways that they skip those candidates that are unlikely to have small errors.

2-D–Log Search Method   One popular fast search algorithm is the *2-D–log search* [19], illustrated in Figure 6.9. It starts from the position corresponding to zero displacement. Each step tests five search points in a diamond arrangement. In the next step, the diamond search is repeated with the center moved to the best matching point resulting from the previous step. The search stepsize (the radius of the diamond) is reduced if the best matching point is the center point, or is on the border of the maximum search range. Otherwise, the stepsize remains the same. The final step is reached when the stepsize is reduced to 1 pel, and nine search points are examined at this last step. The initial stepsize is usually set to half of the maximum search range. With this method, one cannot predetermine the number of steps and the total number of search points, as these depends on the actual MV. But the best case (requiring the minimum number of search points) and the worst case (requiring the maximum number) can be analyzed.

Three-Step Search Method   Another popular fast algorithm is the *three-step search* algorithm [20]. As illustrated in Figure 6.10, the search starts with a stepsize

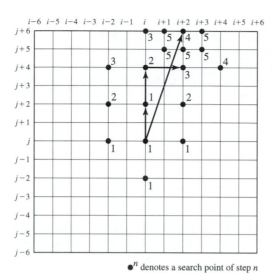

$\bullet^n$ denotes a search point of step n

Figure 6.9 The 2-D–log search method. The search points in a target frame are shown with respect to a block center at (i,j) in the anchor frame. In this example, the best matching MVs in steps 1–5 are (0,2), (0,4), (2,4), (2,6), and (2,6). The final MV is (2,6). Reprinted from H. G. Mussmann, P. Pirsch, and H. J. Grallert, Advances in picture coding, *Proceedings of the IEEE* (April 1985), 73(4):523–48. Copyright 1985 IEEE.

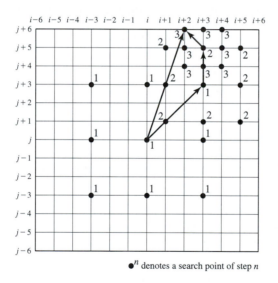

$\bullet^n$ denotes a search point of step n

Figure 6.10 The three-step search method. In this example, the best matching MVs in steps 1–3 are (3,3), (3,5), and (2,6). The final MV is (2,6). Reprinted from H. G. Mussmann, P. Pirsch, and H. J. Grallert, Advances in picture coding, *Proceedings of the IEEE* (April 1985), 73(4):523–48. Copyright 1985 IEEE.

equal to or slightly larger than half of the maximum search range. In each step, nine search points are compared. They consist of the central point of the search square, and eight search points located on the search area boundaries. The stepsize is reduced by half after each step, and the search ends with a stepsize of 1 pel. At each new step, the search center is moved to the best matching point resulting from the previous step. Let R_0 represent the initial search stepsize; then there are at most $L = \lfloor \log_2 R_0 + 1 \rfloor$ search steps, where $\lfloor x \rfloor$ represents the lower integer of x. If $R_0 = R/2$, then $L = \lfloor \log_2 R \rfloor$. At each search step, eight points are searched, except in the very beginning, when nine points must be examined. Therefore, the total number of search points is $8L + 1$. For

TABLE 6.1 COMPARISON OF FAST SEARCH ALGORITHMS FOR A SEARCH RANGE OF $R = 7$. DATA FROM [14, TABLE 1]

Search algorithm	Number of search points		Number of search steps	
	Minimum	Maximum	Minimum	Maximum
EBMA	225	225	1	1
2-D–log [19]	13	26	2	8
Three-step [20]	25	25	3	3

example, for a search range of $R = 32$, with EBMA, the total number of search points is 4225, whereas with the three-step method, the number is reduced to 41, a savings factor of more than 100. Unlike the 2-D–log search method, the three-step method has a fixed, predictable number of search steps and search points. In addition, it has a more regular structure. These features make the three-step method more amenable to VLSI implementation than the 2-D–log method and some other fast algorithms.

Comparison of Fast Algorithms   Table 6.1 compares the minimum and maximum numbers of search points and the number of search steps required by three different search algorithms. As can be seen, some algorithms have a more regular structure and hence a fixed number of computations, whereas others have very different best case and worst case numbers. For VLSI implementation, structural regularity is more important, whereas for software implementation, the average-case complexity (which is closer to the best case in general) is more important. For an analysis of the implementation complexity and cost of using VLSI circuits for these and some other fast algorithms, see [14].

The preceding discussions assume that the search accuracy is integer-pel. To achieve half-pel accuracy, one can add a final step in any fast algorithm, which searches with a half-pel stepsize in a ± 1 pel neighborhood of the best matching point found from the integer-pel search.

6.4.4 Imposing Motion Smoothness Constraints

From Figure 6.8(c), we can see that the motion field obtained using EBMA is quite chaotic. This is because no constraints have been imposed on the spatial variation of the block MVs. Several approaches have been developed to smooth the estimated motion field so that it is closer to the physical motion field. One effective approach is to use a multiresolution approach, which estimates the MVs in a coarser spatial resolution first, and then continuously refines the MVs in successively finer resolutions. The propagation of the MVs from a coarser to a finer resolution is accomplished by spatial interpolation, which induces a certain degree of spatial continuity on the resulting motion field. (This technique will be treated in more detail in Section 6.9.) Another approach is to explicitly impose a smoothness constraint by adding a smoothing term in the error criterion in Equation (6.4.2), which measures the variation of the MVs of adjacent blocks. The resulting overall error function will be similar to that in Equation (6.2.9), except that the motion vectors will be defined over blocks and the prediction error will be summed over a block. The challenge is to determine a proper weighting between the prediction error term and the smoothing term, so that the resulting motion field is not over-smoothed. Ideally, the weighting should be adaptive: it should not be applied near object boundaries. A more difficult task is to identify object boundaries where motion discontinuity should be allowed.

6.4.5 Phase Correlation Method*

Instead of minimizing the DFD, another motion estimation method identifies peaks in the phase correlation function. Assume that two image frames are related by a pure translation, so that

$$\psi_1(\mathbf{x}) = \psi_2(\mathbf{x} + \mathbf{d}). \tag{6.4.3}$$

Taking the Fourier transform on both sides and using the Fourier shift theorem, we obtain

$$\Psi_1(\mathbf{f}) = \Psi_2(\mathbf{f}) \cdot e^{j2\pi \mathbf{d}^T \mathbf{f}}. \tag{6.4.4}$$

The normalized cross-power spectrum between $\psi_1(\mathbf{x})$ and $\psi_2(\mathbf{x})$ is

$$\tilde{\Psi}(\mathbf{f}) = \frac{\Psi_1(\mathbf{f}) \cdot \Psi_2^*(\mathbf{f})}{|\Psi_1(\mathbf{f}) \cdot \Psi_2^*(\mathbf{f})|} = e^{j2\pi \mathbf{d}^T \mathbf{f}}, \tag{6.4.5}$$

where the superscript * indicates complex conjugation. Taking the inverse Fourier transform results in the *phase correlation function* (PCF):[†]

$$\text{PCF}(\mathbf{x}) = \mathcal{F}^{-1}\{\tilde{\Psi}(\mathbf{f})\} = \delta(\mathbf{x} + \mathbf{d}). \tag{6.4.6}$$

We see that *the PCF between two images that are translations of each other is an impulse function, with the impulse located at a position exactly equal to the translation between the two images.* By identifying the peak location of the PCF, one can estimate the translation between two images. This approach was first used by Kuglin [23] for image alignment.

The preceding derivation assumes that both images are continuous in space and have infinite size. On the other hand, an actual image signal is discrete and finite. In practice, we apply the DSFT over the available image domain, which is equivalent to the periodic extension of the CSFT over an infinite image that is zero outside the given image domain (see Sections 2.1–2). In order to suppress the aliasing effect caused by sampling, a frequency-domain weighting function $W(\mathbf{f})$ is usually applied when computing Equation (6.4.5). (In [13], a Kaiser window with $\alpha = 0.2$ is used as the weighting function.) To reduce the effect of boundary samples, a space-domain weighting function $w(\mathbf{x})$ can also be applied to $\psi_1(\mathbf{x})$ and $\psi_2(\mathbf{x})$ before computing the DSFT.

The preceding phase correlation method is used extensively in image registration, where entire images must be aligned [33], [10]. For motion estimation, the two under-

---

*Sections marked with an asterisk may be skipped or left for further exploration.

[†]The name comes from the fact that it is the cross-correlation between the phase portions of the functions $\psi_1(\mathbf{x})$ and $\psi_2(\mathbf{x})$, respectively.

lying frames are usually not related by a global translation. To handle this situation, the phase correlation method is more often applied at the block level. For motion estimation over nonoverlapping blocks of size $N \times N$, both frames are usually divided into overlapping range blocks of size $L \times L$. For a search range of $\pm R$, the range block size should be $L \geq N + 2R$. To determine the MV of a block in $\psi_1(\mathbf{x})$, a size $L \times L$ discrete Fourier transform (DFT) is applied to both this block and its corresponding block in $\psi_2(\mathbf{x})$. Then the PCF is computed using the inverse DFT of the same size, and the peak location is identified. To enable the use of fast Fourier transform (FFT) algorithms, L is usually chosen to be a power of two. For example, if $N = 16$, $R = 16$, then $L = 64$ would be appropriate.

This method assumes that there is a global translation between the two corresponding range blocks. This assumption does not hold for general video sequences. When there are several patches in the range block in $\psi_1(\mathbf{x})$ that undergo different motions, we will observe several peaks in the PCF. Each peak corresponds to the motion of one patch. The location of the peak indicates the MV of the patch, whereas the amplitude of the peak is proportional to the size of the patch [40]. In this sense, the PCF reveals similar information as the MV histogram does over a block. To estimate the dominant MV of the block, we first extract local maxima of the PCF. We then examine the DFD at the corresponding MVs. The MV yielding the minimum DFD will be considered the block MV. Since only a small number of candidate MVs are examined, significant savings in computational complexity may be achieved, compared to the full-search method.

This approach can be extended to motion estimation with fractional-pel accuracy. In [13], the integer-pel candidate motion vectors are augmented by varying the length of the candidate motion vectors by up to ± 1 pel. In [37] and [40], alternative methods are suggested.

An advantage of the phase correlation method for motion estimation is its insensitivity to illumination changes (see Section 5.2). This is because changes in the mean value of an image, or multiplication of an image by a constant, do not affect the phase information. This is not true for the DFD-based methods.

6.4.6 Binary Feature Matching*

In this scheme, known as the hierarchical feature-matching motion estimation scheme (HFM-ME) [25], a sign-truncated feature (STF) is defined and used for block matching, as opposed to the pixel intensity values used in conventional block-matching methods. Using the STF definition, a data block is represented by a mean and a binary bit pattern. The block-matching motion estimation is divided into mean matching and binary phase matching. This technique enables a significant reduction in computational complexity compared with EBMA, because binary phase matching involves only Boolean logic operations. The use of STF also significantly reduces the data transfer time between the frame buffer and motion estimator. Tests have shown that HFM-ME can achieve similar

---

*Sections marked with an asterisk may be skipped or left for further exploration.

prediction accuracy as EBMA under the same search ranges, but can be implemented about 64 times faster. When the search range is doubled for HFM-ME, it can achieve significantly more accurate prediction than EBMA, still with nontrivial time savings [25].

The STF vector of a block of size $2^N \times 2^N$ consists of two parts. The first part consists of the multiresolution mean vectors, and the second part consists of the sign-truncated binary vectors. The mean vectors are determined recursively as follows:

$$\text{Mean}^n(i, j) = \left\lfloor \frac{1}{4} \sum_{p=0}^{1} \sum_{q=0}^{1} \text{Mean}^{n+1}(2i + p, 2j + q) \right\rfloor, \quad 0 \le i, j \le 2^n - 1,$$

$$0 \le n \le N - 1,$$
$$\text{Mean}^N(i, j) = \psi(i, j), \qquad\qquad\qquad 0 \le i, j \le 2^N - 1,$$

(6.4.7)

where $\{\psi(i, j); \ 0 \le i, j \le 2^N - 1\}$ are the pixel intensity values of the original block. The sign-truncated vectors are obtained by

$$\text{ST\_pattern}^n(i, j) = \begin{cases} 0 & \text{if } \text{Mean}^n(i,j) \ge \text{Mean}^{n-1}\left(\lfloor \frac{i}{2} \rfloor, \lfloor \frac{j}{2} \rfloor\right), \\ 1 & \text{otherwise.} \end{cases}$$

(6.4.8)

The STF vectors, decomposed to the nth level for a $2^N \times 2^N$ block, can then be represented as

$$\text{STFV}_n^N = \{\text{ST\_pattern}^N, \text{ST\_pattern}^{N-1}, \dots, \text{ST\_pattern}^{N-n+1}, \text{mean}^{N-n}\}. \quad (6.4.9)$$

When $n = N$, a block is fully decomposed with the following STF vector:

$$\text{STFV}_N^N = \{\text{ST\_pattern}^N, \text{ST\_pattern}^{N-1}, \dots, \text{ST\_pattern}^1, \text{mean}^0\}. \quad (6.4.10)$$

All the intermediate mean vectors are used only to generate ST_patterns and can be discarded. Therefore, the final STF representation consists of a multiresolution binary sequence with $\frac{4}{3}(4^N - 1)$ bits and a one-byte mean. This represents a much reduced data set compared to the original 4^N bytes of pixel values. Also, this feature set allows binary Boolean operations for block matching.

As an example, let us consider how to form the STF vector for a 4×4 block with two layers. First, the mean pyramid is formed as

$$\begin{bmatrix} 158 & 80 & 59 & 74 \\ 80 & 69 & 59 & 74 \\ 87 & 86 & 65 & 62 \\ 116 & 100 & 72 & 58 \end{bmatrix} \Longrightarrow \begin{bmatrix} 97 & 67 \\ 97 & 64 \end{bmatrix} \Longrightarrow 81.$$

The STF vector is then obtained as:

$$\begin{pmatrix} 0 & 1 & 1 & 0 \\ 1 & 1 & 1 & 0 \\ 1 & 1 & 0 & 1 \\ 0 & 0 & 0 & 1 \end{pmatrix}, \begin{pmatrix} 0 & 1 \\ 0 & 1 \end{pmatrix}, \quad 81.$$

The STF vector decomposed to one layer for the preceding example is $\{0110\ 1110\ 1101\ 0001, (97, 67, 97, 64)\}$. The completely decomposed STF vector is $\{0110\ 1110\ 1101\ 0001, 0101, 81\}$. It consists of a twenty-bit binary pattern, which includes a 4×4 first-layer ST_pattern, a 2×2 second-layer ST_pattern, and a mean value. In practical implementations, either completely decomposed or mixed-layer STF vectors can be used.

Comparison of two STF vectors is accomplished by two parallel decision procedures: (1) calculating the absolute error between the mean values, and (2) determining the Hamming distance between the binary patterns. The latter can be accomplished extremely rapidly by using an XOR Boolean operator. Therefore, the main computational load of the HFM-ME lies in the computation of the mean pyramid for the current and all candidate matching blocks. This computation can however be done in advance, only once for every possible block. For a detailed analysis of its computational complexity, and a fast algorithm using logarithmic search, see [25].

6.5 DEFORMABLE BLOCK-MATCHING ALGORITHMS

In the block-matching algorithms introduced previously, each block is assumed to undergo a pure translation. This model is not appropriate for blocks undergoing rotation, zooming, and so on. In general, a more sophisticated model, such as the affine, bilinear, or projective mapping, can be used to describe the motion of each block. (Obviously, this will still cover the translational model as a special case.) With such models, a block in the anchor frame is in general mapped to a nonsquare quadrangle, as shown in Figure 6.11. Therefore, we refer to the class of block-based motion estimation methods using higher order models as *deformable block-matching algorithms* (DBMA) [24]

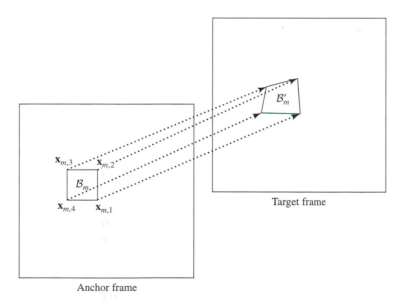

Figure 6.11   The deformable block-matching algorithm finds the best matching quadrangle in the target frame for each block in the anchor frame. The allowed block deformation depends on the motion model used for the block.

(also known as *generalized block-matching algorithms* [36]). In the following, we first discuss how to interpolate the MV at any point in a block using only the MVs at the block corners (called nodes), and then we present an algorithm for estimating nodal MVs.

6.5.1 Node-Based Motion Representation

In Section 5.5, we described several 2-D motion models corresponding to different 3-D motions. All of these models can be used to characterize the motion within a block. In Section 5.5.4, we showed how the most general model, the projective mapping, can be approximated by a polynomial mapping of different orders. In this section, we introduce a node-based block motion model [24], which can characterize the same types of motion as the polynomial model, but is easier to interpret and specify.

In this model, we assume that a selected number of control nodes in a block can move freely, and that the displacement of any interior point can be interpolated from nodal displacements. Let the number of control nodes be denoted by K, and the MVs of the control nodes in $\mathcal{B}_m$ by $\mathbf{d}_{m,k}$; then the motion function over the block is described by

$$\mathbf{d}_m(\mathbf{x}) = \sum_{k=1}^{K} \phi_{m,k}(\mathbf{x})\mathbf{d}_{m,k}, \quad x \in \mathcal{B}_m. \tag{6.5.1}$$

Equation (6.5.1) expresses the displacement at any point in a block as an interpolation of nodal displacements, as shown in Figure 6.12. The interpolation kernel $\phi_{m,k}(\mathbf{x})$ depends on the desired contribution of the kth control point in $\mathcal{B}_m$ to $\mathbf{x}$. One way to design the interpolation kernels is to use the shape functions associated with the corresponding nodal structure. (We discuss further the design of shape functions in Section 6.6.1.)

The translational, affine, and bilinear models introduced previously are special cases of the node-based model with one, three, and four nodes, respectively. A model with more nodes can characterize more complex deformations. The interpolation kernel in the one-node case (at the block center or a chosen corner) is a pulse function, corresponding to nearest-neighbor interpolation. The interpolation functions in the three-node (any three corners in a block) and four-node (the four corners) cases are affine and bilinear functions, respectively. Usually, to use an affine model with a rectangular block, the block is first divided into two triangles, and then each triangle is modeled by the three-node model.

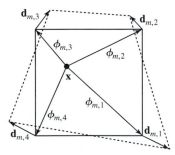

Figure 6.12 Interpolation of motion in a block from nodal MVs.

Compared to the polynomial-based representation introduced previously, the node-based representation is easier to visualize. Can you picture in your head a deformed block, given eight coefficients of a bilinear function? But you certainly can, given the locations of the four corner points in the block! Furthermore, the nodal MVs can be estimated more easily and specified with a lower precision than the polynomial coefficients. It is easier to determine appropriate search ranges and search stepsizes for the nodal MVs than for the polynomial coefficients, based on the a priori knowledge about the dynamic range of the underlying motion and the desired estimation accuracy. In addition, all the motion parameters in the node-based representation are equally important, whereas those in the polynomial representation cannot be treated equally. (For example, the estimation of the high-order coefficients is much harder than that of the constant terms.) Finally, specification of the polynomial coefficients requires a high degree of precision: a small change in a high-order coefficient can generate a very different motion field. On the other hand, to specify nodal MVs, integer- or half-pel accuracy is usually sufficient. These advantages are important for video coding applications.

6.5.2 Motion Estimation Using the Node-Based Model

Because the estimation of nodal movements is independent from block to block, we omit the subscript m that indicates which block is being considered. The following derivation applies to any block $\mathcal{B}$. With the node-based motion model, the motion parameters for any block are the nodal MVs; that is, $\mathbf{a} = [\mathbf{d}_k, k \in \mathcal{K}]$, where $\mathcal{K} = \{1, 2, \ldots, K\}$. They can be estimated by minimizing the prediction error over this block; that is,

$$E(\mathbf{a}) = \sum_{\mathbf{x} \in \mathcal{B}} |\psi_2(\mathbf{w}(\mathbf{x}; \mathbf{a})) - \psi_1(\mathbf{x})|^p, \qquad (6.5.2)$$

where

$$\mathbf{w}(\mathbf{x}; \mathbf{a}) = \mathbf{x} + \sum_{k \in \mathcal{K}} \phi_k(\mathbf{x}) \mathbf{d}_k. \qquad (6.5.3)$$

As with the BMA, there are many ways to minimize the error in Equation (6.5.2), including exhaustive search and a variety of gradient-based search methods. The computational load required of exhaustive search, however, can be unacceptable in practice, because of the high dimension of the search space. Gradient-based search algorithms are more feasible in this case. In the following, we derive a Newton-Ralphson search algorithm, following the approach in [24].

Define $\mathbf{a} = [\mathbf{a}_x^T, \mathbf{a}_y^T]^T$, with $\mathbf{a}_x = [d_{x,1}, d_{x,2}, \ldots, d_{x,K}]^T$, $\mathbf{a}_y = [d_{y,1}, d_{y,2}, \ldots, d_{y,K}]^T$. One can show that

$$\frac{\partial E}{\partial \mathbf{a}}(\mathbf{a}) = \begin{bmatrix} \dfrac{\partial E}{\partial \mathbf{a}_x}(\mathbf{a}) \\[2mm] \dfrac{\partial E}{\partial \mathbf{a}_y}(\mathbf{a}) \end{bmatrix},$$

with

$$\frac{\partial E}{\partial \mathbf{a}_x}(\mathbf{a}) = 2 \sum_{\mathbf{x} \in \mathcal{B}} e(\mathbf{x}; \mathbf{a}) \frac{\partial \psi_2(\mathbf{w}(\mathbf{x}; \mathbf{a}))}{\partial x} \phi(\mathbf{x}),$$

$$\frac{\partial E}{\partial \mathbf{a}_y}(\mathbf{a}) = 2 \sum_{\mathbf{x} \in \mathcal{B}} e(\mathbf{x}; \mathbf{a}) \frac{\partial \psi_2(\mathbf{w}(\mathbf{x}; \mathbf{a}))}{\partial y} \phi(\mathbf{x}).$$

In the preceding equations, $e(\mathbf{x}; \mathbf{a}) = \psi_2(\mathbf{w}(\mathbf{x}; \mathbf{a})) - \psi_1(\mathbf{x})$ and $\phi(\mathbf{x}) = [\phi_1(\mathbf{x}), \phi_2(\mathbf{x}), \ldots, \phi_K(\mathbf{x})]^T$.

By dropping the terms involving second-order gradients, the Hessian matrix can be approximated as

$$[\mathbf{H}(\mathbf{a})] = \begin{bmatrix} [\mathbf{H}_{xx}(\mathbf{a})] & [\mathbf{H}_{xy}(\mathbf{a})] \\ [\mathbf{H}_{xy}(\mathbf{a})] & [\mathbf{H}_{yy}(\mathbf{a})] \end{bmatrix},$$

with

$$[\mathbf{H}_{xx}(\mathbf{a})] = 2 \sum_{\mathbf{x} \in \mathcal{B}} \left(\frac{\partial \psi_2}{\partial x} \right)^2 \bigg|_{\mathbf{w}(\mathbf{x};\mathbf{a})} \phi(\mathbf{x}) \phi(\mathbf{x})^T,$$

$$[\mathbf{H}_{yy}(\mathbf{a})] = 2 \sum_{\mathbf{x} \in \mathcal{B}} \left(\frac{\partial \psi_2}{\partial y} \right)^2 \bigg|_{\mathbf{w}(\mathbf{x};\mathbf{a})} \phi(\mathbf{x}) \phi(\mathbf{x})^T.$$

$$[\mathbf{H}_{xy}(\mathbf{a})] = 2 \sum_{\mathbf{x} \in \mathcal{B}} \frac{\partial \psi_2}{\partial x} \frac{\partial \psi_2}{\partial y} \bigg|_{\mathbf{w}(\mathbf{x};\mathbf{a})} \phi(\mathbf{x}) \phi(\mathbf{x})^T.$$

The Newton-Ralphson update algorithm is:

$$\mathbf{a}^{(l+1)} = \mathbf{a}^{(l)} - \alpha \left[\mathbf{H}(\mathbf{a}^{(l)}) \right]^{-1} \frac{\partial E}{\partial \mathbf{a}}(\mathbf{a}^{(l)}). \tag{6.5.4}$$

The update at each iteration thus requires the inversion of the $2K \times 2K$ symmetric matrix $[\mathbf{H}]$.

To reduce numerical computations, we can update the displacements in the x and y directions separately. Similar derivation will yield:

$$\mathbf{a}_x^{(l+1)} = \mathbf{a}_x^{(l)} - \alpha \left[\mathbf{H}_{xx}(\mathbf{a}^{(l)}) \right]^{-1} \frac{\partial E}{\partial \mathbf{a}_x}(\mathbf{a}^{(l)}), \tag{6.5.5}$$

$$\mathbf{a}_y^{(l+1)} = \mathbf{a}_y^{(l)} - \alpha \left[\mathbf{H}_{yy}(\mathbf{a}^{(l)}) \right]^{-1} \frac{\partial E}{\partial \mathbf{a}_y}(\mathbf{a}^{(l)}). \tag{6.5.6}$$

In this case, we need only invert two $K \times K$ matrices in each update. For the four-node case, $[\mathbf{H}]$ is an 8×8 matrix, whereas $[\mathbf{H}_{xx}]$, and $[\mathbf{H}_{yy}]$ are 4×4 matrices.

As with all gradient-based iterative processes, the preceding update algorithm may reach a bad local minimum that is far from the global minimum, if the initial solution is not chosen properly. A good initial solution can often be provided by the

EBMA. For example, consider the four-node model, with a node at each corner of each block. One can use the average of the motion vectors of the four blocks attached to each node as the initial estimate of the MV for that node. This initial estimate can then be successively updated by using Equation (6.5.4).

Note that this algorithm can also be applied to polynomial-based motion representation. In that case, $\mathbf{a}_x$ and $\mathbf{a}_y$ would represent the polynomial coefficients associated with horizontal and vertical displacements, respectively, and $\phi_k(\cdot)$ would correspond to the elementary polynomial basis functions. However, it is difficult to set the search range for $\mathbf{a}_x$ and $\mathbf{a}_y$, and to check the feasibility of the resulting motion field.

6.6 MESH-BASED MOTION ESTIMATION*

With the block-based model used in either BMA or DBMA, motion parameters in individual blocks are independently specified. Unless motion parameters of adjacent blocks are constrained to vary smoothly, the estimated motion field is often discontinuous and sometimes chaotic, as sketched in Figure 6.13(a). One way to overcome this problem is by using mesh-based motion estimation. As illustrated in Figure 6.13(b), the anchor frame is covered by a mesh, and the motion estimation problem is to find the motion of each node, so that the image pattern within each element in the anchor frame matches well with that in the corresponding deformed element in the target frame. The motion within each element is interpolated from nodal MVs. *As long as the nodes in the target frame still form a feasible mesh, mesh-based motion representation is guaranteed to be continuous and thus be free from the blocking artifacts associated with block-based representation.* Another benefit of mesh-based representation is that it enables continuous tracking of the same set of nodes over consecutive frames, which is desirable in applications requiring object tracking. As shown in Figure 6.13(c), one can generate a mesh for an initial frame, and then estimate the nodal motions between every two frames. At each new frame (the anchor frame), the mesh generated in the previous step is used, so that the same set of nodes is tracked over all frames. This is not possible with block-based representation, because it requires that each new anchor frame be reset to a partition consisting of regular blocks.

Note that the inherent continuity of mesh-based representation is not always desired. The type of motion that can be captured by this representation can be visualized as the deformation of a rubber sheet, which is continuous everywhere. In real-world video sequences, there are often motion discontinuities at object boundaries. A more accurate representation would use separate meshes for different objects. As with block-based representation, the accuracy of mesh-based representation depends on the number of nodes. A very complex motion field can be reproduced, as long as a sufficient number of nodes are used. To minimize the number of nodes required, the mesh should be adapted to the imaged scene, so that the actual motion within each element is smooth (i.e., can be interpolated accurately from the nodal motions). If a regular, nonadaptive

---

*Sections marked with an asterisk may be skipped or left for further exploration.

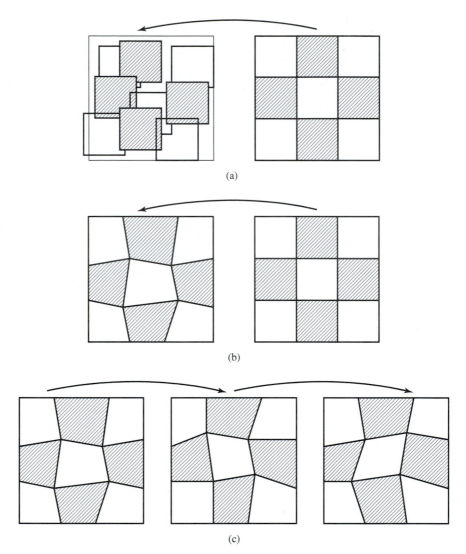

Figure 6.13   Comparison of block-based and mesh-based motion representations: (a) block-based motion estimation between two frames, using a translational model within each block in the anchor frame; (b) mesh-based motion estimation between two frames, using a regular mesh at the anchor frame; (c) mesh-based motion tracking, using the tracked mesh for each new anchor frame.

mesh is used, a large number of nodes are usually needed to approximate the motion field accurately.

In the following, we first describe how to specify a motion field using a mesh-based representation. We then present algorithms for estimating nodal motions in a mesh.

6.6.1 Mesh-Based Motion Representation

In mesh-based motion representation, the underlying image domain in the anchor frame is partitioned into nonoverlapping polygonal elements. Each element is defined by a few nodes and links between the nodes, as shown in Figure 6.14. Such a mesh is also known as a *control grid*. In mesh-based representation, the motion field over the entire frame is described by MVs at the nodes only. The MVs at the interior points of an element are interpolated from the MVs at the nodes of this element. The nodal MVs are constrained so that the nodes in the target frame still form a feasible mesh, with no flip-over elements.

Let the number of elements and nodes be denoted by M and N respectively, and the number of nodes defining each element by K. For convenience, we define the following index sets: $\mathcal{M} = \{1, 2, \ldots, M\}, \mathcal{N} = \{1, 2, \ldots, N\}, \mathcal{K} = \{1, 2, \ldots, K\}$. Let the mth element and nth node in frame t ($t = 1$ for the anchor frame and $t = 2$ for the target frame) be denoted by $\mathcal{B}_{t,m}, m \in \mathcal{M}$ and $\mathbf{x}_{t,n}, n \in \mathcal{N}$, and the MV of the nth node by $\mathbf{d}_n = \mathbf{x}_{2,n} - \mathbf{x}_{1,n}$. The motion field in element $\mathcal{B}_{1,m}$ is related to the nodal MVs $\mathbf{d}_n$ by:

$$\mathbf{d}_m(\mathbf{x}) = \sum_{k \in \mathcal{K}} \phi_{m,k}(\mathbf{x})\mathbf{d}_{n(m,k)}, \quad \mathbf{x} \in \mathcal{B}_{1,m}, \tag{6.6.1}$$

where $n(m, k)$ specifies the global index of the kth node in the mth element (see

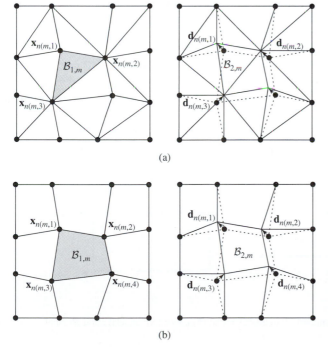

(a)

(b)

Figure 6.14 Illustration of mesh-based motion representation: (a) using a triangular mesh, with three nodes attached to each element, (b) using a quadrilateral mesh, with four nodes attached to each element. In this example, the two meshes have the same number of nodes, but the triangular mesh has twice the number of elements. The left column shows the initial mesh over the anchor frame, the right column the deformed mesh in the target frame.

Figure 6.14). The function $\phi_{m,k}(\mathbf{x})$ is the *interpolation kernel* associated with node k in element m. It depends on the desired contribution of the kth node in $\mathcal{B}_{1,m}$ to the MV at $\mathbf{x}$. This interpolation mechanism has been shown previously in Figure 6.12. To guarantee continuity across element boundaries, the interpolation kernels should satisfy:

$$0 \le \phi_{m,k}(\mathbf{x}) \le 1, \quad \sum_k \phi_{m,k}(\mathbf{x}) = 1, \quad \forall \mathbf{x} \in \mathcal{B}_{1,m},$$

and

$$\phi_{m,k}(\mathbf{x}_l) = \delta_{k,l} = \begin{cases} 1 & k = l, \\ 0 & k \ne l. \end{cases}$$

In finite element method (FEM) analysis, these functions are called *shape functions* [45]. If all the elements have the same shape, then all the shape functions are equal, that is, $\phi_{m,k}(\mathbf{x}) = \phi_k(\mathbf{x})$.

Standard triangular and quadrilateral elements are shown in Figure 6.15. The shape functions for the standard triangular element are:

$$\phi_1^t(x, y) = x, \quad \phi_2^t(x, y) = y, \quad \phi_3^t(x, y) = 1 - x - y. \tag{6.6.2}$$

The shape functions for the standard quadrilateral element are:

$$\phi_1^q(x, y) = (1+x)(1-y)/4, \quad \phi_2^q(x, y) = (1+x)(1+y)/4,$$
$$\phi_3^q(x, y) = (1-x)(1+y)/4, \quad \phi_4^q(x, y) = (1-x)(1-y)/4. \tag{6.6.3}$$

We see that the shape functions for these two cases are affine and bilinear functions, respectively. The readers are referred to [41] for the shape functions for arbitrary triangular elements. The coefficients of these functions depend on the node positions.

Note that the representation of the motion within each element in Equation (6.6.1) is the same as the node-based motion representation introduced in Equation (6.5.1), except that the nodes and elements are denoted using global indices. This is necessary because the nodal MVs are not independent from element to element. It is important not

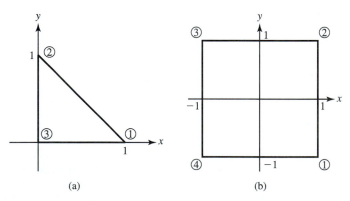

(a)   (b)

Figure 6.15   (a) A standard triangular element; (b) a standard quadrilateral element (a square).

to confuse the mesh-based model with the node-based model introduced in the previous section. There, although several adjacent blocks may share the same node, the nodal MVs are determined independently in each block. Going back to Figure 6.14(b), in the mesh-based model, node n is assigned a single MV, which will affect the interpolated motion functions in four quadrilateral elements attached to this node. In the node-based model, node n can have four different MVs, depending on which block it is considered to be in.

6.6.2 Motion Estimation Using the Mesh-Based Model

With mesh-based motion representation, there are in general two sets of problems to be solved: (1) Given a mesh (or, equivalently, nodes) in the anchor frame, how to determine the nodal positions in the target frame—this is essentially a motion estimation problem. (2) How to set up the mesh in the anchor frame, so that the mesh conforms to the object boundaries. Note that a mesh in which each element corresponds to a smooth surface patch in a single object can lead to more accurate motion estimation than an arbitrarily configured mesh (e.g., a regular mesh). An object-adaptive mesh would also be more appropriate for motion tracking over a sequence of frames. In this book, we consider the first problem only; for solutions to the mesh-generation problem, see, for example [42, 3].

With the mesh-based motion representation described by Equation (6.6.1), the motion parameters include the nodal MVs, that is, $\mathbf{a} = \{\mathbf{d}_n, n \in \mathcal{N}\}$. To estimate them, we can again use an error-minimization approach. Under the mesh-based motion model, the DFD error in Equation (6.2.1) becomes

$$E(\mathbf{d}_n, n \in \mathcal{N}) = \sum_{m \in \mathcal{M}} \sum_{\mathbf{x} \in \mathcal{B}_{1,m}} |\psi_2(\mathbf{w}_m(\mathbf{x})) - \psi_1(\mathbf{x})|^p, \qquad (6.6.4)$$

where, following Equation (6.6.1),

$$\mathbf{w}_m(\mathbf{x}) = \mathbf{x} + \sum_{k \in \mathcal{K}} \phi_{m,k}(\mathbf{x})\mathbf{d}_{n(m,k)}, \quad \mathbf{x} \in \mathcal{B}_{1,m}.$$

In general, the error function in Equation (6.6.4) is difficult to calculate because of the irregular shape of $\mathcal{B}_{1,m}$. To simplify the calculation, we can think of $\mathcal{B}_{t,m}, t = 1, 2$, as being deformed from a *master element* with a regular shape. In general, the master element for different elements could differ. Here, we consider only the case in which all the elements have the same topology that can be mapped from the same master element, denoted by $\tilde{\mathcal{B}}$. Figure 6.16 illustrates such a mapping.

Let $\tilde{\phi}_k(\mathbf{u})$ represent the shape function associated with the kth node in $\tilde{\mathcal{B}}$; then the mapping functions from $\tilde{\mathcal{B}}$ to $\mathcal{B}_{t,m}$ can be represented as

$$\tilde{\mathbf{w}}_{t,m}(\mathbf{u}) = \sum_{k \in \mathcal{K}} \tilde{\phi}_k(\mathbf{u})\mathbf{x}_{t,n(m,k)}, \quad \mathbf{u} \in \tilde{\mathcal{B}}, \quad t = 1, 2. \qquad (6.6.5)$$

The shape functions corresponding to the standard triangular and quadrilateral elements are given in Equations (6.6.2–3). The error in Equation (6.6.4) can be calculated over

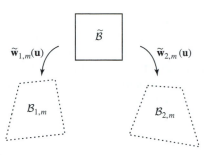

Figure 6.16 Mapping from a master element $\tilde{\mathcal{B}}$ to two corresponding elements in the anchor and target frames, $\mathcal{B}_{1,m}$ and $\mathcal{B}_{2,m}$.

the master element as

$$E(\mathbf{d}_n, n \in \mathcal{N}) = \sum_{m \in \mathcal{M}} \sum_{\mathbf{u} \in \tilde{\mathcal{B}}} |\tilde{e}_m(\mathbf{u})|^p |J_m(\mathbf{u})|, \tag{6.6.6}$$

where

$$\tilde{e}_m(\mathbf{u}) = \psi_2(\tilde{\mathbf{w}}_{2,m}(\mathbf{u})) - \psi_1(\tilde{\mathbf{w}}_{1,m}(\mathbf{u})) \tag{6.6.7}$$

represents the error between the two image frames at points that are both mapped from $\mathbf{u}$ in the master element (see Figure 6.16). The function $J_m(\mathbf{u})$ denotes the Jacobian of the transformation $\tilde{\mathbf{w}}_{1,m}(\mathbf{u}) = [\mathbf{w}_x(\mathbf{u}) \ \mathbf{w}_y(\mathbf{u})]^T$:[*]

$$J_m(\mathbf{u}) = \det\left(\frac{\partial \tilde{\mathbf{w}}_{1,m}(\mathbf{u})}{\partial \mathbf{u}}\right) = \det \begin{bmatrix} \dfrac{\partial \mathbf{w}_x}{\partial x} & \dfrac{\partial \mathbf{w}_y}{\partial x} \\ \dfrac{\partial \mathbf{w}_x}{\partial y} & \dfrac{\partial \mathbf{w}_y}{\partial y} \end{bmatrix}$$

$$= \left(\sum_{k \in K} \frac{\partial \tilde{\phi}_k}{\partial x} x_n\right)\left(\sum_{k \in K} \frac{\partial \tilde{\phi}_k}{\partial y} y_n\right) - \left(\sum_{k \in K} \frac{\partial \tilde{\phi}_k}{\partial x} y_n\right)\left(\sum_{k \in K} \frac{\partial \tilde{\phi}_k}{\partial y} x_n\right), \tag{6.6.8}$$

where $\mathbf{u} = [x \ y]^T$, $\mathbf{x}_{1,n(m,k)} = [x_n, y_n]^T$.

For motion tracking over a set of frames, because the mesh used in each new anchor frame is the tracked mesh resulting from a previous step, the shape of $\mathcal{B}_{1,m}$ is in general irregular (see Figure 6.13(c)). Consequently, the mapping function $\tilde{\mathbf{w}}_1(\mathbf{u})$ and the Jacobian $J_m(\mathbf{u})$ depend on the nodal positions in $\mathcal{B}_{1,m}$. On the other hand, for motion estimation between two frames, to reduce the complexity, one can use a regular mesh for the anchor frame, so that each element is itself equivalent to the master element (see Figure 6.13(b)). In this case, the mapping function in the anchor frame is trivial, that is, $\mathbf{w}_{1,m}(\mathbf{u}) = \mathbf{u}$, and $J_m(\mathbf{u}) = 1$.

The gradient of the error function in Equation (6.6.6) is, when $p = 2$,

$$\frac{\partial E_p}{\partial \mathbf{d}_n} = 2 \sum_{k \in K} \sum_{\mathbf{u} \in \tilde{\mathcal{B}}} \tilde{e}_{m(n,k)}(\mathbf{u}) \tilde{\phi}_k(\mathbf{u}) \left.\frac{\partial \psi_2(\mathbf{x})}{\partial \mathbf{x}}\right|_{\tilde{\mathbf{w}}_{2,m(n,k)}(\mathbf{u})} |J_{m(n,k)}(\mathbf{u})|, \tag{6.6.9}$$

where $m(n, k)$ specifies the global index of the kth element attached to the nth node. Figure 6.17 illustrates the neighboring elements and shape functions attached to node n in the quadrilateral mesh case.

---

[*]Strictly speaking, the use of the Jacobian is correct only when the error is defined as an integral over $\tilde{\mathcal{B}}$. Here we assume that the sampling over $\tilde{\mathcal{B}}$ is sufficiently dense when using the sum to approximate the integration.

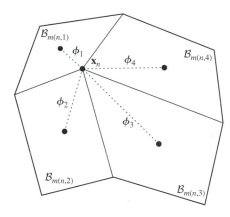

Figure 6.17 Neighborhood structure in a quadrilateral mesh: for a given node n, there are four elements attached to it, each with one shape function connected to this node.

It can be seen that the gradient with respect to one node only depends on the errors in the several elements attached to it. Ideally, in each iteration of a gradient-based search algorithm, to calculate the gradient function associated with any node, one should assume the other nodes are fixed at the positions obtained in the previous iteration. All the nodes should be updated at once at the end of the iteration, before going to the next iteration. But in reality, to speed up the process, one can update one node at a time, while fixing its surrounding nodes. Of course, this suboptimal approach could lead to divergence, or convergence to a local minimum that is worse than the one obtained by updating all the nodes simultaneously. Instead of updating the nodes in the usual raster order, to improve the accuracy and convergence rate, one can order the nodes so that the nodes whose motion vectors can be estimated more accurately are updated first. Because of the uncertainty in motion estimation in smooth regions, it may be better to update the nodes with large edge magnitudes and small motion-compensation errors first. This is known as *highest confidence first* [7], and this approach was taken in [2]. Another possibility is to divide all the nodes into several groups such that the nodes in the same group do not share the same elements, and therefore are independent in their impact on the error function. Sequential update of the nodes in the same group is then equivalent to simultaneous updates of these nodes. (This is the approach adopted in [42].) Either the first-order gradient-descent method, or the second-order Newton-Ralphson type of update algorithm could be used. The second-order method will converge much faster, but it is also more liable to converge to bad local minima.

The newly updated nodal positions based on the gradient function can lead to overly deformed elements (including flip-over and obtuse elements). In order to prevent such things from happening, one should limit the search range into which the updated nodal position can fall. If the updated position goes beyond this region, then it should be projected back to the nearest point in the defined search range. Figure 6.18 shows the legitimate search range for the case of a quadrilateral mesh.

The preceding discussion applies not only to gradient-based update algorithms, but also to exhaustive search algorithms. In this case, one can update one node at a time, by searching for the nodal position that will minimize the prediction errors in elements attached to it in the search range illustrated in Figure 6.18. For each candidate

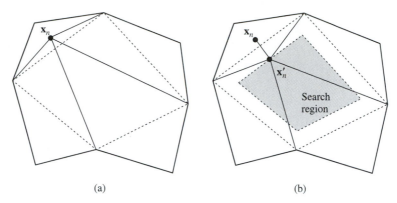

Figure 6.18   The search range for node n, given the positions of other nodes: the diamond region (dashed line) is the theoretical limit, the inner diamond region (shaded) is used in practice. When $\mathbf{x}_n$ falls outside the diamond region as shown in (a), at least one element attached to it becomes obtuse. By projecting $\mathbf{x}_n$ onto the inner diamond as illustrated in (b), all four elements would not be overly deformed.

position, one calculates the error using Equation (6.6.6), but accumulating only over the elements attached to that node. The optimal position is the one with the minimal error. Here again, the search order is very important.

Example 6.2

Figure 6.8(e) and (f) showed the motion estimation result obtained by an exhaustive search approach for backward motion estimation using a rectangular mesh at each new frame [43]. Figure 6.8(e) is the deformed mesh overlaid on top of the target frame, and Figure 6.8(f) is the predicted image for the anchor frame. Note that each deformed quadrangle in Figure 6.8(e) corresponds to a square block in the anchor frame. Thus a narrow quadrangle in the right side of the face indicates that it is expanded in the anchor frame. We can see that the mesh is deformed smoothly, which corresponds to a smooth motion field. The predicted image does not suffer from the blocking artifact associated with the EBMA (Figure 6.8(d) versus Figure 6.8(f)), and appears to be a more successful prediction of the original. A careful comparison between the predicted image (Figure 6.8(f)) and the actual image (Figure 6.8(b)), however, will reveal that the eye closing and mouth movement are not accurately reproduced, and there are some artificial warping artifacts near the jaw and neck. In fact, the PSNR of the predicted image is lower than that obtained by the EBMA.

Up to this point, we have assumed that a single mesh is generated (or propagated from the previous frame in the forward-tracking case) for the entire current frame, and that every node in this mesh is tracked to one and only one node in the target frame, so that the nodes in the target frame still form a mesh that covers the entire frame. In order to handle newly appearing or disappearing objects in a scene, one should allow for the deletion of nodes corresponding to disappeared objects, and the creation of new nodes in newly appearing objects. For a solution to such problems, see [3].

6.7 GLOBAL MOTION ESTIMATION

In Section 5.5, we showed that, depending on the camera and object motion and the object surface geometry, the motion field between two images of the same imaged object can be described by a translation, a geometric transformation, an affine mapping, or a projective mapping. Such a model can be applied to the entire frame if the entire motion field is caused by a camera motion, or if the imaged scene consists of a single object that is undergoing a rigid 3-D motion.*

In practice, one can hardly find a video sequence that contains a single object. There are usually at least two objects: a stationary background and a moving foreground. More often, there is more than one foreground object. Fortunately, when the foreground object motion is small compared to the camera motion, and the camera does not move in the Z-direction, the motion field can be approximated well by a global model. This will be the case, for example, when the camera pans over a scene or zooms in to a particular subject at a relatively fast speed. Such camera motions are quite common in sports videos and movies. Even when the actual 2-D motion field cannot be represented accurately by a single global motion, as long as the effect of the camera motion dominates over other motions (motions of individual small objects), determination of this dominant global motion is still very useful. In this section, we discuss the estimation of global motions.

There are in general two approaches for estimating global motion. One is to estimate the global motion parameters directly by minimizing the prediction errors under a given set of motion parameters. The other is to first determine pixelwise or blockwise motion vectors, using the techniques described previously, and then use a regression method to find the global motion model that best fits the estimated motion field. The latter method can also be applied to motion vectors at selected feature points, such as points with strong edges.

6.7.1 Robust Estimators

A difficulty in estimating global motion is that a pixel may not experience only the global motion. Usually, the motion at any pixel can be decomposed into a global motion (caused by camera movement) and a local motion caused by the movement of the underlying object. Therefore, the prediction error obtained by using the global motion model alone may not be small, even if the correct global motion parameters are available. In other instances, not all the pixels in the same frame experience the global motion, and ideally one should not apply the same motion model to the entire frame. These problems can be overcome by a *robust estimation* method [15], if the global motion is dominant over other local motions—in the sense that the pixels that experience the same global motion and only that global motion occupy a significantly larger portion of the underlying image domain than those pixels that do not.

---

*Recall that in the case that the camera or the object moves in the Z-direction, the motion field can be represented by a projective mapping only if the object surface is planar (see Section 5.5.3). When the object surfaces are spatially varying, the mapping function at any point also depends on the surface depth of that point and cannot be represented by a global model.

The basic idea in robust estimation is to consider the pixels that are governed by the global motion as *inliers,* and the remaining pixels as *outliers.* Initially, one assumes that all the pixels undergo the same global motion, and estimates the motion parameters by minimizing the prediction or fitting error over all the pixels. This will yield an initial set of motion parameters. With this initial solution, one can then calculate the prediction or fitting error over each pixel. The pixels where the errors exceed a certain threshold will be classified as outliers and be eliminated from the next iteration. This process is then repeated for the remaining inlier pixels, and iterates until no outlier pixels exist. This approach is called a *hard-threshold robust estimator.*

Rather than simply labeling a pixel as either an inlier or an outlier at the end of each iteration, one can also assign a different weight to each pixel, with a large weight for a pixel with a small error, and vice versa. At the next minimization or fitting iteration, a weighted error measure is used, so that the pixels with larger errors in the previous iteration will have less impact than those with smaller errors. This approach is known as a *soft-threshold robust estimator.*

6.7.2 Direct Estimation

In either the hard- or soft-threshold robust estimator, each iteration involves the minimization of an error function. Here we derive the form of the function when the model parameters are directly obtained by minimizing the prediction error. We consider only the soft-threshold case, as the hard-threshold case can be considered as a special case in which the weights are either one or zero. Let the motion field from the anchor frame to the target frame be denoted by $\mathbf{d}(\mathbf{x}; \mathbf{a})$, where $\mathbf{a}$ is the vector that contains all the global motion parameters. The prediction error can be written following Equation (6.2.1), as:

$$E_{\text{DFD}} = \sum_{n \in \mathcal{N}} w_n |\psi_2(\mathbf{x}_n + \mathbf{d}(\mathbf{x}_n; \mathbf{a})) - \psi_1(\mathbf{x}_n)|^p \qquad (6.7.1)$$

where $\mathbf{x}_n, n \in \mathcal{N} = \{1, 2, \ldots, N\}$ represent all the pixels over which a global motion is estimated, and w_n are the weighting coefficients for pixels $\mathbf{x}_n$. Within each iteration of the robust estimation process, the parameter vector $\mathbf{a}$ is estimated by minimizing this error, using either gradient-based or exhaustive search methods. The weighting factors w_n in a new iteration will be adjusted based on the DFD at $\mathbf{x}_n$, which is calculated based on the motion parameters estimated in the previous iteration.

6.7.3 Indirect Estimation

For indirect estimation, we assume that the motion vectors $\mathbf{d}_n$ have been estimated at a set of sufficiently dense points $\mathbf{x}_n, n \in \mathcal{N}$. This can be accomplished, for example, using either the block-based or mesh-based approaches previously described. One can also choose to estimate the MVs at selected feature points only, where the estimation accuracy is high. The task here is to determine $\mathbf{a}$ so that the model $\mathbf{d}(\mathbf{x}; \mathbf{a})$ can approximate well the preestimated MVs $\mathbf{d}_n, n \in \mathcal{N}$. This can be accomplished by minimizing the following fitting error:

$$E_{\text{fit}} = \sum_{n \in \mathcal{N}} w_n |\mathbf{d}(\mathbf{x}_n; \mathbf{a}) - \mathbf{d}_n|^p \qquad (6.7.2)$$

As shown in Section 5.5.4, a global motion can usually be described or approximated by a polynomial function. In this case, $\mathbf{a}$ consists of the polynomial coefficients and $\mathbf{d}(\mathbf{x}_n; \mathbf{a})$ is a linear function of $\mathbf{a}$, that is, $\mathbf{d}(\mathbf{x}_n; \mathbf{a}) = [\mathbf{A}(\mathbf{x}_n)]\mathbf{a}$. If we choose $p = 2$, then the minimization problem becomes a weighted least-squares problem. By setting $\partial E_{\text{fit}}/\partial \mathbf{a} = 0$, we obtain the following solution:

$$\mathbf{a} = \left(\sum_{n \in \mathcal{N}} w_n [\mathbf{A}(\mathbf{x}_n)]^T [\mathbf{A}(\mathbf{x}_n)] \right)^{-1} \left(\sum_{n \in \mathcal{N}} w_n [\mathbf{A}(\mathbf{x}_n)]^T \mathbf{d}_n \right). \qquad (6.7.3)$$

As an example, consider the affine motion model given in Equation (5.5.16). The motion parameter vector is $\mathbf{a} = [a_0, a_1, a_2, b_0, b_1, b_2]^T$, and the matrix $[\mathbf{A}(\mathbf{x}_n)]$ is

$$[\mathbf{A}(\mathbf{x}_n)] = \begin{bmatrix} 1 & x_n & y_n & 0 & 0 & 0 \\ 0 & 0 & 0 & 1 & x_n & y_n \end{bmatrix}.$$

In fact, the parameters for the x and y dimensions are not coupled and can be estimated separately, which will reduce the matrix sizes involved. For example, to estimate the x-dimensional parameter $\mathbf{a}_x = [a_0, a_1, a_2]$, the associated matrix is $[\mathbf{A}_x(\mathbf{x}_n)] = [1, x_n, y_n]$, and the solution is:

$$\mathbf{a}_x = \left(\sum_{n \in \mathcal{N}} w_n [\mathbf{A}_x(\mathbf{x}_n)]^T [\mathbf{A}_x(\mathbf{x}_n)] \right)^{-1} \left(\sum_{n \in \mathcal{N}} w_n [\mathbf{A}_x(\mathbf{x}_n)]^T d_{n,x} \right). \qquad (6.7.4)$$

6.8 REGION-BASED MOTION ESTIMATION*

As pointed out in the previous section, there are usually multiple types of motions in an imaged scene, which correspond to motions associated with different objects. In region-based motion estimation, we segment the underlying image frame (the anchor frame) into multiple regions and estimate the motion parameters of each region. The segmentation should be such that a single parametric motion model can represent well the motion in each region. Obviously, region segmentation is dependent on the motion model used for characterizing each region. The simplest approach is to require that each region undergoes a single translational motion. This requirement, however, can result in too many small regions, because the 2-D motion in a region corresponding to a physical object can rarely be modeled by a simple translation. Such a region would have to be split into many small subregions for each subregion to have a single translational motion. For a more efficient motion representation, an affine, bilinear, or perspective motion model should be used.

---

*Sections marked with an asterisk may be skipped or left for further exploration.

In general, there are three approaches for accomplishing region-based motion estimation. With the first approach, one first segments the image frame into different regions—based on texture homogeneity, edge information, and sometimes the motion boundary obtained by analyzing the difference image between two frames—and then estimates the motion in each region. The latter can be accomplished by applying the global motion estimation method described in Section 6.7 to each region. We call such a method *region-first*. With the second approach, one first estimates the motion field over the entire frame, and then segments the resulting motion field so that the motion in each region can be modeled by a single parametric model. We call this method *motion-first*. The resulting regions can be further refined, subject to some spatial connectivity constraints. The first step in this approach can be accomplished using the various motion estimation methods described previously, including the pixel-, block-, and mesh-based methods. The second step involves motion-based segmentation, which is discussed further in Section 6.8.1. The third approach is to jointly estimate the region partitioning and motion in each region. In general, this is accomplished by an iterative process, which performs region segmentation and motion estimation alternately; this approach is described in Section 6.8.2.

6.8.1 Motion-Based Region Segmentation

As described, motion-based segmentation refers to the partitioning of a motion field into multiple regions such that the motion within each region can be described by a single set of motion parameters. Here we present two approaches for accomplishing this task. The first approach uses a clustering technique to identify similar motion vectors. The second approach uses a layering technique to estimate the underlying region and associated motions sequentially, starting from the region with the most dominant motion.

Clustering   Consider the case in which the motion model for each region is a pure translation. Then the segmentation task is to group all spatially connected pixels with similar MVs into one region. This can be easily accomplished by an automated clustering method, such as the K-means or the ISODATA method [8]. This is an iterative process: Starting from an initial partitioning, the mean motion vector, known as the centroid, of each region is calculated. Then each pixel is reclassified into the region whose centroid is closest to the motion vector of this pixel. This results in a new partition, and the preceding two steps can be repeated until the partition no longer changes. In this process, spatial connectivity is not considered. Therefore, the resulting regions may contain pixels that are not spatially connected. Postprocessing steps may be applied at the end of the iterations to improve the spatial connectivity of the resulting regions. For example, a single region may be split into several subregions so that each region is a spatially connected subset, isolated pixels may be merged into their surrounding regions, and finally region boundaries can be smoothed using morphological operators. When the motion model for each region is not a simple translation, motion-based clustering is not as straightforward. This is because one cannot use the similarity between MVs as the criterion for performing clustering. One solution is to find a set of motion parameters for each pixel by fitting the MVs in its neighborhood into a specified model. Then one can employ the clustering method described previously, by replacing the raw MV with the motion parameter vector. If the original motion field is given in the block-based representation using a higher-order model, then one can cluster blocks with similar motion parameters into the same region. Similarly, with the mesh-based motion representation, one can derive a set of motion parameters for each element based on its nodal displacements, and then cluster the elements with similar parameters into the same region. This is the parallel approach described in [44].

Layering   Very often, the motion field in a scene can be decomposed into layers, with the first layer representing the most dominant motion, the second layer the next most dominant, and so on. Here, the dominance of a motion is determined by the area of the region undergoing the corresponding motion. The most dominant motion is often a reflection of the camera motion, which affects the entire imaged domain. For example, in a video clip of a tennis play, the background will be the first layer, which usually undergoes a coherent global camera motion; the player the second layer (which usually contains several subobject-level motions corresponding to the movements of different parts of the body); the racket the third; and the ball the fourth layer. To extract the motion parameters in different layers, one can recursively use the robust estimator method described in Section 6.7.1. First, we try to model the motion field of the entire frame by a single set of parameters, and continuously eliminate outlier pixels from the remaining inlier group, until all the pixels within the inlier group can be modeled well. This will yield the first dominant region (corresponding to the inlier region) and its associated motion. The same approach can then be applied to the remaining pixels (the outlier region) to identify the next dominant region and its motion. This process continues until no outlier pixels exist. As before, postprocessing may be invoked at the end of the iterations to improve the spatial connectivity of the resulting regions. This is the sequential approach described in [44].

For this scheme to work well, the inlier region must be significantly larger than the outlier region at any iteration. This means that the largest region must be greater than the combined area of all other regions, the second largest region must be greater than the combined area of the remaining regions, and so on. This condition is satisfied in most video scenes, which often contain a stationary background that covers a large portion of the underlying image, and different moving objects with varying sizes.

6.8.2 Joint Region Segmentation and Motion Estimation

Theoretically, one can formulate the joint estimation of the region segmentation map and the motion parameters of each region as an optimization problem. The function to be minimized could be a combination of the motion-compensated prediction error and a region smoothness measure. The solution of the optimization problem is difficult, however, because of the very high dimension of the parameter space and the complicated interdependence between these parameters. In practice, a suboptimal approach is often taken, which alternates between the estimation of the segmentation and motion parameters. Based on an initial segmentation, the motion of each region is first estimated. In the next iteration, the segmentation is refined—for example, by eliminating outlier pixels in each region where the prediction errors are large, and by merging pixels sharing similar motion models. The motion parameters for each refined region are then reestimated. This process continues until no more changes in the segmentation map occur.

An alternative approach is to estimate the regions and their associated motions in a layered manner, similar to the layered approach described previously. There, we assume that the motion vector at every point is already known, and the identification of the region with the most dominant motion (i.e., the inlier) is accomplished by examining the fitting error induced by representing individual MVs using a set of motion parameters. This is essentially the indirect robust estimator presented in Section 6.7.3. In the joint region segmentation and motion estimation approach, to extract the next dominant region and its associated motion from the remaining pixels, we can use the direct robust estimator. That is, we directly estimate the motion parameters, by minimizing the prediction errors at these pixels. Once the parameters are determined, we determine whether a pixel belongs to the inlier group by examining the prediction error at this pixel. We then reestimate the motion parameters by minimizing the prediction errors at the inlier pixels only. This approach has been taken by Hsu, Anandan, and Peleg [18].

6.9 MULTIRESOLUTION MOTION ESTIMATION

As can be seen from previous sections, various motion estimation approaches can be reduced to solving an error-minimization problem. Two major difficulties associated with obtaining the correct solution are: (1) the minimization function usually has many local minima, and it is not easy to reach the global minimum, unless it is close to the chosen initial solution; and (2) the amount of computation involved in the minimization process is very large. Both problems can be combated by the multiresolution approach, which searches the solution for an optimization problem in successively finer resolutions. By first searching the solution in a coarse resolution, one can usually obtain a solution that is close to the true motion. In addition, by limiting the search in each finer resolution to a small neighborhood of the solution obtained in the previous resolution, the total number of searches can be reduced, compared to that required by directly searching in the finest resolution over a large range.

In this section, we first describe the multiresolution approach for motion estimation in a general setting, which is applicable to any motion model. We then focus on the block-translation model, and describe a hierarchical block matching algorithm.

6.9.1 General Formulation

As illustrated in Figure 6.19, pyramidal representations of the two raw image frames are first derived, in which each level is a reduced-resolution representation of the lower level, obtained by spatial low-pass filtering and subsampling. The bottom level is the original image. Then, the motion fields between corresponding levels of the two pyramids are estimated, starting from the top (coarsest) level, and progressing to the next finer level repeatedly. At each new finer-resolution level, the motion field obtained at the previous coarser level is interpolated to form the initial solution for the motion at the current level.

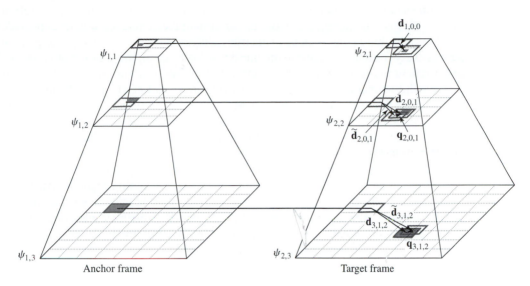

Figure 6.19   Illustration of the hierarchical block matching algorithm.

The most common pyramid structure is one in which the resolution is reduced by half, both horizontally and vertically, between successive levels. Usually, a simple 2×2 averaging filter is used for low-pass filtering; for better performance, a Gaussian filter can be employed.

Assume that the number of levels is L, with the Lth level being the original image. Let the lth level images of the anchor and target frames be represented by $\psi_{t,l}(\mathbf{x}), \mathbf{x} \in \Lambda_l, t = 1, 2$, where Λ_l is the set of pixels at level l. Denote the total motion field obtained from levels 1 to $l - 1$ by $\mathbf{d}_{l-1}(\mathbf{x})$. At the lth level, we first interpolate $\mathbf{d}_{l-1}(\mathbf{x})$ to the resolution of level l, to produce an initial motion estimate $\tilde{\mathbf{d}}_l(\mathbf{x}) = \mathcal{U}(\mathbf{d}_{l-1}(\mathbf{x}))$, where $\mathcal{U}$ represents the interpolation operator. We then determine the update $\mathbf{q}_l(\mathbf{x})$ at this level such that the error

$$\sum_{\mathbf{x} \in \Lambda_l} |\psi_{2,l}(\mathbf{x} + \tilde{\mathbf{d}}_l(\mathbf{x}) + \mathbf{q}_l(\mathbf{x})) - \psi_{1,l}(\mathbf{x})|^p \tag{6.9.1}$$

is minimized. The new motion field obtained after this step is $\mathbf{d}_l(\mathbf{x}) = \mathbf{q}_l(\mathbf{x}) + \tilde{\mathbf{d}}_l(\mathbf{x})$. Upon completion of successive refinements, the total motion at the finest resolution is

$$\mathbf{d}(\mathbf{x}) = \mathbf{q}_L(\mathbf{x}) + \mathcal{U}(\mathbf{q}_{L-1}(\mathbf{x}) + \mathcal{U}(\mathbf{q}_{L-2}(\mathbf{x}) + \cdots + \mathcal{U}(\mathbf{q}_1(\mathbf{x}) + \mathbf{d}_0(\mathbf{x})) \cdots)). \tag{6.9.2}$$

The initial condition for this procedure is $\mathbf{d}_0(\mathbf{x}) = \mathbf{0}$. One can either directly specify the total motion $\mathbf{d}(\mathbf{x})$, or the motion updates at all levels, $\mathbf{q}_l(\mathbf{x}), l = 1, 2, \ldots, L$. The latter represents the motion in a layered structure, which is desired in applications requiring progressive retrieval of the motion field.

The benefits of the multiresolution approach are twofold. First, the minimization problem at a coarser resolution is better-posed than at a finer resolution; therefore, the solution obtained at a coarser level is more likely to be close to the true solution at that resolution. The interpolation of this solution to the next resolution level provides a good initial solution that is close to the true solution at that level. By repeating this step successively from the coarsest to the finest resolution level, the solution obtained at the finest resolution is more likely to be close to the true solution (the global minimum). Second, the estimation at each resolution level can be confined to a significantly smaller search range than the true motion range at the finest resolution, so that the total number of searches to be conducted is smaller than the number of searches required in the finest resolution directly. The actual number of searches will depend on the search ranges set at different levels.

The use of multiresolution representation for image processing was first introduced by Burt and Adelson [6]. Application to motion estimation depends on the motion model used. In the preceding presentation, we have assumed that MVs at all pixels are to be estimated. The algorithm can be easily adapted to estimate block-based, mesh-based, global- or object-level motion parameters. Because the blockwise translational motion model is the most popular for practical applications, we consider this special case in more detail in the following.

6.9.2 Hierarchical Block Matching Algorithm

As indicated in Section 6.4.1, using an exhaustive search scheme to derive block MVs requires an extremely large number of computations. In addition, the estimated block MVs often lead to a chaotic motion field. In this section, we introduce a hierarchical block matching algorithm (HBMA), which is a special case of the multiresolution approach just presented. Here, the anchor and target frames are each represented by a pyramid, and the EBMA or one of its fast variants is employed to estimate MVs of blocks in each level of the pyramid. Figure 6.19 illustrated the process when the spatial resolution is reduced by half, both horizontally and vertically, at each increasing level of the pyramid. Here, we assume that the same block size is used at different levels, so that the number of blocks is reduced by half in each dimension as well. Let the MV for block (m, n) at level l be denoted by $\mathbf{d}_{l,m,n}$. Starting from level 1, we first find the MVs for all blocks in this level, $\mathbf{d}_{1,m,n}$. At each new level $l > 1$, for each block, its initial MV $\tilde{\mathbf{d}}_{l,m,n}$ is interpolated from a corresponding block in level $l - 1$ by

$$\tilde{\mathbf{d}}_{l,m,n} = \mathcal{U}\left(\mathbf{d}_{l-1,\lfloor m/2\rfloor,\lfloor n/2\rfloor}\right) = 2\mathbf{d}_{l-1,\lfloor m/2\rfloor,\lfloor n/2\rfloor}. \tag{6.9.3}$$

Then a correction vector $\mathbf{q}_{l,m,n}$ is searched, yielding the final estimated MV

$$\mathbf{d}_{l,m,n} = \tilde{\mathbf{d}}_{l,m,n} + \mathbf{q}_{l,m,n}. \tag{6.9.4}$$

Example 6.3

In Figure 6.20, we show two video frames, of size 32×32, in which a gray block in the anchor frame is moved by a displacement of $(13,11)$. We show how to use a three level HBMA to estimate the block motion field. The block size used at each level is 4×4, and the search stepsize is 1 pixel. Starting from level 1, for block $(0,0)$ in the anchor frame, the MV is found to be $\mathbf{d}_{1,0,0} = \mathbf{d}_1 = (3, 3)$. When going to level 2, for block $(0, 1)$, it is initially assigned the MV $\tilde{\mathbf{d}}_{2,0,1} = \mathcal{U}(\mathbf{d}_{1,0,0}) = 2\mathbf{d}_1 = (6, 6)$. Starting with this initial

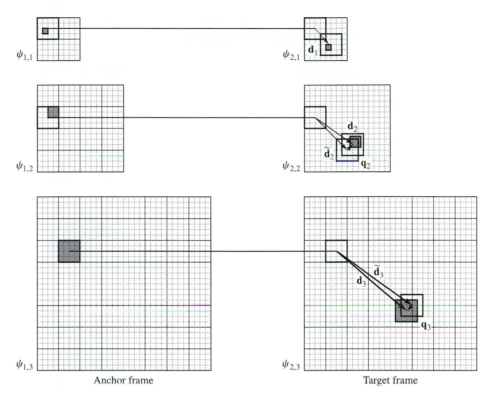

Figure 6.20 An example of a three-level HBMA for block motion estimation (see Example 6.3).

MV, the correction vector is found to be $\mathbf{q}_2 = (1, -1)$, leading to the final estimated MV $\mathbf{d}_{2,0,1} = \mathbf{d}_2 = (7, 5)$. Finally, at level 3, block (1,2) is initially assigned an MV of $\tilde{\mathbf{d}}_{3,1,2} = \mathcal{U}(\mathbf{d}_{2,0,1}) = 2\mathbf{d}_2 = (14, 10)$. With a correction vector of $\mathbf{q}_3 = (-1, 1)$, the final estimated MV is $\mathbf{d}_{3,1,2} = \mathbf{d}_3 = (13, 11)$.

Note that using a block width N at level l corresponds to a block width of $2^{L-l}N$ at the full resolution. The same scaling applies to the search range and stepsize. Therefore, by using the same block size, search range, and stepsize at different levels, we actually use a larger block size, search range, and stepsize in the beginning of the search, and then gradually reduce these (by half) in later steps.

The number of operations involved in the HBMA depends on the search range at each level. If the desired search range is R in the finest resolution, then with an L-level pyramid, one can set the search range to be $R/2^{L-1}$ at the first level. For the remaining levels, because the initial MV interpolated from the previous level is usually quite close to the true MV, the search range for the correction vector need not be very large. However, for simplicity, we assume that every level uses a search range of $R/2^{L-1}$. If the image size is $M \times M$ and the block size is $N \times N$ at every level, the number of blocks at the lth level is $(M/2^{L-l}N)^2$, and the number of searches is

$(M/2^{L-l}N)^2 * (2R/2^{L-1} + 1)^2$. Because the number of operations required for each search is N^2, the total number of operations is

$$\sum_{l=1}^{L}(M/2^{L-l})^2(2R/2^{L-1} + 1)^2 = \frac{4}{3}(1 - 4^{-L})M^2(2R/2^{L-1} + 1)^2$$

$$\approx \frac{1}{3}4^{-(L-2)}4M^2R^2.$$

Recall that the operation count for EBMA is $M^2(2R+1)^2 \approx 4M^2R^2$ (see Section 6.4.1). Therefore, the hierarchical scheme using the preceding parameter selection will reduce the computation by a factor of $3 \cdot 4^{L-2}$. Typically, the number of levels L is 2 or 3.

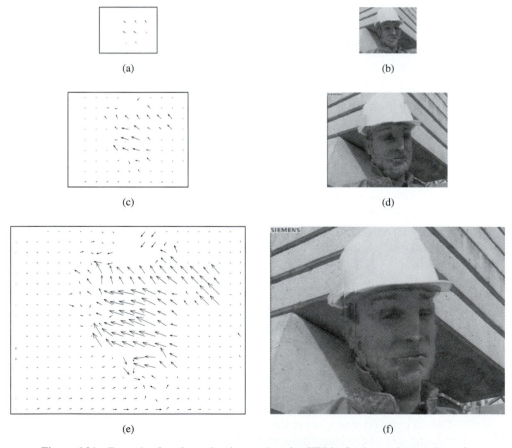

(a)   (b)

(c)   (d)

(e)   (f)

Figure 6.21 Example of motion estimation results using HBMA for the two images shown in Figure 6.8: (a–b) the motion field and predicted image at level 1; (c–d) the motion field and predicted image at level 2; (e–f) the motion field and predicted image at the final level (PSNR = 29.32). A three-level HBMA algorithm is used; the block size is 16×16 at all levels. The search range is 4 at all levels with integer-pel accuracy. The result in the final level is further refined by a half-pel accuracy search in the range of ± 1.

Example 6.4

Figure 6.21 shows estimation results obtained with the HBMA approach, for the same pair of video frames given in Figure 6.8. For this example, a three-level pyramid is used. The search range in each level is set to 4, so that the equivalent search range in the original resolution is $R = 16$. Integer-pel accuracy search is used in all the levels. The final integer-accuracy solution is further refined to half-pel accuracy by using a half-pel accuracy search in a search range of one pixel. Comparing the result in the final level with those shown in Figures 6.8(c) and (d), we can see that the multiresolution approach indeed yields a smoother motion field than the EBMA. Visual observation also reveals that this motion field represents more truthfully the motion between the two image frames in Figures 6.8(a) and (b). This is true in spite of the fact that the EBMA yields a higher PSNR. In terms of computational complexity, the half-pel accuracy EBMA used for Figure 6.8(c–d) requires $352 \times 288(4R + 1)^2 = 4.3\text{E} + 8$ operations, while the three-level algorithm here uses only $\frac{4}{3} \cdot 352 \times 288(\frac{R}{2} + 1)^2 = 1.1\text{E} + 7$ operations, if we neglect the final refinement step using half-pel search.

There are many variant implementations of HBMA. Bierling was the first to apply this idea to the block-based motion model [5]. A special case of hierarchical BMA is known as variable size or quad-tree BMA, which starts with a larger block size, and then repeatedly divides a block into four if the matching error for this block is still larger than a threshold. In this case, all the processing is done on the original image resolution.

6.10 APPLICATION OF MOTION ESTIMATION IN VIDEO CODING

Motion estimation is a key element in any video coding system. As will be discussed in Section 9.3.1, an effective video coding method is to use blockwise temporal prediction, in which a block in a frame to be coded is predicted from its corresponding block in a previously coded frame, called the reference frame,* and then the prediction error is coded. To minimize the bit rate for coding the prediction error, the appropriate criterion for motion estimation is to minimize the prediction error. The fact that the estimated motion field does not necessarily resemble the actual motion field is not problematic in such applications. Therefore, the block-matching algorithms (EBMA and its fast variants including HBMA) offer simple and effective solutions. Instead of using the MV estimated for each block directly for the prediction of that block, one can use a weighted average of the predicted values based on the MVs estimated for its neighboring blocks. This is known as overlapped block-motion compensation, which will be discussed in Section 9.3.2.

Note that in this video coding method, the MVs also must be coded, in addition to the prediction error. Therefore, minimizing the prediction error alone is not the best criterion to use. Since a smoother motion field requires fewer bits to code, imposing smoothness in the estimated motion field, if done properly, can help improve the overall coding efficiency. More advanced motion estimation algorithms therefore operate by minimizing the total bit rate used for coding the MVs and the prediction errors. This subject is discussed further in Section 9.3.3.

*See the footnote on page 145.

To overcome the blocking artifacts produced by block-based motion estimation methods, high-order block-based (DBMA), mesh-based, or a combination of block-based, mesh-based, and/or DBMA can be applied. However, these more complicated schemes usually do not lead to significant gain in coding efficiency.

In more advanced video coding schemes (to be covered in Chapter 10), global motion estimation (Section 6.7) is usually applied to the entire frame, prior to block-based motion estimation, to compensate for the effect of camera motion. Moreover, an entire frame is usually segmented into several regions or objects, and the motion parameters for each region or object are estimated, using the region-based motion estimation method discussed in Section 6.8.

6.11 SUMMARY

Relation between Image Intensity and Motion (Section 6.1)

- Almost all motion estimation algorithms are based on the constant intensity assumption (Equation (6.1.1) or (5.2.11)), or the optical flow equation (Equation (6.1.3)) derived based on that assumption. This enables us to estimate motion by identifying pixels with similar intensity, subject to some motion models. Note that this assumption is valid only when the illumination source is ambient and temporally invariant, and the object surface is diffusely reflecting (Section 5.2).

- When the motion direction is orthogonal to the image intensity gradient, or if the image gradient is zero, motion does not induce changes in image intensity. This is the inherent limitation of intensity-based motion estimation methods.

Key Components in Motion Estimation

- *Motion Representation (Section 6.2.1):*   This depends on the partition used to divide a frame (pixel-based, block-based, mesh-based, region-based, global), the motion model used for each region of the partition (block, mesh-element, object-region, or entire frame), and the constraint between motions in adjacent regions. Different motion representations led to different motion estimation methods, including pixel-based (Section 6.3), block-based (Sections 6.4–5), mesh-based (Section 6.6), region-based (Section 6.8), and global motion estimation (Section 6.7).

- *Motion Estimation Criterion (Section 6.2.2):*   We presented three criteria for estimating the motion parameters over each region: (1) minimizing the DFD error (when the motion is small, this is equivalent to the method based on the optical flow equation); (2) making the resulting motion field as smooth as possible across regions, while minimizing the DFD error; and (3) maximizing the a posteriori probability of the motion field, given the observed frames. We showed that (3) essentially requires (1) and (2). Instead of minimizing the DFD, one can also detect peaks in the phase correlation function, when the motion in a region is a pure translation.

- *Optimization Methods (Section 6.2.3):* For a chosen representation and criterion, the motion estimation problem is usually converted to an optimization (minimization or maximization) problem, which can be accomplished by using exhaustive or gradient-based search. To speed up the search and avoid being trapped in local minima, a multiresolution procedure can be used (Section 6.9).

6.12 PROBLEMS

6.1 Describe the advantages and disadvantages of different motion representation methods (pixel-based, block-based, mesh-based, region-based, and global).

6.2 Describe the advantages and disadvantages of the exhaustive search and gradient-descent methods. Also, compare first- and second order gradient-descent methods.

6.3 What are the main advantages of the multiresolution estimation method, compared to an approach using a single resolution? Are there any disadvantages?

6.4 In Section 6.3.2, we derived the multipoint neighborhood method using the gradient-descent method. Can you find a closed-form solution using the optical flow equation? Under what condition will your solution be valid?

6.5 In Section 6.4.1, we described an exhaustive search algorithm for determining block MVs in the block-based motion representation. Can you find a closed-form solution using the optical flow equation? Under what condition will your solution be valid?

6.6 In Equation (6.2.7), we showed that, if the motion field is a constant, one can use the optical flow equation to set up a least-squares problem, and obtain a closed-form solution. Suppose that the motion field is not a constant, but can be modeled by a polynomial mapping. Can you find a closed-form solution for the polynomial coefficients? (*Hint:* any polynomial mapping function can be represented as $\mathbf{d}(\mathbf{x}; \mathbf{a}) = [\mathbf{A}(\mathbf{x})]\mathbf{a}$, where $\mathbf{a}$ contains all the polynomial coefficients.)

6.7 In Section 6.4.5, we said that when there are several patches in a range block in $\phi_1(\mathbf{x})$ that undergo different motions, there will be several peaks in the PCF. Each peak corresponds to the motion of one patch. The location of the peak indicates the MV of the patch, whereas the amplitude of the peak is proportional to the size of the patch. Can you prove this statement, at least qualitatively? You can simplify your derivation by considering the 1-D case only.

6.8 With the EBMA, does the computational requirement depends on the block size?

6.9 In Section 6.9.2, we derived the number of operations required by the HBMA, if the search range at every level is $R/2^{L-1}$. What will be the number if one uses a search range of ± 1 pel in every level, except at the first level, where the search range is set to $R/2^{L-1}$? Is this parameter set-up appropriate?

6.10 Consider a BT.601-format video, with Y-component frame size of 720×480. Compare the computations required by an EBMA algorithm (integer-pel) with block size 16×16 and that by a two-level HBMA algorithm. Assume that the maximum motion range is ± 32. You can compare the computations by the operation number, with each operation including one subtraction, one absolute value computation, and one addition. You can make your own assumption about the search range at different levels with the HBMA. For simplicity, ignore the computations required for generating the pyramid and assume only integer-pel search.

6.11 Repeat Problem 6.10 for a three-level HBMA algorithm.

6.12 Write a C or MATLAB code for implementing an EBMA with integer-pel accuracy. Use a block size of 16×16. The program should allow the user to choose the search range, so that you can compare the results obtained with different search ranges. (Note that the proper search range depends on the extent of the motion in your test images). Apply the program to two adjacent frames of a video sequence. Your program should produce and plot the estimated motion field, the predicted image and the prediction-error image. It should also calculate the PSNR of the predicted frame compared to the original anchor frame. With MATLAB, you can plot the motion field using the function *quiver*.

6.13 Repeat Problem 6.12 for an EBMA with half-pel accuracy. Compare the PSNR of the predicted image obtained using integer-pel to that using half-pel accuracy estimation. Which method gives you more accurate prediction? Which requires more computation time?

6.14 You can obtain a dense (i.e., pixel-based) motion field from a block-based one by spatial interpolation. Write a C or MATLAB code that can interpolate the motion field resulting from Problem 6.12, by assuming that the MV for each block is actually the MV of the block center. Use bilinear interpolation. Using the interpolated pelwise motion field, you can again produce the predicted image and the prediction-error image. Compare the motion field, predicted image, and prediction-error image obtained in Problems 6.12 and 6.13 with those obtained here. Which method gives you more accurate prediction? Which requires more computation time?

6.15 Implement the HBMA method in C or MATLAB. You can choose to use either two or three levels of resolution. You can use integer-pel search at all levels, but refine your final result by half-pel accuracy search within a ± 1 neighborhood. Use a block size of 16×16 at all levels. The search range should be chosen so that the equivalent search range in the original resolution is ± 32. Compare the results with those obtained in Problems 6.12 and 6.13, in terms of both accuracy and computation time.

6.16 In Section 6.7, we say that the fitting error in Equation (6.7.2) is minimized with the solution given in Equation (6.7.3). Prove this result.

6.17 Assume that the motion between two frames can be modeled by a global affine model. We want to estimate the affine parameters directly based on the DFD criterion. Set up the optimization problem, and derive an iterative algorithm for solving the optimization problem. You can use either the first-order gradient-descent or the Newton-Ralphson method. Write a C or MATLAB code for implementing your algorithm. Apply it to two video frames that are undergoing predominantly camera motion. Compare the resulting motion field and predicted frame with that obtained with the EBMA.

6.18 Repeat Problems 6.17, but use an indirect method to derive the affine parameters from given block MVs. Derive the regression equation and the closed-form solution. Write a C or MATLAB code for implementing your algorithm. You can use your previous code for integer-pel EBMA to generate block MVs. Compare the result obtained here with that from the direct method (Problem 6.17).

6.13 BIBLIOGRAPHY

[1] Aggarwal, J. K., and N. Nandhahumar. On the computation of motion from sequences of images—a review. *Proceedings of the IEEE* (1988), 76:917–35.

[2] Altunbasak, Y., and A. M. Tekalp. Closed-form connectivity-preserving solutions for motion compensation using 2-D meshes. *IEEE Trans. Image Process.* (Sept. 1997), 6:1255–69.

[3] Altunbasak, Y., and A. M. Tekalp. Occlusion-adaptive, content-based mesh design and forward tracking. *IEEE Trans. Image Process.* (Sept. 1997), 6:1270–80.

[4] Barron, J. L., D. J. Fleet, and S. S. Beauchemin. Performance of optical flow techniques. *International Journal of Computer Vision* (1994), 12:43–77.

[5] Bierling, M., Displacement estimation by hierarchical block matching. In *SPIE Conf.: Visual Commun. Image Processing* (Nov. 1988), SPIE-1001:942–51.

[6] Burt, P. J., and E. H. Adelson. The Laplacian pyramid as a compact image code. *IEEE Trans. Commun.*(1983), COM-31:532–40.

[7] Chou, P., and C. Brown. The theory and practice of bayesian image labeling. *International Journal of Computer Vision* (1990), 4:185–210.

[8] Duda, R. O., and P. E. Hart. *Pattern Classification and Scene Analysis*. New York: John Wiley & Sons, 1973.

[9] Fleet, D. J., and A. D. Jepson. Computation of component image velocity from local phase information. *International Journal of Computer Vision* (1990), 5:77–104.

[10] Fleet, D. J. Disparity from local weighted phase-correlation. In *IEEE International Conference on Systems, Man, and Cybernetics: Humans, Information and Technology* (1994), 48–54.

[11] Geman, S., and D. Geman. Stochastic relaxation, Gibbs distributions, and the Bayesian restoration of images. *IEEE Trans. Pattern Anal. Machine Intell.* (Nov. 1984), 6:721–41.

[12] Girod, B. Motion compensation: Visual aspects, accuracy, and fundamental limits. In M. I. Sezan and R. L. Lagendijk, eds., *Motion Analysis and Image Sequence Processing*. Boston: Kluwer Academic Publishers, 1993, 126–52.

[13] Girod, B. Motion-compensating prediction with fractional-pel accuracy. *IEEE Transactions on Communications* (April 1993), 41(4):604–12.

[14] Hang, H.-M., Y.-M. Chou, and S.-C. Cheng. Motion estimation for video coding standards. *Journal of VLSI Signal Processing Systems for Signal, Image, and Video Technology* (Nov. 1997), 17(2/3):113–36.

[15] Haralick, R. M., and J. S. Lee. The facet approach to optical flow. In *Image Understanding Workshop* (1993).

[16] Horn, B. K. P., and B. G. Schunck. Determining optical flow. *Artificial intelligence* (1981), 17(1–3):185–203.

[17] Horn, B. K. P. *Robot Vision*. Cambridge, MA: MIT Press, 1986.

[18] Hsu, S., P. Anandan, and S. Peleg. Accurate computation of optical flow using layered motion representations. In *IEEE Int. Conf. Patt. Recog.* (Oct. 1994), 743–46, Jerusalem, Israel.

[19] Jain, J. R., and A. K. Jain. Displacement measurement and its application in interframe image coding. *IEEE Trans. Commun.* (Dec. 1981), COM-29:1799–808.

[20] Koga, T., et al. Motion-compensated interframe coding for video conferencing. In *Nat. Telecommun. Conf.* (Nov. 1981), G5.3.1–5, New Orleans, LA.

[21] Komarek, T., and P. Pirsch. Array architecture for block matching algorithms. *IEEE Trans. Circuits and Systems* (Oct. 1989), 36:269–77.

[22] Konrad, J., and E. Dubois. Bayesian estimation of motion vector fields. *IEEE Trans. Pattern Anal. Machine Intell.* (Sept. 1992), 14:910–27.

[23] Kuglin, C., and D. Hines. The phase correlation image alignment method. In *IEEE Int. Conf. Cybern. Soc.* (1975), 163–65.

[24] Lee, O., and Y. Wang. Motion compensated prediction using nodal based deformable block matching. *Journal of Visual Communications and Image Representation* (March 1995), 6:26–34.

[25] Lee, X., and Y.-Q. Zhang. A fast hierarchical motion compensation scheme for video coding using block feature matching. *IEEE Trans. Circuits Syst. for Video Technology* (Dec. 1996), 6:627–35.

[26] Mitiche, A., Y. F. Wang, and J. K. Aggarwal. Experiments in computing optical flow with gradient-based, multiconstraint method. *Pattern Recognition* (June 1987), 20(2):173–79.

[27] Musmann, H. G., M. Hotter, and J. Ostermann. Object oriented analysis-synthesis coding of moving images. *Signal Processing: Image Commun.* (Oct. 1989), 1:119–38.

[28] Musmann, H. G., P. Pirsch, and H.-J. Grallert. Advances in picture coding. *Proceedings of the IEEE* (April 1985), 73(4):523–48.

[29] Nagel, H. H. Displacement vectors derived from second-order intensity variations in images sequences. *Computer Graphics and Image Processing* (1983), 21:85–117.

[30] Nagel, H. H., and W. Enklemann. An investigation of smoothness constraints for the estimation of displacement vector fields from image sequences. *IEEE Trans. Pattern Anal. Machine Intell.* (Sept. 1986), 8:565–93.

[31] Netravali, A. N., and J. D. Robbins. Motion-compensated coding: some new results. *Bell System Technical Journal* (Nov. 1980), 59(9):1735–45.

[32] Pirsch, P., N. Demassieux, and W. Gehrke. VLSI architecture for video compression—a survey. *IEEE* (Feb. 1995), 83:220–46.

[33] Reddy, B. S., and B. N. Chatterji. An FFT-based technique for translation, rotation, and scale-invariant image registration. *IEEE Trans. Image Process.* (Aug. 1996), 5(8):1266–71.

[34] Barron, A., J. Rissanen, and Y. Bin. The minimum description length principle in coding and modeling. *IEEE Trans. Information Theory* (Oct. 1998), 44(6):2743–60.

[35] Rousseeuw, P. J., and A. M. Leroy. *Robust Regression and Outlier Detection*. New York: John Wiley & Sons, 1987.

[36] Seferidis, V., and M. Ghanbari. General approach to block matching motion estimation. *Optical Engineering* (July 1993), 32:1464–74.

[37] Shekarforoush, H., M. Berthod, and J. Zerubia. Subpixel image registration by estimating the polyphase decomposition of cross power spectrum. In *IEEE Computer Society Conference on Computer Vision and Pattern Recognition* (1996), 532–37.

[38] Stiller, C., and J. Konrad. Estimating motion in image sequences. *IEEE Signal Processing Magazine* (July 1999), 16:70–91.

[39] Tekalp, A. M., *Digital Video Processing*. Upper Saddle River, NJ: Prentice Hall PTR, 1995.

[40] Thomas, G. A. Television motion measurements for DATV and other applications. Research report 1987/11 (September), British Broadcasting Corporation.

[41] Wang, Y., and O. Lee. Use of 2-D deformable mesh structures for video compression. Part I—the synthesis problem: Mesh based function approximation and mapping. *IEEE Trans. Circuits Syst. for Video Technology* (Dec. 1996), 6:636–46.

[42] Wang, Y., and O. Lee. Active mesh—a feature seeking and tracking image sequence representation scheme. *IEEE Trans. Image Process.* (Sept. 1994), 3:610–24.

[43] Wang, Y., and J. Ostermann. Evaluation of mesh-based motion estimation in H.263 like coders. *IEEE Trans. Circuits Syst. for Video Technology* (June 1998), 8:243–52.

[44] Wang, Y., X.-M. Hsieh, J.-H. Hu, and O. Lee. Region segmentation based on active mesh representation of motion: comparison of parallel and sequential approaches. In *IEEE International Conference on Image Processing (ICIP'95)* (Oct. 1995), 185–88, Washington, DC.

[45] Zienkewicz, O. C., and R. L. Taylor. *The Finite Element Method*, Vol. 1. 4th ed. Upper Saddle River, NJ: Prentice Hall, 1989.

7

Three-Dimensional Motion Estimation

Three-dimensional motion estimation allows for the description of the motion of an object in 3-D space—the motion is estimated from images of the moving object. It is used in computer vision applications for object tracking (as in vehicle guidance and robotics) as well as in object-based video coding for object tracking and motion compensation. In some applications, a moving camera moves through a static scene; in others, a static camera observes moving objects. For some applications, such as surveillance from an airplane, both camera and objects move.

As with 2-D motion estimation, 3-D motion estimation is an ill-posed problem. In order to ensure useful solutions, we make simplifying assumptions—such as that of rigid objects, usually of known shape. The task is to estimate the six motion parameters of an object given in Equation (5.3.9). Most algorithms assume that an object shape is planar or parabolic, if the shape of the object is not known.

Both orthographic and perspective projections can be used as camera models for 3-D motion estimation. Naturally, the use of orthographic projection simplifies the estimation process. However, orthographic projection does not allow us to observe depth, and causes estimation errors if the object moves along the optical axes of the camera or rotates around an axis other than the optical axis.

Another classification of 3-D motion estimation algorithms distinguishes between direct methods, which use the image signal as their input, and indirect methods, which rely on feature correspondences that have been established with a separate algorithm (like block- or feature-matching). In Section 7.1, we discuss algorithms that estimate motion from feature correspondences. The precision of the estimated motion parameters depends to a large extent on the precision of the feature correspondences. Given these correspondences, the algorithms can easily estimate large motion. Since motion is not estimated directly from the image signal but from features derived from the image signal, we call these algorithms *indirect motion estimation* methods. In Section 7.2, we present *direct motion estimation methods*. They have the advantage of not relying on feature correspondences. However, these algorithms can only estimate small motion. Therefore, they are used in an iterative estimation loop.

7.1 FEATURE-BASED MOTION ESTIMATION

Feature-based motion estimation assumes that identical features on an object have been identified in two images. In order to allow for reliable feature detection, these features usually correspond to physical landmarks on the object for which motion must be estimated. Features can be described as points, lines, or corners (connection between lines) in the image plane. In the image plane, displacement vectors may be used to establish the correspondences between feature points. Feature correspondences may be identified using simple block matching. More reliable results can be achieved if we limit the matching to image areas with high spatial gradients. Therefore, we first extract edges in the two images and then try to match the edges. In this section, we will limit our discussion to algorithms that rely on feature correspondences. For a discussion of line and corner correspondences, see [2], or the overview in [11].

In the following sections, we will discuss four algorithms for feature-based motion estimation. Each algorithm assumes rigid 3-D objects. The first (Section 7.1.1) assumes a camera model with orthographic projection and known object shape. The second assumes a camera model with perspective projection and known shape (Section 7.1.2). The third (Section 7.1.3) is suitable for objects that can be approximated by planar surfaces. Finally, we present an algorithm that is based on perspective projection and does not require any knowledge about object shape (Section 7.1.4).

7.1.1 Objects of Known Shape under Orthographic Projection

In this section, we assume that we know the shape of the moving object. Therefore, we know for each point $\mathbf{x} = [x, y]^T$ in image k the 3-D coordinates $\mathbf{X} = [X, Y, Z]^T$ of the object surface point associated with $\mathbf{x}$. We use a camera model with orthographic projection, according to Figure 5.2(b) and Equation (5.1.3). With this camera model and Equation (5.3.9) for 3-D motion, the mapping of a point $\mathbf{X}$ onto the image plane before and after motion becomes

$$\mathbf{x}' = \begin{pmatrix} x' \\ y' \end{pmatrix} = \begin{pmatrix} r_1 x + r_2 y + (r_3 Z + T_x) \\ r_4 x + r_5 y + (r_6 Z + T_y) \end{pmatrix} \qquad (7.1.1)$$

with the rotation matrix

$$[\mathbf{R}] = \begin{bmatrix} r_1 & r_2 & r_3 \\ r_4 & r_5 & r_6 \\ r_7 & r_8 & r_9 \end{bmatrix} \qquad (7.1.2)$$

according to Equation (5.3.6) and the translation vector $\mathbf{T} = [T_x, T_y, T_z]^T$. This is an affine mapping of point $\mathbf{x}$ in image k onto image point $\mathbf{x}'$ in image $k+1$. Linearizing the elements r_i of the rotation matrix according to Equation (5.3.6) to Equation (5.3.10), we have an affine relation between $\mathbf{x}$ and $\mathbf{x}'$:

$$\begin{pmatrix} x' \\ y' \end{pmatrix} = \begin{pmatrix} x - \theta_z y + (\theta_y Z + T_x) \\ \theta_z x + y - (\theta_x Z + T_y) \end{pmatrix}. \tag{7.1.3}$$

Given at least three point correspondences, the solution for the five unknown motion parameters θ_x, θ_y, θ_z, and T_x, T_y can be estimated using the least-squares method. In order to increase the reliability of the estimated motion parameters, it is recommended to use at least six correspondences.

Equation (7.1.3) uses the linearized rotation matrix. Therefore, the estimated motion parameters can only be approximations of the real motion parameters, and the algorithm must be used in an iterative fashion in order to estimate large motion parameters. With each iteration, the estimated motion parameters decrease; at the same time, the error due to the use of the linearized rotation matrix decreases. The iteration should stop as soon as the motion parameter updates approach zero. We present a more detailed example of iterative motion estimation in Section 7.3.

In Equation (7.1.3), the motion parameters θ_x and θ_y are multiplied by the distance Z between the camera center and the object surface point $\mathbf{X}$. Hence, θ_x and θ_y can be estimated accurately only if the object shape and hence Z is known precisely. Since the object shape (or its depth) is not always known, several proposals for estimating the object shape in conjunction with this motion estimation method have been proposed. In [1], a two-step approach for each iteration is proposed that estimates the motion parameters and then updates the depth estimates of the object. In [4], a further improvement is suggested, that of updating the depth estimates Z according to the gradient of an error function. If reasonable depth estimates are not available, Equation (7.1.3) can be used for objects that move translationally mainly parallel to the image plane and rotate mainly around the camera axis. That is, we assume $\theta_x = \theta_y = 0$ and estimate θ_z, T_x, and T_y only.

7.1.2 Objects of Known Shape under Perspective Projection

In order to estimate the arbitrary motion of a 3-D object with arbitrary but known shape, we assume that a known point $\mathbf{X}$ moves to the unknown location $\mathbf{X}'$. Projecting $\mathbf{X}'$ onto the image plane using a camera model with perspective projection in Equation (5.1.2) gives

$$\mathbf{x}' = \begin{bmatrix} x' \\ y' \end{bmatrix} = F \cdot \frac{1}{Z'} \begin{bmatrix} X' \\ Y' \end{bmatrix}. \tag{7.1.4}$$

If we now replace $\mathbf{X}'$ with $\mathbf{X}$ using the 3-D motion equation (Equation (5.3.9)), we get

$$\mathbf{x}' = F \cdot \begin{bmatrix} \dfrac{r_1 X + r_2 Y + r_3 Z + T_x}{r_7 X + r_8 Y + r_9 Z + T_z} \\ \dfrac{r_4 X + r_5 Y + r_6 Z + T_y}{r_7 X + r_8 Y + r_9 Z + T_z} \end{bmatrix}. \tag{7.1.5}$$

The problem is to solve for the motion parameters θ_x, θ_y, θ_z, T_x, T_y, T_z, assuming that $\mathbf{x}$, $\mathbf{x}'$, and $\mathbf{X}$ are known. In order to solve this equation using a system of linear equations, we assume small rotation angles and use the linearized rotation matrix according to Equation (5.3.10). This yields

$$\mathbf{x}' = F \cdot \begin{bmatrix} X - \theta_z Y + \theta_y Z + T_x \\ -\theta_y X + \theta_x Y + Z + T_z \\ \theta_z X + Y - \theta_x Z + T_y \\ -\theta_y X + \theta_x Y + Z + T_z \end{bmatrix}. \tag{7.1.6}$$

After further computation, we get

$$\begin{bmatrix} x' \cdot \frac{Z}{F} - X \\ y' \cdot \frac{Z}{F} - Y \end{bmatrix} = \begin{bmatrix} -Y\frac{x'}{F} & Z + X\frac{x'}{F} & -Y & 1 & 0 & -\frac{x'}{F} \\ -Z - Y\frac{y'}{F} & X\frac{y'}{F} & X & 0 & 1 & -\frac{y'}{F} \end{bmatrix} \cdot \begin{bmatrix} \theta_x \\ \theta_y \\ \theta_z \\ T_x \\ T_y \\ T_z \end{bmatrix} \tag{7.1.7}$$

where the points $\mathbf{x}'$ and $\mathbf{X}$ are known and the six motion parameters are unknown. As before, each point correspondence provides two equations. In order to allow for inaccuracies in the feature point estimation, we should solve Equation (7.1.7) for more than three points using a least-squares algorithm. Due to the use of the linearized rotation matrix, we must estimate the motion parameters iteratively.

7.1.3 Planar Objects

Estimation of the motion parameters of a plane moving arbitrarily in space is an important problem. Often, we can approximate an object surface with one or more planes. For example, if we approximate an object surface using a wireframe, each triangle on the wireframe represents a planar patch. This algorithm does not assume any knowledge of the orientation of the plane in space. Therefore, we will estimate eight parameters that determine the plane orientation and motion.

Assuming the camera model with perspective projection according to Equation (5.1.2), rigid objects moving according to Equation (5.3.9), and the object plane

$$aX + bY + cZ = 1, \tag{7.1.8}$$

we can describe the mapping of point $\mathbf{x}$ from image k to image $k + 1$ as a projective mapping as given in Equation (5.5.14) [17, 5]:

$$[x', y']^T = \left[\frac{a_0 + a_1 x + a_2 y}{1 + c_1 x + c_2 y}, \frac{b_0 + b_1 x + b_2 y}{1 + c_1 x + c_2 y} \right]^T \tag{7.1.9}$$

with the eight unknown motion and structure parameters a_i, b_i, c_i (sometimes referred to as *pure parameters* [18]).

We can solve for the pure parameters by having at least four point correspondences. If we limit ourselves to four point correspondences, no three of the four points are allowed to be colinear [6]. In order to increase the reliability of the results, we suggest using more than eight correspondences, and using a least-squares method to solve for the pure parameters.

In [20], Tsai and Huang estimated the 3-D motion parameters and the orientation of the object plane from the eight pure parameters using singular value decomposition (SVD). This is only useful if the pure parameters can be estimated reliably. Please note that this algorithm for estimating the pure parameters cannot easily be used for object tracking. After estimation of the parameters from two images, the object orientation is fixed. Therefore, there are only six parameters to be estimated for the next image. If we estimate again eight pure parameters, as done in [5], we allow the image plane to have two different orientations for one image frame. Normally, this algorithm is used for backward motion estimation only; that is, for any small region in frame $k + 1$, we assume that it corresponds to a planar patch in 3-D but with unknown plane orientation, and estimate its motion from frame $k+1$ to frame k together with the plane orientation. This does not allow us to continuously track the same 3-D patch over multiple frames.

7.1.4 Objects of Unknown Shape Using the Epipolar Line

In this section, we describe a motion estimation algorithm that allows us to estimate 3-D motion and shape without any knowledge of the object shape [8]. Without loss of generality, we set the camera focal length F to unity. We assume a rigid object and perspective projection. Starting with Equation (7.1.5), and replacing $\mathbf{X}$ with its projection $\mathbf{x}$ onto the image plane and $F = 1$, we get

$$\mathbf{x}' = \begin{pmatrix} \dfrac{r_1 x + r_2 y + r_3 + T_x/Z}{r_7 x + r_8 y + r_9 + T_z/Z} \\[2ex] \dfrac{r_4 x + r_5 y + r_6 + T_y/Z}{r_7 x + r_8 y + r_9 + T_z/Z} \end{pmatrix}. \tag{7.1.10}$$

Note that this equation is the same as that presented in Equation (5.5.13) with $F = 1$. This equation does not change if we multiply $\mathbf{T}$ and Z by a constant value. Therefore, we will only be able to determine the direction of $\mathbf{T}$ but not its absolute length. Its length depends on the object size and the distance of the object from the camera.

In Section 7.1.2, we solved this nonlinear equation using iterative methods and linearizations, assuming that the object shape was known. Using an intermediate matrix allows for estimating the motion without knowing the shape [8, 19]. This intermediate matrix is known as the *E-matrix* or *essential matrix* with its nine *essential parameters*. Eliminating Z in Equation (7.1.10), we can determine the relationship between $\mathbf{x}$ and $\mathbf{x}'$ as

$$[x', y', 1] \cdot [\mathbf{E}] \cdot [x, y, 1]^T = 0, \tag{7.1.11}$$

with

$$[\mathbf{E}] = \begin{bmatrix} e_1 & e_2 & e_3 \\ e_4 & e_5 & e_6 \\ e_7 & e_8 & e_9 \end{bmatrix} = [\mathbf{T}][\mathbf{R}]. \tag{7.1.12}$$

The matrix $[\mathbf{T}]$ is defined as

$$[\mathbf{T}] = \begin{bmatrix} 0 & T_z & -T_y \\ -T_z & 0 & T_x \\ T_y & -T_x & 0 \end{bmatrix} \qquad (7.1.13)$$

and the rotation matrix $[\mathbf{R}]$ is defined according to Equation (5.3.6). Multiplying Equation (7.1.11) with $Z \cdot Z'$, we get

$$[X', Y', Z'] \cdot [\mathbf{E}] \cdot [X, Y, Z]^T = 0. \qquad (7.1.14)$$

Equation (7.1.14) can only be used for motion estimation for $\mathbf{T} \neq 0$; that is, if the object rotates only, we cannot estimate its motion. Therefore we assume that the object of interest also moves translationally.

Epipolar Line   Equation (7.1.11) defines a linear dependency between the corresponding image points $\mathbf{x}$ and $\mathbf{x}'$. Consequently, the possible positions $\mathbf{x}'$ of a point $\mathbf{x}$ after motion lie on a straight line. This line is called the *epipolar line* and is defined by the motion parameters, according to

$$a(\mathbf{x}, [\mathbf{E}])x' + b(\mathbf{x}, [\mathbf{E}])y' + c(\mathbf{x}, [\mathbf{E}]) = 0 \qquad (7.1.15)$$

with

$$\begin{aligned} a(\mathbf{x}, [\mathbf{E}]) &= e_1 x + e_2 y + e_3, \\ b(\mathbf{x}, [\mathbf{E}]) &= e_4 x + e_5 y + e_6, \\ c(\mathbf{x}, [\mathbf{E}]) &= e_7 x + e_8 y + e_9. \end{aligned} \qquad (7.1.16)$$

Figure 7.1 shows the epipolar line for a point $\mathbf{x}$ with rotation $[\mathbf{R}]$ and translation $\mathbf{T}$.

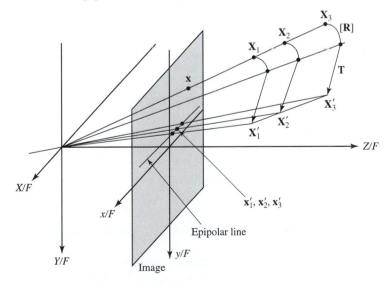

Figure 7.1   The epipolar line for a point $\mathbf{x}$ as defined by object motion.

Motion estimation is carried out in two steps. First the E-matrix is estimated, then it is decomposed into the rotation matrix and the translation vector.

E-Matrix Estimation   It is obvious that the E-matrix can only differ from zero for $\mathbf{T} \neq 0$. With Equation (7.1.11), we define one equation for each point correspondence. Since Equation (7.1.11) is a homogeneous equation, we can set an arbitrary parameter of the matrix to one; for instance,

$$e_9 = 1. \tag{7.1.17}$$

With this constraint, we only need a minimum of eight equations in order to estimate the E-matrix. For each measured point correspondence between $\mathbf{x}^{(j)}$ and $\mathbf{x}'^{(j)}$, we can set up the following equation according to Equation (7.1.11):

$$\mathbf{a}_j^T \mathbf{e}' = r_j \tag{7.1.18}$$

with $\mathbf{a}_j = [x_j' x_j, x_j' y_j, x_j', y_j' x_j, y_j' y_j, y_j', x_j, y_j, 1]^T$ and $\mathbf{e}' = [e_1, e_2, e_3, e_4, e_5, e_6, e_7, e_8, 1]^T$. All the point correspondences will lead to the following system of equations:

$$[\mathbf{A}]\mathbf{e}' = \mathbf{r} \tag{7.1.19}$$

with

$$[\mathbf{A}] = [\mathbf{a}_1, \mathbf{a}_2, \ldots, \mathbf{a}_J]^T. \tag{7.1.20}$$

We can then solve the overdetermined system in Equation (7.1.19) using a least-squares method minimizing $\|\mathbf{r}\|^2$.

It can be shown that setting one of the components of the E-matrix to a constant value as in Equation (7.1.17) may not result in the minimal achievable residual when solving for the E-matrix. An alternative approach is to require the vector $\mathbf{e} = [e_1, e_2, e_3, e_4, e_5, e_6, e_7, e_8, e_9]^T$ to have unit norm, that is,

$$\|\mathbf{e}\| = 1. \tag{7.1.21}$$

Then, we can solve for the E-matrix using a constrained minimization problem

$$\|[\mathbf{A}]\mathbf{e}\| \xrightarrow[\mathbf{e}]{} \min \qquad \text{subject to } \|\mathbf{e}\| = 1. \tag{7.1.22}$$

This will result in the minimal residual and a better solution for the E-matrix. The solution is the unit eigenvector of $[\mathbf{A}]^T[\mathbf{A}]$ associated with the smallest eigenvalue. In order to avoid numerical instabilities, eigenvalues and eigenvectors of $[\mathbf{A}]$ can be computed using SVD according to

$$[\mathbf{A}] = [\mathbf{U}][\mathbf{\Sigma}][\mathbf{V}]^T. \tag{7.1.23}$$

The matrix $[\mathbf{\Sigma}]$ contains only the positive square roots of the nonnegative eigenvalues of $[\mathbf{A}]^T[\mathbf{A}]$ [14].

The influence of an error in a measured point correspondence $(\mathbf{x}, \mathbf{x}')$ depends on the actual image coordinates $\mathbf{x}$ (see Problem 7.6). Errors in correspondences close to the image center do not influence the solution as much as those at the image boundaries.

Estimation of Rotation and Translation   Since the E-matrix is defined to describe the 3-D motion of a rigid object, it has five degrees of freedom, considering that we can only recover the orientation of the translation vector. However, during the estimation of the E-matrix we allow for eight degrees of freedom. Therefore, the extraction of rotation and translation from the E-matrix requires the solution of an optimization problem.

Following [21], we first estimate the translation vector. With $[\mathbf{T}]$ as defined in Equation (7.1.13), we have

$$[\mathbf{T}]^T = -[\mathbf{T}] \quad \text{and} \quad [\mathbf{T}]\mathbf{T} = \mathbf{0}. \tag{7.1.24}$$

Multiplying Equation (7.1.12) by $\mathbf{T}$, we get

$$[\mathbf{E}]^T \cdot \mathbf{T} = [\mathbf{R}]^T[\mathbf{T}]^T \cdot \mathbf{T} = -[\mathbf{R}]^T[\mathbf{T}] \cdot \mathbf{T} = 0. \tag{7.1.25}$$

Hence, all row vectors of $[\mathbf{E}]$ are linearly dependent ($\det([\mathbf{E}]) = 0$), and $\mathbf{T}$ is orthogonal to the plane defined by the column vectors of $[\mathbf{E}]$. If the E-matrix is estimated according to Equation (7.1.22) for noisy point correspondences, we cannot assume Equation (7.1.25) to hold exactly. Therefore, we estimate the translation vector with

$$\|[\mathbf{E}]^T\mathbf{T}\| \xrightarrow[\mathbf{T}]{} \min \quad \text{subject to } \|\mathbf{T}\| = 1, \tag{7.1.26}$$

preferably using SVD. Similar to the solution for $\mathbf{e}$ in Equation (7.1.22), $\mathbf{T}$ is the unit eigenvector of $[\mathbf{E}][\mathbf{E}]^T$ with minimal eigenvalue. From Equation (7.1.25), it is clear that the sign of $\mathbf{T}$ cannot be determined. Similarly, the solution for $[\mathbf{E}]$ is not unique. Given an estimated E-matrix $[\hat{\mathbf{E}}]$, we know that $-[\hat{\mathbf{E}}]$ is also a valid solution. In [22], it is shown that the correct sign for $\mathbf{T}$ can be determined by selecting $\mathbf{T}$ such that the following condition holds for all $\mathbf{x}$:

$$\sum_{\mathbf{x}}[\mathbf{T}[x', y', 1]]^T \cdot [\mathbf{E}] \cdot [x, y, 1]^T > 0. \tag{7.1.27}$$

This equation assures that $\mathbf{X}$ and $\mathbf{X}'$ are located in front of the camera before and after motion.

In a second step, we estimate the rotation matrix $[\mathbf{R}]$. Equation (7.1.12) can also be written as

$$[\mathbf{E}]^T = [\mathbf{R}]^T[\mathbf{T}]^T = -[\mathbf{R}]^T[\mathbf{T}]. \tag{7.1.28}$$

Therefore, $[\mathbf{R}]$ can be found by the following optimization problem

$$\|[\mathbf{R}]^T[-\mathbf{T}] - \mathbf{E}^T\| \xrightarrow[[\mathbf{R}]]{} \min \text{ subject to } [\mathbf{R}]^T = [\mathbf{R}]^{-1} \quad \text{and} \quad \det([\mathbf{R}]) = 1. \tag{7.1.29}$$

The constraints are valid for any rotation matrix $[\mathbf{R}]$. Equation (7.1.29) is a special case of the general problem

$$\|[\mathbf{F}][\mathbf{C}] - [\mathbf{D}]\| \xrightarrow[[\mathbf{F}]]{} \min \tag{7.1.30}$$

where $[\mathbf{F}]$ must satisfy the properties of a rotation matrix, and $[\mathbf{C}]$ and $[\mathbf{D}]$ are arbitrary

3×3 matrices. The solution is computed using the 4×4 matrix $[\mathbf{B}]$ [21]:

$$[\mathbf{B}] = \sum_{i=1}^{3} [\mathbf{B}_i]^T [\mathbf{B}_i] \tag{7.1.31}$$

with

$$[\mathbf{B}_i] = \begin{bmatrix} 0 & c_{1i} - d_{1i} & c_{2i} - d_{2i} & c_{3i} - d_{3i} \\ d_{1i} - c_{1i} & 0 & -(c_{3i} + d_{3i}) & c_{2i} + d_{2i} \\ d_{2i} - c_{3i} & c_{3i} + d_{3i} & 0 & -(c_{1i} + d_{1i}) \\ d_{3i} - c_{3i} & -(c_{2i} + d_{2i}) & c_{1i} + d_{1i} & 0 \end{bmatrix} \tag{7.1.32}$$

where c_{ji}, d_{ji} are elements of $[\mathbf{C}]$ and $[\mathbf{D}]$. First, we determine the unit eigenvector $\mathbf{g} = [g_1, g_2, g_3, g_4]^T$ of $[\mathbf{B}]$ with the smallest eigenvalue, preferably using SVD. Then the matrix $[\mathbf{F}]$ can be obtained by

$$[\mathbf{F}] = \begin{bmatrix} g_1^2 + g_2^2 + g_3^2 + g_4^2 & 2(g_2 g_3 - g_1 g_4) & 2(g_2 g_4 + g_1 g_3) \\ 2(g_2 g_3 + g_1 g_4) & g_1^2 - g_2^2 + g_3^2 - g_4^2 & 2(g_3 g_4 - g_1 g_2) \\ 2(g_2 g_4 - g_1 g_3) & 2(g_3 g_4 + g_1 g_2) & g_1^2 - g_2^2 - g_3^2 + g_4^2 \end{bmatrix}. \tag{7.1.33}$$

Setting $[\mathbf{C}] = -[\mathbf{T}]$, $[\mathbf{D}] = [\mathbf{E}]^T$, then the rotation matrix is related to the preceding solution by $[\mathbf{R}] = [\mathbf{F}]^T$.

Estimation Accuracy   In this section, we investigate how well the estimated parameters represent the feature point correspondences. A suitable measure for the accuracy of the derived motion parameters is to compute the epipolar line for each feature point $\mathbf{x}$ according to Equation (7.1.15), using the estimated rotation and translation parameters. Then we measure the distance d^2 between $\mathbf{x}'$ and its epipolar line:

$$d^2 = \frac{(\mathbf{x}'[\mathbf{E}]\mathbf{x})^2}{a(\mathbf{x}, [\mathbf{E}])^2 + b(\mathbf{x}, [\mathbf{E}])^2} \tag{7.1.34}$$

The average distance $\bar{d}$ indicates how well the E-matrix approximates the given correspondences. We have two possibilities for computing $\bar{d}$. We can use the estimated E-matrix according to Equation (7.1.22), or we can use an E-matrix $[\hat{\mathbf{E}}] = [\mathbf{T}][\mathbf{R}]$, which we compute from the estimated translation and rotation parameters according to Equations (7.1.26) and (7.1.29), which are in turn calculated from the estimated E-matrix. It is important to note that $\bar{d}$ computed using the second option is often an order of magnitude higher than $\bar{d}$ computed using the first option. This indicates that motion estimation using the E-matrix is very sensitive to noise in the correspondences. Advanced algorithms use Equation (7.1.22) as a starting point for an iterative algorithm that estimates the E-matrix, subject to the constraints that $[\mathbf{E}]$ is defined by $[\mathbf{R}]$ and $[\mathbf{T}]$ through Equations (7.1.12) and (7.1.13).

Estimation of Object Shape   Knowing the corresponding points $\mathbf{x}$ and $\mathbf{x}'$ as well as the motion $[\mathbf{R}]$ and $\mathbf{T}$, we can compute Z for a point according to Equa-

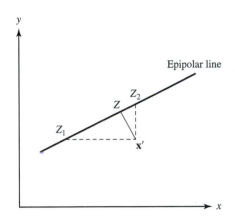

Figure 7.2 The projection of a point $\mathbf{x}'$ onto the epipolar line at $\mathbf{z}$ determines the location of the corresponding object point in the 3-D space. Z_1 and Z_2 are computed using Equations (7.1.35) and (7.1.36).

tion (7.1.10) in two ways:

$$Z = \frac{T_y - y' T_z}{(xr_7 + yr_7 + r_9)y' - (xr_4 + yr_5 + r_6)},\tag{7.1.35}$$

or

$$Z = \frac{T_x - x' T_z}{(xr_7 + yr_8 + r_9)x' - (xr_1 + yr_2 + r_3)}.\tag{7.1.36}$$

If $\mathbf{x}'$ is not located on the epipolar line, these two equations give different estimates for Z. Figure 7.2 shows how Equations (7.1.35) and (7.1.36) project $\mathbf{x}'$ onto the epipolar line. A better approach to determining Z is to choose Z at the orthogonal projection of $\mathbf{x}'$ onto the epipolar line. With Z known, X and Y can be determined from $\mathbf{x}$ using inverse perspective projection.

7.2 DIRECT MOTION ESTIMATION

In the previous section, we assumed a small number of accurate feature point correspondences for features that are visible in two images. Since such accurate feature point correspondences are not always available, we now introduce motion estimation algorithms that use the image signal and its gradients directly. Since the image signal tends to be noisy, we expect to use a rather large number of image points for determining 3-D motion. We call these points *observation points*. We do not know the position of an observation point in a previous or following image. If the assumptions that we make with respect to object shape, illumination, and camera model hold, these algorithms can measure large motion. An obvious consequence of the assumptions is that the projections of an object in two consecutive images must overlap. Therefore, the motion estimation range of direct motion estimation algorithms may be smaller than that of indirect motion estimation algorithms.

We will start by deriving the basic relationship between images and motion (Section 7.2.1). Then, we will present an algorithm similar to that in Section 7.1.2 that allows us to estimate the 3-D motion of a 3-D object with known shape without feature points (Section 7.2.2). Furthermore, we will apply this knowledge to extend the algorithm of Section 7.1.3 to work without feature points (Section 7.2.3). In Section 7.2.4, we present several methods by which the robustness of these estimation algorithms with respect to estimation accuracy can be increased.

7.2.1 Image Signal Models and Motion

In order to derive a direct motion estimation algorithm, we assume that differences between two consecutive luminance images $\psi_k(\mathbf{x})$ and $\psi_{k+1}(\mathbf{x})$ are due to object motion only. This implies that objects have diffuse reflecting surfaces and that the scene is illuminated by ambient light sources (see Section 5.2). (In [16] and [4], illumination effects are considered as well.) We now relate the frame difference between the two video frames $\psi_k(\mathbf{x})$ and $\psi_{k+1}(\mathbf{x})$ to motion.

First, we need an analytical description of the image signal. Using the first order Taylor expansion of $\psi(\mathbf{x} + \Delta\mathbf{x})$ at $\mathbf{x}$, we obtain

$$\psi(\mathbf{x} + \Delta\mathbf{x}) \approx \psi(\mathbf{x}) + \frac{\partial \psi}{\partial x}\Delta x + \frac{\partial \psi}{\partial y}\Delta y. \tag{7.2.1}$$

Obviously, this linear signal model is only a crude approximation of a real image. Furthermore, the approximation is only valid for small $\Delta\mathbf{x}$. In Appendix A, we describe several methods for computing the image gradient $\mathbf{g}(\mathbf{x}) = [\partial\psi(\mathbf{x})/\partial x, \partial(\mathbf{x})\psi/\partial y]^T$ at a location $\mathbf{x}$ in an image.

Let us consider an observation point on an object surface, which is located at $\mathbf{X}$ at time instant k and at $\mathbf{X}'$ at time instant $k + 1$. With $\mathbf{X}$ projected onto $\mathbf{x}$ in image ψ_k, and the same point $\mathbf{X}'$ projected onto $\mathbf{x}'$ in image ψ_{k+1} after motion, the luminance difference between images k and $k + 1$ at position $\mathbf{x}$ is

$$\begin{aligned}\Delta\psi_k(\mathbf{x}) &= \psi_{k+1}(\mathbf{x}) - \psi_k(\mathbf{x}) \\ &= \psi_{k+1}(\mathbf{x}) - \psi_{k+1}(\mathbf{x}').\end{aligned} \tag{7.2.2}$$

Equation (7.2.2) holds because we assume that all changes are due to object motion according to the constant intensity assumption $\psi_{k+1}(\mathbf{x}') = \psi_k(\mathbf{x})$ (Equation (5.2.11)). With the linear signal model of Equation (7.2.1) and $\Delta x = \mathbf{x}' - \mathbf{x}$, we can rewrite Equation (7.2.2) as

$$\Delta\psi_k(\mathbf{x}) = -\mathbf{g}_{k+1}(\mathbf{x}) \cdot (\mathbf{x}' - \mathbf{x}), \tag{7.2.3}$$

where "$\cdot$" denotes the vector inner product. Due to the linear signal model, $\mathbf{g}_{k+1}(\mathbf{x}) = \mathbf{g}_k(\mathbf{x})$ holds. Therefore, Equation (7.2.3) can also be represented as

$$\Delta\psi_k(\mathbf{x}) = -\mathbf{g}_k(\mathbf{x}) \cdot (\mathbf{x}' - \mathbf{x}). \tag{7.2.4}$$

Note that this equation is equivalent to the discrete version of the optical flow equation given in Equation (6.2.3).

Equation (7.2.4) gives us a relationship between an observation point $\mathbf{x}$ in image k and its displaced position $\mathbf{x}'$ in image $k + 1$. Figure 7.3 illustrates this relationship between image gradient, frame difference, and 2-D motion. For Figure 7.3(a), Equation (7.2.4) holds exactly, because the image signals ψ_k and ψ_{k+1} are linear. In this

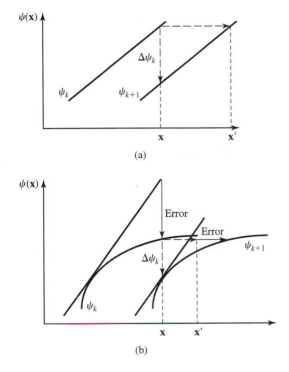

$\psi(\mathbf{x})$

ψ_k

ψ_{k+1}

$\Delta\psi_k$

$\mathbf{x}$ $\mathbf{x}'$

(a)

$\psi(\mathbf{x})$

Error

Error

ψ_{k+1}

$\Delta\psi_k$

ψ_k

$\mathbf{x}$ $\mathbf{x}'$

(b)

Figure 7.3   Two examples showing how Equation (7.2.4) relates the image signal to 2-D motion of an observation point: (a) no approximation error, because the linear signal model holds; (b) since the linear signal model cannot approximate the image signal very well, we measure a luminance difference that is smaller than required. As a result, we estimate a displacement that is too small.

example, the luminance difference $\Delta\psi_k(\mathbf{x})$ is negative, and the image gradient $\mathbf{g}_k(\mathbf{x})$ is positive, resulting in a positive displacement $(\mathbf{x}' - \mathbf{x}) > 0$. The situation in Figure 7.3(b) is similar; however, the image signal is no longer linear. In this example, the measured frame difference $\Delta\psi_k(\mathbf{x})$ is smaller than the linear signal model would suggest. Hence, we will estimate a displacement with the correct sign but a too small amplitude. Due to this effect, we will have to estimate motion iteratively.

Looking at Figure 7.3(b), we can imagine that we would measure a luminance difference with an unpredictable sign if the image signal contains very high frequencies. In this case, the displacement $(\mathbf{x}' - \mathbf{x})$ may be larger than the period of the image signal around $\mathbf{x}$. This may result in estimating displacements with an incorrect sign. Therefore, we recommend the use of a low-pass filtered version of the image for the first couple of iterations.

The motion estimation algorithms described in the following two sections are based on an image signal model, in order to derive a relationship between frame differences and 3-D motion. They relate the motion of a point $\mathbf{x}$ to the motion in 3-D space using the camera model, motion equation, and object shape. Since Equation (7.2.4) is based on a linear approximation of the image signal, the estimation algorithm will be used within an iterative estimation process until the correct motion is estimated. In order to speed up the convergence of the algorithms, a higher-order approximation of the image signal is desirable. According to [3], we can approximate the image signal by a quadratic image signal model, using a Taylor series expansion of second order without explicitly computing the second derivatives. Using this model, the frame difference is approximated by

$$\Delta \psi_k(\mathbf{x}) = -\bar{\mathbf{g}}(\mathbf{x}) \cdot (\mathbf{x}' - \mathbf{x}) \tag{7.2.5}$$

with

$$\bar{\mathbf{g}}(\mathbf{x}) = \frac{1}{2}(\mathbf{g}_k(\mathbf{x}) + \mathbf{g}_{k+1}(\mathbf{x})). \tag{7.2.6}$$

The averaging gradient $\bar{\mathbf{g}}(\mathbf{x})$ takes into account both the first-order and second-order gradients of $\psi_{k+1}(\mathbf{x})$. Since this quadratic image signal model enables a better image signal approximation, yields better motion estimates, and is simple to compute, we recommend this model whenever appropriate.

7.2.2 Objects of Known Shape

Here we extend the algorithm of Section 7.1.2 to work without feature points. During motion estimation, we use a large set of observation points to represent the object. Since we assume that we know the object shape, the 3-D positions of these points are known. Replacing image coordinates $\mathbf{x}$ in Equation (7.2.5) with world coordinates $\mathbf{X}$, using the pinhole camera model according to Equation (5.1.2), we get

$$\Delta \psi_k(\mathbf{x}) = F \cdot \bar{g}_x \left(\frac{X'}{Z'} - \frac{X}{Z} \right) + F \cdot \bar{g}_y \left(\frac{Y'}{Z'} - \frac{Y}{Z} \right). \tag{7.2.7}$$

where $\bar{g}_x$ and $\bar{g}_y$ are the x and y components of $\bar{g}(\mathbf{x})$. Assuming small rotation angles and object rotation around its center $\mathbf{C} = [\mathbf{C_x}, \mathbf{C_y}, \mathbf{C_z}]^{\mathrm{T}}$, Equations (5.3.10) and (5.3.11) result in

$$\mathbf{X}' = [\mathbf{R}'] \cdot (\mathbf{X} - \mathbf{C}) + \mathbf{C} + \mathbf{T}. \tag{7.2.8}$$

Substituting Equation (7.2.8) for $\mathbf{X}'$ in Equation (7.2.7) results in

$$
\begin{aligned}
\Delta \psi_k(\mathbf{x}) = {}& F \cdot \bar{g}_x / Z \cdot T_x + F \cdot \bar{g}_y / Z \cdot T_y \\
& - [(X\bar{g}_x + Y\bar{g}_y)F/Z^2 + \Delta \psi_k(\mathbf{x})/Z] \cdot T_z \\
& - [[X\bar{g}_x(Y - C_y) + Y\bar{g}_y(Y - C_y) + Z\bar{g}_y(Z - C_z)]F/Z^2 \\
& + \Delta \psi(\mathbf{x})/Z(Y - C_y)] \cdot \theta_x + [[Y\bar{g}_y(X - C_x) + X\bar{g}_1(X - C_x) \\
& + Z\bar{g}_x(Z - C_z)]F/Z^2 + \Delta \psi(\mathbf{x})/Z(X - C_x)] \cdot \theta_y \\
& - [\bar{g}_x(Y - C_y) - \bar{g}_y(X - C_x)]F/Z \cdot \theta_z
\end{aligned}
\tag{7.2.9}
$$

where $T_x, T_y, T_z, \theta_x, \theta_y,$ and θ_z are the six unknown motion parameters, $\mathbf{X} = [X, Y, Z]^T$ is the point on the object surface, $\bar{\mathbf{g}} = [\bar{g}_x, \bar{g}_y]^T$ is the image gradient according to Equation (7.2.6), F is the camera focal length, and $\mathbf{x}$ is the location of the observation point in frame k prior to motion.

In order to obtain reliable estimates for the six motion parameters, Equation (7.2.9) must be established for many observation points $\mathbf{x}_j, j = 1, 2, \ldots, J$, on the surface of the moving object, resulting in an overdetermined system of linear equations

$$[\mathbf{A}] \cdot \phi - \mathbf{b} = \mathbf{r} \tag{7.2.10}$$

with the residual $\mathbf{r} = [r_1, \ldots, r_J]^T$, $\phi = [T_x, T_y, T_z, \theta_x, \theta_y, \theta_z]^T$, $\mathbf{b} = [\Delta \psi_k(\mathbf{x}_1), \ldots,$ $\Delta \psi_k(\mathbf{x}_J)]^T$, $[\mathbf{A}] = (\mathbf{a}_1, \ldots, \mathbf{a}_J)^T$, and $\mathbf{a}_j^T$ defined according to Equation (7.2.9) for the jth observation point. This system of linear equations can be solved by the following optimization problem:

$$\|[\mathbf{A}]\phi - \mathbf{b}\|^2 = \|\mathbf{r}\|^2 \xrightarrow{\phi} \min, \tag{7.2.11}$$

which has the effect of minimizing the motion-compensated frame difference measured at the observation points $\mathbf{x}_1, \ldots, \mathbf{x}_J$ of the object. The motion parameters are given by

$$\hat{\phi} = ([\mathbf{A}]^T [\mathbf{A}])^{-1} [\mathbf{A}]^T \mathbf{b}. \tag{7.2.12}$$

In order to avoid the inversion of large matrices, we do not compute $[\mathbf{A}]$, but immediately compute the 6×6 matrix $[\mathbf{A}]^T [\mathbf{A}]$ and the vector $[\mathbf{A}]^T \mathbf{b}$.

Due to the linearizations in Equations (7.2.8) and (7.2.5), motion parameters must be estimated iteratively. After every iteration, the model object and its observation points are moved according to the nonlinear 3-D motion equation (Equation (5.3.11)) using the estimated motion parameters $\hat{\phi}$. Then, a new set of motion equations is established, giving new motion parameter updates. The motion parameter updates decrease during the iterations, assuming that the motion estimation algorithm converges. Since the motion parameter updates approach zero during the iterations, the error due to the linearization of the rotation matrix also approaches zero. The iteration process terminates if the decrease of the residual $\|[\mathbf{A}]\phi - \mathbf{b}\|^2$ becomes negligible.

This algorithm estimates object motion from frame k to $k+1$. Due to this forward motion estimation, we can track an object as it is moving forward in a video scene.

7.2.3 Planar Objects

Planar objects are often used as an approximation for real objects (Section 5.5.3). Since these approximations are usually used not for tracking but for motion compensation in video coding, we now extend the algorithm presented in Section 7.1.3 to a direct estimation method [5] that estimates motion backward. Starting with Equation (7.1.9) defining the mapping $\mathbf{A}(x, y)$ of a point $\mathbf{x}$ in image $k+1$ to $\mathbf{x}'$ in image k, with

$$
\begin{aligned}
\mathbf{x}' &= [x', y']^T \\
&= [A_x(\mathbf{x}), A_y(\mathbf{x})]^T \\
&= \left[\frac{a_1 x + a_2 y + a_3}{a_7 x + a_8 y + 1}, \frac{a_4 x + a_5 y + a_6}{a_7 x + a_8 y + 1} \right]^T \\
&= \mathbf{A}(\mathbf{x})
\end{aligned} \tag{7.2.13}
$$

and assuming again that all changes in the image sequence are due to object motion, we get

$$\psi_{k+1}(\mathbf{x}) = \psi_k(\mathbf{x}') = \psi_k(\mathbf{A}(\mathbf{x})). \tag{7.2.14}$$

The frame difference, similarly to Equation (7.2.2), becomes

$$\Delta\psi_k(\mathbf{x}) = \psi_{k+1}(\mathbf{x}) - \psi_k(\mathbf{x})$$

$$= \psi_k(\mathbf{A}(\mathbf{x})) - \psi_k(\mathbf{x}). \tag{7.2.15}$$

With a Taylor series expansion, we express the luminance function at an observation point $\mathbf{x}$ with respect to the motion parameters a_i:

$$\psi_k(\mathbf{A}(\mathbf{x})) = \psi_k(\mathbf{x}_k) + \sum_{i=1}^{8} \frac{\partial \psi_k(\mathbf{x})}{\partial a_i} \Delta a_i$$

$$+ \frac{1}{2} \sum_{i=1}^{8} \sum_{j=1}^{8} \frac{\partial \psi_k(\mathbf{x})}{\partial a_i} \frac{\partial \psi_k(\mathbf{x})}{\partial a_j} \Delta a_i \Delta a_j + R_2(\mathbf{x}). \tag{7.2.16}$$

with $\mathbf{a} = (a_1, \dots, a_8)^T$, $\Delta\mathbf{a} = \mathbf{a} - \mathbf{e}$, $\mathbf{e} = (1, 0, 0, 0, 1, 0, 0, 0)^T$, $\frac{\partial \psi_k(\mathbf{x})}{\partial a_i} = \frac{\partial \psi_k(\mathbf{x})}{\partial x} \frac{\partial A_x(\mathbf{x})}{\partial a_i} + \frac{\partial \psi_k(\mathbf{x})}{\partial y} \frac{\partial A_y(\mathbf{x})}{\partial a_i}$, and $R_2(\mathbf{x})$ denoting the higher-order terms of the Taylor series expansion. Using this in Equation (7.2.15) results in

$$\Delta\psi_k(\mathbf{x}) = \sum_{i=1}^{8} \frac{\partial \psi_k(\mathbf{x})}{\partial a_i} \Delta a_i + \frac{1}{2} \sum_{i=1}^{8} \sum_{j=1}^{8} \frac{\partial \psi_k(\mathbf{x})}{\partial a_i} \frac{\partial \psi_k(\mathbf{x})}{\partial a_j} \Delta a_i \Delta a_j + R_2(\mathbf{x}). \tag{7.2.17}$$

If the quadratic image signal model is valid, residual $R_2(\mathbf{x})$ is zero, simplifying Equation (7.2.17) to

$$\Delta\psi_k(\mathbf{x}) = \bar{g}_x x \Delta a_1 + \bar{g}_x y \Delta a_2 + \bar{g}_x \Delta a_3 + \bar{g}_y x \Delta a_4 + \bar{g}_y y \Delta a_5$$

$$+ \bar{g}_y \Delta a_6 - x(\bar{g}_x x + \bar{g}_y y)\Delta a_7 - y(\bar{g}_x x + \bar{g}_y y)\Delta a_8$$

$$= \mathbf{h}^T \Delta\mathbf{a}. \tag{7.2.18}$$

with $\bar{\mathbf{g}}(\mathbf{x}) = [\bar{g}_x, \bar{g}_y]^T$ according to Equation (7.2.6).

Equation (7.2.18) defines an equation for one observation point of the object with the eight unknown motion parameters. In order to estimate the motion, we set up Equation (7.2.18) for J observation points $\mathbf{x}$, and solve the following system of J linear equations:

$$[\mathbf{H}]\Delta\mathbf{a} = \Delta\psi. \tag{7.2.19}$$

The row vectors of matrix $[\mathbf{H}]$ are the $\mathbf{h}^T$ determined according to Equation (7.2.18), and $\Delta\psi$ is a vector containing frame differences between the two images according to Equation (7.2.18). Solving this system of linear equations with the least-squares method, we get, similarly to Equation (7.2.12),

$$\Delta\mathbf{a} = ([\mathbf{H}]^T[\mathbf{H}])^{-1}[\mathbf{H}]^T\Delta\psi. \tag{7.2.20}$$

Due to the approximations used in the image signal model, this estimation process must be carried out iteratively. After each iteration, we use the estimated motion parameters to motion-compensate frame $k + 1$ using

$$\hat{\psi}_k(\mathbf{x}) = \psi_{k+1}(\mathbf{A}'(\mathbf{x})). \tag{7.2.21}$$

The displaced frame difference becomes

$$\mathrm{DFD}(\mathbf{x}) = \psi_{k+1}(\mathbf{A}'(\mathbf{x})) - \psi_k(\mathbf{x}) \tag{7.2.22}$$

where the inverse mapping $\mathbf{A}'(\mathbf{x})$ is defined as $\mathbf{A}'(\mathbf{A}(\mathbf{x})) = \mathbf{x}$. During the iteration, Equation (7.2.22) replaces Equation (7.2.15). Furthermore, the image gradients in Equation (7.2.18) must be computed using the motion-compensated image and image k. If we are interested in the overall motion parameter, the estimated parameters during each iteration must be concatenated appropriately (see Problem 7.9).

7.2.4 Robust Estimation

In order to enable motion estimation that is not easily affected by image noise or violation of our model assumptions, we must take care in the selection and use of observation points. Two types of errors affect motion estimation: *measurement errors* due to image noise occurring when we sample the image signal, and *modeling errors* occurring when the model assumptions of our estimation algorithm are not valid. In this section, we deal with measurement errors first, and then with modeling errors and how to minimize their effect on the estimated motion parameters.

Measurement Errors   The algorithms in the previous two sections are based on a quadratic image signal model. Since this is valid only for a small area around the location of an observation point at most, the selection of appropriate observation points is an important factor in developing a robust direct motion estimator. Here we present concepts that are valid for any direct motion estimation algorithm.

Considering image noise, we can derive a basic criterion for selecting observation points. Assuming that the camera noise in two consecutive frames is i.i.d and the noise variance is σ_n^2, the noise of the image difference signal will be

$$\sigma_{\Delta \psi_k}^2 = 2 \cdot \sigma_n^2. \tag{7.2.23}$$

Since we use the frame difference for computing motion, we must take care that we consider only observation points where the influence of the camera noise is small; that is, we prefer observation points at which the frame difference exceeds a certain minimum.

According to Equation (7.2.4), we represent the local displacement $\mathbf{x}' - \mathbf{x}$ as a function of the noise-free luminance signal. In a noisy environment, the gradient $\mathbf{g}(\mathbf{x})$ should have a large absolute value in order to limit the influence of camera noise. Hence, we select as observation points only points with a gradient larger than a threshold T_G:

$$|g_x(\mathbf{x})| > T_G$$

$$|g_y(\mathbf{x})| > T_G \tag{7.2.24}$$

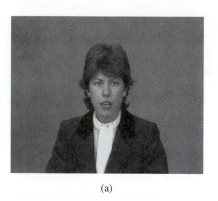
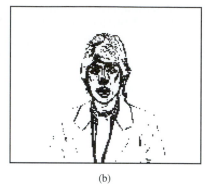

(a)   (b)

Figure 7.4   (a) Image of the test sequence "Claire." (b) Observation points for this image.

Relatively large gradients allow for a precise estimation of the motion parameters as well, because a slight mismatch would immediately result in a high DFD or $\Delta\psi(\mathbf{x})$. If we choose an observation point that has a zero gradient or a very small gradient, the point could be moved by several pels without causing any significant DFD. Figure 7.4 shows an image of the test sequence "Claire" and its observation points. We selected all points $\mathbf{x}$ with $|g_x(\mathbf{x})| > 32$ and $|g_y(\mathbf{x})| > 32$ as observation points. With this initial selection of observation points, we look now into methods of improving the performance of the motion estimation algorithms.

Modeling Errors   Equations (7.2.10) and (7.2.19) are solved such that the variances of the residual errors are minimized. However, this approach is sensitive to modeling errors [10]. Modeling errors occur because Equations (7.2.10) and (7.2.19) are based on several model assumptions and approximations that tend to be valid for the majority of observation points but not for all. Observation points that violate these assumptions are named *outliers* [15]. When using a least-squares method for solving Equations (7.2.10) and (7.2.19), outliers have a significant influence on the solution. Therefore, we must take measures that limit the influence of these outliers on the estimation process. The following assumptions are sometimes not valid:

1. rigid real object,
2. quadratic image signal model,
3. small deviations of model object shape from real object shape,
4. diffuse illumination and diffuse reflecting surfaces.

In the following, we discuss the influence of each of these assumptions on motion estimation accuracy.

Assumption 1:   In the case that parts of the real object are nonrigid (i.e. the object is flexible), we have image areas that cannot be described by the motion, and shape, and color parameters of the object. These image areas are named *model failures* and can be detected due to their potentially high frame difference $\Delta\psi(\mathbf{x})$. Observation points with a high frame difference can be classified as outliers. During an iteration, we will consider only those observation points $\mathbf{x}$ for which

$$\Delta\psi(\mathbf{x}) < \sigma_{\Delta\psi} \cdot T_{\text{ST}} \tag{7.2.25}$$

with

$$\sigma_{\Delta\psi}^2 = \frac{1}{J} \sum_{j=0}^{J} \left(\Delta\psi\left(\mathbf{x}^{(j)}\right)\right)^2 \tag{7.2.26}$$

holds. The threshold T_{ST} is used to remove the outliers from consideration. The variance $\sigma_{\Delta\psi}^2$ should be recalculated after each iteration.

Assumption 2:   According to Equation (7.2.6), motion estimation is based on a quadratic image signal model. This is at most valid for a small area around an observation point $\mathbf{x}$. It allows for estimating only small displacements $(\mathbf{x}' - \mathbf{x})$ in one iteration step [12]. Given an image gradient $\mathbf{g}(\mathbf{x})$ and a maximum displacement d_{max} that we want to allow for estimation, we can limit the allowable frame difference $\Delta\psi(\mathbf{x})$ for each observation point and if necessary exclude an observation point for a given iteration. We assume that this point does not conform to the image signal model.

Assumption 3:   The shape difference between the model object and the real object can be modeled by means of a spatial uncertainty of an observation point along the line of sight. This can be considered using a Kalman filter during motion estimation [9].

Assumption 4:   Although the influences of nondiffuse illumination and reflections disturb image analysis, they are usually not modeled separately. To some extent, these image areas are automatically detected when detecting model failures (see assumption 1).

Summarizing these observations, we conclude that we should select observation points with large absolute image gradients according to Equation (7.2.24). Only observation points that pass this initial test will be considered during motion estimation. Equation (7.2.25) provides a selection criterion for the observation points that we will use for a given iteration step. In [13], further details on robust direct motion estimation and maximum-likelihood estimation are discussed. In the following, we will briefly focus on an efficient estimator described in [13] and [7].

Instead of using the binary selection criteria for observation points according to Equation (7.2.25), we can use continuous cost functions to control the influence of an observation point on the parameter estimation. We use the residual r_j according to Equation (7.2.10) as a measure for the influence of an observation point [14, 23]. A critical issue for selecting an appropriate measure is the treatment of outliers. Ideally, we would know the probability density function of the outliers. Furthermore, we want outliers to have no influence on the estimated motion parameters. However, the probability density function of outliers is often unknown. Therefore, heuristic solutions using the cost function $w = (1 - (r^2/b^2))^2$ as suggested by Tukey were found [7, 23]:

$$r'_j = \begin{cases} r_j \cdot \left(1 - \frac{r_j^2}{b^2}\right) & \text{if } |r_j| < b \\ 0 & \text{otherwise.} \end{cases} \tag{7.2.27}$$

Instead of $\sum_j |r_j|^2 \to \min$ we now minimize

$$\sum_j |r_j|^2 |w_j|^2 \to \min. \tag{7.2.28}$$

The cost fuction $w = (1 - (r^2/b^2))^2$ increases to 1 when $|r|$ decreases. Observation points with $|r_j| \geq b$ are excluded from the next iteration. The constant b is the threshold for detecting outliers. In order to adapt the outlier detection to the image difference signal $\Delta\psi$, we select b proportional to $\sigma_{\Delta\psi}$ in Equation (7.2.26). During motion estimation, we multiply the equation of each observation point $\mathbf{x}^{(j)}$ with its individual cost $w_j = (1 - (r_j^2/b^2))^2$, therefore adapting its influence on the estimated motion parameters according to Equations (7.2.11) and (7.2.20).

7.3 ITERATIVE MOTION ESTIMATION

Direct motion estimation algorithms can estimate small motion only, mainly due to the limited range for which the assumptions of the image signal model hold. In order to estimate larger motion, we must use an iterative algorithm.

Let us assume that we estimate the motion parameters, starting with a set of observation points and equations as described in one of the previous sections. During this first iteration we solve for the motion parameters $[\mathbf{R}_i]$ and $\mathbf{T}_i$ with $i = 1$, giving the residual r_i. We use these motion parameters to motion-compensate the object and its observation points. If appropriate, we use the nonlinear motion equations for motion compensation, even if the motion parameters are solved by using a linearized rotation matrix. We now set up a new set of equations following the same procedure as for the first iteration, resulting in the motion parameters $[\mathbf{R}_i]$ and $\mathbf{T}_i$ with $i = 2$. We continue these iterations as long as the new motion parameters decrease the residual; that is, $(r_j^2 - r_{j+1}^2) > T$ with threshold T chosen appropriately. The final estimated motion parameters from image k to image $k+1$ can be computed by concatenating the motion parameters of each iteration.

To show the principle of motion parameter concatenation, we compute the concatenation for the algorithm presented in Section 7.2.2. Assuming motion according to Equation (5.3.9), the estimated rotation matrix $[\hat{\mathbf{R}}]$ becomes

$$[\hat{\mathbf{R}}] = \Pi_{i=1}^{I}[\mathbf{R}_{I+1-i}] \tag{7.3.1}$$

with the rotation matrix $[\mathbf{R}_i]$ estimated in iteration i. The computation of the translation vector $\hat{\mathbf{T}}$ must consider the center of rotation. Assuming that we compute rotation around $\mathbf{C}_i$ during iteration i, we get

$$\hat{\mathbf{T}} = \mathbf{T}_I + \mathbf{C}_I + [\mathbf{R}_I]\,(\mathbf{T}_{I-1} + \mathbf{C}_{I-1} - \mathbf{C}_I + [\mathbf{R}_{I-1}](\mathbf{T}_{I-2} + \mathbf{C}_{I-2} - \mathbf{C}_{I-1} + \cdots))$$

$$(7.3.2)$$

assuming rotation around the coordinate center; that is, the object moves according to

$$\mathbf{X}' = [\hat{\mathbf{R}}]\mathbf{X} + \hat{\mathbf{T}}. \qquad (7.3.3)$$

Compared to indirect motion estimation, direct motion estimation usually requires iterative algorithms that tend to be more computationally intensive. However, direct motion estimation has the advantage of considering the image signal during estimation.

7.4 SUMMARY

In this chapter, we discussed several methods of estimating 3-D motion. We presented two estimation methods, indirect and direct motion estimation.

- **Indirect Motion Estimation Algorithms (Section 7.1)** depend on feature points on the moving object. The feature points must be reliably marked on two images. Then we can estimate 3-D motion from these correspondences. The algorithms allow for estimating large motion; however, the estimation algorithms tend to be sensitive to errors in the feature point correspondences. Different algorithms can be used depending on whether we want to assume orthographic (Section 7.1.1) or perspective (Sections 7.1.2, 7.1.4) projection, and whether we know the object shape. Using an orthographic camera model is acceptable if the object does not move significantly in the direction of the camera axis and object rotation is mainly within the camera plane. Otherwise, we use an algorithm with a perspective camera model, which requires more computations than an orthographic model. Estimating the essential matrix from feature point correspondences (Section 7.1.4) enables us to determine the motion of the object as well as the 3-D location of these feature points. In order to gain robustness against noise, we suggest the use of the epipolar line constraint and the minimization of the distance of the feature points from this line (Equation 7.1.34) when determining the motion and shape parameters from the essential matrix.
- **Direct Motion Estimation Algorithms (Section 7.2)** approximate the image signal using a Taylor series expansion (Section 7.2.1) and derive pel-based motion (optical flow) from the frame difference signal. This pel-based motion is used directly to compute the 3-D motion parameters. These algorithms are used if we have small object motion from one frame to another. Usually, we are required to make some assumption about the 3-D object shape. The algorithm described in Section 7.2.3 assumes a planar object shape. Key to the successful use of direct

motion estimation algorithms is the careful selection of those observation points at which we measure the frame difference signal. In Section 7.2.4, we describe robust methods for selecting these points during motion estimation.

- **Iterative Motion Estimation (Section 7.3)**   Direct motion estimation usually allows only for estimation of small motion. The motion estimation range can be extended using an iterative approach.

The selection of the estimation method depends to a large extent on what measurements are available. Indirect motion estimation is the method of choice if we have only a few highly accurate feature point locations. Direct motion estimation should be used if we can reliably estimate motion from the image signal using optical flow. As always, a combination of these approaches might give the best results for some applications.

7.5 PROBLEMS

7.1 What are the differences between a direct and an indirect motion estimator? What are the potential advantages and disadvantages?

7.2 Derive Equation (7.1.9). Start with the general 3-D motion equation $\mathbf{X}' = [\mathbf{R}]\mathbf{X} + \mathbf{T}$ and the plane equation in Equation (7.1.8).

7.3 Show that Equation (7.1.10) can be written as Equation (7.1.11).

7.4 Show that the solution to the optimization problem in Equation (7.1.22) is the unit eigenvector of $[\mathbf{A}]^T[\mathbf{A}]$ associated with the smallest eigenvalue.

7.5 Consider an image point $\mathbf{x}$ that is the projection of $\mathbf{X}$. This point moves to $\mathbf{x}'$ and $\mathbf{X}'$, respectively. Derive the epipolar line for this point $\mathbf{x}$ for the motion parameters $\mathbf{T}$ and $[\mathbf{R}]$. For an image point $\mathbf{x}$ and its epipolar line, illustrate how the distance of a point $\mathbf{X}$ from the camera center (that is, the Z coordinate) determines its location on the epipolar line, by locating the points corresponding to $Z = \infty$ and $Z = 1$.

7.6 When estimating the E-matrix, the influence of a measurement error δ in the point correspondence depends on the absolute coordinates of these correspondences. Show how the residual of a point correspondence depends on the coordinates of the corresponding points. In Equations (7.1.11) and (7.1.19), replace $\mathbf{x}'$ by $\mathbf{x}' + \delta$. What would be the weighting factors for each equation in Equation (7.1.22), such that the image coordinates of a correspondence do not influence the estimation of the E-matrix? (*Hint:* Minimize the distance of each point from the epipolar line.)

7.7 Present a polynomial representation of the luminance signal $\psi(\mathbf{x})$ around $\mathbf{x}$ using a first and second order Taylor expansion.

7.8 Derive Equations (7.2.6) and (7.2.5) using a second-order Taylor series expansion for the image signal.

7.9 In order to implement the direct motion estimation algorithm for planar objects (Section 7.2.3), the inverse mapping of $\mathbf{A}$ must be known. Show that the mapping

A fulfills the four axioms of a group (namely, closure, existence of an inverse, existence of an identity, and associativity). Compute the inverse of a mapping **A** and the sum of the motion parameter estimates after two iterations.

7.10 Discuss measures to increase the robustness of a direct motion estimation algorithm.

7.6 BIBLIOGRAPHY

[1] Aizawa, K., H. Harashima, and T. Saito. Model-based analysis-synthesis image coding (MBASIC) system for a person's face. *Signal Processing: Image Commun.* (Oct. 1989), 1(2):139–52.

[2] Beardsley, P., P. H. S. Torr, and A. Zisserman. 3-D model aquisition from extended image sequences. In B. Buxton and Cipolla R., eds., *Proceedings of the 4th European Conference on Computer Vision, LNCS 1065,* 683–95. Cambridge, England: Springer-Verlag, 1996.

[3] Bierling, M. Displacement estimation by hierarchical block matching. In *SPIE: Visual Commun. Image Processing* (Nov. 1988), SPIE-1001:942–51.

[4] Bozdagi, G., A. M. Tekalp, and L. Onural. An improvement to MBASIC algorithm for 3-D motion and depth estimation. *IEEE Trans. Image Process.* (June 1994), 3:711–16.

[5] Hötter, M., and R. Thoma. Image segmentation based on object oriented mapping parameter estimation. *Signal Processing* (Oct. 1988), 15(3):315–34.

[6] Huang, T. S., and A. N. Netravali. Motion and structure from feature correspondences: A review. *Proceedings of the IEEE* (Feb. 1994), 82(2):252–68.

[7] Li, H., and R. Forchheimer. Two-view facial movement estimation. *IEEE Trans. on Circuits and Systems for Video Technology* (June 1994), 4(3):276–87.

[8] Longuet-Higgins, H. C. A computer algorithm for reconstructing a scene from two projections. *Nature* (Sept. 1981), 293(5828):133–35.

[9] Martinez, G. Shape estimation of articulated 3-D objects for object-based analysis-synthesis coding (OBASC). *Signal Processing: Image Communication* (March 1997), 9(3):175–99.

[10] Meer, P., D. Mintz, and D. Y. Kim. Robust regression methods in computer vision: a review. *International Journal of Computer Vision* (April 1991), 6(1):59–70.

[11] Moons, T. A guided tour through multiview relations. In R. Koch and L. van Gool, eds., *Proceedings SMILE Workshop, LNCS 1506,* 304–46. Berlin: Springer-Verlag, 1998.

[12] Musmann, H. G., P. Pirsch, and H.-J. Grallert. Advances in picture coding. *Proceedings of the IEEE* (April, 1985), 73(4):523–48.

[13] Ostermann, J. Object-based analysis-synthesis coding based on moving 3-D objects. In L. Guan, S. Y. Kung, and J. Larsen, eds., *Multimedia Image and Video Processing*, 289–329. New York: CRC Press, 2000.

[14] Press, W. H., S. A. Teukolsky, W. T. Vetterling, and B. P. Flannery. *Numerical Recipes in C: The Art of Scientific Computing*. Cambridge University Press, 1992.

[15] Rousseeuw, P. J., and A. M. Leroy. *Robust Regression and Outlier Detection*. New York: John Wiley & Sons, 1987.

[16] Stauder, J. Estimation of point light source parameters for object-based coding. *Signal Processing: Image Communication* (Nov. 1995), 46(7):355–79.

[17] Tsai, R. Y., and T. S. Huang. Uniqueness and estimation of three-dimensional motion parameters of rigid objects with curved surfaces. *IEEE Trans. Pattern Anal. Machine Intell.* (Jan. 1981), 6:13–16.

[18] Tsai, R. Y., and T. S. Huang. Estimating three-dimensional motion parameters of a rigid planar patch. *IEEE Transactions on Acoustics, Speech and Signal Processing* (Dec. 1981), 29(6):1147–52.

[19] Tsai, R. Y., and T. S. Huang. Uniqueness and estimation of three-dimensional motion parameters of rigid objects with curved surfaces. *IEEE Transactions on Pattern Analysis and Machine Intelligence* (Jan. 1984), 6(1):13–27.

[20] Tsai, R. Y., T. S. Huang, and W.-L. Zhu. Estimating three-dimensional motion parameters of a rigid planar patch. II. Singular value decomposition. *IEEE Transactions on Acoustics, Speech and Signal Processing* (Aug. 1982), 30(4):525–34.

[21] Weng, J., T. S. Huang, and N. Ahuja. Motion and structure from two perspective views: algorithms, error analysis, and error estimation. *IEEE Transactions on Pattern Analysis and Machine Intelligence* (May 1989), 11(5):451–76.

[22] Zhuang, X. A simplification to linear 2-view motion algorithms. *Computer Vision Graphics and Image Processing* (1989), 46(2):175–78.

[23] Zhuang, X., T. Wang, and P. Zhang. A highly robust estimator through partially likelihood function modeling and its application in computer vision. *IEEE Transactions on Pattern Analysis and Machine Intelligience* (Jan. 1992), 14(1):19–34.

8

Foundations of Video Coding

Video coding is an important application of digital video processing. The goal of video coding is to reduce the data rate of a video sequence, so that it is feasible to transmit the video in real time over a given communication channel. The channel bandwidth varies, depending on the application and the transmission media. For video phone applications using regular telephone lines, 20 kbps may be all that is available for video coding. For broadcasting of standard-definition television signals over satellite, a data rate of 6 mbps may be available. In addition to communications applications, video coding is also necessary for storage and retrieval, where different storage media have different capacity and access rates, thus demanding varying amount of compression. Due to this wide range of data rates, different types of algorithms have been developed. One class of algorithms allows efficient coding of arbitrary video signals without trying to analyze the video content. A second class of algorithms identifies regions and objects in a video sequence and codes those. We call the former *waveform-based* video coders and the latter *content-dependent* video coders.

In this chapter, we first give an overview of different waveform-based and content-dependent video coders in Section 8.1. Then, following a review of basic probability and information theory concepts in Section 8.2, information theory used to optimize lossless coding, and rate distortion theory used to optimize lossy coding are introduced in Section 8.3. This is followed by descriptions of lossless coding (Section 8.4), and the most basic lossy coding techniques, including scalar and vector quantization (Sections 8.5 and 8.6). Chapters 9 and 10 will discuss techniques for waveform-based and content-dependent video coding, respectively.

8.1 OVERVIEW OF CODING SYSTEMS

8.1.1 General Framework

The components of a video coding algorithm are determined to a large extent by the *source model* that is adopted for modeling the video sequences. The video coder seeks to describe the contents of a video sequence by means of its source model. The source model may make assumptions about the spatial and temporal correlation between pixels of a sequence. It might also consider the shape and motion of objects or illumination effects. In Figure 8.1, we show the basic components in a video coding system. In the encoder, the digitized video sequence is first described using the parameters of the source model. If we use a source model of statistically independent pixels, then the parameters of this source model would be the luminance and chrominance amplitudes of each pixel. On the other hand, if we use a model that describes a scene as several objects, the parameters would be the shape, texture, and motion of individual objects. In the next step, the parameters of the source model are quantized into a finite set of symbols. The quantization parameters depend on the desired trade off between the bit rate and distortion. The quantized parameters are finally mapped into binary codewords using lossless coding techniques, which further exploit the statistics of the quantized parameters. The resulting bit stream is transmitted over the communication channel. The decoder retrieves the quantized parameters of the source model by reversing the binary encoding and quantization processes of the encoder. Then, the image synthesis algorithm at the decoder computes the decoded video frame using the quantized parameters of the source model.

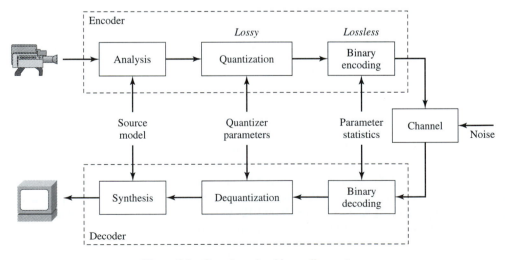

Figure 8.1   Overview of a video coding system.

8.1.2 Categorization of Video Coding Schemes

In this subsection, we provide an overview of several popular video coding algorithms, and put them into context by looking at their underlying source models. The source model of a coding algorithm can be determined by looking at the coded parameter set and at the image synthesis algorithm that composes the decoded image based on the decoded parameters.

Waveform-Based Coding   The techniques in this first category all attempt to represent the color values of individual pixels as accurately as possible, without considering the fact that a group of pixels may represent a physical object.

Assuming statistically independent pixels results in the simplest source model (Table 8.1). The related coding technique is named *pulse coded modulation* (PCM). A PCM representation of the image signal is not usually used for video coding due to its inefficiency compared to other source models.

In most images, we find that the colors of neighboring pixels are highly correlated. For bit-rate reduction, this property is best exploited using *transforms* like the Karhunen-Loeve (KLT), discrete cosine (DCT), or wavelet transforms (discussed in Chapters 9 and 11). The transform serves to decorrelate the original sample values and concentrate the energy of the original signal into a few coefficients. The parameters to be quantized and coded are the coefficients of the transform. Another way to exploit the correlation among adjacent samples is *predictive coding,* by which the sample value to be coded is first predicted from previously coded samples. The prediction error, which has less correlation and lower energy than the original signal, is then quantized and coded. Both transform and predictive coding can be considered as special cases of *vector quantization,* which quantizes a block of samples (a vector) at a time. Essentially, it finds the typical block patterns occurring in the signal and approximates any block by one of the typical patterns.

TABLE 8.1   COMPARISON OF SOURCE MODELS, PARAMETER SETS, AND CODING TECHNIQUES.

Source model	Encoded parameters	Coding technique
Statistically independent pels	Color of each pel	PCM
Statistically dependent pels	Color of each block	Transform coding, predictive coding, and vector quantization
Translationally moving blocks	Color and motion vector of each block	Block-based hybrid coding
Moving unknown objects	Shape, motion, and color of each object	Analysis-synthesis coding
Moving known object	Shape, motion, and color of each known object	Knowledge-based coding
Moving known object with known behavior	Shape, color, and behavior of each object	Semantic coding

Today's video coding standards, H.261, H.263, MPEG-1, MPEG-2, and MPEG-4 (discussed in Chapter 13), all employ the approach of *block-based hybrid coding,* which combines predictive coding with transform coding (Chapter 9). This coding technique subdivides each image into fixed-size blocks. Each block of a frame k is synthesized using a block of the same size at a displaced position of the previous frame $k-1$. This is done for all blocks of frame k. The resulting image is called the *predicted image.* The encoder transmits the 2-D motion vectors for all blocks to the decoder so that the decoder can compute the same predicted image. The encoder subtracts this predicted image from the original image, resulting in the *prediction error image.* If the synthesis of a block using the predicted image is not sufficiently accurate—that is, if the prediction error of a block exceeds a threshold—then the encoder uses a transform coder for transmitting the prediction error of this block to the decoder. The decoder adds the prediction error to the predicted image and thus synthesizes the decoded image. Therefore, block-based hybrid coding is based on the source model of translationally moving blocks. In addition to the color information encoded as transform coefficients of the prediction error, motion vectors must be transmitted. It is worth noting that such a coder can *switch* to the simpler source model of statistically dependent pixels. This is done whenever the coding of a block is achieved more efficiently without referring to the previous frame.

Content-Dependent Coding   The preceding block-based hybrid coding techniques essentially approximate the shape of objects in a scene with square blocks of fixed size. Therefore, they generate high prediction errors in blocks on object boundaries. These boundary blocks contain two objects with different motions, which cannot be accounted for by a single motion vector. Content-dependent coders recognize such problems, and attempt to segment a video frame into regions corresponding to different objects and code these objects separately. For each object, shape information must be transmitted in addition to motion and texture information.

In *object-based analysis-synthesis coding* (OBASC; discussed in Chapter 10), each moving object of a video scene is described by a model object. In order to describe the shape of an object, OBASC employs a segmentation algorithm. Further, motion and texture parameters of each object are estimated. In the simplest case, a 2-D silhouette describes the shape of an object, a motion vector field describes its motion, and a color waveform represents its texture. Other approaches describe an object with a 3-D wireframe. An object in frame k is described by the shape and color of an object in frame $k-1$ and an update of shape and motion parameters. The decoder synthesizes the object using the current motion and shape parameters as well as the color parameters of the preceding frame. Only for those image regions where the image synthesis fails is color information transmitted.

In the case that the object types in a video sequence belong to known object classes, *knowledge-based coding* can be employed, which uses a wireframe specially designed to describe the recognized object type (Chapter 10). Several approaches have been developed to code human heads using predefined wireframes. Using a predefined wireframe increases coding efficiency because it is adapted to the shape of the object. Sometimes, this technique is also referred to as *model-based coding.*

When possible object types as well as their behaviors are known, semantic coding may be used. For example, for a human face object, a "behavior" refers to the temporal trajectory of a set of facial feature points associated with a particular facial expression. The possible behaviors of a human face include typical facial expressions such as happiness, sadness, anger, and so on. In this case, the parameters for describing the behavior of an object are estimated and transmitted to the decoder. This coding technique has the potential for achieving very high coding efficiency, because the number of possible behaviors for an object like a face is very small, and the number of bits required to specify the behavior is much smaller than that for describing the actual action with conventional motion and color parameters.

8.2 BASIC NOTIONS IN PROBABILITY AND INFORMATION THEORY

In video coding or coding of any signal, we treat a given signal as the realization of a random process. In source coding theory, a random process is referred to as a source. The efficiency of a coding technique depends on how fully the source statistics are exploited. Before moving into the theory and techniques for source coding, we first review how to characterize a random process using probability distributions, and introduce the concepts of entropy and mutual information from information theory.

8.2.1 Characterization of Stationary Sources

We consider only sources defined over a discrete index set, so that a source is a random sequence. We use $\mathcal{F} = \{\mathcal{F}_n\}$ to represent a source, in which $\mathcal{F}_n$ represents the random variable (RV) corresponding to the nth sample. The actual value that $\mathcal{F}_n$ takes in a given realization is denoted by f_n. Note that for a discrete signal over a multidimensional space, we can either order it into a 1-D sequence, so that n refers to the 1-D index of a sample, or we can assume that n is a multidimensional index. Also, $\mathcal{F}_n$ can be a scalar or a vector. In the latter case, it is also called a random vector. If $\mathcal{F}_n$ can only take symbols from a finite alphabet $\mathcal{A} = \{a_1, a_2, \ldots, a_L\}$, then we call $\mathcal{F}_n$ a discrete RV and $\mathcal{F}$ a *discrete source*. If $\mathcal{F}_n$ takes value over a continuous supporting range or the union of several continuous ranges, denoted by $\mathcal{B}$, then $\mathcal{F}_n$ is called a continuous RV and $\mathcal{F}$ a *continuous source*.[*]

As an example, consider a digital color video. In this case, the index n represents a particular combination of pixel location and frame number, and $\mathcal{F}_n$ is a 3-D random vector, representing the three color values taken by the nth pixel. Because each color value is quantized into 256 levels, a digital video is a discrete source with an alphabet size of 256^3. Any given video sequence is a particular realization of this discrete random process. An analog video, on the other hand, is a continuous-space continuous-amplitude random process. The sampling process converts it to a discrete-space continuous-amplitude process. Only after quantization does it become a discrete-space discrete-amplitude process (i.e., a discrete source). In the real world,

---

[*]Note that in this chapter, the words "discrete" and "continuous" refer to the sample amplitude, whereas in earlier chapters, they referred to the space over which the signal was defined. A digital signal is a discrete-space signal in which each sample takes only discrete values.

a discrete source is often obtained from a continuous-space continuous-amplitude source by sampling and quantization.

In this book, we consider only *stationary* processes, in which the probability distribution of $\mathcal{F}_n$ does not depend on the index n, and the joint distribution of a group of N samples is invariant with respect to a common shift in the index. We use $p_{\mathcal{F}_n}(f)$ to represent the *probability mass function* (pmf), in the case of a discrete source, or the *probability density function* (pdf), in the case of a continuous source, of any sample $\mathcal{F}_n$. In addition, we use $p_{\mathcal{F}_{n+1}, \mathcal{F}_{n+2}, \dots, \mathcal{F}_{n+N}}(f_1, f_2, \dots, f_N)$ to represent the joint pmf or pdf of any N successive samples in $\mathcal{F}$. We also use $p_{\mathcal{F}_n | \mathcal{F}_{n-1}, \mathcal{F}_{n-2}, \dots, \mathcal{F}_{n-M}}(f_{M+1} \mid f_M, f_{M-1}, \dots, f_1)$ to represent the conditional pmf or pdf of any sample $\mathcal{F}_n$ given its previous M samples. When it is clear from the underlying context, we simply use $p(f)$, $p(f_1, f_2, \dots, f_N)$, and $p(f_{M+1} \mid f_M, f_{M-1}, \dots, f_1)$ to denote the preceding functions.

An important class of stationary sources is known as *independent and identically distributed* (i.i.d.) sources, which satisfy $p(f_1, f_2, \dots, f_N) = p(f_1)p(f_2) \cdots p(f_N)$, and $p(f_{M+1} \mid f_M, f_{M-1}, \dots, f_1) = p(f_{M+1})$. An i.i.d. source is also referred to as *memoryless*. Another important class is the *Markov process*, in which a sample depends only on its immediate predecessor, that is, $p(f_{M+1} \mid f_M, f_{M-1}, \dots, f_1) = p(f_{M+1} \mid f_M)$. More generally, an *Mth order Markov process* is one in which a sample depends only on its previous M samples. A process in which any N samples follow an Nth order Gaussian distribution is called a *Gaussian process*. A Gaussian process is Markov if the covariance between two samples $\mathcal{F}_n$ and $\mathcal{F}_m$ has the form $C(\mathcal{F}_n, \mathcal{F}_m) = \sigma^2 \rho^{-(n-m)}$. Such a process is called a *Gauss-Markov process*, or a *Gauss-Markov field (GMF)* when the underlying process is 2-D. In image and video processing, a real image or video frame is often modeled as a GMF.

8.2.2 Entropy and Mutual Information for Discrete Sources

In this section, we introduce two very important notions from information theory: *entropy* and *mutual information*. As will be seen, they provide bounds on the minimal bit rates achievable using lossless and lossy coding, respectively. We first define entropy and mutual information for discrete RVs, and then apply these definitions to samples in a discrete source.

Definition 8.1 (Entropy of a Discrete RV).   The entropy of a discrete RV $\mathcal{F}$ with an alphabet $\mathcal{A}$ and a pmf $p_{\mathcal{F}}(f)$ is defined by

$$H(\mathcal{F}) = - \sum_{f \in \mathcal{A}} p_{\mathcal{F}}(f) \log_2 p_{\mathcal{F}}(f). \tag{8.2.1}$$

In general, the log can be taken to any base. Here we consider only the base-two case, so that entropy is expressed in bits. Also, it is assumed that $0 \log 0 = 0$, so that adding a symbol with zero probability to the alphabet does not change the entropy.

The entropy of a discrete RV is always nonnegative, because $0 \le p(f) \le 1$ for any pmf. Among all RVs with an alphabet size L, the uniform distribution, $p(f) = 1/L$, $f \in \mathcal{A}$, achieves the maximum entropy, $H_{\max} = \log_2 L$.

Entropy is a measure of the uncertainty about a RV $\mathcal{F}$. It depends on the pmf of $\mathcal{F}$, not the actual alphabet used by $\mathcal{F}$. When $\mathcal{F}$ can take on any value in $\mathcal{A}$ with equal probability, it is most uncertain, hence it has maximum entropy. On the other hand, if $\mathcal{F}$ takes a particular symbol in $\mathcal{A}$ with probability one, it has no uncertainty, and hence the entropy is zero. The uncertainty about a RV can in turn be considered as the information that may be conveyed by a RV: when one is shown the actual value taken by $\mathcal{F}$, one is told much more in the first case than in the second.

Definition 8.2 (Joint Entropy of Two Discrete RVs).   Let $\mathcal{F}$ and $\mathcal{G}$ represent two discrete RVs, with a joint pmf $p_{\mathcal{F},\mathcal{G}}(f, g)$, $f \in \mathcal{A}_f, g \in \mathcal{A}_g$. Their joint entropy is defined as

$$H(\mathcal{F}, \mathcal{G}) = -\sum_{f \in \mathcal{A}_f} \sum_{g \in \mathcal{A}_g} p_{\mathcal{F},\mathcal{G}}(f, g) \log_2 p_{\mathcal{F},\mathcal{G}}(f, g). \tag{8.2.2}$$

Definition 8.3 (Conditional Entropy between Two Discrete RVs).   Let $\mathcal{F}$ and $\mathcal{G}$ represent two discrete RVs, with marginal pmf $p_{\mathcal{G}}(g)$ and conditional pmf $p_{\mathcal{F}|\mathcal{G}}(f \,|\, g)$, $f \in \mathcal{A}_f, g \in \mathcal{A}_g$. The conditional entropy of $\mathcal{F}$ given $\mathcal{G}$ is defined as

$$H(\mathcal{F} \,|\, \mathcal{G}) = \sum_{g \in \mathcal{A}_g} p_{\mathcal{G}}(g) H(\mathcal{F} \,|\, g)$$

$$= -\sum_{g \in \mathcal{A}_g} p_{\mathcal{G}}(g) \sum_{f \in \mathcal{A}_f} p_{\mathcal{F}|\mathcal{G}}(f \,|\, g) \log_2 p_{\mathcal{F}|\mathcal{G}}(f \,|\, g). \tag{8.2.3}$$

Definition 8.4 (Mutual Information between Two Discrete RVs).   Let $\mathcal{F}$ and $\mathcal{G}$ represent two discrete RVs, with a joint pmf $p_{\mathcal{F},\mathcal{G}}(f, g)$, $f \in \mathcal{A}_f, g \in \mathcal{A}_g$, and marginal distributions $p_{\mathcal{F}}(f)$ and $p_{\mathcal{G}}(g)$. The mutual information between $\mathcal{F}$ and $\mathcal{G}$ is defined as

$$I(\mathcal{F}; \mathcal{G}) = \sum_{f \in \mathcal{A}_f} \sum_{g \in \mathcal{A}_g} p_{\mathcal{F},\mathcal{G}}(f, g) \log_2 \frac{p_{\mathcal{F},\mathcal{G}}(f, g)}{p_{\mathcal{F}}(f) p_{\mathcal{G}}(g)}. \tag{8.2.4}$$

The mutual information between $\mathcal{F}$ and $\mathcal{G}$ specifies the amount of information provided by $\mathcal{G}$ about $\mathcal{F}$. In other words, it measures the reduction in the number of bits required to specify $\mathcal{F}$, given the realization of $\mathcal{G}$. The mutual information is always nonnegative. It is easy to show that

$$I(\mathcal{F}; \mathcal{G}) = I(\mathcal{G}; \mathcal{F}) \geq 0. \tag{8.2.5}$$

Relations between Entropy, Conditional Entropy, and Mutual Information   We summarize some important relations between entropy, conditional entropy and mutual information in the following. These relations can be proved quite easily (Problem 8.2).

$$H(\mathcal{F}, \mathcal{G}) = H(\mathcal{G}) + H(\mathcal{F} \,|\, \mathcal{G}) = H(\mathcal{F}) + H(\mathcal{G} \,|\, \mathcal{F}) \tag{8.2.6}$$

$$I(\mathcal{F}; \mathcal{G}) = H(\mathcal{F}) - H(\mathcal{F} \,|\, \mathcal{G}) = H(\mathcal{G}) - H(\mathcal{G} \,|\, \mathcal{F}) \tag{8.2.7}$$

$$I(\mathcal{F}; \mathcal{G}) = H(\mathcal{F}) + H(\mathcal{G}) - H(\mathcal{F}, \mathcal{G}) \tag{8.2.8}$$

$$I(\mathcal{F}; \mathcal{G}) \leq H(\mathcal{F}), \qquad I(\mathcal{F}; \mathcal{G}) \leq H(\mathcal{G}) \tag{8.2.9}$$

$$H(\mathcal{F}) \geq H(\mathcal{F} \mid \mathcal{G}), \qquad H(\mathcal{G}) \geq H(\mathcal{G} \mid \mathcal{F}) \tag{8.2.10}$$

$$H(\mathcal{F}, \mathcal{G}) \leq H(\mathcal{F}) + H(\mathcal{G}) \tag{8.2.11}$$

$$I(\mathcal{F}; \mathcal{F}) = H(\mathcal{F}) = H(\mathcal{F}, \mathcal{F}) \tag{8.2.12}$$

Equation (8.2.10) says that conditioning reduces the entropy (i.e., uncertainty) of a RV. From Equation (8.2.7), we can see that $I(\mathcal{F}; \mathcal{G})$ describes the reduction in the uncertainty (or information) of $\mathcal{F}$ due to the knowledge of $\mathcal{G}$.

The preceding definitions are given for a RV or between two RVs. They can be similarly defined for random vectors. By applying these definitions to random vectors consisting of successive samples in a discrete stationary source, we arrive at the following definitions.

Definition 8.5 (Entropy of a Discrete Source). The Nth order entropy of a discrete stationary source $\mathcal{F}$ with Nth order joint pmf $p(f_1, f_2, \ldots, f_N)$ is the joint entropy among N successive samples $\mathcal{F}_1, \mathcal{F}_2, \ldots, \mathcal{F}_N$ of $\mathcal{F}$:

$$H_N(\mathcal{F}) = H(\mathcal{F}_1, \mathcal{F}_2, \ldots, \mathcal{F}_N)$$

$$= - \sum_{[f_1, f_2, \ldots, f_N] \in \mathcal{A}^N} p(f_1, f_2, \ldots, f_N) \log_2 p(f_1, f_2, \ldots, f_N), \tag{8.2.13}$$

where $\mathcal{A}^N$ represents the N-fold Cartesian product of $\mathcal{A}$.

Definition 8.6 (Conditional Entropy of a Discrete Source). The Mth order conditional entropy of a discrete stationary source $\mathcal{F}$ with Mth order joint pmf $p(f_1, f_2, \ldots, f_M)$ and Mth order conditional pmf $p(f_{M+1} \mid f_M, f_{M-1}, \ldots, f_1)$ is the conditional entropy of a sample $\mathcal{F}_{M+1}$ given its previous M samples $\mathcal{F}_M, \mathcal{F}_{M-1}, \ldots, \mathcal{F}_1$:

$$H_{C,M}(\mathcal{F}) = H(\mathcal{F}_{M+1} \mid \mathcal{F}_M, \mathcal{F}_{M-1}, \ldots, \mathcal{F}_1)$$

$$= \sum_{[f_1, f_2, \ldots, f_M] \in \mathcal{A}^M} p(f_1, f_2, \ldots, f_M) H(\mathcal{F}_{M+1} \mid f_M, f_{M-1}, \ldots, f_1) \tag{8.2.14}$$

where

$$H(\mathcal{F}_{M+1} \mid f_M, f_{M-1}, \ldots, f_1)$$

$$= - \sum_{f_{M+1} \in \mathcal{A}} p(f_{M+1} \mid f_M, f_{M-1}, \ldots, f_1) \log_2 p(f_{M+1} \mid f_M, f_{M-1}, \ldots, f_1). \tag{8.2.15}$$

Definition 8.7 (Entropy Rate of a Discrete Source). It can be shown that $\frac{1}{N} H_N(\mathcal{F})$ and $H_{C,N}(\mathcal{F})$ are both nonincreasing functions of N. Furthermore, $\lim_{N \to \infty} \frac{1}{N} H_N(\mathcal{F})$ and $\lim_{N \to \infty} H_{C,N}(\mathcal{F})$ both exist and are equal. This limit is defined as the *entropy rate* of the source:

$$\bar{H}(\mathcal{F}) = \lim_{N\to\infty} \frac{1}{N} H_N(\mathcal{F}) = \lim_{N\to\infty} H_{C,N}(\mathcal{F}). \qquad (8.2.16)$$

Relation between Joint Entropy, Conditional Entropy and Entropy Rate
It can be shown (Problem 8.3) that, for any finite N and M,

$$H_N(\mathcal{F}) = \sum_{m=1}^{N-1} H_{C,m}(\mathcal{F}) + H_1(\mathcal{F}) \qquad (8.2.17)$$

$$\bar{H}(\mathcal{F}) \le H_{C,N-1}(\mathcal{F}) \le \frac{1}{N} H_N(\mathcal{F}) \le H_1(\mathcal{F}), \qquad (8.2.18)$$

If $\mathcal{F}$ is i.i.d., then

$$\bar{H}(\mathcal{F}) = \frac{1}{N} H_N(\mathcal{F}) = H_{C,M}(\mathcal{F}) = H_1(\mathcal{F}). \qquad (8.2.19)$$

As will be shown in Section 8.3.1, the entropy rate is the lower bound on the minimal bit rate required to code a discrete source losslessly. The bound can be achieved only when an infinite number of samples are coded together. On the other hand, the first-order entropy and conditional entropy provide lower bounds on the achievable bit rates when one sample is coded independently, or conditioned on the previous samples, respectively.

Definition 8.8 (Mutual Information between Discrete Sources).   Let $\mathcal{F}$ and $\mathcal{G}$ represent two discrete stationary sources, with Nth order pmfs $p(f_1, f_2, \ldots, f_N)$ and $p(g_1, g_2, \ldots, g_N)$, respectively. Further, let $p(f_1, f_2, \ldots, f_N, g_1, g_2, \ldots, g_N)$ represent the joint pmf of N samples each from $\mathcal{F}$ and $\mathcal{G}$. The Nth order mutual information between $\mathcal{F}$ and $\mathcal{G}$ is defined as

$$I_N(\mathcal{F}; \mathcal{G}) = \sum_{[f_1, f_2, \ldots, f_N] \in \mathcal{A}_f^N} \sum_{[g_1, g_2, \ldots, g_N] \in \mathcal{A}_g^N} p(f_1, f_2, \ldots, f_N, g_1, g_2, \ldots, g_N)$$

$$\cdot \log_2 \frac{p(f_1, f_2, \ldots, f_N, g_1, g_2, \ldots, g_N)}{p(f_1, f_2, \ldots, f_N) p(g_1, g_2, \ldots, g_N)} \qquad (8.2.20)$$

In lossy source coding, a source $\mathcal{F} = \{\mathcal{F}_n\}$ is represented by a quantized version $\mathcal{G} = \{\mathcal{G}_n\}$. The first-order mutual information $I_1(\mathcal{F}; \mathcal{G})$ measures the amount of information provided by a quantized sample $\mathcal{G}_n$ about the original sample $\mathcal{F}_n$. On the other hand, $I_N(\mathcal{F}; \mathcal{G})$ measures the information provided by a block of N quantized samples about the original N samples. As will be shown in Section 8.3.2, $\lim_{N\to\infty} \min \frac{1}{N} I_N(\mathcal{F}; \mathcal{G})$ provides a lower bound on the minimal bit rate required for a desired distortion D between $\mathcal{F}$ and $\mathcal{G}$, where the minimum is taken over all possible $\mathcal{G}$s that have distortion D from $\mathcal{F}$.

8.2.3 Entropy and Mutual Information for Continuous Sources

For a continuous source, the pmf for any possible value in its support region is zero. Direct application of the previously defined entropy would yield an infinite entropy. Instead, we use the pdf of the source to define differential entropy.*

Definition 8.9 (Differential Entropy of a Continuous RV).   The differential entropy of a continuous RV $\mathcal{F}$ with support range $\mathcal{B}$ and pdf $p_{\mathcal{F}}(f)$ is defined by

$$h(\mathcal{F}) = - \int_{f \in \mathcal{B}} p_{\mathcal{F}}(f) \log_2 p_{\mathcal{F}}(f) \, df. \tag{8.2.21}$$

As with entropy, differential entropy measures the uncertainty or information content of a continuous RV. Unlike entropy, differential entropy can take on negative values and in fact can vary in the range of $(-\infty, \infty)$, depending on the pdf.

All other definitions (including joint entropy, conditional entropy, and mutual information) given in Section 8.2.2 for discrete sources can be similarly given for continuous sources, by replacing any involved pmf with a corresponding pdf. Similarly, the relations between entropy, conditional entropy, and mutual information are still valid, with entropy replaced by differential entropy.

Gaussian Sources   A very important continuous source is the Gaussian source, in which each sample follows a Gaussian distribution and each group of N samples follows a joint Gaussian distribution. Here, we give the differential entropy for a Gaussian RV and a Gaussion random vector.

A Gaussian RV with mean η and variance σ^2 has a pdf

$$p(f) = \frac{1}{\sqrt{2\pi}\sigma} e^{-(f-\eta)^2/2\sigma^2}, \quad f \in (-\infty, \infty). \tag{8.2.22}$$

It can be shown (Problem 8.4) that its differential entropy is

$$h_{\text{Gaussian}} = \frac{1}{2} \log_2(2\pi e \sigma^2). \tag{8.2.23}$$

An N-dimensional Gaussian random vector $\mathcal{F}$ with mean η and covariance matrix $[\mathbf{C}]$ has a joint pdf

$$p(\mathbf{f}) = \frac{1}{(\sqrt{2\pi})^N |\det[\mathbf{C}]|^{1/2}} \exp\left(-\frac{1}{2}(\mathbf{f} - \eta)^T [\mathbf{C}]^{-1} (\mathbf{f} - \eta)\right). \tag{8.2.24}$$

It can be shown that the differential entropy of this random vector (or joint differential entropy among the N components) is

$$h_{\text{Gaussian}-N} = \frac{1}{2} \log_2((2\pi e)^N |\det[\mathbf{C}]|) = \frac{1}{2} \log_2\left((2\pi e)^N \prod_n \lambda_n\right), \tag{8.2.25}$$

where $\lambda_n, n = 1, 2, \ldots, N$, are the eigenvalues of $[\mathbf{C}]$.

---

*The previously defined entropy based on the pmf is also known as *absolute entropy*.

An important property of the Gaussian distribution is that the Gaussian RV has the greatest entropy among all continuous RVs with the same variance. Similarly, a Gaussian random vector has the greatest entropy among all random vectors with the same covariance matrix. This makes the Gaussian source most difficult to code, in that it requires a higher bit rate to represent than any other source, for a given distortion criterion.

8.3 INFORMATION THEORY FOR SOURCE CODING

The theoretical foundation for source coding is established by several important results of information theory.* These results, due to Shannon, the father of information theory, establish bounds on the minimal bit rate required to realize lossless and lossy coding. A rigorous and comprehensive treatment of this subject is beyond the scope of this book. The intent of this section is to help the reader to understand the theoretical limits of source coding and how they may be used to guide the design of practical coding techniques. The reader is referred to Cover and Thomas [3] and Berger [1] for more in-depth exposition.

8.3.1 Bound for Lossless Coding

Scalar Lossless Coding   For a particular realization $\{f_n\}$ of a discrete source $\mathcal{F}$ with an alphabet $\mathcal{A} = \{a_1, a_2, \ldots, a_L\}$, scalar lossless coding refers to the assignment of a binary codeword c_n for each sample f_n. This requires that one predesign a codebook $\mathcal{C} = \{c(a_1), c(a_2), \ldots, c(a_L)\}$, where $c(a_i)$ is the codeword for symbol a_i. Then the codeword for f_n will be $c_n = c(f_n)$. For a codebook to be useful, a coded sequence must be *uniquely decodable;* that is a sequence of codewords must correspond to one and only one possible sequence of source symbols. Note that this is a stronger requirement than one that the mapping between the alphabet and the codebook be one-to-one. Let $l(a_i)$ denote the length (i.e., the number of bits) of $c(a_i)$; then the *bit rate,* defined as the average number of bits per sample,[†] will be equal to the average length per codeword, that is,

$$R = \sum_{a_i \in \mathcal{A}} p(a_i) l(a_i). \qquad (8.3.1)$$

Theorem 8.1 (Bound for Scalar Lossless Coding).   The minimum bit rate $\bar{R}_1(\mathcal{F})$ required to represent a discrete stationary source $\mathcal{F}$ by assigning one codeword to each sample satisfies

$$H_1(\mathcal{F}) \leq \bar{R}_1(\mathcal{F}) \leq H_1(\mathcal{F}) + 1. \qquad (8.3.2)$$

---

*Source coding refers to the process of converting samples from a source into a stream of binary bits, whereas channel coding refers to further manipulating the source bits to add protection against transmission errors.

[†]Elsewhere in this book, we also define bit rate as the number of bits per second (bps). For a source outputting f_s sample/s and representing each sample with R bits, the bit rate in terms of bps is $f_s R$.

The lower bound can be achieved when the pmf of the source is a function of power two. That is, there exist a set of integers $\{m_1, m_2, \ldots, m_L\}$ such that $p(a_i) = 2^{-m_i}$. In this case, $l(a_i) = -\log_2 p(a_i) = m_i$.   ∎

Theorem 8.1 tells us that the first-order entropy of a discrete source $H_1(\mathcal{F})$ determines the range of the minimum bit rate for scalar coding of the source. This is not surprising, as $H_1(\mathcal{F})$ measures the information content (in bit/sample) carried by each new sample in a source.

Vector Lossless Coding   The preceding theorem tells us that, depending on the source, the minimal achievable rate can differ from the first-order entropy by as much as one bit/sample. To further reduce the bit rate, one can treat each successive N samples in a discrete sequence as a vector sample, and assign one codeword for each possible vector symbol in $\mathcal{A}^N$. The first order entropy of the vector source would be the Nth order entropy of the original source. If $\bar{R}^N$ represent the minimal number of bits per vector, then applying Theorem 8.1 to the preceding vector source yields

$$H_N(\mathcal{F}) \leq \bar{R}^N(\mathcal{F}) \leq H_N(\mathcal{F}) + 1 \qquad (8.3.3)$$

If we let $\bar{R}_N = \bar{R}^N/N$ represent the minimal number of bits per sample, we arrive at the following theorem.

Theorem 8.2 (Bound for Vector Lossless Coding).   The minimum bit rate $\bar{R}_N(\mathcal{F})$ required to represent a discrete stationary source $\mathcal{F}$ by assigning one codeword to each group of N successive samples satisfies

$$H_N(\mathcal{F})/N \leq \bar{R}_N(\mathcal{F}) \leq H_N(\mathcal{F})/N + 1/N. \qquad (8.3.4)$$

In the limit when $N \to \infty$, from Equation (8.2.16), we have

$$\lim_{N \to \infty} \bar{R}_N(\mathcal{F}) = \bar{H}(\mathcal{F}). \qquad (8.3.5)$$

∎

Theorem 8.2 tells us that the bit rate can be made arbitrarily close to the source entropy rate by coding many samples together.

Conditional Lossless Coding   We have seen that one can improve upon the efficiency of scalar coding by vector coding. An alternative is conditional coding, also known as *context-based coding*. With Mth order conditional coding, the codeword for a current sample depends on the pattern formed by the previous M samples. Such a pattern is called the *context*. Specifically, a separate codebook is designed for each possible context, based on the conditional distribution of the output sample given this context. If the source alphabet size is L, then the maximum number of contexts, and consequently the maximum number of codebooks, is L^M. Applying Theorem 8.1 to the conditional distribution under context m, the minimal bit rate under this context satisfies

$$H_{C,M}^m(\mathcal{F}) \leq \bar{R}_{C,M}^m(\mathcal{F}) \leq H_{C,M}^m(\mathcal{F}) + 1, \qquad (8.3.6)$$

where $H^m_{C,M}(\mathcal{F})$ represents the entropy of the Mth order conditional distribution under context m. Let p_m represent the probability of context m, then the average minimal bit rate with a conditioning order M is

$$\bar{R}_{C,M}(\mathcal{F}) = \sum_m p_m \bar{R}^m_{C,M}(\mathcal{F}). \tag{8.3.7}$$

Substituting Equation (8.3.6) into the above yields the following result.

Theorem 8.3 (Bound for Conditional Lossless Coding).   The minimum bit rate $\bar{R}_{C,M}(\mathcal{F})$ required to represent a discrete stationary source $\mathcal{F}$ using Mth order conditional coding satisfies

$$H_{C,M}(\mathcal{F}) \leq \bar{R}_{C,M}(\mathcal{F}) \leq H_{C,M}(\mathcal{F}) + 1. \tag{8.3.8}$$

In the limit when $M \to \infty$, we have

$$\bar{H}(\mathcal{F}) \leq \lim_{M \to \infty} \bar{R}_{C,M}(\mathcal{F}) \leq \bar{H}(\mathcal{F}) + 1. \tag{8.3.9}$$

∎

Compared to Equation (8.3.2), conditional coding can achieve a lower bit rate than scalar coding, because $\bar{H}(\mathcal{F}) < H_1(\mathcal{F})$—unless the source is i.i.d. However, because one still codes one sample at a time, the upper bound still differs from the entropy rate by one bit, even when the conditioning order M goes to infinity. Comparing Nth order vector lossless coding with $(N-1)$th order conditional coding, the former always has the same or a larger lower bound, according to Equation (8.2.18). On the other hand, vector coding will have a smaller upper bound if $\frac{1}{N}H_N(\mathcal{F}) - H_{C,N-1}(\mathcal{F}) < (N-1)/N$. Therefore, which method is more efficient depends on the actual source statistics.

8.3.2 Bound for Lossy Coding*

When a source is continuous, it is impossible to represent it exactly with a finite number of bits. Recall that the absolute entropy rate of a continuous source is infinite. Therefore, applying the bound on lossless coding (Equation (8.3.5)) to a continuous source leads to a minimal bit rate that is infinite. To describe a continuous source with a finite number of bits, one must in some way quantize the source so that each sample or group of samples can only take symbols from a finite alphabet. Such quantization will inevitably induce distortion between the original source and the reconstructed source based on the quantized representation. One can directly quantize the source samples, either one at a time (scalar quantization) or a group of samples together (vector quantization), or transform the samples into some other parameters and quantize the resulting parameters. The quantized symbols can then be encoded losslessly into binary codewords. At the decoder, the codewords are first mapped back to the quantized symbols, from which each sample is reconstructed.

In general, a lossy coding process can be considered as mapping every input vector $\mathbf{f} = \{f_1, f_2, \ldots, f_N\}$ consisting of N samples from the original source $\mathcal{F}$ into a

---

*Sections marked with an asterisk may be skipped or left for further exploration.

quantized vector $\mathbf{g} = Q(\mathbf{f})$. The vector $\mathbf{g}$ must belong to a predesigned *reconstruction codebook* of a finite size $L, C = \{\mathbf{g}_1, \mathbf{g}_2, \ldots, \mathbf{g}_L\}$. We will assume that each component of $\mathbf{f}$ and $\mathbf{g}$ can only take values from the same support range $\mathcal{B}$. Using fixed-length coding, each quantized vector is represented by $\log_2(L)$ bits, so that the bit rate (bit/sample) of the coded sequence is

$$R_N = \frac{1}{N} \log_2(L). \qquad (8.3.10)$$

Given a predesigned reconstruction codebook, the bit rate can be further reduced from Equation (8.3.10) by applying variable-length coding to the indices of codewords. However, in the remaining discussions, we assume that the bit rate is determined according to Equation (8.3.10).

To meet a given distortion criterion between $\mathcal{F}$ and $\mathcal{G}$, the required codebook size L depends on the coding scheme (i.e., how the codebook is designed). Intuitively, the bit rate can be reduced by increasing the vector length N, especially when the source is not i.i.d. Rate distortion theory defines the lower bound on the bit rate required to achieve a given distortion, when N approaches infinity.

Note that lossy coding can be applied to both continuous and discrete sources. Recall that for a discrete source, the minimal bit rate required for lossless (zero-distortion) coding is bounded by the source entropy rate. To reach an even lower rate, one must quantize the samples to a smaller alphabet, thereby incurring a certain degree of distortion. The following discussions apply to both continuous and discrete sources.

Distortion Measure   Up to this point, we have used the term "distortion" without formally defining it. To measure the distortion between an original source $\mathcal{F}$ and a reconstructed source $\mathcal{G}$, we first define a distortion measure $d_1(f, g)$ between two scalars f and g. When f and g are both real values, the most common distortion measures are the squared error, $d_1(f, g) = (f - g)^2$, and absolute error, $d_1(f, g) = |f - g|$. Then the average distortion per sample between an original vector $\mathbf{f}$ and a reconstructed vector $\mathbf{g}$ is

$$d_N(\mathbf{f}, \mathbf{g}) = \frac{1}{N} \sum_n d_1(f_n, g_n). \qquad (8.3.11)$$

This distortion is defined for a given vector and its reconstruction vector. To evaluate the average performance over all possible input vectors, we define the following expected distortion between $\mathcal{F}$ and $\mathcal{G}$ when the vector length is N:

$$E\{d_N(\mathcal{F}, \mathcal{G})\} = \int_{\mathbf{f} \in \mathcal{B}^N} \int_{\mathbf{g} \in \mathcal{B}^N} p_N(\mathbf{f}) q_N(\mathbf{g}\,|\,\mathbf{f}) d_N(\mathbf{f}, \mathbf{g})\, d\mathbf{g}\, d\mathbf{f}, \qquad (8.3.12)$$

where $p_N(\mathbf{f})$ represents the joint pdf of N successive samples in $\mathcal{F}$, $q_N(\mathbf{g}\,|\,\mathbf{f})$ represents the conditional probability that $\mathbf{f}$ is reconstructed as $\mathbf{g}$, and $\mathcal{B}^N$ represents the N-fold Cartesian product of $\mathcal{B}$. Note that $q_N(\mathbf{g}\,|\,\mathbf{f})$ essentially characterizes the coding scheme, although in practice, a deterministic mapping between $\mathbf{f}$ and $\mathbf{g}$ is used. When $d_1(f, g)$ is the squared error, the preceding measure is known as *mean squared error* (MSE), which is widely used, primarily because of its mathematical tractability.

In the preceding, we have assumed that the source is continuous, and the functions $p_N(\cdot)$, $q_N(\cdot)$ represent pdfs. When the source is discrete, the integration over $\mathcal{B}^N$ should be replaced by a sum over $\mathcal{A}^N$, and $p_N(\cdot)$, $q_N(\cdot)$ would represent corresponding pmfs.

Rate-Distortion Bound   The performance of a lossy coder is measured by its *operational rate-distortion (RD) function,* denoted by $R(D)$, which describes the bit rate R required to achieve a given distortion D, for given source characteristics. The rate-distortion theorem (Theorem 8.4) establishes the bound on the minimal rate required among all possible coders, when the vector length N reaches infinity. We call the function relating the minimal rate and the distortion the *rate-distortion bound,* denoted by $\bar{R}(D)$. Let $R_N(D; q_N(\mathbf{g}\,|\,\mathbf{f}))$ represent the bit rate required to achieve distortion D by a coding scheme described by $q_N(\mathbf{g}\,|\,\mathbf{f})$. The RD bound is defined as

$$\bar{R}(D) = \lim_{N\to\infty} \min_{q_N(\mathbf{g}\,|\,\mathbf{f})\in Q_{D,N}} R_N(D; q_N(\mathbf{g}\,|\,\mathbf{f})) \tag{8.3.13}$$

where

$$Q_{D,N} = \{q_N(\mathbf{g}\,|\,\mathbf{f}) : E\{d_N(\mathcal{F}, \mathcal{G})\} \le D\} \tag{8.3.14}$$

is the set of conditional probabilities (i.e., coding schemes) that satisfy the specified distortion constraint.

Instead of representing R as a function of D, one can also represent D in terms of R, which will lead to the *operational distortion-rate function $D(R)$* or *distortion-rate bound $\bar{D}(R)$.*

Theorem 8.4 (Bound on Lossy Coding).   The RD bound for coding a stationary source $\mathcal{F}$ at a distortion D is given by[*]

$$\bar{R}(D) = \lim_{N\to\infty} \min_{q_N(\mathbf{g}\,|\,\mathbf{f})\in Q_{D,N}} \frac{1}{N} I_N(\mathcal{F}; \mathcal{G}). \tag{8.3.15}$$

When the source is i.i.d., the RD bound becomes

$$\bar{R}(D) = \min_{q_1(\mathbf{g}\,|\,\mathbf{f})\in Q_{D,1}} I_1(\mathcal{F}; \mathcal{G}). \tag{8.3.16}$$

■

Recall that $I_N(\mathcal{F}; \mathcal{G})$ describes the amount of information provided by N samples of $\mathcal{G}$ about the corresponding N samples of $\mathcal{F}$. Therefore, it is not surprising that $-I_N(\mathcal{F}; \mathcal{G})$ determines the number of bits per sample required to specify $\mathcal{F}$ with distortion D when $N \to \infty$.

An important result from information theory is that, even when the source is i.i.d., one can reduce the bit rate R by coding many samples together. In fact, the RD bound is achievable only when the vector length N reaches infinity.

In general, the bound in Theoem 8.4 is difficult to calculate for an arbitrary source. A more useful bound is described by the following theorem.

---

[*]In information theory, the RD function of a source $\mathcal{F}$ is defined as the right-hand side of Equation (8.3.15). Here we call it the RD bound. The Shannon source-coding theorem states that there exists a coding scheme by which the bit rate required to achieve a distortion D approaches $\bar{R}(D)$ [1, 3].

Theorem 8.5. Under the MSE distortion criterion, the RD bound for any stationary source $\mathcal{F}$ satisfies

$$\bar{R}_L(D) \le \bar{R}(D) \le \bar{R}_G(D), \tag{8.3.17}$$

where $\bar{R}_G(D)$ is the RD bound for a Gaussian source with the same variance as $\mathcal{F}$, and $\bar{R}_L(D)$ is known as the Shannon lower bound, with

$$\bar{R}_L(D) = \bar{h}(\mathcal{F}) - \frac{1}{2} \log_2 2\pi e D = \frac{1}{2} \log_2 \frac{Q(\mathcal{F})}{D}, \tag{8.3.18}$$

where $\bar{h}(\mathcal{F})$ is the entropy rate of $\mathcal{F}$, and $Q(\mathcal{F}) = (1/2\pi e)2^{2\bar{h}(\mathcal{F})}$ is called the entropy power of $\mathcal{F}$. For a Gaussian source, $\bar{R}_L(D) = \bar{R}_G(D)$. ∎

Theorem 8.5 tells us that among all sources with the same variance, the Gaussian source requires the highest bit rate, to satisfy the same distortion criterion. In fact, for any source, the optimal coding scheme that achieves the RD bound is the one by which the quantization error sequence is an i.i.d. Gaussian source with variance D, with a differential entropy $\frac{1}{2} \log_2 2\pi e D$. Therefore, the lower bound given in Equation (8.3.18) is equal to the difference between the differential entropy of the original source and that of the quantization error.

8.3.3 Rate-Distortion Bounds for Gaussian Sources*

For most sources, a closed-form solution for the RD bound is difficult to derive. One group of exceptions is the various types of Gaussian sources, for which closed-form solutions have been found, when the distortion criterion is MSE. We summarize these results below.

i.i.d. Gaussian Source The RD bound for an i.i.d. Gaussian source with variance σ^2 is

$$\bar{R}(D) = \begin{cases} \frac{1}{2} \log_2 \frac{\sigma^2}{D}, & 0 \le D < \sigma^2, \\ 0, & D \ge \sigma^2. \end{cases} \tag{8.3.19}$$

Alternatively, one can write

$$\bar{D}(R) = \sigma^2 2^{-2R}. \tag{8.3.20}$$

i.i.d. Vector Gaussian Source with Independent Components Consider an i.i.d. vector Gaussian source, in which each vector sample consists of N independent Gaussian RVs with variances $\sigma_n^2, n = 1, 2, \ldots, N$. The RD bound for this source is given by

$$\begin{cases} R(\alpha) = \frac{1}{N} \sum_n \max\{0, \frac{1}{2} \log_2 \frac{\sigma_n^2}{\alpha}\}, \\ D(\alpha) = \frac{1}{N} \sum_n \min\{\alpha, \sigma_n^2\}, \end{cases} \tag{8.3.21}$$

*Sections marked with an asterisk may be skipped or left for further exploration.

where α is an intermediate parameter. The RD bound for the feasible distortion region can be obtained by letting α traverse the range $(0, \max\{\sigma_n^2\})$.

In the case when D is sufficiently small that $D \leq \min\{\sigma_n^2\}$, the preceding reduces to

$$\bar{R}(D) = \frac{1}{2}\log_2 \frac{\left(\prod_n \sigma_n^2\right)^{1/N}}{D}, \tag{8.3.22}$$

or

$$\bar{D}(R) = \left(\prod_n \sigma_n^2\right)^{1/N} 2^{-2R}. \tag{8.3.23}$$

This RD bound is achieved by coding each component in a vector sample independently,* with $R_n = \max\{0, \frac{1}{2}\log_2(\sigma_n^2/\alpha)\}$ bits used for component n. For a component with $\sigma_n^2 \leq \alpha$, it can be always represented by zero, thus $R_n = 0$. For the remaining components, the bits are allocated so that they all have equal distortion $D_n = \alpha$. This result is obtained by optimally allocating the total bit rate NR among the N components so that the average distortion D is minimized. Essentially, it requires that the bit rates R_n are chosen such that the slopes of the distortion-rate functions for different components, $\partial D_n(R_n)/\partial R_n$, are equal at these rates.

i.i.d. Vector Gaussian Source with Correlated Components   Let $[\mathbf{C}]$ represent the covariance matrix of the N components of each vector sample, and let $\lambda_n, n = 1, 2, \ldots N$, represent the eigenvalues of $[\mathbf{C}]$. The RD bound for this source is given by

$$\begin{cases} R(\alpha) = \frac{1}{N}\sum_n \max\left\{0, \frac{1}{2}\log_2 \frac{\lambda_n}{\alpha}\right\}, \\ D(\alpha) = \frac{1}{N}\sum_n \min\{\alpha, \lambda_n\}, \end{cases} \tag{8.3.24}$$

where $\alpha \in (0, \max\{\lambda_n\})$ is an intermediate parameter.

In the case when D is sufficiently small that $D \leq \min\{\lambda_n\}$, the preceding reduces to

$$\bar{R}(D) = \frac{1}{2}\log_2 \frac{\left(\prod_n \lambda_n\right)^{1/N}}{D} = \frac{1}{2}\log_2 \frac{|\det[\mathbf{C}]|^{1/N}}{D} \tag{8.3.25}$$

or

$$\bar{D}(R) = \left(\prod_n \lambda_n\right)^{1/N} 2^{-2R} = |\det[\mathbf{C}]|^{1/N} 2^{-2R}. \tag{8.3.26}$$

The preceding result is obtained by transforming the original vector using a transform matrix consisting of the eigenvectors of $[\mathbf{C}]$, and then applying the RD bound in Equation (8.3.21) to the transformed vector, which now has independent components.

---

*But successive samples in the nth component must be coded together to achieve the RD bound.

General Gaussian Source   An arbitrary stationary Gaussian source with mean η can be described by its auto-covariance function $C(k) = E\{(\mathcal{F}_n - \eta)(\mathcal{F}_{n+k} - \eta)\}$, $k = 0, \pm 1, \ldots$ or its power spectrum $S(e^{j\omega})$.* The RD bound for such a source is the limit of the RD bound given in Equation (8.3.24) when $N \to \infty$. The result is

$$\begin{cases} R(\alpha) = \frac{1}{2\pi} \int_{-\pi}^{\pi} \max\{0, \frac{1}{2} \log_2 \frac{S(e^{j\omega})}{\alpha}\} \, d\omega, \\ D(\alpha) = \frac{1}{2\pi} \int_{-\pi}^{\pi} \min\{\alpha, S(e^{j\omega})\} \, d\omega. \end{cases} \tag{8.3.27}$$

When D is sufficiently small that the region of ω where $D > S(e^{j\omega})$ can be neglected, we have

$$\bar{R}(D) = \frac{1}{4\pi} \int_{-\pi}^{\pi} \log_2 \frac{S(e^{j\omega})}{D} \, d\omega. \tag{8.3.28}$$

Equation (8.3.27) tells us that to represent a source with a spectrum $S(e^{j\omega})$ at distortion α, the frequency range over which $S(e^{j\omega}) < \alpha$ need not be coded, whereas the remaining range should be coded so that the error signal has a power equal to α over this range. This bit-allocation procedure is commonly known as *reverse water filling*.

In the special case of a first-order Gauss-Markov process with $C(k) = \sigma^2 |\rho|^k$, we have

$$R(D) = \frac{1}{2} \log_2 \frac{(1 - \rho^2)\sigma^2}{D}, \tag{8.3.29}$$

or

$$D(R) = (1 - \rho^2)\sigma^2 2^{-2R}. \tag{8.3.30}$$

8.4 BINARY ENCODING

Binary encoding is the process of representing each possible symbol from a finite alphabet source by a sequence of binary bits, called a *codeword*. The codewords for all possible symbols form a *codebook* or simply *code*. A symbol may correspond to one or several original or quantized pixel values or model parameters. Because the mapping from symbols to codewords is one-to-one, this process is also called lossless coding.

For a code to be useful, it should satisfy the following properties: (1) it should be uniquely decodable, which means that there is a one-to-one mapping between the codeword and the symbol; and (2) the code should be *instantaneously decodable,* which means that one can immediately decode a group of bits if they match a codeword, without examining subsequent bits in the coded sequence (see Figure 8.2). This second property requires that no prefix of any codeword is another valid codeword; such code is called *prefix code*. Although instantaneous decodability is a stronger requirement than uniqueness and enables fast decoding, it does not limit coding efficiency. It can be shown that a prefix code can produce the minimal bit rate among all uniquely decodable codes, for the same source [4]. All practical coding methods produce prefix codes.

Obviously, the simplest binary code is the fixed-length binary representation of all possible symbols. If the number of symbols is L, then the bit rate will be $\lceil \log_2 L \rceil$

---

*$S(e^{j\omega})$ is the discrete-time Fourier transform of $C(k)$.

Codebook 1
(a prefix code)

Symbol	Codeword
a_1	"0"
a_2	"10"
a_3	"110"
a_4	"111"

Codebook 2
(not a prefix code)

Symbol	Codeword
a_1	"0"
a_2	"01"
a_3	"100"
a_4	"011"

Bitstream: 0 0 1 1 0 1 0 1 1 0 1 0 0

Decoded string based on codebook 1: $0|0|1\ 1\ 0|1\ 0|1\ 1\ 0|1\ 0|0 \rightarrow a_1\ a_1\ a_3\ a_2\ a_3\ a_2\ a_1$
(can decode instantaneously)

Decoded string based on codebook 2: $0|0\ 1\ 1|0\ 1|0\ 1\ 1|0|1\ 0\ 0 \rightarrow a_1\ a_4\ a_2\ a_4\ a_1\ a_3$
(must look ahead to decode)

Figure 8.2   With the codebook on the left (a prefix code), the given bit stream can be instantaneously decoded upon the detection of a complete codeword. With the codebook on the right (a nonprefix code), one does not know whether the first "0" in the bit stream corresponds to the codeword "0", or to the first bit in the second or fourth codeword. One must examine additional bits to make the decision. Because the second bit is still "0", yet there is no codeword that starts with two "0"s, the first "0" must correspond to the codeword "0".

bit/symbol. From Section 8.3.1, we know that the lowest possible bit rate by any codebook is the entropy rate of the source. Unless the source has a uniform distribution, the fixed-length coding scheme will be very inefficient in that the bit rate is much higher than the source entropy. To reduce the bit rate, *variable-length coding* (VLC) is needed, which assigns a shorter codeword to a higher probability symbol, so that the average bit rate is low. Because the bit rate of an appropriately designed variable-length coder can approach the entropy of the source, VLC is also referred to as *entropy coding*.

There are three popular VLC methods. Huffman coding converts a fixed number of symbols into a variable-length codeword; the LZW method converts a variable number of symbols into a fixed-length codeword; and, finally, arithmetic coding converts a variable number of symbols into a variable-length codeword. Huffman and arithmetic methods are probability-model–based, and both can reach the entropy bound asymptotically. The arithmetic coding method can more easily achieve asymptotic performance and can easily adapt to variations in signal statistics, but it is also more complex than Huffman coding. The LZW method [12, 10] does not require knowledge of the signal statistics and thus is universally applicable, but it is less efficient than the other two methods. Huffman and arithmetic coding methods have been employed in various video coding standards. We describe these two methods in the following two subsections.

8.4.1 Huffman Coding

Huffman coding is the most popular approach for lossless coding of a discrete source with alphabet $\mathcal{A} = \{a_1, a_2, \ldots, a_L\}$ and pmf $p(a_l)$. It designs a codebook for all possible symbols, so that a symbol appearing more frequently is assigned a shorter codeword. The basic procedure for codebook design using Huffman coding is as follows:

Step 1: Arrange the symbol probabilities $p(a_l), l = 1, 2, \ldots, L$, in a decreasing order and consider them as leaf nodes of a tree.

Step 2: While there is more than one node:

(a) Find the two nodes with the smallest probability and arbitrarily assign 1 and 0 to these two nodes.

(b) Merge the two nodes to form a new node whose probability is the sum of the two merged nodes. Go back to Step 1.

Step 3: For each symbol, determine its codeword by tracing the assigned bits from the corresponding leaf node to the top of the tree. The bit at the leaf node is the last bit of the codeword.

Example 8.1 (Scalar Huffman Coding)

An example of Huffman coding is shown in Figure 8.3, in which the source consists of four symbols. The symbols and their probabilities are shown in the left two columns. The resulting codewords and the codeword lengths for these symbols are shown in the right two columns. Also given are the average bit rate R and the first-order entropy of the source H_1. We can see that, indeed, $H_1 < R < H_1 + 1$.

A notorious disadvantage of Huffman coding, when applied to individual samples, is that at least one bit must be used for each sample. To further reduce the bit rate, one can treat each group of N samples as one entity and give each group a codeword. This leads to vector Huffman coding, a special case of vector lossless coding. This is shown in the next example.

Example 8.2 (Vector Huffman Coding)

Consider a source with the same alphabet and the pmf as in Example 8.1. Suppose that we further know that the conditional distribution of a sample $\mathcal{F}_n$ given its previous sample

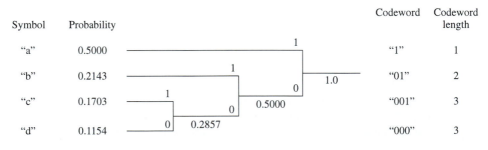

Bit rate $R = 1.7857$ Entropy $H_1 = 1.7707$

Figure 8.3 Example of scalar Huffman coding (see Example 8.1).

$\mathcal{F}_{n-1}$ is described by the following matrix:

$$[Q] = \begin{bmatrix} 0.6250 & 0.3750 & 0.3750 & 0.3750 \\ 0.1875 & 0.3125 & 0.1875 & 0.1875 \\ 0.1250 & 0.1875 & 0.3125 & 0.1250 \\ 0.0625 & 0.1250 & 0.1250 & 0.3125 \end{bmatrix}$$

(8.4.1)

where the element in row i and column j specifies the conditional probability $q(i \mid j)$, which is the probability that $\mathcal{F}_n$ takes the ith symbol, given that $\mathcal{F}_{n-1}$ equals the jth symbol. We can determine the joint pmf of every two samples by

$$p(\hat{f}_{n-1}, \hat{f}_n) = p(\hat{f}_{n-1})q(\hat{f}_n \mid \hat{f}_{n-1}).$$

(8.4.2)

Applying this to all possible combination of two symbols, we obtain the probabilities of all possible 2-D vector symbols, which are given in the first column of Figure 8.4. Then we can apply Huffman coding to this new source; the resulting codebook is also shown in Figure 8.4. In this case, the bit rate per vector sample is $R^2 = 3.5003$, and bit rate per sample is $R_2 = R^2/2 = 1.75015$. On the other hand, $H_2/2 = 1.7314$. Indeed, we have $\frac{1}{2}H_2 < R_2 < \frac{1}{2}H_2 + 1/2$. Compared to the result in Example 8.1, we see that the bit rate is reduced by using vector coding.

Instead of coding two samples together, we can also use conditional Huffman coding, which uses different codebooks depending on the symbols taken by previous samples. This is shown in the next example.

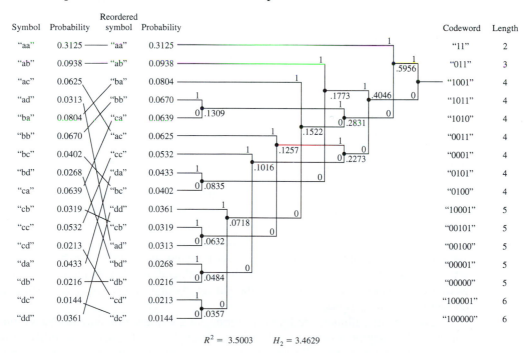

$R^2 = 3.5003 \qquad H_2 = 3.4629$

Figure 8.4   Example of vector Huffman coding (see Example 8.2).

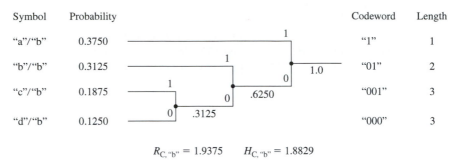

$$R_{C,\,\text{"b"}} = 1.9375 \qquad H_{C,\,\text{"b"}} = 1.8829$$

Figure 8.5   Example of conditional Huffman coding, when the conditioning context is "b" (see Example 8.3).

Example 8.3 (Conditional Huffman Coding)

We continue with the situation of previous example, but we now design a separate codebook for each possible context consisting of one previous sample. In this case, there are only four possible contexts. Figure 8.5 shows the codebook design for the context "b" (i.e., the previous symbol being "b"), based on the conditional probabilities given in the second column of the matrix given in Equation (8.4.1). Following the same procedure, one can design the codebook for the other three possible contexts. The resulting bit rates for the four contexts "a", "b", "c", and "d" are 1.5625, 1.9375, 1.9375, and 1.9375, and the average bit rate over all possible contexts is $R_{C,1} = 1.7500$. On the other hand, the conditional entropies for the four contexts are 1.5016, 1.8829, 1.8829, and 1.8829, respectively, with an average conditional entropy being $H_{C,1} = 1.6922$. As expected, we have $H_{C,1} < R_{C,1} < H_{C,1} + 1$. In this particular example, $R_{C,1} < R_2$, although this may not be the case in general.

If the source is a Markov-1 process, then the entropy rate of the source is equal to its first-order conditional entropy. Hence, $H_{C,1}$ is the lowest possible bit rate achievable by vector coding even when the vector length is infinite.

8.4.2 Arithmetic Coding

A disadvantage of Huffman coding is that it cannot closely approach the entropy bound unless many samples are coded together. This is because each sample (or group of samples) uses at least one bit. Therefore, even for a very small alphabet, the bit rate cannot be lower than one bit/sample, in the case of scalar coding, or one bit/vector-sample, in the case of vector coding. One way to avoid such problems is by converting a variable number of samples into a variable-length codeword. This method is known as *arithmetic coding*. It can approach the entropy bound more closely than Huffman coding for common signals.

The idea in arithmetic coding is to represent a sequence of symbols by an interval in the line segment from zero to one, with length equal to the probability of the sequence. Because the probabilities of all possible sequences sum to one, the intervals corresponding to all possible sequences will fill up the entire line segment. The coded bits for a sequence essentially became the binary representation of any point in the interval corresponding to the sequence. Instead of waiting for the entire sequence to appear before deciding on the interval and its binary representation, one starts from an initial interval determined based on the first symbol, and then recursively divides the previous interval after each new symbol appears. To specify an interval, the lower and upper boundaries of the interval are represented in binary. Whenever the most significant bit (MSB) of the lower boundary is the same as that of the upper boundary, this bit is shifted out. At the end of the source sequence, all the bits that have been collected would be the binary representation of an intermediate point in the interval corresponding to the sequence. The more likely a sequence is, the longer will be the interval, and the fewer bits will be needed to specify the interval.

Let $a_l, l = 1, 2, \ldots, L$, represent the L possible symbols of the source, p_l the probability of symbol a_l, and $q_l = \sum_{k=1}^{l} p_l$ the cumulative probability up to the lth symbol. Let d_n, l_n, and u_n represent the length, lower boundary, and upper boundary, respectively, of the interval at step n, with $l_0 = 0$, $u_0 = 1$, and $d_0 = 1$. Upon receiving the nth symbol of the source sequence, if the symbol is a_l, the lower and upper boundaries are computed using

$$d_n = d_{n-1} * p_l; \quad l_n = l_{n-1} + d_{n-1} * q_{l-1}; \quad u_n = l_n + d_n. \tag{8.4.3}$$

Let the binary representation of l_n and u_n be $(b_1^l, b_2^l, \ldots, b_K^l)$ and $(b_1^u, b_2^u, \ldots, b_K^u)$, respectively.* If the first few b_k^l and b_k^u are the same, then they are shifted out; otherwise, all the bits are retained. The algorithm then proceeds by examining the next symbol.

Example 8.4

Figure 8.6(a) shows an example of arithmetic encoding. The source alphabet contains 3 symbols, "a", "b", and "c", with probability distribution $p(\text{"a"}) = 1/2$, $p(\text{"b"}) = 1/4$, $p(\text{"c"}) = 1/4$. The intervals corresponding to these symbols are shown in the first column in Figure 8.6(a). The first source symbol is "a", which corresponds to the first interval in the first column. The lower and upper boundaries are $l_1 = 0 = (0000...)$ and $u_1 = 1/2 = (1000...)$. Because the MSBs of l_1 and u_1 are different, no bits are shifted out. The next source symbol is "b"; therefore, we take the second interval in the second column. The boundaries of this interval are, using Equation (8.4.3), $l_2 = 1/4 = (01000...)$ and $u_2 = 3/8 = (011000...)$. The first two bits are common; therefore, we can shift out "01" in the coded bit stream. Following this procedure, we can determine the coded bits for all the following source symbols.

To decode a received bit stream, one determines the lower and upper boundaries of an interval corresponding to the received bits. If the interval falls in a range that can only be associated with a particular source symbol, that symbol is decoded. If not, additional bits are examined.

Example 8.4 (continued)

The decoding process is shown in Figure 8.6(b). The first received bit is "0", and the possible lower and upper boundaries both with "0" as the MSB are $l = (0000...) = 0$ and $u = (0111...) = (1000...) = \frac{1}{2}$. From the first column in Figure 8.6(a), this interval

---

*The binary representation of a number f between 0 and 1 is given by $(b_1, b_2, \ldots, b_K)$ if $f = \sum_{k=1}^{K} b_k 2^{-k}$.

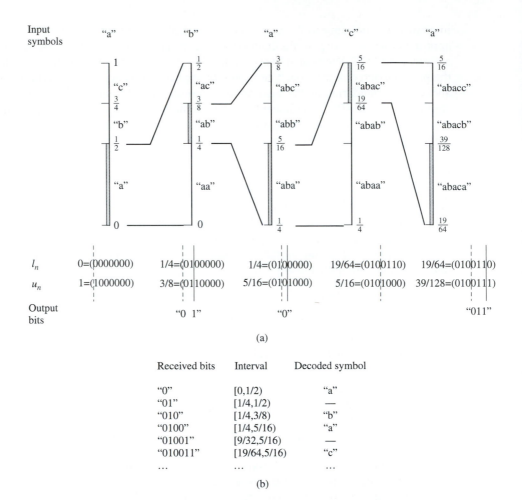

Figure 8.6 Example of arithmetic coding. (a) Encoding a source sequence
"abaca...". The shaded interval in the nth column indicates the interval corresponding
to the sequence up to the nth symbol. (b) Decoding a bit stream "010011...".

corresponds to "a", therefore, one can output "a" as a decoded symbol. With the next received bit being "1", the possible lower and upper boundaries both with "01" as the first two
bits are $l = (0100...) = \frac{1}{4}$ and $u = (0111...) = (1000...) = \frac{1}{2}$. From the second column
in Figure 8.6(a), the range $(\frac{1}{4}, \frac{1}{2})$ can come from either "ab" or "ac". Therefore, one cannot
decode a symbol at this step. With the third bit being "0", the lower and upper boundaries
are $l = (010000...) = \frac{1}{4}$ and $u = (010111...) = (011000...) = \frac{3}{8}$. From the second column in Figure 8.6(a), this interval corresponds to "ab". Therefore, one can output "b" in
this step. Continuing this process, one can decode all the symbols for the given bit stream.

The preceding description of arithmetic coding takes place on a conceptual level
only, which assumes infinite precision in calculation of the interval boundaries. As

more source symbols are processed, the interval corresponding to the input sequence becomes increasingly small. To process a reasonably long sequence, one would need a computer with an extremely high precision to specify the boundary values. Fortunately, it is not necessary to record the absolute boundary values for the encoding and decoding process to work. A previously determined interval can be rescaled to length one once a common bit of the lower and upper boundaries is shifted out. Such implementation is known as *finite-precision arithmetic coding*. Instead of using fractional arithmetic for the lower and upper bound calculation, integer implementation, which is simpler, has also been developed. A full description of practical encoding and decoding algorithms is beyond the scope of this book; the reader is referred to the excellent tutorial by Witten, Neal, and Cleary [11]. A good coverage of both Huffman and arithmetic coding can also be found in Sayood [9].

The bit rate of arithmetic coding is bounded [9] by

$$H_N(\mathcal{F})/N \le R \le H_N(\mathcal{F})/N + 2/N, \tag{8.4.4}$$

where N is the number of symbols in the sequence being coded and $H_N(\mathcal{F})$ is the Nth order entropy of the sequence. Recall that if we code the entire length-N sequence using vector Huffman coding, the bit rate will be bounded as in Equation (8.3.4). Therefore, both methods can approach the entropy rate closely when N is sufficiently large. However, with Huffman coding, one must design and store the codebook for all possible sequences of length N, which has complexity that grows exponentially with N. This limits the sequence length that can be used in practice. With arithmetic coding, no codebook need be designed in advance for each possible source sequence. Rather, bits are sequentially obtained whenever the lower and upper boundaries of the identified interval have common MSB. The coded sequence length can be as long as the length of the source. Therefore, in practice, arithmetic coding can closely approach the entropy rate.

Another advantage of arithmetic coding is that one can adapt to changes in the source statistics simply by updating the symbol probability table. One can also easily realize conditional coding, by using different probability tables for different contexts. With Huffman coding, one would have to redesign the codebook based on an updated probability table, or have multiple codebooks for different conditioning states.

Because of the higher coding efficiency and ease of adaptation, arithmetic coding is a better alternative than Huffman coding, as long as the computation involved is acceptable.

8.5 SCALAR QUANTIZATION

8.5.1 Fundamentals

The most basic lossy coding method is scalar quantization (SQ), which quantizes each sample in a source signal to one of the reconstruction values in a predesigned reconstruction codebook. In general, the original source can be either continuous or discrete. Here we consider only the first case, where the original sample can take on any value in a continuous support range B over the real line. This support range is divided into a number of regions, $B_l, l = 1, 2, \ldots, L$, so that values in the same region B_l are mapped into the same reconstruction value g_l. When the source is discrete, but all possible symbols can be ordered along the real line, we can treat each sample as the output from an fictitious continuous source, which only produces certain discrete values over a continuous range. With this approach, the theory and techniques discussed in the following for continuous sources can be applied to discrete sources as well.

A quantizer is described by the *number of reconstruction values, L*, the *boundary values, $b_l, l = 0, 1, \ldots, L$*, and *reconstruction values, $g_l, l = 1, 2, \ldots, L$*. The boundary values can alternatively be described by the *partition regions $B_l = [b_{l-1}, b_l)$*.* Letting $\mathcal{L} = \{1, 2, \ldots, L\}$, the quantization function is described by:

$$Q(f) = g_l \quad \text{if } f \in B_l, \ \ l \in \mathcal{L}. \tag{8.5.1}$$

This function is illustrated in Figure 8.7. In a digital computer, the reconstruction value g_l is simply specified by the integer index l, which, in a fixed-length binary representation, requires $R = \lceil \log_2 L \rceil$ bits. Here, the symbol $\lceil x \rceil$ denotes the upper integer of x, that is, the smallest integer that is equal to or greater than x.

Distortion of a Quantizer   In Section 8.3.2, we defined a distortion measure between original and quantized sources, when N samples are quantized together. In the case of a scalar quantizer, $N = 1$, the distortion measure given in Equation (8.3.12) becomes

$$D_q = E\{d_1(\mathcal{F}, Q(\mathcal{F}))\} = \int_{f \in B} d_1(f, Q(f)) p(f) \, df \tag{8.5.2}$$

$$= \sum_{l \in \mathcal{L}} P(B_l) D_{q,l} \tag{8.5.3}$$

with

$$D_{q,l} = \int_{f \in B_l} d_1(f, g_l) p(f \mid f \in B_l) \, df. \tag{8.5.4}$$

In the preceding equations, $\mathcal{F}$ represents the RV corresponding to any sample in a source $\mathcal{F}$; $p(f), f \in B$, the pdf of $\mathcal{F}$; $P(B_l) = \int_{B_l} p(f) \, df$ the probability that $f \in B_l$; and $p(f \mid f \in B_l) = p(f)/P(B_l)$ the conditional pdf of $\mathcal{F}$ given $\mathcal{F} \in B_l$. Obviously, $D_{q,l}$ denotes the mean distortion incurred in region B_l.

When $d_1(f, g) = (f - g)^2$, the preceding distortion reduces to the MSE, which is equal to the variance of the quantization error $\mathcal{F} - Q(\mathcal{F})$, denoted by σ_q^2. Specifically,

$$\sigma_q^2 = E\{|\mathcal{F} - Q(\mathcal{F})|^2\} = \sum_{l \in \mathcal{L}} P(B_l) \int_{b_{l-1}}^{b_l} (f - g_l)^2 p(f \mid B_l) \, df. \tag{8.5.5}$$

---

*The use of a closed left-hand boundary is arbitrary. One can also use $B_l = (b_{l-1}, b_l]$.

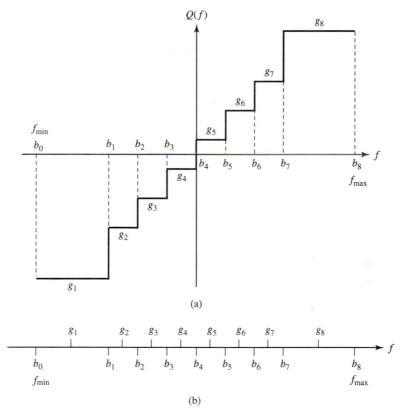

Figure 8.7 Illustration of a scalar quantizer: (a) representation as a function; (b) representation as a line partition.

8.5.2 Uniform Quantization

The simplest scalar quantizer is a uniform quantizer, which has equal distances between adjacent boundary values and between adjacent reconstruction values:

$$b_l - b_{l-1} = g_l - g_{l-1} = q \tag{8.5.6}$$

where q is called the *quantization stepsize*. Such a quantizer is only applicable to a signal with a finite dynamic range B. Let f_{min} and f_{max} represent the minimal and maximal values, then $B = f_{max} - f_{min}$. The quantizer is specified by the number of quantization levels L, or the bit rate R, or the quantization stepsize q. These parameters are related by $q = B/L = B2^{-R}$. The boundary and reconstruction values are given by

$$b_l = l * q + f_{min}, \quad g_l = (l-1) * q + q/2 + f_{min}. \tag{8.5.7}$$

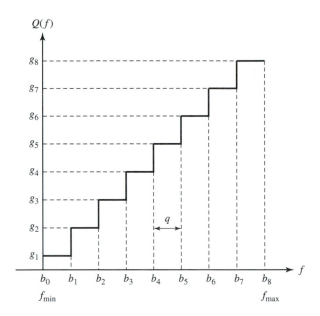

Figure 8.8   A uniform quantizer.

The quantization function can be described in a closed form:

$$Q(f) = \left\lfloor \frac{f - f_{min}}{q} \right\rfloor * q + \frac{q}{2} + f_{min}, \tag{8.5.8}$$

where, $\lfloor x \rfloor$ represents the lower integer of x, that is, the largest integer that is equal or smaller than x (see Figure 8.8).

If the source is uniformly distributed, that is, if

$$p(f) = \begin{cases} 1/B & f \in (f_{min}, f_{max}) \\ 0 & \text{otherwise} \end{cases} \tag{8.5.9}$$

then substituting Equations (8.5.7) and (8.5.9) into Equation (8.5.5) yields

$$\sigma_q^2 = \frac{q^2}{12} = \sigma_f^2 \, 2^{-2R} \tag{8.5.10}$$

where $\sigma_f^2 = B^2/12$ is the variance of the original signal $\mathcal{F}$. The SNR of the quantizer is

$$\text{SNR} = 10 \log_{10} \frac{\sigma_f^2}{\sigma_q^2} = (20 \log_{10} 2) \; R = 6.02R \; \text{(dB)} \tag{8.5.11}$$

Therefore, *every additional bit in a uniform quantizer leads to a 6.02 dB gain in SNR for a uniform source.* This is a well known result in quantization theory.

8.5.3 Optimal Scalar Quantizer

The uniform quantizers introduced thus far are simple to implement, but may not be the most efficient in representing a source. In this section, we examine how to design an optimal scalar quantizer, so that the distortion is minimized for a fixed bit rate, given the source pdf $p(f)$.

Minimum Mean Square Error (MMSE) Quantizer   In an MMSE quantizer, b_l, g_l are chosen such that the MSE σ_q^2 is minimized for a given L. From the calculus of variations, the necessary conditions are $\partial \sigma_q^2 / \partial b_l = 0$ and $\partial \sigma_q^2 / \partial g_l = 0$, $\forall l$. From Equation (8.5.5),

$$\frac{\partial \sigma_q^2}{\partial b_l} = \frac{\partial}{\partial b_l} \left(\int_{b_{l-1}}^{b_l} (f - g_l)^2 p(f) \, df + \int_{b_l}^{b_{l+1}} (f - g_{l+1})^2 p(f) \, df \right)$$

$$= (b_l - g_l)^2 p(b_l) - (b_l - g_{l+1})^2 p(b_l) = 0,$$

$$\frac{\partial \sigma_q^2}{\partial g_l} = \frac{\partial}{\partial g_l} \int_{b_{l-1}}^{b_l} (f - g_l)^2 p(f) \, df = -\int_{b_{l-1}}^{b_l} 2(f - g_l) p(f) \, df = 0.$$

Simplification of the preceding equations yields:

$$b_l = \frac{g_l + g_{l+1}}{2}, \tag{8.5.12}$$

$$g_l = E\{\mathcal{F} \mid \mathcal{F} \in \mathcal{B}_l\} = \int_{\mathcal{B}_l} f \, p(f \mid f \in \mathcal{B}_l) \, df. \tag{8.5.13}$$

The conditional mean $E\{\mathcal{F} \mid \mathcal{F} \in \mathcal{B}_l\}$ is referred to as the *centroid of $\mathcal{B}_l$*.

The results in Equations (8.5.12–13) state that *the optimum boundary values lie halfway between the optimal reconstruction values, which, in turn, lie at the centroids of the regions between the boundary values*. Note that requiring $b_l = (g_l + g_{l+1})/2$ is equivalent to quantizing any $f \in (g_l, g_{l+1})$ to g_l if f is closer to g_l than to g_{l+1}. A more general statement is that any value f is quantized to the reconstruction value that is closest to f. Therefore, Equation (8.5.12) can be equivalently written as

$$\mathcal{B}_l = \{ f : d_1(f, g_l) \le d_1(f, g_l'), \forall l' \ne l \}. \tag{8.5.14}$$

The condition given by Equation (8.5.12) or (8.5.14) is called the *nearest-neighbor condition*. On the other hand, equation (8.5.13) is known as the *centroid condition*.

It can be shown that the nearest-neighbor condition given in Equation (8.5.14) is still valid with a distortion measure rather than the MSE. The centroid condition in Equation (8.5.13), however, must be changed to

$$g_l = \operatorname{argmin}_g \{ E\{d_1(\mathcal{F}, g) \mid \mathcal{F} \in \mathcal{B}_l\} \}, \tag{8.5.15}$$

where $\operatorname{argmin}_g \{E(g)\}$ represents the argument g that minimizes the function $E(g)$. The g_l defined in Equation (8.5.15) is called the *generalized centroid of $\mathcal{B}_l$*, and its solution depends on the definition of $d_1(f, g)$. For example, if $d_1(f, g) = |f - g|$, then g_l is the median value in $\mathcal{B}_l$ (see Problem 8.12).

An important property of an MMSE quantizer is that it equalizes the quantization error in different partition regions, that is,

$$P(\mathcal{B}_l) D_{g,l} = D_q, \quad \forall l \in \mathcal{L} \tag{8.5.16}$$

For other important statistical properties of an optimal quantizer, see [5, 4]; also see Problem 8.8.

MMSE Quantizer for a Uniform Source   Substituting the uniform distribution in Equation (8.5.9) into Equations (8.5.12–13) will yield the solution given in Equation (8.5.7). Thus, as expected, *the MMSE quantizer for a uniform source is uniform.*

MMSE Quantizer for a Nonuniform Source   For an arbitrary pdf $p(f)$, the conditions in Equations (8.5.12–13) do not always have a closed-form solution. Numerical procedures must be used to determine the optimal set of b_l, g_l. Table 4.3 in [5] gives the optimal $\tilde{b}_l$ and $\tilde{g}_l$ for uniform, Gaussian, Laplacian, and Gamma sources with unit variance and zero mean. It can be shown (see Problem 8.10) that the optimal b_l and g_l for a source with arbitrary mean μ_f and variance σ_f^2 can be obtained by:

$$b_l = \sigma_f \tilde{b}_l + \mu_f, \qquad g_l = \sigma_f \tilde{g}_l + \mu_f. \tag{8.5.17}$$

The MMSE quantizer for a nonuniform source is in general nonuniform, that is, the partition regions are unequal in length. In this case, the quantized value cannot be determined by a closed-form formula as with a uniform quantizer. In general, one must compare the given value f with all boundary values, until the region for which $b_{l-1} \leq f < b_l$ is found. Alternatively, one can find g_l that is closest to f. A sequential search would require up to L comparisons. Using a binary tree search procedure, the number of comparisons can be reduced to $\lceil \log_2 L \rceil$.

Asymptotic Performance of MMSE Quantizer   For a source with an arbitrary pdf, it is difficult to derive a closed-form solution for the quantizer parameters b_l, g_l and, consequently, the quantization error σ_q^2, using the nearest-neighbor and centroid conditions. However, when the bit rate R and, consequently, the quantization level $L = 2^R$ are very high, a closed-form solution for the MMSE quantizer in terms of the pdf can be derived [5, 4]. The associated quantization error can be expressed as

$$\sigma_q^2 = \epsilon^2 \sigma_f^2 2^{-2R} \tag{8.5.18}$$

with

$$\epsilon^2 = \frac{1}{12} \left(\int_{-\infty}^{\infty} \tilde{p}(f)^{1/3}\, df \right)^3, \tag{8.5.19}$$

where $\tilde{p}(f) = \sigma_f p(\sigma_f f)$ denotes the pdf of a normalized source with unit variance. Equation (8.5.18) is known as the high-rate approximation of the scalar quantizer performance. Comparing the preceding with Equation (8.5.10), we see that $\epsilon^2 = 1$ for a uniform source. For a nonuniform source, $\epsilon^2 > 1$ in general. In particular, for a Gaussian source, $\epsilon^2 = 2.71$. Recall that the distortion-rate bound for an i.i.d. Gaussian source is given in Equation (8.3.20). Therefore, the best scalar quantizer still differs from the bound by $\frac{1}{2} \log_2 \epsilon^2 = 0.7191$ bit/sample. For the values of ϵ^2 for several other sources, see [5, Table 4.8]. To be closer to the bound, multiple samples must be quantized together. The factor ϵ^2 can be reduced by applying variable-length coding to the quantized indices; this is discussed further in Section 8.6.4.

Lloyd Algorithm for Optimal Scalar Quantizer Design Based on Training Data   When the distribution of the signal to be quantized is unknown, the quantizer can be designed based on a training set containing representative samples to be quantized. A popular method for designing a quantizer based on training data is the Lloyd algorithm [7, 4]. As shown in Fig. 8.9, the algorithm updates the reconstruction and boundary

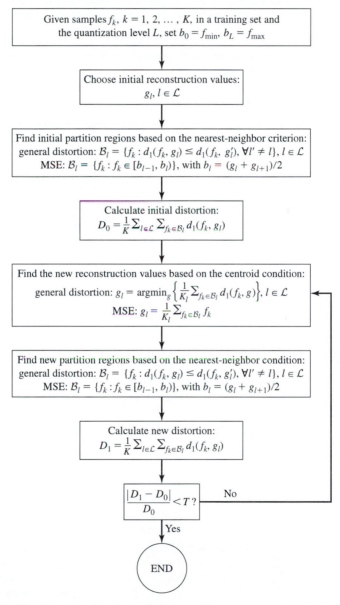

Figure 8.9   Lloyd algorithm for designing an optimal scalar quantizer from training data.

values iteratively. At each iteration, one first recalculates the reconstruction values based on the centroid condition, and then partitions all the training samples based on the nearest-neighbor condition. The statistical expectation required in the calculation of the generalized centroid and distortion are approximated by sample averages. Under the MSE criterion, the reconstruction value for a partition is simply the mean value of training samples belonging to that partition, whereas the boundary values are simply the averages of two adjacent reconstruction values. In the figure, K_l denotes the number of samples in B_l.

8.6 VECTOR QUANTIZATION

8.6.1 Fundamentals

Rather than quantizing one sample at a time, one can quantize a group of N samples together. Each sample group is known as a vector, and the process of quantizing a vector at a time is called vector quantization (VQ). In the image or video case, a vector usually corresponds to a square block of pixels. The motivation for using VQ primarily stems from the observation that in a typical image or video frame, samples in a block are correlated such that there are certain block patterns that are more likely than others. The task of VQ is essentially to find the L most popular patterns, and replace any given block by one of the representative patterns that most closely resembles the original pattern.

   The preceding discussion assumes that a vector corresponds to a group of sample values. In a source in which each sample is described by multiple values, each sample can itself be quantized using a vector quantizer. For example, each sample in a color image contains three components. To display a 24-bit color image on a computer monitor using an eight-bit graphics card, one must find 256 colors that best represent all possible $2^{24} = 16$ million colors.

   Consider each possible ND vector $\mathbf{f}$ as a point in an ND space; the VQ problem is to partition the ND space into L regions $\mathcal{B}_l$, and represent all the points in region $\mathcal{B}_l$ by a representative point $\mathbf{g}_l$. This process is illustrated in Figure 8.10 for the case of $N = 2$.

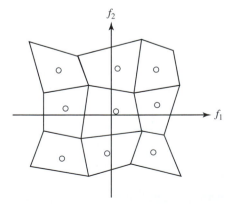

Figure 8.10   Illustration of vector quantization with vector dimension $N = 2$. The open circle inside each partition region represents the reconstruction codeword of that region.

We call $\mathcal{B}_l$ the *partition regions* and $\mathbf{g}_l$ the *reconstruction vectors* or *codewords*. The set containing all codewords $C = \{\mathbf{g}_l, l \in \mathcal{L}\}$ is called the *codebook*.* The quantization function can be described as

$$Q(\mathbf{f}) = \mathbf{g}_l, \quad \text{if } \mathbf{f} \in \mathcal{B}_l. \tag{8.6.1}$$

If the reconstruction codewords are converted into binary bits using fixed-length binary coding, then each group of N samples requires $\lceil \log_2 L \rceil$ bits, and the bit rate (bit/sample) is

$$R = \frac{1}{N} \lceil \log_2 L \rceil. \tag{8.6.2}$$

Comparing Equation (8.6.1) with Equation (8.5.1) and Figure 8.10 with Figure 8.7(b), it can be seen that the scalar quantizer is just a special case of the vector quantizer, where $\mathbf{f}$ is a scalar and $\mathcal{B}_l$ is a line segment. In general, the partition region $\mathcal{B}_l$ cannot be described simply by a few decision values, as in the scalar case.

To evaluate the quantization error introduced by a vector quantizer, let $\mathcal{F}$ and $Q(\mathcal{F})$ represent the original and the quantized vector, and let $p_N(\mathbf{f})$ represent the joint pdf of the components in $\mathcal{F}$. Using the notation in Section 8.3.2, the conditional probability $q_N(\mathbf{g} \,|\, \mathbf{f})$ for the vector quantizer described by Equation (8.6.1) is

$$q_N(\mathbf{g} \,|\, \mathbf{f}) = \begin{cases} 1, & \text{if } \mathbf{f} \in \mathcal{B}_l, \mathbf{g} = \mathbf{g}_l, \\ 0, & \text{otherwise.} \end{cases} \tag{8.6.3}$$

The distortion criterion in Equation (8.3.12) reduces, in this case, to

$$D_q = E\{d_N(\mathcal{F}, Q(\mathcal{F}))\} = \int_{\mathcal{B}} p_N(\mathbf{f}) d_N(\mathbf{f}, Q(\mathbf{f})) \, d\mathbf{f} \tag{8.6.4}$$

$$= \sum_{l=1}^{L} P(\mathcal{B}_l) D_{q,l} \tag{8.6.5}$$

with

$$D_{q,l} = E\{d_N(\mathcal{F}, Q(\mathcal{F})) \,|\, \mathcal{F} \in \mathcal{B}_l\} = \int_{\mathbf{f} \in \mathcal{B}_l} p_N(\mathbf{f} \,|\, \mathbf{f} \in \mathcal{B}_l) d_N(\mathbf{f}, \mathbf{g}_l) \, d\mathbf{f}. \tag{8.6.6}$$

Nearest-Neighbor Quantizer and Complexity of VQ   In general, a vector quantizer is specified by both the codewords $\mathbf{g}_l$ and the partition regions $\mathcal{B}_l$. With a *nearest-neighbor quantizer*, for any input vector $\mathbf{f}$, the quantized vector is determined by comparing $\mathbf{f}$ with all codewords and finding the one that is closest to it under the distance measure $d_N(\mathbf{f}, \mathbf{g}_l)$. In other words, the partition regions are implicitly determined by the codewords through

$$\mathcal{B}_l = \{\mathbf{f} \in \mathcal{R}^N : d_N(\mathbf{f}, \mathbf{g}_l) \leq d_N(\mathbf{f}, \mathbf{g}_{l'}), \forall l' \neq l\}. \tag{8.6.7}$$

---

*Note that each reconstruction codeword must be mapped into a binary codeword, to represent the quantized signal using binary bits. "Codeword" and "codebook" in this section refer to the reconstruction codeword and codebook, whereas in Sections 8.3.1 and 8.4 they referred to binary codeword and codebook.

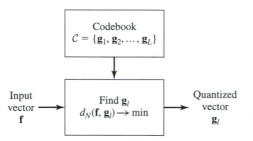

Figure 8.11   Operation of a nearest-neighbor quantizer.

As will be shown in Section 8.6.3, this is a necessary condition for minimizing the mean quantization error. Essentially, all quantizers used in practice belong to this category; Figure 8.11 illustrates the operation of a nearest-neighbor quantizer.

With a nearest-neighbor quantizer, for each input vector, one must evaluate the distortion between this vector and all L reconstruction codewords. With the squared error measure,

$$d_N(\mathbf{f}, \mathbf{g}) = \frac{1}{N} \sum_{n=1}^{N} (f_n - g_n)^2, \tag{8.6.8}$$

each evaluation requires N operations (an operation consists of a subtraction, a multiplication, and an addition). Hence, the total number of operations required to quantize one vector is NL. From Equation (8.6.2), we have $L = 2^{NR}$; hence, the number of operations is $N2^{NR}$. In terms of the storage requirement, each codeword requires N basic memory units and all L codewords require $NL = N2^{NR}$ memory units. *Thus, both the complexity and storage requirements of VQ increase exponentially with the vector dimension N.* Recall that Theorem 8.4 tells us that the RD bound for a source is typically achieved only when N reaches infinity. In general, the larger N, the more efficient the coder. However, the exponential growth of the complexity with N limits the vector dimension that can be used in practice. Consider the quantization of video frames, by which each block of N pixels is quantized using a size-L codebook. Assuming the frame rate is f_t and the frame size is $N_x \times N_y$, then the number of operations per second will be $(f_t N_x N_y / N) \cdot N2^{NR} = f_t N_x N_y 2^{NR}$. For a BT.601 video Y component, $f_t = 30$, $N_x = 720$, $N_y = 480$, a vector size of 4×4 and bit rate of 1 bit/sample lead to an operation count of $6.8E + 11$ per second! For image and video coding, N is usually limited to 4×4 or below.

To reduce the complexity, similarly to binary tree search in the SQ case, one need not do exhaustive search if each partition region is a polytope. In this case, different partition regions are defined by intersecting hyperplanes. One can compare a given vector with one hyperplane at a time, each time eliminating codewords on one side of the plane. This way, the number of comparisons can be reduced to a number significantly lower than L, but may not be as low as $\log_2 L$, unless the codebook has a special tree structure. To further reduce the complexity, various fast VQ schemes have been developed, which impose certain constraints on the codebook structure, so that the number of comparisons can be reduced. Gersho and Gray [4] provide extensive coverage of various VQ techniques.

8.6.2 Lattice Vector Quantizer

As we learned in Section 8.5.2, the simplest scalar quantizer is a uniform quantizer, in which a certain range of the real line is partitioned into equal-length quantization bins. An analogue of the uniform quantizer in the VQ case is a lattice quantizer, in which all the partition regions have the same shape and size, and in fact are all shifted versions of a basic region, which is the Voronoi region of the lattice. Figure 8.12 shows two lattice quantizers, using rectangular and hexagonal lattices, respectively. In Chapter 3, we introduced the concept of lattices for uniform sampling over a multidimensional space. Here we see that lattices can also be used to define uniform quantizers over a multidimensional space.

Recall that a lattice Λ in an ND real space $\mathcal{R}^N$ is defined by a set of basis vectors $\mathbf{v}_n, n = 1, 2, \ldots, N$; or, equivalently, a generating matrix $[\mathbf{V}] = [\mathbf{v}_1, \mathbf{v}_2, \ldots, \mathbf{v}_N]$. The points in a lattice are those vectors that can be represented as

$$\mathbf{g}_l = \sum_{n=1}^{N} m_{l,n}\mathbf{v}_n = [\mathbf{V}]\mathbf{m}_l, \tag{8.6.9}$$

where $m_{l,n} \in \mathcal{Z}$ are integers or $\mathbf{m}_l \in \mathcal{Z}^N$ is an ND integer vector. Usually a codeword $\mathbf{g}_l$ is indexed by its corresponding $\mathbf{m}_l$. The Voronoi region $\mathcal{V}$ of Λ is the region in which all points are closer to the origin than any other nonzero lattice points. With a lattice quantizer, the codebook consists of all the points in a lattice, or in a coset of a lattice, which is a shifted version of a lattice. Such quantizers have infinite numbers of

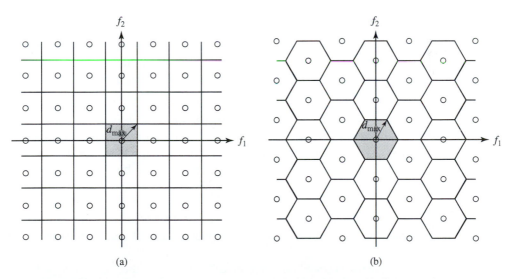

(a)   (b)

Figure 8.12 Lattice quantizers using (a) rectangular and (b) hexagonal lattices. The shaded regions indicate the Voronoi regions; $d_{\max}$ is the maximal distance between an input vector and its quantized version.

codewords. In reality, a truncated version of a lattice or its coset is more often used, which only take sample points that are within a certain support range in $\mathcal{R}^N$.

For a uniformly distributed source over a certain finite region in $\mathcal{R}^N$, if we assume that all the partition regions fit in the support region perfectly,* then all the partition regions are equally likely, and the distortion in each region is the same as that over the Voronoi region. Recall that the volume of the Voronoi region $\mathcal{V}$ is equal to $|\det[\mathbf{V}]|$. Thus, the conditional pdf $p(\mathbf{f}\,|\,\mathbf{f} \in \mathcal{V}) = 1/|\det[\mathbf{V}]|$. The mean quantization error is therefore

$$D_q = D_{q,l} = \frac{1}{|\det[\mathbf{V}]|} \int_{\mathcal{V}} d_N(\mathbf{f}, \mathbf{0})\, d\mathbf{f}. \tag{8.6.10}$$

With the MSE criterion, we have

$$D_q = \frac{1}{|\det[\mathbf{V}]|} \int_{\mathcal{V}} \|\mathbf{f}\|^2\, d\mathbf{f}. \tag{8.6.11}$$

The values of D_q for several high-dimensional lattices have been tabulated by Conway and Sloane [2].

It is worth noting that even for an i.i.d. source, VQ can improve the coding efficiency over SQ. This is because VQ allows more flexibility in packing a space using partition regions. Consider the two examples given in Figure 8.12. In either case, the maximum distortion in representing a point in any region by its center (the codeword) is $d_{\max} = 1$, so that the two quantizers will yield the same maximum distortion. But to fill the same 2-D space, fewer hexagons are needed, because the Voronoi region for the hexagonal lattice is larger than that for the rectangular lattice. Specifically, the areas of the Voronoi regions for the rectangular and hexagonal lattices in this example are $2d_{\max}^2$ and $2.59d_{\max}^2$, respectively. In the entire support region considered in the figure, the rectangular lattice has more than 36 partition regions, whereas the hexagonal lattice has 30 regions.[†] Now, the rectangular lattice quantizer can also be realized by applying a uniform scalar quantizer in each dimension. This means that even when two samples are independent and both are uniformly distributed, quantizing them together as a vector with a well-designed vector quantizer can reduce the bit rate.

Implementation of a Lattice Quantizer   As described previously, for an arbitrary nearest-neighbor quantizer, the quantization of a given vector $\mathbf{f}$ involves an exhaustive search. This is not necessary with a lattice quantizer. In this case, one can first determine a real index vector of the quantized vector by

$$\mathbf{m} = [\mathbf{V}]^{-1}\mathbf{f}. \tag{8.6.12}$$

Then one can evaluate the distortions associated with all the integer index vectors neighboring $\mathbf{m}$, by taking the lower and upper integer of each component of $\mathbf{m}$, and

---

*This is not true in general unless a hypercubic lattice is used. But when L is large, the number of partition regions on the boundary of the support region is small, and the impact due to those regions can be ignored.

[†]Recall from Chapter 3 that we know that the hexagonal lattice is more efficient in covering a space.

determine which gives the minimal distortion. To further reduce the complexity, one can also simply round **m** to the nearest integer vector. The reader is referred to [2] for a more detailed discussion of lattice encoding algorithms and performances.

8.6.3 Optimal Vector Quantizer

Although lattice quantizers are simple to implement, they may not be the most efficient when the source is not uniformly distributed. For a given vector dimension N and source pdf $p_N(\mathbf{f})$, it is desirable to design an *optimal* quantizer that minimizes a given distortion criterion for a fixed bit rate R or, equivalently, codebook size L. Ideally, one should determine the codewords $\mathbf{g}_l$ and the partition regions $\mathcal{B}_l$ simultaneously to minimize the average distortion D_q. The optimal solution, however, is difficult to find. Instead, one can proceed in two stages. First, for given codewords $\mathbf{g}_l, l \in \mathcal{L}$, the partition regions $\mathcal{B}_l, l \in \mathcal{L}$, should be such that the average distortion over all regions, D_q, is minimized. Second, for a given partition region $\mathcal{B}_l$, the codeword $\mathbf{g}_l$ should be chosen to minimize the distortion within $\mathcal{B}_l$, $D_{q,l}$.

For the first problem, it is easy to see from Equation (8.6.4) that D_q will be minimized if $Q(\mathbf{f})$ is chosen so that $d_N(\mathbf{f}, Q(\mathbf{f}))$ is minimized for any $\mathbf{f}$. That is,

$$Q(\mathbf{f}) = \mathbf{g}_l = \text{argmin}_{\mathbf{g}'_l, l' \in \mathcal{L}}\{d_N(\mathbf{f}, \mathbf{g}'_l)\}. \tag{8.6.13}$$

In other words, $\mathbf{f}$ should be quantized to the codeword nearest to it. This assignment is equivalent to setting the partition region $\mathcal{B}_l$ according to

$$\mathcal{B}_l = \{\mathbf{f} : d_N(\mathbf{f}, \mathbf{g}_l) \leq d_N(\mathbf{f}, \mathbf{g}'_l), \forall l' \neq l\}. \tag{8.6.14}$$

We call either Equation (8.6.13) or Equation (8.6.14) the nearest-neighbor condition.

Now, we consider the second problem. Obviously, we should choose $\mathbf{g}_l$ such that $D_{q,l}$ in Equation (8.6.6) is minimized, that is,

$$\mathbf{g}_l = \text{argmin}_{\mathbf{g}} E\{d_N(\mathcal{F}, \mathbf{g}) \,|\, \mathcal{F} \in \mathcal{B}_l\}. \tag{8.6.15}$$

We call this $\mathbf{g}_l$ the generalized centroid of $\mathcal{B}_l$, and Equation (8.6.15) the centroid condition.

When the distortion measure is MSE,

$$D_{q,l}(\mathbf{g}) = \frac{1}{N} E\{\|\mathcal{F} - \mathbf{g}\|^2 \,|\, \mathcal{F} \in \mathcal{B}_l\} = \frac{1}{N} \int_{\mathcal{B}_l} \|\mathbf{f} - \mathbf{g}\|^2 p(\mathbf{f} \,|\, \mathbf{f} \in \mathcal{B}_l) \, d\mathbf{f}. \tag{8.6.16}$$

The $\mathbf{g}_l$ that minimizes $D_{q,l}$ must satisfy $\partial D_{q,l}/\partial \mathbf{g} = 0$. This yields

$$\mathbf{g}_l = \int_{\mathcal{B}_l} p(\mathbf{f} \,|\, \mathbf{f} \in \mathcal{B}_l)\mathbf{f} \, d\mathbf{f} = E\{\mathcal{F} \,|\, \mathcal{F} \in \mathcal{B}_l\}. \tag{8.6.17}$$

That is, $\mathbf{g}_l$ is the conditional mean or centroid of $\mathcal{B}_l$. Equations (8.6.14–15) define an optimal vector quantizer for an arbitrary distortion measure, and Equations (8.6.14) and (8.6.17) together define a MMSE vector quantizer.

Note that the nearest-neighbor and centroid conditions are necessary but not sufficient conditions for minimizing D_q. A quantizer satisfying both conditions still may

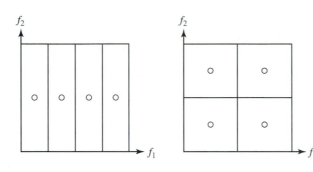

Figure 8.13   Two vector quantizers that each satisfy the nearest-neighbor and centroid conditions, yet the quantizer on the right has a lower distortion.

not achieve the global minimum of D_q. Figure 8.13 shows two 2-D vector quantizers for a uniformly distributed signal. It is easy to verify that each of the quantizers satisfies both conditions, yet clearly the quantizer on the right will have a lower MSE. Therefore, the quantizer on the left is only locally optimal.

Generalized Lloyd Algorithm for Designing an Optimal Vector Quantizer
For an arbitrary source, it is difficult to derive closed-form solutions for $\mathbf{g}_l, \mathcal{B}_l, l \in \mathcal{L}$. A locally optimal solution can be obtained by the *generalized Lloyd algorithm*.* Like the Lloyd algorithm for SQ, the generalized Lloyd algorithm for VQ iteratively determines the optimal $\mathbf{g}_l$ and $\mathcal{B}_l$ based on training data. Figure 8.14 shows the flowchart of this algorithm. Comparing this figure to Figure 8.9, one can easily see the analogy between these two algorithms.

Example 8.5 (Vector Quantizer Design)
Figure 8.15 shows an example of quantizer design using the generalized Lloyd algorithm under the MSE criterion. Within each iteration, we first recalculate the centroid vector (indicated by a circle) for each previously determined partition region, by averaging all the training vectors in that region, and then we repartition (indicated by ovals) all the training vectors based on their distances to the newly updated centroid vectors. Results for the first three iterations are shown. Because the partition after the third iteration remains the same as that after the second iteration, the fourth iteration will not change the codewords or, consequently, the partition regions. Therefore, the result given after the third iteration is the final solution.

The quantizer resulting from the generalized Lloyd algorithm depends on the initial codewords. When they are not chosen properly, the algorithm may converge to a bad local minimum that is far from the global minimum. One simple and yet quite effective way to choose the initial codewords is to use the codewords that would have resulted from applying uniform scalar quantization in each dimension. For an extensive coverage of VQ design algorithms, including the selection of initial codewords, see [4].

---

*This algorithm is also known as the *LBG algorithm* following the authors who first generalized the Lloyd algorithm for vector quantization [6]. The same algorithm has also been used in pattern recognition for automatic clustering of data points into a few classes; there, it is known as the *K*-means algorithm [8].

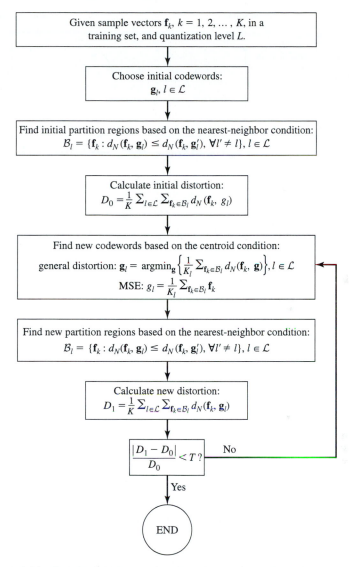

Figure 8.14   Generalized Lloyd algorithm for designing an optimal vector quantizer from training data.

8.6.4 Entropy-Constrained Optimal Quantizer Design

Up to this point, we have assumed that each codeword in a size-L codebook takes $\lceil \log_2 L \rceil$ bits to represent. Because the reconstruction codewords usually appear with different probabilities, we can further reduce the bit rate by using Huffman coding or other variable-length coding methods. In the methods described for optimal quantizer design in Sections 8.5.3 and 8.6.3, we tried to minimize the distortion for a given bit rate R or codebook size L by assuming that R is related to L by Equation (8.6.2). The

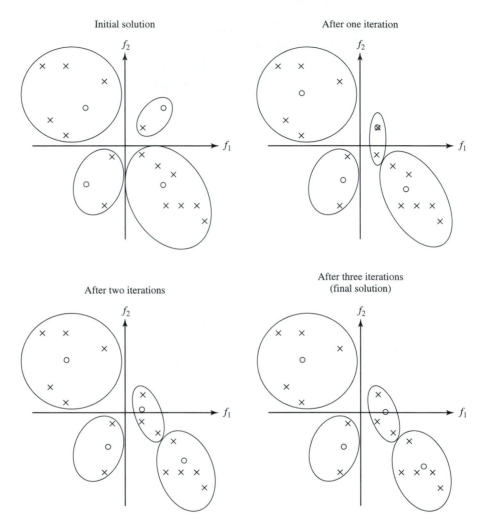

Figure 8.15   An example of vector quantizer design using the generalized Lloyd algorithm. Training vectors are denoted by × and codewords by ○. (see Example 8.5).

resulting solutions in terms of $\mathbf{g}_l$, $\mathcal{B}_l$ are no longer optimal, if the quantized vectors are coded using variable-length coding. With entropy-constrained optimal quantizer design, we minimize the distortion subject to a rate constraint, where the rate is assumed to be equal to the entropy of the quantized source. If we code one codeword at a time, then the bit rate is bounded by

$$\bar{R}_N = -\frac{1}{N} \sum_{l \in \mathcal{L}} P(\mathcal{B}_l) \log_2 P(\mathcal{B}_l), \qquad (8.6.18)$$

because the probability for codeword $\mathbf{g}_l$ is $P(\mathcal{B}_l)$. Given the desired bit rate R and vector dimension N, the optimization problem is to minimize D_q subject to

$$-\sum_{l \in \mathcal{L}} P(\mathcal{B}_l) \log_2 P(\mathcal{B}_l) \leq RN. \tag{8.6.19}$$

Using the Lagrange multiplier method, this constrained optimization problem can be converted into an unconstrained one:

$$\text{minimize } J(\lambda) = D_q + \lambda \left(-\sum_{l \in \mathcal{L}} P(\mathcal{B}_l) \log_2 P(\mathcal{B}_l) - NR \right) \tag{8.6.20}$$

where λ must be chosen such that the constraint in Equation (8.6.19) is satisfied. For each fixed λ, the solution for $\mathbf{g}_l$ and $\mathcal{B}_l$ can be obtained by using the generalized Lloyd algorithm, with the distortion criterion changed from D_q to Equation (8.6.20). Then one can calculate the bit rate required by this solution using Equation (8.6.18). Essentially, different λ values yield solutions that are optimal for different bit rates.

Under the entropy constraint, the optimal quantizers for a nonuniform source tend to approach uniformity when L is large. This way, the maximum distortion in each partition region is similar. However, because the codewords that correspond to regions with higher pdfs will be more likely than others, they will be coded with shorter codewords.

Note that the preceding discussion is equally applicable to scalar quantization. Recall that the distortion-rate function of any non–entropy-constrained MMSE quantizer can be written in the form of Equation (8.5.18) in the high-rate case. The net effect of applying entropy coding is to reduce the factor ϵ^2. For a Gaussian source, using an entropy-constrained MMSE quantizer and scalar lossless coding can reduce ϵ^2 from 2.71 to 1.42, which corresponds to a saving of 0.467 bit/sample for the same distortion. Compared to the RD bound for an i.i.d. Gaussian source, this represents a gap of only 0.255 bit/sample from the bound [5, Section 4.6.2].

8.7 SUMMARY

General Framework of Coding System (Section 8.1)

- A coding system usually consists of three components (Figure 8.1): analysis based on the source model, which converts the signal samples into model parameters; quantization of the parameters (for lossy coding systems); and binary encoding of quantized parameters.
- The main difference between different coding systems lies in the source model used. Video coding methods can be classified into two groups, those using source models that directly characterize the pixel values (waveform-based), and those using models that describe the underlying object structures (content-dependent).

Bounds for Lossless and Lossy Coding (Section 8.3)

- The minimal rate required for lossless coding for a discrete source is determined by the source entropy rate (Theorem 8.2).

- The minimal rate required for coding at a given distortion (the RD bound) is determined by the minimal mutual information between the original and the quantized source (Theorem 8.4).
- The preceding bounds are achievable only if an infinite number of samples are coded together.
- Among all the sources with the same variance, the Gaussian source requires the highest bit rate.

Binary Encoding (Section 8.4)

- Binary encoding can be applied to a single sample at a time or a group of samples, with or without conditioning based on previous samples.
- Huffman and arithmetic coding are the two most popular methods for variable length coding. Huffman coders are simpler to design and operate, but cannot reach the lossless coding bound with reasonable complexity. It is also difficult to adapt a Huffman coder to variations in signal statistics. Arithmetic coders can more easily approach the entropy bound and are more effective for nonstationary signals, but they are also more complex to implement.

Scalar Quantization (Section 8.5)

- The necessary conditions for MSE optimality are the centroid condition, Equations (8.5.13) or (8.5.15), and the nearest-neighbor condition, Equation (8.5.12) or (8.5.14). A locally optimal quantizer can be designed using the Lloyd algorithm (Figure 8.9).
- With an MMSE quantizer, the operational RD function of a source has the general form of Equation (8.5.18) when the rate is sufficiently high. The factor ϵ^2 depends on the pdf of the source.
- The MMSE quantizer assumes that the quantized indices are coded using fixed-length coding. With variable-length coding and when the rate is sufficiently high, uniform quantization is nearly optimal. For image and video coding, uniform quantization followed by VLC is preferred over nonuniform quantization.

Vector Quantization (Section 8.6)

- VQ can exploit the correlation between samples, and its efficiency increases as the vector dimension increases. It has the potential of reaching the RD bound when the vector dimension reaches infinity.
- The computational complexity and storage requirement of a general unconstrained VQ encoder grow exponentially with the vector dimension. Practical image and video coders use vector dimensions of 4×4 or lower.
- The necessary conditions for MSE optimality are generalized version of the centroid condition, Equations (8.6.17) or (8.6.15), and the nearest-neighbor condition, Equation (8.6.14). A locally optimal vector quantizer can be designed using the generalized Lloyd algorithm (Figure 8.14).

8.8 PROBLEMS

8.1 Consider a discrete source with an alphabet $\mathcal{A} = \{a_1, a_2, \ldots, a_L\}$. Compute the entropy of the source for the following two cases:

(a) The source is uniformly distributed, with $p(a_l) = 1/L, \forall l \in \mathcal{L}$.

(b) For a particular $k \in \mathcal{L}$, $p(a_k) = 1$ and $p(a_l) = 0, l \neq k$.

8.2 Prove the relations given in Equations (8.2.6–12).

8.3 Prove the relations given in Equations (8.2.17–19).

8.4 Prove that the differential entropy of a Gaussian RV with mean η and variance σ^2 is as given in Equation (8.2.23).

8.5 Show that Nth-order vector coding is worse than $(N - 1)$th-order conditional coding, if

$$\frac{1}{N} H_N(\mathcal{F}) - H_{c,N-1}(\mathcal{F}) > \frac{N-1}{N}.$$

8.6 A Markov source with three symbols $\mathcal{A} = \{a_1, a_2, a_3\}$ has the following probability distributions:

$$p(a_i) = \frac{1}{3}; \quad i = 1, 2, 3,$$

$$P\{a_i/a_j\} = \begin{cases} \frac{1}{2} & \text{if } i = j; \\ \frac{1}{4} & \text{otherwise.} \end{cases}$$

(a) Calculate the first-order entropy, second-order entropy, first-order conditional entropy, and the entropy rate of this source.

(b) Design the first-order, second-order, and first-order conditional Huffman codes for this source. Calculate the resulting bit rate for each case. Compare it to the corresponding lower and upper bounds defined by the entropy.

(c) What is the minimal bit rate achievable with this source? How could you achieve the minimal rate?

8.7 Encode and decode the following sequence using arithmetic coding. Use the occurrence frequency of each symbol in the sequence as the estimate of the probability of the symbol.

$$\text{source sequence: a c b a a b a c a c b a}$$

What is the bit rate of the coded sequence? Compare the result to scalar Huffman coding.

8.8 Show that for an MMSE quantizer, the original RV $\mathcal{F}$, the quantized RV $\mathcal{G}$, and the quantization error $\mathcal{Q} = \mathcal{F} - \mathcal{G}$ satisfy the following statistical relations:

(a) The quantized value is an unbiased estimate of the original value: $E\{\mathcal{G}\} = E\{\mathcal{F}\}$.

(b) The quantized value is orthogonal to the quantization error: $E\{\mathcal{G}\mathcal{Q}\} = 0$.

(c) The quantization process reduces the signal variance: $\sigma_{\mathcal{G}}^2 = \sigma_{\mathcal{F}}^2 - \sigma_{\mathcal{Q}}^2$.

8.9 Show that the MMSE quantizer for a uniform source is given by Equation (8.5.7).

8.10 Prove that for a source with arbitrary mean η and variance σ^2, the optimal reconstruction and boundary values can be obtained from those derived for a source with zero mean and unit variance according to Equation (8.5.17).

8.11 Consider a RV $\mathcal{F}$ with pdf $p(f) = (\lambda/2)e^{-\lambda|f|}$. A three-level quantizer is defined as

$$Q(f) = \begin{cases} b & f > a \\ 0 & -a \le f \le a \\ -b & f < -a \end{cases}$$

(a) Find b for a given a such that the centroid condition is satisfied when the distortion measure is the MSE.

(b) Find a for a given b such that the nearest-neighbor condition is met.

(c) Find an optimal set of a, b in terms of λ such that both conditions are satisfied. Derive the final MSE.

8.12 The optimality conditions given in Equations (8.5.12–13) are derived by minimizing the MSE. In the following, we consider other distortion measures.

(a) Derive a similar set of equations if the distortion criterion is the mean absolute error (MAE), that is, $D_q = E\{|\mathcal{F} - Q(\mathcal{F})|\}$. Show that the generalized centroid g_l for the decision region $\mathcal{B}_l = [b_{l-1}, b_l)$ is

$$\int_{b_{l-1}}^{g_l} p(f \mid f \in \mathcal{B}_l)\, df = \int_{g_l}^{b_l} p(f \mid f \in \mathcal{B}_l)\, df.$$

That is, g_l equally splits $\mathcal{B}_l$ in terms of probability. Such a g_l is also called the *median* of the region $\mathcal{B}_l$. What is the form of the nearest-neighbor condition under this distortion measure?

(b) For a source with Laplacian distribution, $p(f) = \frac{1}{2}e^{-|f|}$, determine the generalized centroid for the region $\mathcal{B}_l$ under the MAE criterion. Compare this centroid to that determined based on the MSE criterion $D_q = E\{|\mathcal{F} - Q(\mathcal{F})|^2\}$.

8.13 Consider a quantizer for a discrete vector source $\mathcal{F} = \{\mathbf{f}_k, k = 1, 2, \ldots, K\}$, where $\mathbf{f}_k$ is an ND vector. Suppose that the distortion measure between two vectors is the Hamming distance, defined as

$$d(\mathbf{f}, \mathbf{g}) = \begin{cases} 0 & \text{if } \mathbf{f} = \mathbf{g}; \\ 1 & \text{otherwise.} \end{cases}$$

Assume that these vectors are to be quantized into $L < K$ codewords.

(a) For a given partition region of the source, $\mathcal{B}_l = \{\mathbf{f}_k, k = 1, 2, \ldots, K_l\}$, determine a new codeword for this region based on the centroid rule.

(b) For a given set of codewords $\mathcal{C} = \{\mathbf{g}_1, \mathbf{g}_2, \ldots, \mathbf{g}_L\}$, determine a new partition of the source based on the nearest-neighbor rule. Is the partition unique?

(c) Starting from an arbitrary initial partition of the source, how many iterations are needed to arrive at a locally optimal codebook?

    (*Hint:* Consider two separate cases: 1) when all the training vectors are different; 2) when some of the training vectors are the same.)

8.14 A 2-D vector quantizer has two codewords: $\mathbf{g}_1 = [1/2, 1/2]^T$; $\mathbf{g}_2 = [-1/2, -1/2]^T$. Suppose that the input vectors $\mathbf{f} = [f_1, f_2]^T$ are uniformly distributed in the square defined by $-1 < f_1 < 1$ and $-1 < f_2 < 1$. Illustrate the partition regions associated with the two codewords, and determine the MSE of this quantizer (it is sufficient to write the integral formula).

8.15 A 2-D lattice quantizer over a finite region is composed of the codewords (g_1, g_2) defined by $g_1 = m$ and $g_2 = m + 2n$, where m, n are integers such that (g_1, g_2) fall in the specified region.

(a) Sketch the codewords and partition regions near the origin.

(b) Design an algorithm that will identify the nearest codeword for any given input point in the plane.

(c) Find the nearest codeword for the point $(123.4, 456.7)$ using the algorithm in part (b).

(d) Determine the MSE of this quantizer if the input is uniformly distributed in the space spanned by all the partition regions.

8.16 Write a C or MATLAB code to implement the generalized Lloyd algorithm, based on training data generated from a 2-D i.i.d. Gaussian source. Assume that the two components in each sample are independent and that both have zero mean and unit variance. You can choose the number of samples used and the codebook size. Indicate the training samples and the codewords after each iteration using crosses and circles in a 2-D plot as in Figure 8.15.

8.17 Repeat Problem 8.16, but assume that the two vector components are correlated with a correlation coefficient ρ. Compare the results obtained with different values of ρ.

8.9 BIBLIOGRAPHY

[1] Berger, T. *Rate Distortion Theory*. Englewood Cliffs, NJ: Prentice Hall, 1971.

[2] Conway, J. H., and N. J. A. Sloane. Fast quantizing and decoding algorithms for lattice quantizers and codes. *IEEE Trans. Inform. Theory* (March 1982), IT-28:227–32.

[3] Cover, T. M., and J. A. Thomas. *Elements of Information Theory*. New York: John Wiley & Sons, 1991.

[4] Gersho, A., and R. M. Gray. *Vector Quantization and Signal Compression*. Boston: Kluwer Academic Press, 1992.

[5] Jayant, N. S., and P. Noll. *Digital Coding of Waveforms*. Englewood Cliffs, NJ: Prentice-Hall, 1984.

[6] Linde, Y., A. Buzo, and R. M. Gray. An algorithm for vector quantizer design. *IEEE Trans. Commun.* (Jan. 1980), COM-28:84–95.

[7] Lloyd, P. S. Least squares quantization in PCM. *IEEE Trans. Inform. Theory* (March 1982), IT-28:127–35. (Originally published as Bell Laborataries Technical Note, 1957.)

[8] MacQueen, J. Some methods for classification and analysis of multivariate observations. In *Fifth Berkeley Symposium on Math. Stat. and Prob.* (1967), 1:281–96.

[9] Sayood, K. *Introduction to Data Compression*. San Francisco: Morgan Kaufmann, 1996.

[10] Welch, T. A. A technique for high performance data compression, *IEEE Computer Magazine* (June 1984), 8–19.

[11] Witten, I. H., R. M. Neal, and J. G. Cleary. Arithmetic coding for data compression. *Commun. ACM* (1987), 30:520–40.

[12] Ziv, J., and A. Lempel. A universal algorithm for sequential data compression. *IEEE Trans. Inform. Theory* (May 1977), IT-23(3):337–43.

9

Waveform-Based Video Coding

In Chapter 8, we introduced the fundamental techniques for video coding common to both waveform-based and content-dependent coders. In this chapter, we focus on techniques for waveform-based video coding. First, we describe block-based transform coding techniques (Section 9.1), by which each block of pixels is converted into a set of uncorrelated coefficients, which are then quantized and coded. We introduce the important Karhunen-Loeve transform (KLT) and its approximation, the discrete cosine transform (DCT), which is used in most modern image coding standards. (A coding method using another important transform, wavelet transform, is introduced in Chapter 11, when we describe scalable coding.) In Section 9.2, we introduce coding techniques using spatial and temporal prediction. These techniques reduce the data rate of a video sequence by exploiting spatial correlation between neighboring pixels as well as temporal correlation between consecutive frames of the sequence. (The spatial correlation is due to the fact that color values of adjacent pixels in the same video frame usually change smoothly; the temporal correlation refers to the fact that consecutive frames of a sequence usually show the same physical scene, occupied by the same objects that may have moved.) We conclude this chapter by presenting a coding scheme that combines transform and predictive coding (Section 9.3), a popular and effective technique that has been adopted by all international video coding standards.

9.1 BLOCK-BASED TRANSFORM CODING

Transform coding has proven to be especially effective for compression of still images and video frames. Ideally, a transform should be applied to an entire image or video frame, to fully exploit the spatial correlation among pixels. But to reduce computational complexity, block-based transform coding—which divides an image into nonoverlapping blocks, and applies the transform over each block—is more often used in practice. In this chapter, we describe only block-based transform coding, with a focus on the type of algorithm used in standard image and video coders. For a more comprehensive treatment of transform coding, the reader is referred to Clark [6].

We start with an overview of transform coding, describing on a conceptual level how and why it works (Section 9.1.1). We then formally define 1-D and 2-D unitary transforms (Sections 9.1.2–3) and, as an example, the DCT (Section 9.1.4). Section 9.1.5 discusses how to optimally allocate bits among the transform coefficients, so as to minimize the mean square error. This section also derives the gain of transform coding over scalar quantization. These results are obtained by assuming an arbitrary unitary transform. Section 9.1.6 derives the necessary conditions for a transform to achieve the highest coding gain among all unitary transforms, and shows that the KLT is an optimal transform in this sense. The KLT, however, depends on the statistics of the signals and is difficult to compute; fortunately, the DCT is a good approximation of the KLT for most image signals. For this reason, DCT-based coding has been adopted by all image and video coding standards. Section 9.1.7 presents the DCT-based image coding algorithm, as used by the JPEG standard. The modification of this algorithm for video coding is described in Section 9.3.1 and in Chapter 13. We conclude this section by briefly describing vector transforms in Section 9.1.8.

9.1.1 Overview

One of the most popular waveform-based coding schemes for images is transform coding. In block-based transform coding, one divides an image into nonoverlapping blocks, each of which is transformed into a set of coefficients. These coefficients are quantized separately using scalar quantizers. The quantized coefficient indices are finally converted into binary bits using variable-length coding. In the decoder, the image block is recovered from the quantized coefficients through an inverse transform. Figure 9.1 shows the encoder and decoder operations of a typical transform coder.

Transform as Projection onto Selected Basis Functions   One can think of the transform process as representing an image block as a linear combination of a set of basic patterns (known as transform basis functions), as illustrated in Figure 9.2. The contribution from each basic pattern is the transform coefficient corresponding to that

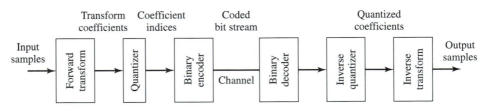

Figure 9.1   Block diagram for the encoder and decoder of a block-transform coding system.

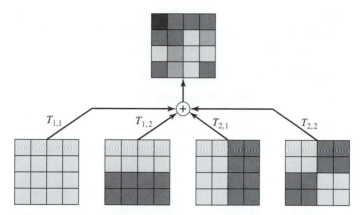

Figure 9.2   A 2-D transform is a projection of an image onto a set of basic block patterns, each known as a transform basis.

transform basis function. The process of deriving the transform coefficients for a given image block is the *forward transform,* whereas reconstruction of the image block from the transform coefficients is the *inverse transform.*

Transform Design Criterion   Obviously, the performance of a transform coder depends on the basis functions that are used. A good transform should: (1) decorrelate the signal to be quantized, so that scalar quantization over individual values can be effectively used without losing too much coding efficiency (in comparison with vector quantization); and (2) compact the energy of the original pixel block into as few coefficients as possible. The latter property allows one to represent the original block by a few coefficients with large magnitudes. As will be shown in Section 9.1.6, the best transform under these criteria is the KLT. But, because the KLT depends on the second-order statistics of the signal, and is difficult to calculate, in practice, one uses a fixed transform to approximate the KLT. The transform that best approximates the KLT for common image signals is the DCT; therefore, in almost all transform-based image coders, the DCT is used.

Transform Coding versus Vector Quantization   Because adjacent pixels in an image are usually correlated, independent representation of each pixel value is not efficient. One way to exploit the correlation among adjacent pixels is to quantize a block of pixels together using vector quantization, which replaces each image block by one of the typical block patterns that is closest to the original block. The larger the block size, the more completely the correlation among pixels can be exploited, and the higher the compression gain that can be achieved. Unfortunately, as we learned in Section 8.6, the complexity in searching for the best matching pattern also grows exponentially with the block size. Transform coding is one way to implement a constrained vector quantizer without using an exhaustive search. Specifically, the codewords in this quantizer are those representable by linear combinations of basis vectors with quantized coefficients. Because each coefficient depends on the entire image block, scalar quantization of the transform coefficients essentially accomplishes vector quantization of the entire block, but at a reduced complexity.

9.1.2 One-Dimensional Unitary Transform

General 1-D Linear Transform   As described, the transform process can be thought of as decomposing an image block into a set of basic block patterns. This is in fact true for a 2-D linear transform. Before formally defining the 2-D transform, we first define the 1-D transform. In this case, a 1-D signal is divided into vectors of length N, and each vector is represented as a linear combination of a set of basis vectors. Each transform coefficient represents the contribution from one basis vector. Mathematically, these coefficients are obtained by a matrix operation. Let $\mathbf{s} = [s_1, s_2, \ldots, s_N]^T$ represent the vector consisting of the original samples, and $\mathbf{t} = [t_1, t_2, \ldots, t_N]^T$ the vector containing all the transformed coefficients. Further, let $\mathbf{u}_k = [u_{k;1}, u_{k;2}, \ldots, u_{k;N}]^T$ denote the vector corresponding to the kth transform basis and $[\mathbf{U}] = [\mathbf{u}_1, \mathbf{u}_2, \ldots, \mathbf{u}_N]$ the matrix containing all basis vectors. Then the inverse transform represents the sample vector $\mathbf{s}$ as

$$\text{inverse transform:} \quad \mathbf{s} = \sum_{k \in \mathcal{N}} t_k \mathbf{u}_k = [\mathbf{U}]\mathbf{t} \qquad (9.1.1)$$

where $\mathcal{N} = \{1, 2, \ldots, N\}$. In order to be able to represent any N-D vector with N basis vectors, the basis vectors must be linearly independent. In this case, the matrix $[\mathbf{U}]$ must be invertible, and the transform coefficients can be determined by

$$\text{forward transform:} \quad \mathbf{t} = [\mathbf{U}]^{-1}\mathbf{s} = [\mathbf{V}]\mathbf{s} \qquad (9.1.2)$$

For generality, we assume that the basis vectors $\mathbf{u}_k$ may contain complex components, so that $\mathbf{u}_k$ belongs to the N-D complex space $\mathcal{C}^N$. In this case, even if $\mathbf{s}$ is a real vector, $\mathbf{t}$ is in general a complex vector. In the following discussion, we assume that both $\mathbf{s}$ and $\mathbf{t}$ belong to $\mathcal{C}^N$.

1-D Unitary Transform   The preceding discussion does not impose any constraint on the transform basis vectors. A special class of linear transform is the unitary transform, in which the basis vectors are orthonormal to one another.

Before formally defining the unitary transform, we first review the definition of inner product and norm in a complex space $\mathcal{C}^N$. The *inner product* of two vectors $\mathbf{s}_1, \mathbf{s}_2$, both in $\mathcal{C}^N$, is defined as

$$\langle \mathbf{s}_1, \mathbf{s}_2 \rangle = \mathbf{s}_1^H \mathbf{s}_2 = \sum_{n \in \mathcal{N}} s_{1;n}^* s_{2;n}, \qquad (9.1.3)$$

where the superscript H denotes vector transposition followed by element conjugation. Two vectors are said to be *orthogonal* to each other if $\langle \mathbf{s}_1, \mathbf{s}_2 \rangle = 0$. The *norm* $\|\mathbf{s}\|$ of $\mathbf{s} \in \mathcal{C}^N$ is defined as

$$\|\mathbf{s}\|^2 = \langle \mathbf{s}, \mathbf{s} \rangle = \sum_{n \in \mathcal{N}} |s_n|^2. \qquad (9.1.4)$$

The norm $\|\mathbf{s}\|$ represents the length or magnitude of $\mathbf{s}$, and the squared norm $\|\mathbf{s}\|^2$ the *energy* of $\mathbf{s}$. A vector with a unit norm is said to be a *normalized* vector. If two vectors are orthogonal to each other, and each is normalized, they are said to be *orthonormal*.

If the basis vectors in the linear transform defined in Equation (9.1.2) are orthonormal to each other—that is, if

$$\langle \mathbf{u}_k, \mathbf{u}_l \rangle = \sum_{n \in \mathcal{N}} u_{k;n}^* u_{l;n} = \delta_{k,l} = \begin{cases} 1 & \text{if } k = l, \\ 0 & \text{if } k \neq l, \end{cases} \tag{9.1.5}$$

then the matrix $[\mathbf{U}]$ satisfies

$$[\mathbf{U}]^H[\mathbf{U}] = [\mathbf{U}][\mathbf{U}]^H = [\mathbf{I}]_N \tag{9.1.6}$$

where $[\mathbf{I}]_N$ represents an $N \times N$ identity matrix. A matrix satisfying this condition is called *unitary*. A linear transform where the transform matrix $[\mathbf{U}]$ is unitary is called a *unitary transform*. In this case, $[\mathbf{U}]^{-1} = [\mathbf{U}]^H$ and the transform coefficients can be found by $\mathbf{t} = [\mathbf{U}]^H \mathbf{s}$ or $t_k = \langle \mathbf{u}_k, \mathbf{s} \rangle = \mathbf{u}_k^H \mathbf{s}$. That is, the kth transform coefficient t_k can be found simply by the inner product of $\mathbf{s}$ with $\mathbf{u}_k$.

To summarize, a unitary transform pair is specified by

forward transform: $\quad t_k = \langle \mathbf{u}_k, \mathbf{s} \rangle \quad$ or $\quad \mathbf{t} = [\mathbf{U}]^H \mathbf{s} = [\mathbf{V}]\mathbf{s}$

inverse transform: $\quad \mathbf{s} = \sum_{k \in \mathcal{N}} t_k \mathbf{u}_k = [\mathbf{U}]\mathbf{t} = [\mathbf{V}]^H \mathbf{t}. \tag{9.1.7}$

Usually, we refer to a transform by its forward transform matrix $[\mathbf{V}] = [\mathbf{U}]^H$. The basis vectors of the transform are the conjugates of the row vectors in $[\mathbf{V}]$.

Example 9.1   (1-D DFT)

Recall that the N-point DFT is specified by

forward transform: $\quad t_k = \dfrac{1}{\sqrt{N}} \displaystyle\sum_{n=0}^{N-1} s_n e^{-j2\pi kn/N}, \quad k = 0, 1, \ldots, N - 1,$

$$\tag{9.1.8}$$

inverse transform: $\quad s_n = \dfrac{1}{\sqrt{N}} \displaystyle\sum_{k=0}^{N-1} t_k e^{j2\pi kn/N}, \quad n = 0, 1, \ldots, N - 1.$

Using the notation that we have described, we see that it is a unitary transform with basis vectors defined by:

$$u_{k;n} = \frac{1}{\sqrt{N}} e^{j2\pi \frac{kn}{N}}, \quad n = 0, 1, \ldots, N - 1, \quad k = 0, 1, \ldots, N - 1, \tag{9.1.9}$$

or

$$\mathbf{u}_k = \frac{1}{\sqrt{N}} \begin{bmatrix} 1 & e^{j2\pi \frac{k}{N}} & \cdots & e^{j2\pi \frac{k(N-1)}{N}} \end{bmatrix}^T, \quad k = 0, 1, \ldots, N - 1. \tag{9.1.10}$$

Note that, in this example, we follow the convention of using an index range of 0 to $N-1$ for k and n, instead of 1 to N.

Properties of Unitary Transforms   Up to this point, we have assumed that $\mathbf{s}$ is a given sample vector, and $\mathbf{t}$ is its transform. In reality, $\mathbf{s}$ is a realization of a random vector $\boldsymbol{\mathcal{S}} = [\mathcal{S}_1, \mathcal{S}_2, \ldots, \mathcal{S}_N]^T$, which corresponds to any N samples extracted from a source. Similarly, $\mathbf{t}$ is a realization of a random vector $\boldsymbol{\mathcal{T}} = [\mathcal{T}_1, \mathcal{T}_2, \ldots, \mathcal{T}_N]^T = [\mathbf{V}]\boldsymbol{\mathcal{S}}$. In transform coding, we are interested not only in the relations between $\mathbf{s}$ and $\mathbf{t}$, but also in the relations between the statistics of $\boldsymbol{\mathcal{S}}$ and $\boldsymbol{\mathcal{T}}$. With a unitary transform, the following relations exist.

1. The mean vectors $\boldsymbol{\eta}_s = E\{\boldsymbol{\mathcal{S}}\}$ and $\boldsymbol{\eta}_t = E\{\boldsymbol{\mathcal{T}}\}$ are related by

$$\boldsymbol{\eta}_t = [\mathbf{V}]\boldsymbol{\eta}_s, \qquad \boldsymbol{\eta}_s = [\mathbf{V}]^H\boldsymbol{\eta}_t. \tag{9.1.11}$$

The covariance matrices $[\mathbf{C}]_s = E\{(\boldsymbol{\mathcal{S}} - \boldsymbol{\eta}_s)(\boldsymbol{\mathcal{S}} - \boldsymbol{\eta}_s)^H\}$ and $[\mathbf{C}]_t = E\{(\boldsymbol{\mathcal{T}} - \boldsymbol{\eta}_t)(\boldsymbol{\mathcal{T}} - \boldsymbol{\eta}_t)^H\}$ are related by

$$[\mathbf{C}]_t = [\mathbf{V}][\mathbf{C}]_s[\mathbf{V}]^H, \qquad [\mathbf{C}]_s = [\mathbf{V}]^H[\mathbf{C}]_t[\mathbf{V}]. \tag{9.1.12}$$

2. The total energy of the transformed vector equals that of the sample vector. This is true both for a given realization and the ensemble average; that is,

$$\sum_{n \in \mathcal{N}} s_n^2 = \sum_{k \in \mathcal{N}} t_k^2, \tag{9.1.13}$$

$$\sum_{n \in \mathcal{N}} \sigma_{s,n}^2 = \sum_{k \in \mathcal{N}} \sigma_{t,k}^2. \tag{9.1.14}$$

where $\sigma_{s,n}^2 = E\{(\mathcal{S}_n - \eta_{s,n})^2\}$ and $\sigma_{t,k}^2 = E\{(\mathcal{T}_k - \eta_{t,k})^2\}$ are the variances of $\mathcal{S}_n$ and $\mathcal{T}_k$, respectively. (This property is equivalent to Parseval's theorem for the Fourier transform.)

3. Suppose that we use only the first $K < N$ coefficients to approximate $\mathbf{s}$, with the approximated vector being $\hat{\mathbf{s}}_K = \sum_{k=1}^K t_k \mathbf{u}_k$; then the approximation error signal is $\mathbf{e}_K = \mathbf{s} - \hat{\mathbf{s}}_K = \sum_{k=K+1}^N t_k \mathbf{u}_k$. The approximation error energy for a given $\mathbf{s}$ is

$$\|\mathbf{e}_K\|^2 = \sum_{n \in \mathcal{N}} e_n^2 = \sum_{k=K+1}^N t_k^2. \tag{9.1.15}$$

The variance of $\boldsymbol{\mathcal{E}}_K = \boldsymbol{\mathcal{S}} - \hat{\boldsymbol{\mathcal{S}}}_K$ over the ensemble of $\boldsymbol{\mathcal{S}}$ is

$$E\{\|\boldsymbol{\mathcal{E}}_K\|^2\} = \sum_{n \in \mathcal{N}} \sigma_{e,n}^2 = \sum_{k=K+1}^N \sigma_{t,k}^2. \tag{9.1.16}$$

Because the sum of all coefficient squares or variances is a constant (Equation (9.1.13) or (9.1.14)), the approximation error for a particular signal vector is minimized if one chooses K coefficients that have the largest values to approximate the original signal. Similarly, the mean approximation error is minimized by choosing K coefficients with the largest variances.

The proof of the above relations is quite straightforward, and is left as an exercise (Problem 9.1).

9.1.3 Two-Dimensional Unitary Transform

The preceding discussion assumed that the input samples were from a 1-D signal. For a 2-D image block, these descriptions still apply if we think of **s** as a vector consisting of the samples from an image block arranged in a 1-D order, and each $\mathbf{u}_k$ as a vector representation of a basis image, using the same 1-D ordering. However, one can also extend the definition of the 1-D transform to 2-D. In this case, an $M \times N$ image block $[\mathbf{S}] = [S_{m,n}]$ is represented as a linear combination of $M \times N$ basis images $[\mathbf{U}]_{k,l} = [U_{k,l;m,n}]$, $k \in \mathcal{M}, l \in \mathcal{N}$, (see Figure 9.2) as

$$[\mathbf{S}] = \sum_{k \in \mathcal{M}} \sum_{l \in \mathcal{N}} T_{k,l} [\mathbf{U}]_{k,l}, \tag{9.1.17}$$

where $\mathcal{M} = \{1, 2, \ldots, M\}, \mathcal{N} = \{1, 2, \ldots, N\}$.

Let $\mathcal{C}^{M \times N}$ represent the $M \times N$–dimensional complex space. Let $[\mathbf{S}]_1 = [S_{1;m,n}]$, $[\mathbf{S}]_2 = [S_{2;m,n}]$ be two matrices in this space. Then the *inner product* of the two is defined by:

$$\langle [\mathbf{S}]_1, [\mathbf{S}]_2 \rangle = \sum_{m \in \mathcal{M}} \sum_{n \in \mathcal{N}} S_{1;m,n}^* S_{2;m,n}. \tag{9.1.18}$$

The *norm* of a matrix $[\mathbf{S}] = [S_{m,n}]$ is defined as:

$$\|[\mathbf{S}]\|^2 = \langle [\mathbf{S}], [\mathbf{S}] \rangle = \sum_{m \in \mathcal{M}} \sum_{n \in \mathcal{N}} |S_{m,n}|^2. \tag{9.1.19}$$

With these definitions, a set of basis images $[\mathbf{U}]_{k,l}, k \in \mathcal{M}, l \in \mathcal{N}$, are said to be orthonormal with one another if

$$\langle [\mathbf{U}]_{k,l}, [\mathbf{U}]_{i,j} \rangle = \delta_{k,i} \delta_{l,j} = \begin{cases} 1, & \text{if } k = i, l = j; \\ 0 & \text{otherwise.} \end{cases} \tag{9.1.20}$$

With an orthonormal set of basis images, one can easily find the transform coefficients by $T_{k,l} = \langle [\mathbf{U}]_{k,l}, [\mathbf{S}] \rangle$. The forward and inverse transform relations for an $M \times N$ 2-D unitary transform are

$$\begin{aligned} \text{forward transform:} \quad & T_{k,l} = \langle [\mathbf{U}]_{k,l}, [\mathbf{S}] \rangle, \\ \text{inverse transform:} \quad & [\mathbf{S}] = \sum_{k \in \mathcal{M}} \sum_{l \in \mathcal{N}} T_{k,l} [\mathbf{U}]_{k,l}. \end{aligned} \tag{9.1.21}$$

Example 9.2   (2-D DFT)

The 2-D DFT of size $M \times N$ is defined as:

$$\begin{aligned} \text{forward transform:} \quad & T_{k,l} = \frac{1}{\sqrt{MN}} \sum_{m=0}^{M-1} \sum_{n=0}^{N-1} S_{m,n} e^{-j2\pi \left(\frac{km}{M} + \frac{ln}{N} \right)}, \\ & k = 0, 1, \ldots, M-1, \quad l = 0, 1, \ldots, N-1; \\ \text{inverse transform:} \quad & S_{m,n} = \frac{1}{\sqrt{MN}} \sum_{k=0}^{M-1} \sum_{l=0}^{N-1} T_{k,l} e^{j2\pi \left(\frac{km}{M} + \frac{ln}{N} \right)}, \\ & m = 0, 1, \ldots, M-1, \quad n = 0, 1, \ldots, N-1. \end{aligned} \tag{9.1.22}$$

One can see that this is a 2-D unitary transform with basis images defined by:

$$U_{k,l;m,n} = \frac{1}{\sqrt{MN}} e^{j2\pi\left(\frac{km}{M} + \frac{ln}{N}\right)}, \quad m, k = 0, 1, \ldots, M-1, \quad n, l = 0, 1, \ldots, N-1. \quad (9.1.23)$$

Separable Transforms   One particularly important class of 2-D transforms consists of those in which each basis image is the outer product of two basis vectors, each from a 1-D transform. Virtually all of the 2-D transforms in use belong to this type. Specifically, let $\mathbf{h}_k, k \in \mathcal{M}$ and $\mathbf{g}_l, l \in \mathcal{N}$ represent basis vectors in $\mathcal{C}^M$ and $\mathcal{C}^N$, respectively; then one can form basis images in $\mathcal{C}^{M \times N}$ by:

$$[\mathbf{U}]_{k,l} = \mathbf{h}_k \mathbf{g}_l^T \quad \text{or} \quad U_{k,l;m,n} = h_{k;m} g_{l;n}, \quad k \in \mathcal{M}, \quad l \in \mathcal{N}. \quad (9.1.24)$$

It can be shown that $[\mathbf{U}]_{k,l}, k \in \mathcal{M}, l \in \mathcal{N}$, will form an orthonormal basis set in $\mathcal{C}^{M \times N}$ as long as $\mathbf{h}_k, k \in \mathcal{M}$, and $\mathbf{g}_l, l \in \mathcal{N}$, form orthonormal basis sets in $\mathcal{C}^M$ and $\mathcal{C}^N$, respectively.

Example 9.3   (2-D DFT as Separable Transforms)

Recall that the basis images for an $M \times N$–point 2-D DFT are:

$$U_{k,l;m,n} = \frac{1}{\sqrt{MN}} e^{j2\pi\left(\frac{km}{M} + \frac{ln}{N}\right)},$$

whereas the basis vectors for an M-point DFT and an N-point DFT are, respectively,

$$h_{k;m} = \frac{1}{\sqrt{M}} e^{j2\pi\frac{km}{M}}, \qquad g_{l;n} = \frac{1}{\sqrt{N}} e^{j2\pi\frac{ln}{N}}.$$

Obviously, $U_{k,l;m,n} = h_{k;m} g_{l;n}$. Therefore, the $M \times N$–point 2-D DFT is a separable transform constructed from an M-point 1-D DFT and an N-point 1-D DFT.

When the transform is separable, one can first accomplish 1-D transforms for each row using the basis matrix $[\mathbf{G}] = [\mathbf{g}_1, \mathbf{g}_2, \ldots, \mathbf{g}_N]$, and then perform 1-D transforms for each column of the intermediate image using the basis matrix $[\mathbf{H}] = [\mathbf{h}_1, \mathbf{h}_2, \ldots, \mathbf{h}_M]$. This operation can be represented as

$$
\begin{aligned}
\text{forward transform:} \quad & [\mathbf{T}] = [\mathbf{H}]^H [\mathbf{S}] [\mathbf{G}], \\
\text{inverse transform:} \quad & [\mathbf{S}] = [\mathbf{H}] [\mathbf{S}] [\mathbf{G}]^H.
\end{aligned}
\quad (9.1.25)
$$

When a 2-D transform is separable, its equivalent 1-D transform representation will have a transform matrix that is the Kronecker product of the matrices $[\mathbf{G}]$ and $[\mathbf{H}]$. For a more detailed discussion of this topic, including additional properties of separable transforms, see [18, Section 5.2].

Computational Savings of Separable Transforms   An $M \times N$ transform in general takes $M \times N$ calculations for determining each coefficient. To determine all $M \times N$ coefficients, the total number of calculations is $M^2 N^2$, which becomes N^4 when $M = N$. On the other hand, when the transform is separable, one first calculates N-point 1-D transforms along each of M rows, each requiring N^2 calculations. One then calculates M-point 1-D transforms along each of N columns, each requiring M^2

calculations. The total number of calculations is $MN^2 + NM^2$, which is $2N^3$ if $M = N$. When N is large, the reduction from N^4 to $2N^3$ is significant.

If there exists a fast algorithm for calculating the 1-D transform, then further savings are obtainable. For example, if an N-point 1-D transform takes $N \log_2 N$ operations, then an $N \times N$ 2-D separable transform can be done in $2N^2 \log N$ operations.

9.1.4 The Discrete Cosine Transform

Besides the DFT, there are many other transforms that have been developed for signal analysis, including the DCT, discrete sine transform (DST), Hadamard transform, Walsh transform, Haar transform, and slant transforms. (See [12] or [18] for definitions of the basis vectors of these transforms.) All these transforms are originally defined in the 1-D form, and can be used to construct 2-D separable transforms. The DFT is most widely used for performing frequency domain analysis for discrete signals. On the other hand, the DCT has been found to be more useful for source coding, especially image coding. For this reason, we introduce the DCT in this section and show examples of using it to represent images.

The basis vectors of the 1-D N-point DCT are defined by:

$$u_{k;n} = \alpha(k) \cos \frac{(2n+1)k\pi}{2N}, \quad n = 0, 1, \ldots, N-1 \tag{9.1.26}$$

with

$$\alpha(k) = \begin{cases} \sqrt{\frac{1}{N}} & k = 0 \\ \sqrt{\frac{2}{N}} & k = 1, 2, \ldots, N-1. \end{cases} \tag{9.1.27}$$
$$\tag{9.1.28}$$

The forward and inverse transforms are described by:

$$t_k = \sum_{n=0}^{N-1} u_{k;n}^* s_n = \alpha(k) \sum_{n=0}^{N-1} s_n \cos \frac{(2n+1)k\pi}{2N}, \tag{9.1.29}$$

$$s_n = \sum_{k=0}^{N-1} u_{k;n} t_k = \sum_{k=0}^{N-1} \alpha(k) t_k \cos \frac{(2n+1)k\pi}{2N} \tag{9.1.30}$$

As with the DFT, we follow the convention of using an index range of 0 to $N-1$ for k and n.

Notice that the DCT basis vectors are real vectors, varying in a sinusoidal pattern with increasing frequency. Each DCT coefficient specifies the contribution of a sinusoidal pattern at a particular frequency to the actual signal. The lowest coefficient, known as the *DC coefficient,* represents the average value of the signal. The other coefficients, known as *AC coefficients,* are associated with increasingly higher frequencies.

The 2-D $M \times N$–point DCT is constructed from the 1-D M-point DCT basis and N-point DCT basis. That is, each $M \times N$ basis image is the outer product of an M-point DCT basis vector with an N-point DCT basis vector. The basis images corresponding to an 8×8 DCT are shown in Figure 9.3. To obtain a 2-D DCT of an image block, one can

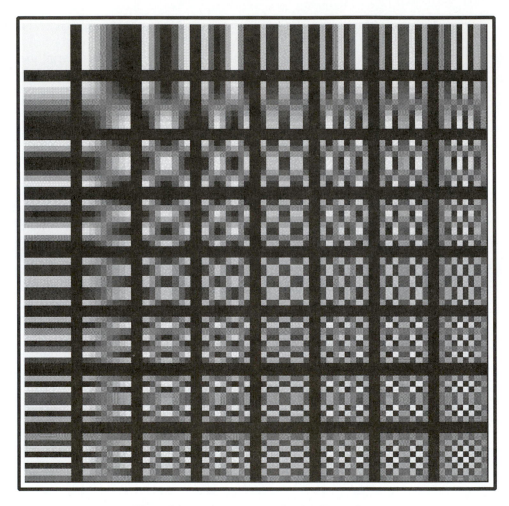

Figure 9.3   Basis images of an 8 × 8 DCT transform.

first apply the corresponding 1-D DCT to each row of the image block, and then apply the 1-D DCT to each column of the row transformed block. There exist several fast algorithms for computing N-point 1-D DCTs with $N \log_2 N$ operations [4, 21].

The reason that the DCT is well suited for image coding is that an image block can often be represented with a few low-frequency DCT coefficients. This is because the intensity values in an image usually vary smoothly, and very high frequency components exist only near edges. Figure 9.4 shows the energy (i.e., the variance) distribution of the 8 × 8 DCT coefficients for a test image. The coefficients are ordered using a zigzag scan (described in Section 9.1.7), so that the coefficients with low indices in both horizontal and vertical directions are put in the front. We can see that the energy of the DCT coefficients drops quickly as the frequency index increases. Figure 9.5 shows the

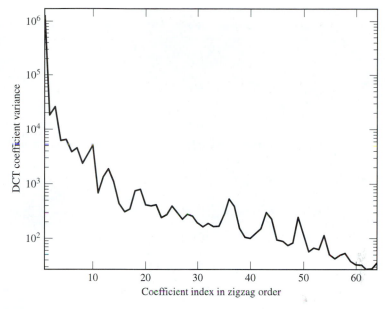

Figure 9.4   Energy distribution of the 8 × 8 DCT coefficients of the test image "flower."

approximation of a test image using four different numbers of DCT coefficients. We can see that with only sixteen out of 64 coefficients per block, we can already represent the original image quite well.

9.1.5 Bit Allocation and Transform Coding Gain

In this section, we consider the rate-distortion performance of a transform coder. Although we are mostly interested in 2-D transforms, we will use the 1-D transform representation in our analysis. Since any 2-D signal can be mapped to 1-D and then transformed, this choice does not limit the applicability of the analysis result. We will assume that the transform is an arbitrary 1-D unitary transform. We further assume that the transform coefficients are quantized separately, each using a scalar quantizer optimized based on the pdf of that coefficient, and that the quantized coefficients are converted into bits using fixed-length coding (see Figure 9.1). Our goal is to derive the relation between the average distortion per sample in a given vector, and the average bit rate per sample. We will also compare the performance achievable by transform coding with that of a PCM system in which a scalar quantizer is applied to each original sample.

Relation between Distortions in the Samples and Coefficients   First, we must relate the quantization errors in the reconstructed samples with those in the transform coefficients. We will use the MSE as the distortion criterion exclusively in this section. As before, we use $\mathcal{S}_n$ and $\mathcal{T}_k$ to denote the RVs corresponding to s_n and t_k, respectively, and $\mathcal{S}$ and $\mathcal{T}$ the random vectors corresponding to **s** and **t**. In addition,

Figure 9.5 Versions of the image "flower" reconstructed with different numbers of DCT coefficients. The top left is the original image with 64 coefficients per block, the top right with sixteen coefficients, the bottom left with eight coefficients, and the bottom right with four coefficients. The DCT coefficients are arranged in zigzag order.

we use $\hat{s}_n$, $\hat{\mathbf{s}}$, $\hat{\mathcal{S}}_n$, $\hat{\boldsymbol{S}}$ to represent the quantized versions of s_n, $\mathbf{s}$, $\mathcal{S}_n$, $\boldsymbol{S}$, and $\hat{t}_k$, $\hat{\mathbf{t}}$, $\hat{\mathcal{T}}_k$, $\hat{\boldsymbol{T}}$ the quantized versions of t_k, $\mathbf{t}$, $\mathcal{T}_k$, $\boldsymbol{T}$. The MSE per sample between the original and the reconstructed samples is

$$D_s = \frac{1}{N} E\{\|\boldsymbol{S} - \hat{\boldsymbol{S}}\|^2\} = \frac{1}{N} \sum_{n \in \mathcal{N}} D_{s,n} \qquad (9.1.31)$$

with

$$D_{s,n} = E\{(\mathcal{S}_n - \hat{\mathcal{S}}_n)^2\}. \qquad (9.1.32)$$

Similarly, the MSE per coefficient between the original and the quantized coefficients is

$$D_t = \frac{1}{N} E\{\|\mathcal{T} - \hat{\mathcal{T}}\|^2\} = \frac{1}{N} \sum_{k \in \mathcal{N}} D_{t,k} \qquad (9.1.33)$$

with

$$D_{t,k} = E\{(\mathcal{T}_k - \hat{\mathcal{T}}_k)^2\}. \qquad (9.1.34)$$

Using the relation

$$\mathcal{S} = [\mathbf{V}]^H \mathcal{T} \quad \text{and} \quad \hat{\mathcal{S}} = [\mathbf{V}]^H \hat{\mathcal{T}}, \qquad (9.1.35)$$

we have

$$D_s = \frac{1}{N} E\{\|\mathcal{S} - \hat{\mathcal{S}}\|^2\} = \frac{1}{N} E\{\|[\mathbf{V}]^H (\mathcal{T} - \hat{\mathcal{T}})\|^2\}$$

$$= \frac{1}{N} E\{(\mathcal{T} - \hat{\mathcal{T}})^H [\mathbf{V}][\mathbf{V}]^H (\mathcal{T} - \hat{\mathcal{T}})\} = \frac{1}{N} E\{\|\mathcal{T} - \hat{\mathcal{T}}\|^2\} = D_t,$$

in which the fourth equality is due to the unitary property of $[\mathbf{V}]$. This result tells us that the average quantization error in the sample domain is equal to the average quantization error in the transform domain.

Let $\sigma_{t,k}^2$ represent the variance of coefficient $\mathcal{T}_k$, and let R_k and $D_{t,k}$ denote the bit rate and distortion for $\mathcal{T}_k$. From Equation (8.5.18), when R_k is sufficiently high, we have

$$D_{t,k}(R_k) = \epsilon_{t,k}^2 \sigma_{t,k}^2 2^{-2R_k} \qquad (9.1.36)$$

where $\epsilon_{t,k}^2$ depends on the pdf of $\mathcal{T}_k$. Therefore, the average distortion using transform coding is

$$D_{\text{TC}} = D_s = D_t = \frac{1}{N} \sum_{k \in \mathcal{N}} \epsilon_{t,k}^2 \sigma_{t,k}^2 2^{-2R_k}. \qquad (9.1.37)$$

Optimal Bit Allocation   Given the desired average bit rate R, the question is how to allocate the total bits RN among the N coefficients so that the error in Equation (9.1.37) is minimized. Using the Lagrange multiplier method, this constrained minimization problem can be converted to the minimization of

$$J(R_k, \forall k \in \mathcal{N}) = \sum_{k \in \mathcal{N}} \epsilon_{t,k}^2 \sigma_{t,k}^2 2^{-2R_k} + \lambda \left(\sum_{k \in \mathcal{N}} R_k - RN \right) \qquad (9.1.38)$$

where λ must be chosen to satisfy

$$\sum_{k \in \mathcal{N}} R_k = RN \qquad (9.1.39)$$

If we let $(\partial J / \partial R_k) = 0$, we obtain

$$\frac{\partial D_{t,k}}{\partial R_k} = -2 \ln 2 D_{t,k} = -(2 \ln 2) \epsilon_{t,k}^2 \sigma_{t,k}^2 2^{-2R_k} = -\lambda, \quad \forall k \in \mathcal{N} \qquad (9.1.40)$$

To determine λ, we take the products of the preceding equations for all k. This yields

$$\lambda^N = (2\ln 2)^N \left(\prod_k \epsilon_{t,k}^2 \sigma_{t,k}^2\right) 2^{-2\sum_k R_k} = (2\ln 2)^N \left(\prod_k \epsilon_{t,k}^2 \sigma_{t,k}^2\right) 2^{-2NR} \quad (9.1.41)$$

or

$$\lambda = (2\ln 2)\left(\prod_k \epsilon_{t,k}^2 \sigma_{t,k}^2\right)^{1/N} 2^{-2R}. \quad (9.1.42)$$

Substituting this into Equation (9.1.40) gives

$$R_k = R + \frac{1}{2}\log_2 \frac{\epsilon_{t,k}^2 \sigma_{t,k}^2}{\left(\prod_k \epsilon_{t,k}^2 \sigma_{t,k}^2\right)^{1/N}}. \quad (9.1.43)$$

With this bit allocation, the distortions incurred on all the coefficients are equal; that is,

$$D_{\text{TC}} = D_t = D_{t,k} = \left(\prod_k \epsilon_{t,k}^2 \sigma_{t,k}^2\right)^{1/N} 2^{-2R}. \quad (9.1.44)$$

The solution given in Equation (9.1.43) implies that a coefficient with a larger variance should be given more bits, whereas a coefficient with a smaller variance should be given fewer bits. The optimal allocation is such that all the coefficients have the same quantization error.

Note that for a coefficient with sufficiently small variance, the rate determined according to Equation (9.1.43) could be negative. In practice, the bit rate for this coefficient must be reset to zero, corresponding to not coding the coefficient at all. The bits allocated to other coefficients would have to be reduced to satisfy the original average bit rate specification. For bit allocation algorithms that explicitly constrain R_k to be nonnegative or nonnegative integers, see [34, 31].

Transform Coding Gain over PCM   Assume that the original samples come from a stationary source so that each sample has the same variance σ_s^2. If we code this source using PCM; that is, we apply optimal scalar quantization to individual samples separately, then the distortion is related to the rate by

$$D_{\text{PCM}} = D_{s,n} = \epsilon_s^2 \sigma_s^2 2^{-2R} \quad (9.1.45)$$

where ϵ_s^2 depends on the pdf of the source sample.

A measure of the performance of transform coding is the *transform coding gain over PCM*, which is defined as

$$G_{\text{TC}} = \frac{D_{\text{PCM}}}{D_{\text{TC}}}. \quad (9.1.46)$$

Substituting Equations (9.1.45) and (9.1.44) into (9.1.46), we have

$$G_{\text{TC}} = \frac{\epsilon_s^2 \sigma_s^2}{\left(\prod_k \epsilon_{t,k}^2 \sigma_{t,k}^2\right)^{1/N}} = \frac{\epsilon_s^2}{\left(\prod_k \epsilon_{t,k}^2\right)^{1/N}} \frac{\frac{1}{N} \sum \sigma_{t,k}^2}{\left(\prod_k \sigma_{t,k}^2\right)^{1/N}}, \qquad (9.1.47)$$

in which the second equality results from the energy preservation property of the unitary transform given in Equation (9.1.14) and the fact that $\sigma_{s,n}^2 = \sigma_s^2$. Therefore, *the transform coding gain is proportional to the ratio of the arithmetic mean of transform coefficient variances to the geometric mean of these variances.* It is a well known result that for any arbitrary set of values, the arithmetic mean is equal to or larger than the geometric mean, the two means being equal only when all the values are the same. In general, we also have $\epsilon_s^2 \geq \left(\prod_k \epsilon_{t,k}^2\right)^{1/N}$. Therefore, $G_{\text{TC}} \geq 1$ or $D_{s,\text{TC}} \leq D_{s,\text{PCM}}$ under the same average bit rate. The more unevenly valued the coefficient variances, the smaller their geometric mean, and the higher the coding gain.

When the source is Gaussian, each sample is Gaussian. Because each transform coefficient is a linear combination of the samples, it follows a Gaussian distribution as well. In this case, $\epsilon_s^2 = \epsilon_{t,k}^2$ and, hence,

$$G_{\text{TC,Gaussian}} = \frac{\sigma_s^2}{\left(\prod_k \sigma_{t,k}^2\right)^{1/N}} = \frac{\frac{1}{N} \sum \sigma_{t,k}^2}{\left(\prod_k \sigma_{t,k}^2\right)^{1/N}}. \qquad (9.1.48)$$

Example 9.4

Consider applying transform coding to each 2×2 block of an image, using a 2×2 DCT. Assume that the image is a stationary Gaussian process with variance σ_s^2 and that the correlation coefficients between horizontally, vertically, and diagonally adjacent pixels are ρ_h, ρ_v, and ρ_d, respectively, as illustrated in Figure 9.6. We want to determine the optimal bit allocation for a given average bit rate of R, and the corresponding distortion; to compare this coder with a PCM coder that directly quantizes each sample using an optimal scalar quantizer; and to give the numerical results for the special case of $\sigma_s^2 = 1$, $\rho_h = \rho_v = \rho = 0.95$, $\rho_d = \rho^2 = 0.9025$, and $R = 2$.

To solve the problem, we arrange the 2×2 pixel array into a 4-D vector, and think of the 2×2 2-D DCT as a four-point 1-D DCT. Using the 1-D ordering $\mathbf{s} = [A, B, C, D]^T$,

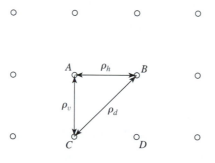

Figure 9.6 The image source considered in Examples 9.4–6.

the covariance matrix of **s**, assuming that **s** has zero mean, is,

$$
[\mathbf{C}]_s = E\left\{ \begin{bmatrix} A \\ B \\ C \\ D \end{bmatrix} [A \ \ B \ \ C \ \ D] \right\} = \begin{bmatrix} C_{AA} & C_{AB} & C_{AC} & C_{AD} \\ C_{BA} & C_{BB} & C_{BC} & C_{BD} \\ C_{CA} & C_{CB} & C_{CC} & C_{CD} \\ C_{DA} & C_{DB} & C_{DC} & C_{DD} \end{bmatrix}
$$

$$
= \sigma_s^2 \begin{bmatrix} 1 & \rho_h & \rho_v & \rho_d \\ \rho_h & 1 & \rho_d & \rho_v \\ \rho_v & \rho_d & 1 & \rho_h \\ \rho_d & \rho_v & \rho_h & 1 \end{bmatrix}. \tag{9.1.49}
$$

The basis images of the 2×2 DCT are:

$$
\frac{1}{2}\begin{bmatrix} 1 & 1 \\ 1 & 1 \end{bmatrix}, \quad \frac{1}{2}\begin{bmatrix} 1 & -1 \\ 1 & -1 \end{bmatrix}, \quad \frac{1}{2}\begin{bmatrix} 1 & 1 \\ -1 & -1 \end{bmatrix}, \quad \frac{1}{2}\begin{bmatrix} 1 & -1 \\ -1 & 1 \end{bmatrix}. \tag{9.1.50}
$$

Rearranging each basis image into a vector, the equivalent 1-D inverse transform matrix is

$$
[\mathbf{U}] = \frac{1}{2}\begin{bmatrix} 1 & 1 & 1 & 1 \\ 1 & -1 & 1 & -1 \\ 1 & 1 & -1 & -1 \\ 1 & -1 & -1 & 1 \end{bmatrix}. \tag{9.1.51}
$$

The covariance matrix of the transformed vector $[\mathbf{C}]_t$ can be obtained using the property in Equation (9.1.12), and the variances of the transformed coefficients are the diagonal element in $[\mathbf{C}]_t$. In the special case of $\rho_h = \rho_v = \rho = 0.95$, $\rho_d = \rho^2 = 0.9025$, and $\sigma_s^2 = 1$, the coefficient variances are

$$
\sigma_{t,k}^2 = \{(1+\rho)^2, (1-\rho^2), (1-\rho^2), (1-\rho)^2\}\sigma_s^2 \tag{9.1.52}
$$

$$
= \{3.8025, 0.0975, 0.0975, 0.0025\}. \tag{9.1.53}
$$

The geometric mean of the coefficient variances is

$$
\sigma_t^2 = \left(\prod_k \sigma_{t,k}^2 \right)^{1/4} = (1-\rho^2)\sigma_s^2 = 0.0975. \tag{9.1.54}
$$

If the desired average bit rate is $R = 2$ bpp, the optimal bit allocation among the four coefficients, using Equation (9.1.43), is

$$
R_k = \{4.64, 2, 2, -0.64\}. \tag{9.1.55}
$$

The average distortion per pixel, using Equation (9.1.44) and $\epsilon_{t,k}^2 = \epsilon_{\text{Gaussian}}^2$, is

$$
D_{\text{TC}} = \epsilon_{\text{Gaussian}}^2 (1-\rho^2)\sigma_s^2 2^{-2R} = 0.0061\epsilon_{\text{Gaussian}}^2 \tag{9.1.56}
$$

If one directly quantizes each pixel, then the average distortion (assuming optimal quantizer), using Equation (9.1.45), is

$$
D_{\text{PCM}} = \epsilon_{\text{Gaussian}}^2 \sigma_s^2 2^{-2R} = 0.0625\epsilon_{\text{Gaussian}}^2 \tag{9.1.57}
$$

The transform coding gain, using Equation (9.1.48), is

$$
G_{\text{TC}} = \frac{\sigma_s^2}{\sigma_t^2} = \frac{1}{1-\rho^2} = 10.25. \tag{9.1.58}
$$

Notice that the optimal bit allocation in Equation (9.1.55) assumes that one can code a variable using a negative number of bits. In practice, one must set $R_4 = 0$, and reallocate the total number of bits among the remaining three variables. A simple solution is to set $R_1 = 4$, so that the modified bit allocation is

$$R_k = \{4, 2, 2, 0\}. \tag{9.1.59}$$

The average distortion in this case is

$$D_{\text{TC}} = \frac{1}{4} \sum_k \epsilon_{\text{Gaussian}}^2 \sigma_{t,k}^2 2^{-2R_k} = 0.0074\, \epsilon_{\text{Gaussian}}^2. \tag{9.1.60}$$

Although this distortion is higher than that indicated by Equation (9.1.56), it is still much lower than that achievable by PCM.

9.1.6 Optimal Transform Design and the KLT

As we have just shown, with optimal bit allocation and optimal scalar quantization, transform coding yields a lower average distortion than the PCM method. This is true with any unitary transform. It is natural then to question whether there exists an optimal transform that will minimize the distortion D_s (or, equivalently, maximize the gain G_{TC}). Because $\epsilon_s^2 \sigma_s^2$ is fixed for a given source, the transform should be such that it yields the minimal geometric mean $(\prod_k \epsilon_{t,k}^2 \sigma_{t,k}^2)^{1/N}$, among all $N \times N$ unitary transforms. Fortunately, when the source is Gaussian, the answer is a definite "yes," and such an optimal transform is the KLT.

Construction of the KLT Bases   With the KLT, the basis vectors are designed based on the covariance matrix of the original signal $[\mathbf{C}]_s$. Let λ_k and ϕ_k represent the kth eigenvalue and normalized eigenvector of $[\mathbf{C}]_s$, respectively. Then they satisfy:

$$[\mathbf{C}]_s \phi_k = \lambda_k \phi_k, \quad \text{with} \quad \langle \phi_k, \phi_l \rangle = \delta_{k,l}.$$

The KLT for a given covariance matrix $[\mathbf{C}]_s$ is the unitary transform that uses ϕ_k as the basis vectors. The corresponding inverse transform matrix is:

$$[\mathbf{\Phi}] = [\phi_0, \phi_1, \ldots, \phi_{N-1}]. \tag{9.1.61}$$

Replacing $[\mathbf{V}]$ in Equation (9.1.12) by $[\mathbf{\Phi}]^H$, we get

$$[\mathbf{C}]_t = [\mathbf{\Phi}]^H [\mathbf{C}]_s [\mathbf{\Phi}] = \begin{bmatrix} \phi_1^H \\ \phi_2^H \\ \cdots \\ \phi_N^H \end{bmatrix} [\mathbf{C}]_s [\phi_1, \phi_2, \ldots, \phi_N] \tag{9.1.62}$$

$$= \begin{bmatrix} \phi_1^H \\ \phi_2^H \\ \cdots \\ \phi_N^H \end{bmatrix} [\lambda_1 \phi_1, \lambda_2 \phi_2, \ldots, \lambda_N \phi_N] = \begin{bmatrix} \lambda_1 & 0 & \cdots & 0 \\ 0 & \lambda_2 & \cdots & 0 \\ \cdots & \cdots & \cdots & \cdots \\ 0 & 0 & \cdots & \lambda_N \end{bmatrix}. \tag{9.1.63}$$

That is, the KLT diagonalizes $[\mathbf{C}]_s$. Further, the variance of the kth transformed coefficient is $\sigma_k^2 = \lambda_k$.

To prove that the KLT will minimize the coding distortion, we make use of the inequality

$$\det[\mathbf{C}]_t \leq \prod_{k \in \mathcal{N}} \sigma_{t,k}^2 \tag{9.1.64}$$

which applies to any covariance matrix $[\mathbf{C}]$ with diagonal entries $\sigma_{t,k}^2$. On the other hand, from Equation (9.1.12), we have

$$\det[\mathbf{C}]_t = \det[\mathbf{C}]_s \tag{9.1.65}$$

for all unitary transforms. Therefore, for any unitary transform, we have

$$\prod_{k \in \mathcal{N}} \sigma_{t,k}^2 \geq \det[\mathbf{C}]_s. \tag{9.1.66}$$

But with the KLT, we have

$$\prod_{k \in \mathcal{N}} \sigma_{t,k}^2 = \det[\mathbf{C}]_t = \det[\mathbf{C}]_s. \tag{9.1.67}$$

Thus, among all unitary transforms, the KLT yields the minimal geometric mean of the transform coefficient variances. Consequently, it maximizes the transform coding gain, if the source is Gaussian. Even for a non-Gaussian source, the effect of the ϵ^2 factors can usually be ignored, and the KLT can be considered the optimal transform.

Another property of the KLT is that it yields the minimal approximation error with $K < N$ coefficients among all unitary transforms, where the error is defined as in Equation (9.1.16). Alternatively, we say that the KLT has the highest energy compaction capability among all transforms. This is because the approximation error is directly related to the geometric mean of the coefficient variances. The lower the geometric mean, the more unequally distributed these variances and, consequently, the more energy that can be packed into a fixed number of coefficients.

With the KLT and a Gaussian source, we have

$$D_{\text{TC}} = \epsilon_{\text{Gaussian}}^2 (\det[\mathbf{C}]_s)^{1/N} 2^{-2R}. \tag{9.1.68}$$

Recall that the RD bound for a correlated Gaussian vector source is given by Equation (8.3.26). Therefore, the operational RD function achievable by transform coding exceeds the bound by the constant factor $\epsilon_{\text{Gaussian}}^2 = 2.71$.

The coding gain using the KLT, from Equations (9.1.47) and (9.1.65), is

$$G_{\text{TC,KLT}} = \frac{\epsilon_s^2}{\left(\prod_k \epsilon_k^2\right)^{1/N}} \frac{\sigma_s^2}{\left(\prod_k \lambda_k\right)^{1/N}} = \frac{\epsilon_s^2}{\left(\prod_k \epsilon_k^2\right)^{1/N}} \frac{\sigma_s^2}{(\det[\mathbf{C}]_s)^{1/N}} \tag{9.1.69}$$

Example 9.5

Consider again the image source in Figure 9.6. Instead of using the DCT, we would like to use the KLT. The covariance matrix $[\mathbf{C}]_s$ of the 4-D vector consisting of any 2×2 pixels is given by Equation (9.1.49). To determine the transform matrix for the KLT, we need to determine the eigenvalues and eigenvectors of $[\mathbf{C}]_s$. The eigenvalues are determined by solving

$$\det([\mathbf{C}]_s - \lambda[\mathbf{I}]) = 0 \qquad (9.1.70)$$

For each eigenvalue λ_k, we can determine the corresponding eigenvector ϕ_k by solving

$$([\mathbf{C}]_s - \lambda[\mathbf{I}])\phi_k = \mathbf{0} \qquad (9.1.71)$$

with the constraint $\|\phi_k\|^2 = 1$. For the special case of $\rho_h = \rho_d = \rho, \rho_d = \rho^2$, the resulting eigenvalues are

$$\lambda_k = \{(1+\rho)^2, (1-\rho^2), (1-\rho^2), (1-\rho)^2\}\sigma_s^2.$$

Comparing this result to Equation (9.1.52), we see that the eigenvalues in this case are the same as the variances obtained by the DCT in Example 9.4. Consequently, the KLT matrix, consisting of the eigenvectors, is the same as the DCT transform matrix, given in Equation (9.1.51).* It can be easily verified that $\det[\mathbf{C}]_s = \prod_k \lambda_k$.

For this particular example, the DCT is equal to the KLT, for any value of ρ. It has been show that, for an arbitrary first-order Markov process, its KLT basis functions resemble closely those of the DCT [1, 6]. In general, however, the DCT is not as efficient as the KLT.

Although the KLT is optimal in terms of its capability in maximizing the coding gain (or, equivalently, energy compaction) and in decorrelating the signal, it is computable only for a stationary source with a known covariance matrix. In reality, the source may by temporally or spatially varying, and one must constantly update the covariance matrix based on previous signal samples and recalculate the eigenvectors, which is computationally very demanding. Furthermore, there exist no fast algorithms for the KLT derived from an arbitrary covariance matrix. Therefore, for practical applications, it is desirable to employ transforms that are signal-independent. It has been shown that the DCT is very close to the KLT for the covariance matrices of common image signals. Therefore, the DCT has been used in place of the KLT for image coding.

9.1.7 DCT-Based Image Coders and the JPEG Standard

The previous sections have introduced the basic principles for transform coding. As shown in Figure 9.1, a transform coder consists of three components: transform, quantization, and binary encoding. In the analysis to this point, we have assumed that each coefficient is quantized using a pdf-optimized scalar quantizer, and the quantization indices are coded using a fixed-length coder. But in practice, the combination of uniform quantization and variable-length coding has been found to be more effective. Here we describe a typical block transform coder, which is the basis for the image coding

---

*The solutions for eigenvectors corresponding to λ_2 and λ_3 are not unique. Another set of solutions, for example, is

$$[\mathbf{\Phi}] = [\phi_1, \phi_2, \phi_3, \phi_4] = \begin{bmatrix} 1/2 & 1/\sqrt{2} & 0 & 1/2 \\ 1/2 & 0 & -1/\sqrt{2} & -1/2 \\ 1/2 & 0 & 1/\sqrt{2} & -1/2 \\ 1/2 & -1/\sqrt{2} & 0 & 1/2 \end{bmatrix}.$$

16	11	10	16	24	40	51	61
12	12	14	19	26	58	60	55
14	13	16	24	40	57	69	56
14	17	22	29	51	87	80	62
18	22	37	56	68	109	103	77
24	35	55	64	81	104	113	92
49	64	78	87	103	121	120	101
72	92	95	98	112	100	103	99

Figure 9.7 An example quantization table for the luminance component [15, 29].

standard known as JPEG* [15, 29] and all the video coding standards to be introduced in Chapter 13. In the JPEG standard, this method is applied to each image block directly, whereas in the video coding standards, it is applied to the error between an original image block in a current video frame and the predicted block based on a previous video frame. Occasionally, it is also applied to the original image block directly, when the prediction is not accurate or when it is desirable to reset the prediction loop.

Given an image, it is first divided into 8×8 nonoverlapping blocks. An 8×8–point DCT is then applied to each block. The coefficients are then quantized using uniform quantizers, each with a different stepsize. The stepsizes for different coefficients are specified in a quantization table. The table is designed based on the visual sensitivity to different frequency components: typically, low-frequency coefficients are assigned smaller stepsizes than high-frequency coefficients, because the human eye is more sensitive to changes in low-frequency components, as described in Chapter 2. The particular table used is specified in the beginning of the compressed bit stream as side information. Typically, different tables are used for the luminance and chrominance components. An example luminance quantization table given in the JPEG standard is shown in Figure 9.7. Instead of using a specified table as it is, one can scale the table to increase or decrease the stepsizes so as to achieve the desired bit rate. The scale factor is called the *quality factor* in JPEG coders, and the *quantization parameter* (QP) in MPEG and H-series video coders.

For binary encoding of the quantized DCT coefficients, the DCT coefficients are arranged into a 1-D array following the zigzag order illustrated in Figure 9.8. This scan order puts the low-frequency coefficients in front of the high-frequency coefficients. Because many quantized coefficients are zeros in a typical image block, it is not efficient to specify the coefficient values individually. Rather, a run-length representation is used, which starts with the DC value, followed by a series of symbols.[†] Each symbol consists

*The JPEG standard refers to the international standard for still image compression recommended by the Joint Photographic Expert Group (JPEG) of the International Standards Organization (ISO).

[†]The run-length coding method used in video coding is slightly different, as will be described in Chapter 13.

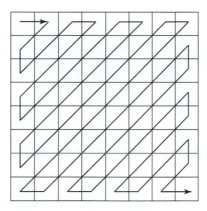

Figure 9.8   The zigzag order of the DCT coefficients.

DCT coefficients: [5 0 0 2 3 0 0 4 0 0 0 0 0 0 1 0 0 0 0 0 … 0]

Coding symbols: 5, (2,2), (0,3), (2,4), (6,1), EOB

Figure 9.9   Converting the 64 quantized DCT coefficients resulting from the zigzag scan into symbols for binary encoding.

of two numbers: the number of zeros (known as the *run-length*) from the last nonzero value, and the following nonzero value. A special symbol, "EOB," is placed after the last nonzero value in a block, to signal the end of the block. Figure 9.9 shows an example of how to convert the DCT coefficients of an 8×8 block into such symbols. Ideally, the DC value and the symbols would be coded independently using two separate VLC tables. In order to avoid the use of very large VLC tables, the dynamic range of the coefficient value is partitioned into several segments. The magnitude of a nonzero coefficient is specified by the segment number and the relative value in that segment. For the DC coefficient, the segment number is Huffman coded, based on the frequencies of different segments, whereas the relative magnitude is coded using a fixed-length codeword. For the AC coefficients, each symbol is further partitioned into two parts: the part consisting of the zero run-length with the segment number of the nonzero value is Huffman coded, whereas the relative magnitude of the nonzero value is coded using a fixed-length code. The VLC tables are usually predesigned based on training data. For improved coding efficiency, arithmetic coding can be used instead of Huffman coding.

When the preceding method is applied to original image blocks directly, the DC values of adjacent blocks are usually similar. To further improve the coding efficiency, the DC coefficient (i.e., the mean value) of a block may be predicted from the DC value of a previously coded block, and then the DC prediction error quantized and coded using the method just described. Thus, the value "5" in Figure 9.9 actually represents the DC prediction error. Also, the stepsize in the top-left entry of the table given in Figure 9.7 is actually for the DC prediction error. For a color image, each color component can be coded separately using this method. Figure 9.10 shows several JPEG-coded images, obtained by choosing different quality factors. These images are obtained by

Figure 9.10   Versions of the image "Lena" coded by the JPEG algorithm at different bit rates. The top left is the original image, at 8 bpp; the top right is coded at 0.59 bpp; the bottom left at 0.37 bpp; and the bottom right at 0.22 bpp.

scaling the quantization table given in Figure 9.7 and using Huffman tables recommended by the standard.

9.1.8 Vector Transform Coding

For the transform coding discussed thus far, we have assumed that transform coefficients are subject to scalar quantization. Although vector quantization can be applied to the transform coefficients, the gain over SQ is not significant, as transform coefficients are already decorrelated. A method that further improves transform coding gain uses the so-called vector transform [22]. Such transforms convert a group of vector samples into a set of vector coefficients, each of which is then vector quantized. To maximize the coding efficiency, a vector transform should introduce correlation within variables in a vector coefficient, but decorrelate successive vector coefficients. For more information on vector transforms and their application in image and video coding, the reader is referred to the review article by Li and Zhang [23].

9.2 PREDICTIVE CODING

In addition to transform coding, predictive coding is another important technique for image and video coding. In fact, temporal predictive coding using motion-compensated prediction is the key to the success of modern video coding standards. We start this section with an overview of predictive coding (Section 9.2.1). We then describe how to design the predictor to minimize the prediction error, and derive the gain of predictive coding over scalar quantization (Section 9.2.2). The last two subsections illustrate how to apply spatial and temporal prediction in image and video coding (Sections 9.2.3–4).

9.2.1 Overview

In predictive coding, a pixel is not coded directly; rather, its value is predicted from those of adjacent pixels in the same frame or in a previous frame. This is motivated by the fact that adjacent pixels usually have similar color values, thus it is wasteful of bits to specify the current value independent of the past. Figure 9.11 shows the block

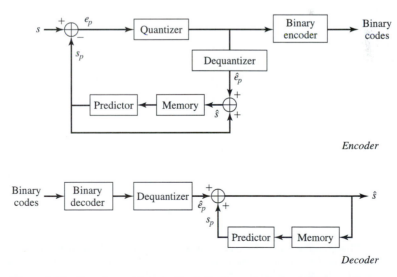

Encoder

Decoder

Figure 9.11   Encoder and decoder block diagrams of a lossy predictive coding system.

diagrams of the encoder and decoder of a generic *lossy predictive coding* system. In the encoder, an input sample is first predicted from a few previously reconstructed samples stored in the memory, then the prediction error is quantized and then coded using a variable-length coder. The reconstructed value at the decoder is the predicted value plus the quantized error. To guarantee that the encoder and the decoder use exactly the same prediction value, the encoder must repeat the same process as the decoder to reproduce reconstructed samples. This is called *closed-loop prediction*. This kind of coder is generally referred to as *differential pulse coded modulation* (DPCM).*

When the prediction error value is coded directly without quantization, the system reduces to *lossless predictive coding,* which is useful for applications that require storage or transmission of original signals with no distortion.

Error Analysis of a Lossy Predictive Coder   Refer to Figure 9.11. Let s and s_p represent the original and predicted sample value, $e_p = s - s_p$ and $\hat{e}_p$ the original and quantized prediction error. Let e_q represent the quantization error for e_p, then $e_p = \hat{e}_p + e_q$. The reconstructed value for s is

$$\hat{s} = s_p + \hat{e}_p = s_p + e_p - e_q = s - e_q. \qquad (9.2.1)$$

Therefore, the error between the original and the reconstructed sample value is $e_s = s - \hat{s} = e_q$, exactly the same as the quantization error for the prediction error. Thus, the distortion in a lossy predictive coder is completely dependent on the quantizer for the prediction error, for a fixed predictor.

For the design of the error quantizer, various quantization techniques discussed in Chapter 8 can be employed. Because the error usually has a nonuniform distribution, a nonuniform quantizer optimized for the distribution of the error signal (typically modeled by a zero-mean Laplacian distribution) is usually used. For improved coding efficiency, one can also use a vector quantizer or transform coder for the error signal.

A key to the efficiency of a predictive coder is the predictor used. From Section 8.5, the distortion introduced by a quantizer at a given bit rate is proportional to the variance of the signal (see Equation (8.5.18)). To maximize the coding efficiency, the predictor should minimize the prediction error variance. How to design a predictor to achieve this objective is the topic of the next subsection.

9.2.2 Optimal Predictor Design and Predictive Coding Gain

In general, one can use various kinds of predictors, linear or nonlinear. But in practice, linear predictors are used almost exclusively for ease of implementation. Let s_0 represent the current pixel, and $s_k, k = 1, 2, \ldots, K$, the previous pixels that are used to predict

---

*Traditionally, the prediction error is uniformly quantized and then transmitted using an amplitude modulation method known as pulse coded modulation. With the advent of digital coding and transmission, the error can be quantized nonuniformly, various variable-length coding methods can be used and, finally, different digital modulation techniques can be used to carry binary bits. So the name DPCM is no longer an accurate description of this coding scheme.

s_0. In a linear predictor, the predicted value for s_0 is described by

$$s_p = \sum_{k=1}^{K} a_k s_k \tag{9.2.2}$$

where the a_k are called *prediction coefficients*. The number of samples used for prediction, K, is usually called the *predictor order*. The key problem in designing a predictive coder is how to determine these coefficients. As previously described, to maximize coding efficiency, one should minimize the variance of the prediction error (which is equivalent to the MSE of the predictor). Letting $\mathcal{S}_k$ represent the RV corresponding to s_k, and $\mathcal{S}_p$ the RV corresponding to s_p, this error is defined as:

$$\sigma_p^2 = E\{|\mathcal{S}_0 - \mathcal{S}_p|^2\} = E\left\{ \left| \mathcal{S}_0 - \sum_{k=1}^{K} a_k \mathcal{S}_k \right|^2 \right\}. \tag{9.2.3}$$

The optimal prediction coefficients should minimize this error. This can be accomplished by letting $\partial \sigma_p^2 / \partial a_l = 0$, which yields

$$E\left\{ \left(\mathcal{S}_0 - \sum_{k=1}^{K} a_k \mathcal{S}_k \right) \mathcal{S}_l \right\} = 0, \quad l = 1, 2 \ldots, K. \tag{9.2.4}$$

Note that this equation is precisely the *orthogonality principle* for the linear minimum mean square error (LMMSE) estimator. It says that the prediction error must be orthogonal to each past sample used for prediction. Let $R(k, l) = E\{\mathcal{S}_k \mathcal{S}_l\}$ represent the correlation between $\mathcal{S}_k$ and $\mathcal{S}_l$. From Equation (9.2.4), one can easily derive the following set of linear equations:

$$\sum_{k=1}^{K} a_k R(k, l) = R(0, l), \quad l = 1, 2 \ldots, K, \tag{9.2.5}$$

or, in matrix form,

$$\begin{bmatrix} R(1, 1) & R(2, 1) & \cdots & R(K, 1) \\ R(1, 2) & R(2, 2) & \cdots & R(K, 2) \\ \cdots & \cdots & \cdots & \cdots \\ R(1, K) & R(2, K) & \cdots & R(K, K) \end{bmatrix} \begin{bmatrix} a_1 \\ a_2 \\ \cdots \\ a_K \end{bmatrix} = \begin{bmatrix} R(0, 1) \\ R(0, 2) \\ \cdots \\ R(0, K) \end{bmatrix} \tag{9.2.6}$$

or

$$[\mathbf{R}]\mathbf{a} = \mathbf{r}. \tag{9.2.7}$$

This equation for solving the LMMSE predictor is usually called the Yule-Walker equation.

From Equation (9.2.7), the optimal prediction coefficients are:

$$\mathbf{a} = [\mathbf{R}]^{-1}\mathbf{r}. \tag{9.2.8}$$

The MSE of this predictor is:

$$\sigma_p^2 = E\{(\mathcal{S}_0 - \mathcal{S}_p)\mathcal{S}_0\} = R(0, 0) - \sum_{k=0}^{K} a_k R(k, 0) \tag{9.2.9}$$

$$= R(0, 0) - \mathbf{r}^T \mathbf{a} = R(0, 0) - \mathbf{r}^T R^{-1} \mathbf{r}. \tag{9.2.10}$$

The first equality results from Equation (9.2.4), the orthogonality principle.

For a stationary source, the autocorrelation of a pixel is a constant, independent of its spatial location; that is, $R(l, l) = R(0, 0), l = 1, 2, \ldots, K$. Furthermore, the correlations between two pixels are symmetrical; that is, $R(k, l) = R(l, k)$. The matrix R is in fact the autocorrelation matrix of the prediction vector $\mathbf{S}_p = [\mathcal{S}_1, \mathcal{S}_2, \ldots, \mathcal{S}_K]^T$, and $\mathbf{r} = E\{\mathcal{S}_0 \mathbf{S}\}$ is the correlation between the current pixel $\mathcal{S}_0$ and $\mathbf{S}_p$. When the source is a 1-D stationary source, $\mathcal{S}_0 = S(n)$ is an arbitrary sample of the source, and the $\mathcal{S}_k = S(n - k)$ are the kth previous samples from $S(n)$, we have $R(k, l) = E\{S(n - k)S(n - l)\} = R_s(k - l)$, the correlation between two samples separated by $k - l$. In this case, the matrix $[\mathbf{R}]$ is Toeplitz in that all the entries on the same diagonal line are identical, and the corresponding upper and lower triangle lines are conjugates of each other. Such a matrix can be inverted efficiently using the Levinson-Durbin algorithm [30]. When the source is 2-D, the matrix $[\mathbf{R}]$ in general does not have such nice structure. But only a limited number of past samples is typically used for prediction, as the correlation between two pixels decreases rapidly as their distance increases. Usually, at most four pixels (to the left, top, top left, and top right of the current one) are used for prediction. In this case, direct solution of the preceding equation is a simple task.

This solution for the optimal predictor assumes that the predicted value s_p is obtained from original past samples s_k. In a lossy predictive coder, s_k must be replaced by reconstructed past samples $\hat{s}_k$. Ideally, the error in Equation (9.2.3) should be replaced by

$$\sigma_p^2 = E\left\{ \left| \mathcal{S}_0 - \sum_k a_k \hat{\mathcal{S}}_k \right|^2 \right\} \tag{9.2.11}$$

But because $\hat{\mathcal{S}}_k$ is dependent on the predictor and the quantizer in a complicated relation, finding a solution that minimizes this error is very difficult. In practice, one simply assumes that the quantization error is negligible and designs the predictor by minimizing the error in Equation (9.2.3).

Predictive Coding Gain   As with transform coding, we can define the gain of predictive coding over PCM as the ratio of the distortions incurred by the two types of coders under the same bit rate. Recall that with predictive coding, it is the prediction error that is quantized and coded, and the distortion on the original signal is equal to the quantization error for the prediction error. Assuming optimal scalar quantization for the prediction error, the distortion is related to the rate by

$$D_{\text{DPCM}} = \epsilon_p^2 \sigma_p^2 2^{-2R} \tag{9.2.12}$$

where ϵ_p^2 depends on the pdf of the prediction error, and σ_p^2 is the variance of the prediction error, which is equal to the MSE of the predictor. On the other hand, the distortion-rate function of PCM is given by Equation (9.1.45). Therefore, the *predictive coding gain* is

$$G_{\text{DPCM}} = \frac{D_{\text{PCM}}}{D_{\text{DPCM}}} = \frac{\epsilon_s^2 \, \sigma_s^2}{\epsilon_p^2 \, \sigma_p^2} \tag{9.2.13}$$

In general, the prediction error σ_p^2 reduces as the predictor order K increases.

With the LMMSE predictor given by Equation (9.2.8), it can be shown [19] that the minimal prediction error when the predictor order goes to infinity is related to the power spectrum density (PSD)* of the signal $S(e^{j\omega})$ by

$$\sigma_{p,\text{min}}^2 = \lim_{K \to \infty} \sigma_p^2 = \exp\left(\frac{1}{2\pi} \int_{-\pi}^{\pi} \log_e S(e^{j\omega}) \, d\omega\right). \tag{9.2.14}$$

The ratio $\gamma_s = \sigma_{p,\text{min}}^2/\sigma_s^2$ is called the *spectral flatness measure* of a signal. The predictive coding gain is thus inversely proportional to the spectral flatness. This is because a signal with a narrowly concentrated spectrum is more amenable to prediction. A signal with a flat spectrum (i.e., white noise) is unpredictable.

It can be shown that the integral in Equation (9.2.14) can be written [19] as

$$\sigma_{p,\text{min}}^2 = \lim_{K \to \infty} \left(\prod_k \lambda_k\right)^{1/N}, \tag{9.2.15}$$

where λ_k is the kth eigenvalue of the Nth order covariance matrix of the signal.† On the other hand, the signal variance equals

$$\sigma_s^2 = \lim_{K \to \infty} \frac{1}{K} \sum_k \lambda_k \tag{9.2.16}$$

Therefore, the predictive coding gain when $K \to \infty$ can be expressed as

$$\lim_{K \to \infty} G_{\text{DPCM}} = \frac{\epsilon_s^2 \, \lim_{K \to \infty} \frac{1}{K} \sum_k \lambda_k}{\epsilon_p^2 \, \lim_{K \to \infty} \left(\prod_k \lambda_k\right)^K}, \tag{9.2.17}$$

which is equal to the transform coding gain achievable with the KLT (see Equation (9.1.47)), when the transform length N goes to infinity. Therefore, the two methods are equivalent when the transform length and the predictor order are both infinite. *However, for any finite N, length-N transform coding is less efficient than order-N predictive coding [19].* This is because a length-N transform involves only N samples at a time, whereas a predictive coder of any order has an effective memory that is infinite, because of the feedback of the quantization error in the closed-loop prediction.

---

*The PSD of a stationary random sequence $\{S(n)\}$ is the Fourier transform of its autocorrelation function $R(k) = E\{S(n)S(n-k)\}$.

†It can be shown that $\lambda_k = S(e^{j\omega_k})$, where ω_k are in general nonequidistant samples over the interval $(-\pi, \pi)$.

9.2.3 Spatial-Domain Linear Prediction

Let $\psi(m, n)$ represent a video frame. In spatial-domain linear prediction, a current pixel $\psi(m, n)$ is predicted from its past neighboring pixels, $\psi(k, l)$, $(k, l) \in \mathcal{B}_{(m,n)}$, where $\mathcal{B}_{(m,n)}$ represents a causal neighborhood of pixel (m, n). The optimal predictor can be determined based on Equation (9.2.6) if the correlations between $\psi(m, n)$ and each of the pixels in $\mathcal{B}_{(m,n)}$ are known.

Example 9.6

Consider again the image source shown in Figure 9.6. This time, we use a predictive coder, which predicts the pixel $s_0 = \psi(m, n)$ (pixel D) from $s_1 = \psi(m, n-1)$ (pixel C), $s_2 = \psi(m-1, n)$ (pixel B), and $s_3 = \psi(m-1, n-1)$ (pixel A). Equation (9.2.6) is reduced to, in this case:

$$\begin{bmatrix} R(C, C) & R(C, B) & R(C, A) \\ R(B, C) & R(B, B) & R(B, A) \\ R(A, C) & R(A, B) & R(A, A) \end{bmatrix} \begin{bmatrix} a_1 \\ a_2 \\ a_3 \end{bmatrix} = \begin{bmatrix} R(D, C) \\ R(D, B) \\ R(D, A) \end{bmatrix}$$

or

$$\begin{bmatrix} 1 & \rho_d & \rho_v \\ \rho_d & 1 & \rho_h \\ \rho_v & \rho_h & 1 \end{bmatrix} \begin{bmatrix} a_1 \\ a_2 \\ a_3 \end{bmatrix} = \begin{bmatrix} \rho_h \\ \rho_v \\ \rho_d \end{bmatrix}.$$

In the special case of $\rho_h = \rho_v = \rho$, $\rho_d = \rho^2$, the optimal predictor is

$$\begin{bmatrix} a_1 \\ a_2 \\ a_3 \end{bmatrix} = \begin{bmatrix} \rho \\ \rho \\ -\rho^2 \end{bmatrix}.$$

The MSE of this predictor, using Equation (9.2.10), is

$$\sigma_p^2 = R(0, 0) - [R(0, 1) \quad R(0, 2) \quad R(0, 3)] \begin{bmatrix} a_1 \\ a_2 \\ a_3 \end{bmatrix} = (1 - \rho^2)^2 \sigma_s^2.$$

With $R = $ two–bit optimal scalar quantization, the quantization error is

$$D = \epsilon_{\text{Gaussian}}^2 \sigma_p^2 2^{-2R} = 0.000594 \, \epsilon_{\text{Gaussian}}^2. \tag{9.2.18}$$

The coding gain over PCM is

$$G_{\text{DPCM}} = \frac{\sigma_s^2}{\sigma_p^2} = \frac{1}{(1 - \rho^2)^2} = 105.19. \tag{9.2.19}$$

Comparing this result with those in Examples 9.4 and 9.5, we see that, among the three methods that all involve a 2×2 pixel block when coding, the predictive coding method is the most efficient. This is consistent with the theoretical result, that predictive coding using an Nth-order predictor is better than transform coding with a length-N transform. Note that the theoretical analysis as well as the examples all assume optimal scalar quantization, which may not be realizable in practice. A major problem with predictive coding is that a transmission error in the coded bit stream that affects a single sample can lead to mismatch between the encoder and decoder, so that all the following reconstructed samples may be in error. This is known as the *error propagation effect*. The effect of transmission errors on coded bit streams, and mechanisms to prevent or suppress such effects, will be discussed in Chapter 14.

Note that spatial prediction can be applied not only to original pixel values, but also to transform coefficients. For example, in a block transform coder, the DC coefficients (i.e., the mean values) of adjacent blocks are often similar, and one may predict the DC value of a current block from those of the block above and to the left. In the JPEG image coding standard discussed in Section 9.1.7, the left block DC value is used for prediction. Usually, the correlations between AC coefficients of adjacent blocks are not strong enough to warrant the use of prediction.

In coders using wavelet transforms (Chapter 11), the coefficients in the same spatial position but different scales are often correlated. Therefore, prediction among these coefficients can also be applied. There, instead of predicting the actual coefficient value, a nonlinear predictor is often used, which predicts locations of nonzero coefficients in a fine scale from the locations of nonzero coefficients in a coarser scale.

9.2.4 Motion-Compensated Temporal Prediction

Unidirectional Temporal Prediction   In addition to applying prediction within a frame, one can predict a pixel value in a current frame from its corresponding pixel in a previous frame. Let $\psi(\mathbf{x}, t)$ represent the pixel value in frame t at pixel $\mathbf{x}$, and let t_- denote the previous frame time. Then the prediction process is described by

$$\psi_p(\mathbf{x}, t) = \psi(\mathbf{x}, t_-). \tag{9.2.20}$$

This is known as *linear temporal prediction*. Such prediction is effective only if the underlying scene is stationary. In a real-world video, the objects in the scene as well as the camera are usually moving, so that the pixels with the same spatial location in two adjacent frames can have very different values. In this case, *motion-compensated prediction* (MCP) is more appropriate, which uses

$$\psi_p(\mathbf{x}, t) = \psi(\mathbf{x} + \mathbf{d}(\mathbf{x}), t_-), \tag{9.2.21}$$

where $\mathbf{d}(\mathbf{x})$ represent the motion vector of pixel $\mathbf{x}$ from time t to t_-. Using the terminology introduced in Chapter 6, frame t is the anchor frame, and frame t_- is the target frame. In video coding literature, frame t_- is usually referred to as the *reference frame,* frame t the *coded frame* or *current frame,* and $\psi_p(\mathbf{x}, t)$ the *predicted frame*. The reference frame must be coded and reconstructed before the coded frame.

In a coder using MCP, both the MV and the prediction error must be specified. Obviously, if one specifies an MV at every pixel plus a prediction error value, one may have to use more bits than specifying the original pixel value for each pixel. In a real coder, more compact motion representation is used to reduce the bit rate for coding the motion. One popular video coding method is to use block-based motion representation, and code only one MV per block. (We discuss block-based video coders in Section 9.3.)

The temporal predictors introduced thus far use only one pixel in a previous frame. This restriction results mainly from a practical implementation constraint, where storage of more than one previous frame is often too costly. Theoretically, however, using pixels from more than one previous frame can improve prediction accuracy. In this case, one can still use the optimal predictor design solution given in Equation (9.2.6) to determine the predictor coefficients associated with different previous pixels.

Bidirectional Temporal Prediction   In fact, the predictor need not rely on pixels in previous frames only. In *bidirectional temporal prediction,* a pixel in a current frame is predicted from a pixel in a previous frame t_- as well as a pixel in a following frame t_+. The predicted value at frame t is described by

$$\psi_p(\mathbf{x}, t) = a_- \psi(\mathbf{x} + \mathbf{d}^-(\mathbf{x}), t_-) + a_+ \psi(\mathbf{x} + \mathbf{d}^+(\mathbf{x}), t_+) \qquad (9.2.22)$$

where $\mathbf{d}^-(\mathbf{x})$ and $\mathbf{d}^+(\mathbf{x})$ represent the MVs at $\mathbf{x}$ from t to t_- and from t to t_+, respectively. Typically, we call the prediction of the current frame from a previous $(t_- < t)$ reference frame *forward motion compensation,* and that from a future $(t_+ > t)$ reference frame *backward motion compensation.** Ideally, coefficients a_- and a_+ should be determined using the optimal predictor solution given in Equation (9.2.6), which requires knowledge of the correlations between corresponding pixels in the three frames involved. In practice, they are typically chosen heuristically. For the predicted value to have the same mean value as the original value, they are chosen so that $a_- + a_+ = 1$.

The use of bidirectional temporal prediction can be very beneficial when there are regions in the coded frame that are uncovered by object motion. Such a region does not have a corresponding region in the previous frame, but only in the following frame. For example, if we look back at Figure 5.12, we see that the uncovered background in frame $k + 1$ does not have a correspondence in frame k, but it will have a correspondence in frame $k + 2$. The prediction coefficients a_- and a_+ are usually adapted based on the prediction accuracy. In the example just considered, one may choose $a_- = 0, a_+ = 1$. When correspondences can be found in both frame t_- and frame t_+, a good choice is $a_- = a_+ = 0.5$. As with unidirectional prediction, bidirectional prediction can be performed at a block level, so that only two MVs must be specified for each block.

Note that the use of bidirectional prediction necessitates the coding of frames in an order that is different from the original temporal order. For example, a subset of frames may be coded first, using unidirectional prediction from the past coded frames only, and then the remaining frames can be coded using bidirectional prediction. This type of out-of-sequence coding is illustrated in Figure 9.12. Although bidirectional prediction can improve prediction accuracy and consequently coding efficiency, it incurs encoding delay and is typically not used in real-time applications such as video phones or video conferencing. For example, the H.261/H.263 standards intended for interactive communications use only unidirectional prediction and a restricted

---

*These should not be confused with the backward and forward motion estimation defined in Section 6.2.

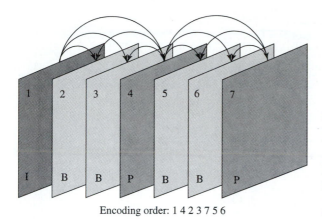

Figure 9.12   Video coding using both uni- and bidirectional temporal prediction. The arrows indicate the reference frames used for predicting a coded frame. Frames labeled I, P, and B are coded without prediction, with unidirectional prediction, and with bidirectional prediction, respectively.

Encoding order: 1 4 2 3 7 5 6

bidirectional prediction (known as the PB-mode). On the other hand, the MPEG standard series, targeted mainly for video distribution, employs both uni- and bidirectional prediction.

9.3 VIDEO CODING USING TEMPORAL PREDICTION AND TRANSFORM CODING

A popular and effective video coding method is the one that uses block-based temporal prediction and transform coding. This method is essentially the core of all the international video coding standards. In this section, we first provide an overview of this coding method (Section 9.3.1). This will be followed by an extension, which uses more sophisticated motion-compensated prediction (Section 9.3.2). Next, we consider the issue of how to choose various coding parameters in this coding method to maximize the coding efficiency under the rate constraint imposed by the underlying application (Sections 9.3.3–4). As an optional advanced topic, we analyze the influence of motion estimation accuracy and loop filtering on coding efficiency (Section 9.3.5).

9.3.1 Block-Based Hybrid Video Coding

In this coder, each video frame is divided into blocks of a fixed size and each block is processed more or less independently, hence the designation "block-based." The word "hybrid" means that each block is coded using a combination of motion-compensated temporal prediction and transform coding. Figure 9.13 shows the key steps in this coding paradigm. First, a block is predicted from a previously coded reference frame using block-based motion estimation. The motion vector specifies the displacement between the current block and the best matching block. The predicted block is obtained from

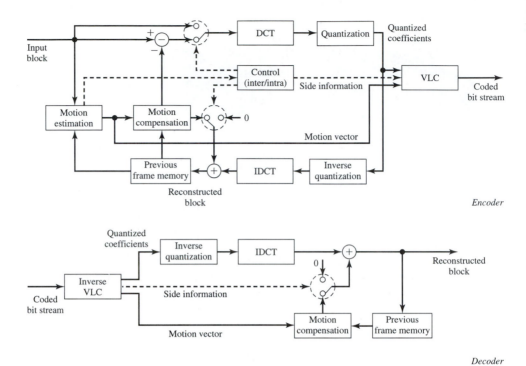

Figure 9.13 Encoding and decoding process for a block in a typical block-based hybrid coding system.

the previous frame based on the estimated MV using motion compensation.* Then, the prediction error block is coded, by transforming it using the DCT,† quantizing the DCT coefficients, and converting them into binary codewords using variable-length coding. The actual coding method is similar in principle to that described in Section 9.1.7. As with the JPEG standard, the quantization of the DCT coefficients is controlled by a quantization parameter, which scales a predefined quantization table. The quantization tables, quantizer implementations, and run length coding methods employed by various video coding standards are described in Chapter 13. Note that the encoder must emulate the decoder operation to deduce the same reconstructed frame as the decoder, so that there will be no mismatch between the reference frames used for prediction.

---

*The phrase "motion compensation," here and subsequently in this book, will refer to motion-compensated prediction.

†Ideally, if the prediction is very accurate, then the error pattern will be close to white noise and can be effectively coded using scalar quantization. In reality, however, the use of a block-based motion model often leads to structured error patterns. Transform coding is used to exploit the correlation between error pixels.

The preceding discussion assumes that temporal prediction is successful, in that the prediction error block requires fewer bits to code than the original image block. This method of coding is called *P-mode*. When this is not the case, the original block will be coded directly using transform coding. This is known as *intra-mode*. Instead of using a single reference frame for prediction, bidirectional prediction can be used, which finds two best matching blocks, one in a previous frame and another in a following frame, and uses a weighted average of the two matches as the prediction for the current block. In this case, two MVs are associated with each block. This is known as the *B-mode*. Both P- and B-modes are generally referred to as *inter-mode*. The mode information and the MVs, as well as other side information regarding picture format, block location, and so on, is also coded using VLC.

In practice, the block size for motion estimation may not be the same as that for transform coding. Typically, motion estimation is done on a larger block known as a *macroblock* (MB), which is subdivided into several blocks for which the DCT is evaluated. For example, in most video coding standards, the MB size is 16×16 pels and the block size is 8×8 pels. If the color subsampling format is 4:2:0, then each MB consists of 4 Y blocks, 1 Cb block, and 1 Cr block. The coding mode (i.e., intra- or inter-mode) is decided at the MB level. Because MVs and DC coefficients of adjacent MBs or blocks are usually similar, they are typically coded predictively, using the MV and DC coefficients of the previous MB or block as the predicted values. In all the video coding standards described in Chapter 13, a number of MBs form a *group of blocks* (GOB) or a *slice,* and several GOBs or slices form a picture. Sizes and shapes of GOBs and slices differ among the various video coding standards and picture sizes, and can often be tailored to the application's needs. Prediction of MVs and DC coefficients are usually restricted within the same GOB or slice. That is, the predicted MV or DC value for the first MB or block in a GOB is set to some default value. This is to suppress error propagation when the compressed bit stream is corrupted by transmission or storage errors.

A frame may be coded entirely in the intra-mode, and such a frame is called an *intra-frame* or I-frame. The first frame of a sequence is always coded as an intra-frame. In applications employing high bit rates, or with relaxed real-time constraints, intra-frames are also used periodically to stop potential error propagation, and to enable random access. (Low-latency applications, such as video conferencing, cannot use such periodic refresh, because the number of bits from an intra-frame is typically much greater than for a predicted frame, and thus will cause a sudden surge in the transmitted data.) A *P-frame* uses only a past frame for prediction, and depending on the prediction accuracy, an MB can be coded in either intra- or P-mode, which is usually determined based on the prediction error. Finally, a *B-frame* uses bidirectional prediction, and an MB in a B-frame can be coded in intra-, P-, or B-mode. A B-frame can only be coded after the surrounding I- or P-frames are coded (see Figure 9.12). Both P- and B-frames are generally referred as *inter-frames.*

In the MPEG-1 and MPEG-2 standards, the frames are partitioned into *groups of pictures* (GOPs), with each GOP starting with an I-frame, followed by interleaving P- and B-frames. (A typical GOP is illustrated in Figure 9.12). This enables random access: one can access any GOP without decoding previous GOPs. The GOP structure also enables fast-forwarding and rewinding: fast-forwarding can be accomplished by decoding only I- or I- and P-frames. Fast rewinding can be realized by decoding only I-frames, in a backward order.

9.3.2 Overlapped Block Motion Compensation

In this and the next subsection, for notational simplicity, we do not specify the time index in the video function. Rather, we use $\psi(\mathbf{x})$, $\psi_r(\mathbf{x})$, and $\psi_p(\mathbf{x})$ to represent the coded frame, the reference frame, and the predicted frame, respectively. In block-based motion compensation, every pixel in a block is predicted by a single pixel in the reference frame, using

$$\psi_p(\mathbf{x}) = \psi_r(\mathbf{x} + \mathbf{d}_m), \quad \mathbf{x} \in \mathcal{B}_m, \tag{9.3.1}$$

where $\mathbf{d}_m$ represents the MV of the block $\mathcal{B}_m$. When $\mathbf{d}_m$ is noninteger, an interpolation is needed to obtain $\psi_r(\mathbf{x} + \mathbf{d}_m)$. As previously described, when the estimated MV is inaccurate, or when the blockwise translational model is not appropriate, the predicted image will suffer from the notorious blocking artifact. To overcome this problem, two approaches have been taken. One is to combine the motion estimates for neighboring blocks to produce the predicted value for a given pixel, and another is to apply filtering to the predicted image. We discuss the first approach in this section, and the filtering approach in Section 9.3.5.

　　With overlapped block motion compensation (OBMC) [25, 26], a pixel is predicted based not only on the estimated MV for the block it belongs to, but also the MVs of its neighboring blocks, as illustrated in Figure 9.14. Let K be the total number of neighboring blocks considered, and let $\mathcal{B}_{m,k}$ and $\mathbf{d}_{m,k}$, $k \in \mathcal{K} = \{0, 1, 2, \ldots, K\}$,

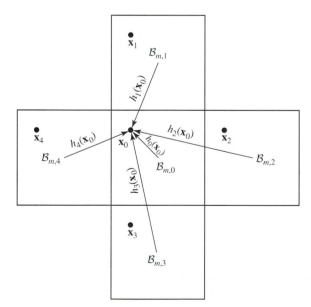

Figure 9.14　An example of overlapped block motion compensation, using four neighboring blocks.

denote the kth neighboring block and its MV, with $\mathcal{B}_{m,0} = \mathcal{B}_m, \mathbf{d}_{m,0} = \mathbf{d}_m$. The predicted value is determined by

$$\psi_p(\mathbf{x}) = \sum_{k\in\mathcal{K}} h_k(\mathbf{x})\psi_r(\mathbf{x}+\mathbf{d}_{m,k}), \quad \mathbf{x} \in \mathcal{B}_m. \tag{9.3.2}$$

The coefficient $h_k(\mathbf{x})$ can be considered as the weight assigned to the estimated value based on $\mathbf{d}_{m,k}$. Intuitively, for a given $\mathbf{x}$, $h_k(\mathbf{x})$ should be inversely proportional to the distance between $\mathbf{x}$ and the center position of $\mathcal{B}_{m,k}$. For example, with the neighborhood sketched in Figure 9.14, for $\mathbf{x}_0$, the weight for $\mathbf{d}_{m,1}$ and $\mathbf{d}_{m,4}$ should be larger than that for $\mathbf{d}_{m,2}$ and $\mathbf{d}_{m,3}$.

From an estimation-theoretic perspective, the predictor in Equation (9.3.2) can be interpreted as a linear estimator of $\psi(\mathbf{x})$ given the observations $\psi_r(\mathbf{x}+\mathbf{d}_{m,k})$. For a given $\mathbf{x}$, the optimal estimator $\mathbf{h}(\mathbf{x}) = [h_k(\mathbf{x}), k \in \mathcal{K}]^T$ can be determined by minimizing the mean square prediction error

$$E\left\{\left|\psi(\mathbf{x}) - \sum_{k\in\mathcal{K}} h_k(\mathbf{x})\psi_r(\mathbf{x}+\mathbf{d}_{m,k})\right|^2\right\}. \tag{9.3.3}$$

To preserve the mean value of the estimated image, the weights must be constrained so that

$$\sum_{k\in\mathcal{K}} h_k(\mathbf{x}) = 1.$$

Incorporating this constraint in the optimization by using the Lagrange multiplier method [26] gives

$$\mathbf{h}(\mathbf{x}) = [\mathbf{R}(\mathbf{x})]^{-1}\left(\mathbf{r}(\mathbf{x}) - \frac{\mathbf{i}^T[\mathbf{R}(\mathbf{x})]^{-1}\mathbf{r}(\mathbf{x}) - 1}{\mathbf{i}^T[\mathbf{R}(\mathbf{x})]^{-1}\mathbf{i}}\mathbf{i}\right), \tag{9.3.4}$$

where $\mathbf{i}$ is a column vector of dimension K with all elements equal to 1, $[\mathbf{R}(\mathbf{x})]$ and $\mathbf{r}(\mathbf{x})$ are the autocorrelation matrix and cross-correlation vector, respectively, with elements

$$R_{k,l}(\mathbf{x}) = E\{\psi_r(\mathbf{x}+\mathbf{d}_{m,k})\psi_r(\mathbf{x}+\mathbf{d}_{m,l})\}, k,l \in \mathcal{K}$$

$$r_k(\mathbf{x}) = E\{\psi(\mathbf{x})\psi_r(\mathbf{x}+\mathbf{d}_{m,k})\}, k \in \mathcal{K}.$$

This solution can be used to derive the weights for different neighboring block MVs for every pixel $\mathbf{x}$ in a block. The expectation in the preceding equations is with respect to uncertainties in both image intensity and motion. The preceding statistics can be estimated from real video data. As indicated by the notation, they are in general spatially variant. However, assuming that $\psi(\mathbf{x})$ is stationary, these statistics are dependent only on the relative position of the pixel in a block, not on the absolute position with respect to the entire image. If the correlation between pixels is symmetric, then the solution $h_k(\mathbf{x})$ will also be symmetric, so that only the solution for pixels in the upper left quarter of the block need be determined.

In general, OBMC can be applied to block MVs estimated by any method, including the simple block-matching method (BMA). However, the prediction error can be further reduced if motion estimation is accomplished by considering OBMC. Ideally, one should minimize the following modified MCP error:

$$\sum_{\mathbf{x}\in\mathcal{B}_m}\left|\psi(\mathbf{x})-\sum_{k\in\mathcal{K}}h_k(\mathbf{x})\psi_r(\mathbf{x}+\mathbf{d}_{m,k})\right|^p. \tag{9.3.5}$$

Because the error for a block depends on the MVs of its adjacent blocks, one cannot find one block MV at a time, as in the BMA case. One approach is to iteratively perform motion estimation and OBMC [26]. First, the conventional BMA is used to generate an initial set of MVs, where prediction error is calculated with the assumption that Equation (9.3.1) is used for MCP. Then, OBMC is applied using these motion estimates. In the next iteration, for each block, the MV is estimated by minimizing the error given in Equation (9.3.5), where the MVs for the neighboring blocks are assumed to be the same as those found in the previous iteration.

Instead of using iterations, a simpler approach is to apply unequal weighting to the pixels in the current block as well as neighboring blocks, when calculating the prediction error using the BMA. To determine the proper weighting function, notice that the MV $\mathbf{d}_m$ of block $\mathcal{B}_m$ affects not only this block, but also its neighboring blocks $\mathcal{B}_{m,k}$ when performing OBMC. Because the contribution of $\mathbf{d}_m$ to pixel $\mathbf{x}$ in $\mathcal{B}_{m,k}$ is weighted by $h_k(\mathbf{x})$, we can minimize the following error

$$\sum_{k\in\mathcal{K}}\sum_{\mathbf{x}\in\mathcal{B}_{m,k}}|\psi(\mathbf{x})-\psi_r(\mathbf{x}+\mathbf{d}_m)|^p h_k(\mathbf{x}). \tag{9.3.6}$$

Let $\bar{\mathcal{B}}_m = \bigcup_{k\in\mathcal{K}}\mathcal{B}_{m,k}$ be the superblock including the current block, $\mathcal{B}_{m,o}$, and its neighboring blocks $\mathcal{B}_{m,k}, k \in K$; then this error can be rewritten as

$$\sum_{\mathbf{x}\in\bar{\mathcal{B}}_m}|\psi(\mathbf{x})-\psi_r(\mathbf{x}+\mathbf{d}_m)|^p\bar{h}(\mathbf{x}) \tag{9.3.7}$$

where the window function $\bar{h}(\mathbf{x})$ is related to $h_k(\mathbf{x})$. For example, in Figure 9.14, $\bar{h}(\mathbf{x}_0) = h_0(\mathbf{x}_0), \bar{h}(\mathbf{x}_1) = h_3(\mathbf{x}_0), \bar{h}(\mathbf{x}_2) = h_4(\mathbf{x}_0), \bar{h}(\mathbf{x}_3) = h_1(\mathbf{x}_0)$, and $\bar{h}(\mathbf{x}_4) = h_2(\mathbf{x}_0)$. The error in Equation (9.37) is a weighted MCP error over the superblock $\bar{\mathcal{B}}_m$, and the window function $\bar{h}(\mathbf{x})$ is a rearrangement of the original weighting coefficients $h_k(\mathbf{x})$. The optimal window function, determined by Orchard and Sullivan [26], is shown in Figure 9.15.

Experimental results have shown that using OBMC can improve the prediction image quality by up to 1 dB, when OBMC is combined with standard BMA. When iterative motion estimation is applied, up to 2 dB improvement can be obtained [26]. Due to its significant improvement in prediction accuracy, OBMC has been incorporated in the ITU-T H.263 standard for video coding, as an advanced option. The standard uses the four-neighbor structure shown in Figure 9.14, but only two of the neighbors have nonzero weights for each given pixel. The weighting coefficients $h_k(\mathbf{x})$ used in the standard, which are chosen to facilitate fast computation [17], are given in Figure 9.16. The equivalent window function $\bar{h}(\mathbf{x})$ is shown in Figure 9.17.

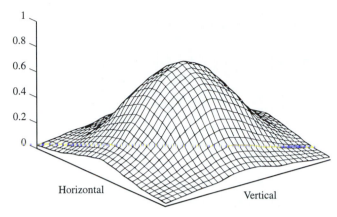

Figure 9.15 The optimal window function for OBMC determined by Orchard and Sullivan. Reprinted from M. T. Orchard and G. J. Sullivan, Overlapped block motion compensation: An estimation-theoretic approach, *IEEE Trans. Image Process.* (1994), 3:693–99. Copyright 1994 IEEE.

(a)

4	5	5	5	5	5	5	4
5	5	5	5	5	5	5	5
5	5	6	6	6	6	5	5
5	5	6	6	6	6	5	5
5	5	6	6	6	6	5	5
5	5	6	6	6	6	5	5
5	5	5	5	5	5	5	5
4	5	5	5	5	5	5	4

(b)

2	2	2	2	2	2	2	2
1	1	2	2	2	2	1	1
1	1	1	1	1	1	1	1
1	1	1	1	1	1	1	1
1	1	1	1	1	1	1	1
1	1	1	1	1	1	1	1
1	1	2	2	2	2	1	1
2	2	2	2	2	2	2	2

(c)

2	1	1	1	1	1	1	2
2	2	1	1	1	1	2	2
2	2	1	1	1	1	2	2
2	2	1	1	1	1	2	2
2	2	1	1	1	1	2	2
2	2	1	1	1	1	2	2
2	2	1	1	1	1	2	2
2	1	1	1	1	1	1	2

Figure 9.16 (a) The weighting coefficients for OBMC, specified in the H.263 video coding standard [17]: (a) for prediction with the motion vector of the current block; (b) for prediction with motion vectors of the blocks on top or bottom of the current block; (c) for prediction with motion vectors of the blocks to the left or right of the current block. The numbers given are 8 × the actual weights. For example, to predict the pixel at the top left corner of the block, the weights associated with the MVs of the current MB, the top MB, and the left MB are 4/8, 2/8, and 2/8, respectively. For the pixel on the first row and the second column, the weights are 5/8, 2/8, 1/8, respectively.

9.3.3 Coding Parameter Selection

In the hybrid coder described in Section 9.3.1, there are multiple choices that the encoder must make, including the mode to use for each MB in an inter-coded frame, the QP for each MB, the motion estimation methods and parameters (e.g., with or without overlapping, block size, search range), and so on. Each combination of these coding parameters will lead to a different trade off between the overall rate and distortion of the coded video. In this subsection, we address the issue of how to select these parameters.

In the early development of hybrid coders, such decisions are usually driven by heuristics. For example, in inter-coded pictures, the decision to code an MB in intra-mode or inter-mode can be based on the variance of the MB itself, σ_{intra}^2, and the variance

				1	1	1	1	1	1	1	1				
				1	1	1	1	1	1	1	1				
				1	1	2	2	2	2	1	1				
				2	2	2	2	2	2	2	2				
1	1	1	2	4	5	5	5	5	5	5	4	2	1	1	1
1	1	2	2	5	5	5	5	5	5	5	5	2	2	1	1
1	1	2	2	5	5	6	6	6	6	5	5	2	2	1	1
1	1	2	2	5	5	6	6	6	6	5	5	2	2	1	1
1	1	2	2	5	5	6	6	6	6	5	5	2	2	1	1
1	1	2	2	5	5	6	6	6	6	5	5	2	2	1	1
1	1	2	2	5	5	5	5	5	5	5	5	2	2	1	1
1	1	1	2	4	5	5	5	5	5	5	4	2	1	1	1
				2	2	2	2	2	2	2	2				
				1	1	2	2	2	2	1	1				
				1	1	1	1	1	1	1	1				
				1	1	1	1	1	1	1	1				

Figure 9.17 The window function corresponding to the weighting coefficients given in Figure 9.16.

of the motion compensation error, σ^2_{inter}.* If $\sigma^2_{intra} \leq \sigma^2_{inter} + c$, then the intra-mode is chosen over the inter-mode. This heuristic is motivated by the observation that the number of bits required to code a block at a given distortion is proportional to the variance of the block. The positive constant c is added to account for the additional bits required to code the motion vector in the inter-mode.

For improved performance, the choice between different parameters can be determined by a rate-distortion optimization approach. Here, the bits required by the use of different parameters, and the resulting distortions (say, the MSE), are determined by actually coding the source using these parameter settings. The setting that yields the best trade off between the rate and distortion is then chosen. Typically, there is a constraint on the average rate of the coded sequence, and the problem is to find the parameter setting that has the minimal distortion while meeting the rate constraint. Such a constrained optimization problem can be solved by either the Lagrangian multiplier method [7] or the dynamic programming approach [2].

Let us consider an example of determining the coding modes for all MBs in a frame, assuming that all other options are fixed and that the desired number of bits for the entire frame is R_d. Denote the distortion for the nth MB using mode m_n by $D_n(m_n)$, and the required number of bits by $R_n(m_k, \forall k)$. The reason that R_n depends on the coding modes for other MBs is because the motion vectors and DC coefficients are predictively coded from neighboring blocks. The problem is to

$$\text{minimize} \quad \sum_n D_n(m_n),$$
$$\text{subject to} \quad \sum_n R_n(m_k, \forall k) \leq R_d. \tag{9.3.8}$$

*In practice, to reduce computation, the variance is often replaced by the sum of the absolute differences between the values (original or prediction error) of each pixel and the block mean.

Using the Lagrangian multiplier method, this problem can be converted to

$$\text{minimize} \quad J(m_n, \forall n) = \sum_n D_n(m_n) + \lambda \sum_n R_n(m_k, \forall k) \qquad (9.3.9)$$

where λ must be chosen to satisfy the rate constraint.

Strictly speaking, the optimal coding modes for different MBs are interdependent. For ease of understanding of the basic concept, let us ignore the dependence of the rate R_n on the coding modes of other MBs; that is, we assume that $R_n(m_k, \forall k) = R_n(m_n)$. Then the coding mode for each MB can be determined individually, by minimizing

$$J_n(m_n) = D_n(m_n) + \lambda R_n(m_n). \qquad (9.3.10)$$

Given that there are only a small number of possible modes to choose from, the optimal mode for each block can be found by an exhaustive search.

Note that if m_n (and, consequently, R_n) is a continuous variable, then minimizing J_n is equivalent to setting $\partial J_n/\partial R_n = 0$, which leads to $\partial D_n/\partial R_n = -\lambda$. *This implies that the optimal mode in each MB is such that the resulting RD slope $\partial D_n/\partial R_n$ is the same among different MBs.* In practice, there are only a finite number of modes to choose from. Each possible mode corresponds to an operating point on a piecewise linear RD curve, so that each mode is associated with a range of RD slopes. The optimal modes for different MBs found by minimizing $J_n(m_n)$ for a given λ will have similar ranges in their RD slopes.

The result that different MBs should operate at the same RD slope is a special case of a variety of RD optimization problem involving multiple independent coding parameters. Recall that in the bit allocation problem in transform coding (Section 9.1.5), the optimal solution is such that the RD slopes for different coefficients are the same (Equation (9.1.40)). In that case, the RD slope is proportional to the distortion; therefore, the optimal bit allocation yields identical distortion among different coefficients.

A difficult problem in this approach is how to determine λ for a given desired rate. For an arbitrarily chosen λ, this method will yield a solution that is optimal at a particular rate, which may or may not be close to the desired rate. A bisection method is commonly employed to search for the λ that will yield the desired rate. A closed-form relation between λ and the QP is derived in [36], which assumes that the QP is chosen based on the target bit rate (see the next subsection).

Rate-distortion optimized selection of coding modes was first considered by Wiegand, et al. [38]. There, to handle the interdependence among the coding modes of different MBs in the same frame and over adjacent frames, a dynamic programming scheme was employed to find the optimal coding modes for a set of MBs simultaneously. The paper by Sullivan and Wiegand [36] includes a good discussion on how to use a similar approach to determine a variety of coding options in the H.263 coder. It is worth noting that the RD-based approach leads to rather marginal gain

---

*By definition, $\partial D_n/\partial R_n = \lim_{\Delta R \to 0} \Delta D/\Delta R$; that is, the RD slope measures the reduction in the distortion by each extra bit. The equal-slope condition is equivalent to requiring all blocks to operate in such a way that an extra bit in each block will reduce the error by the same amount. Otherwise, a block operating at a higher (lower) slope should use more (fewer) bits.

over heuristic approaches within the H.263 framework (about 10% savings in the bit rate, or 0.5 dB in PSNR) [36]. In practice, such gains may not be justified, considering the quite significant increase in complexity. Thus, the RD optimized approach mainly serves as a benchmark for evaluating the performance of a heuristic, yet practical approach.

The computationally most demanding step in the RD-based parameter selection approach is to collect the RD data associated with different parameter settings (coding mode and QP, and possibly different motion estimation methods). This would require actually coding all the MBs using all different parameters. To reduce the computation, some RD models have been proposed that relate the rate and distortion, on the one hand, and the QP and coding modes, on the other [5, 24]. Once the RD data are obtained, one can use either a Lagrangian multiplier method or a dynamic programming approach to find the optimal allocation. The Lagrangian multiplier method is simpler, but has suboptimal performance, because it ignores interdependence between rates of adjacent MBs in the same frame or between neighboring frames.

In addition to coding parameter selection, RD-based methods can be applied in a variety of problems in image and video coding. One important area is motion estimation for video coding. Whereas traditional approaches to motion estimation focus on minimizing motion-compensated prediction error only (see Chapter 6), RD optimized approaches also consider the rate needed to code the resulting MVs. For example, the zero MV may be preferred over a nonzero MV, if it leads to only slightly higher prediction error, considering the extra bits required to code the nonzero MV. Furthermore, because MVs are coded in a predictive manner, smoother motion fields are preferred. Work on RD optimized motion estimation can be found in [35, 11, 20, 33, 3].

For more complete coverage on how to use RD theory to optimize image and video coding, the reader is referred to the review paper by Ortega and Ramchandran [58], which also contains a good exposition of Lagrangian and dynamic programming methods as optimization tools. More extensive coverage of this topic can be found in Schuster and Katsaggelos [32].

9.3.4 Rate Control

One important issue in video coding is rate control, which refers to how to code a video so that the resulting bitstream satisfies a target bit rate. Within the framework of hybrid coding, this reduces to how to choose the coding parameters (frame rate, inter- vs. intra-, QP, etc.) to meet the rate constraint. We alluded to this problem in the previous subsection, where we described an RD optimized approach for selecting the coding mode for each MB, subject to a rate constraint.

For pleasant visual perception, it is better to represent a video so that it has a constant (or smoothly changing) quality. Because the scene activity (in terms of motion as well as texture complexity) changes in time, the required bits will change from frame to frame. Even if we accept changes in video quality, the bit rate cannot be made strictly constant, due to the use of VLC and possibly different coding modes on the frame level. Therefore, the best that one can hope to achieve are constant average rates over short intervals of time. The variability of the bit rate within each interval must be handled by a smoothing buffer following the encoder. The output bits from the encoder first enter the smoothing buffer at a variable rate, and are then moved out of the buffer at a constant rate. The maximum time interval and the allowed variability within this interval depend on the delay requirement of the underlying application and the affordable buffer size.* A larger buffer will enable better video quality, but will also introduce longer delay.

Rate control is typically accomplished in three steps:

1. Update the target average bit rate (in terms of bps) for each short time interval, also referred to as the *rate update interval*;†
2. Determine the coding mode (e.g., I-, P-, or B-frame) and the target bit budget for each frame to be coded in this interval, which is usually based on the target average rate for the interval and the current buffer fullness; and
3. Determine the coding mode and QP for each MB in a frame to meet the target rate for this frame.

Step 1 depends on the underlying applications. For video transport over a constant bit rate (CBR) network (such as a modem line or an ISDN channel), the target average rate should be constant. For networks that can accommodate variable bit rate (VBR), such as an ATM network, the coder should try to regulate the rate to maintain a constant video quality, and yet satisfy the average and peak rate requirements [14]. Finally, for networks where channel capacity and error characteristics vary in time, such as the best-effort Internet and wireless channels, one must update the target average rate based on the channel conditions. The update interval depends on how fast the network conditions change. In all these applications, the target average bit rate per update interval should depend on not only the available bandwidth (which would need to be estimated in time-varying channels), but also the end-to-end delay requirement, and the current encoder buffer size (i.e., how many bits remain to be sent). *Essentially, the target rate should be set so that the bits from the current update interval as well as remaining bits in the buffer can reach the destination within the maximally allowed delay time.* In this sense, the rate constraint is determined by the delay constraint [27]. In lossy networks, one also must take into account the fact that some previously sent data may have to be retransmitted in case they are damaged during transmission [13]. In [27], Ortega provides a good categorization of rate control problems and extensive coverage of how to determine the target rate based on the underlying network characteristics and delay requirements. For rate estimation approaches designed for video streaming over Internet applications, see the discussion in Chapter 15.

---

*Given the continuously diminishing cost of memory, it is reasonable to assume that the buffer can be as large as the end-to-end delay allows and, therefore, this physical limitation of the buffer size can be ignored [27]. The encoder buffer is only one factor that contributes to the end-to-end delay of a video transmission. For a description of different factors contributing to the end-to-end delay, see Chapter 14.

†In practice, this may be done in a sliding window fashion. That is, at each new frame time, the target bit rate for a short period of time following this frame is determined.

Step 2 is typically accomplished by choosing an appropriate frame rate based on the total rate budgeted for an update interval, and assigning the same target rate for all the frames in the interval. Frames can be skipped when the buffer occupancy is in danger of overflow. This could happen when the bits used for the previous frames too greatly exceed the specified targets, or the remaining bit budget for this interval is insufficient to code the video at the specified frame rate. This is the approach typically employed for interactive applications, where all the frames, except the first, are coded in the inter-mode, to maintain a fairly constant delay between frames. For broadcast or streaming of MPEG-2 coded video, an update interval usually corresponds to a GOP, which contains one I-frame followed by interleaving P- and B-frames. Typically, rate allocation within a GOP assumes some constant ratios between the bits used for I-, P-, and B-frames [16]. More sophisticated, RD-based approaches would try to allocate the total bits among frames based on the scene content, so that the average distortion is minimized. The look-ahead involved, however, will incur significant computations and additional coding delay.

Rate control within a frame (i.e., step 3) can be accomplished by adjusting the coding mode and QP for each MB, which can be done either based on heuristic rules or using an RD optimized approach, as described in Section 9.3.3.

The preceding discussions focus on video transmission applications. For storage applications (such as in DVD movies), the total bits used for a video are limited by the capacity of the storage medium; however, the bit rate may change from time to time to accommodate changes in scene activity.

The rate control solutions discussed thus far consider only waveform-based coders. For object-based coders (such as MPEG-4), rate control must consider rate allocation among multiple objects; and, within each object, allocation among shape, motion, and texture. The frame rates of different objects can also vary based on the relative importance of each object. These problems are addressed by Vetro, Sun, and Wang [37].

Instead of producing a bit stream with a fixed target rate, another approach is to produce a scalable bit stream, from which the receiver can choose to extract only partial bits based on the available bandwidth. This approach is more appropriate when the same video is to be accessed by multiple users with different link capacities. Scalable coding is discussed in Chapter 11; the distribution of a scalable stream to users with different bandwidth capacities is considered in Chapter 15.

A problem that is related to rate control is *rate shaping*. This refers to the function of an interface (called a transcoder or filter) between the compression layer and the network transport layer or between two network segments, with which a precompressed video stream is converted so that the rate of the resulting stream matches the available network bandwidth. This subject is discussed in Chapter 15.

It is important to note that, in video coding standards, coding mode selection and rate control are not defined. Instead, the encoder has the flexibility to optimize their implementations as long as the coded bit stream follows the standardized syntax.

9.3.5 Loop Filtering*

The sources for motion-compensated prediction error can be classified into three groups: (1) motion estimation error; (2) noise in the reference image, which can be due to the accumulation of errors caused by motion estimation in prior frames and quantization errors; and (3) changes in the coded frame that are not due to motion, such as occlusion and illumination variation. In order to suppress primarily the second source of noise, *loop filtering* can be applied, which consists of a low-pass filter applied to the predicted image. Note that with noninteger motion estimates, an interpolation filter is used implicitly when one performs motion-compensated prediction. Similarly, loop filtering is accomplished implicitly with OBMC. However, as can be seen in the following, a filter can be specifically designed to reduce the effect of noise in the video signal. Motion compensation with filtering can be generally described by

$$\psi_p(x, y) = g(x, y) * \psi_r(x + d_x(x, y), y + d_y(x, y)), \tag{9.3.11}$$

where $\psi_p(x, y)$ represents the predicted frame, $\psi_r(x, y)$ the reference frame, $g(x, y)$ the impulse response of the loop filter, and $*$ denotes 2-D linear convolution. In the following, we derive the optimal filter following the approach of Girod [10, 8, 9].

To simplify the design, the following assumptions are made: (1) the reference video frame is a stationary random field; (2) the errors in estimated motion vectors at different spatial positions are independent; and (3) the noise term is independent of the reference frame. With these assumptions, the power spectral density of the prediction error can be shown [8] to be

$$S_{e,e}(f_x, f_y) = S_{\psi,\psi}(f_x, f_y)(1 + |G(f_x, f_y)|^2 - 2\mathcal{R}\{G(f_x, f_y)P(f_x, f_y)\})$$
$$+ S_{n,n}(f_x, f_y)|G(f_x, f_y)|^2. \tag{9.3.12}$$

In this relation, $S_{\psi,\psi}(f_x, f_y)$ and $S_{n,n}(f_x, f_y)$ represent the PSDs of the reference frame and the noise signal, respectively; $G(f_x, f_y)$ is the Fourier transform of $g(x, y)$; $P(f_x, f_y)$ is the Fourier transform of the pdf of the motion estimation error; and $\mathcal{R}$ denotes the real part of a complex variable. In this analysis, the signals are assumed to be discrete and the Fourier transforms are DSFT.

By differentiating Equation (9.3.12) with respect to $G(f_x, f_y)$, we arrive at the optimal filter that minimizes the mean squared prediction error:

$$G(f_x, f_y) = P^*(f_x, f_y) \frac{S_{\psi,\psi}(f_x, f_y)}{S_{\psi,\psi}(f_x, f_y) + S_{n,n}(f_x, f_y)}, \tag{9.3.13}$$

where $*$ denotes complex conjugation. Note that, without the multiplication term $P^*(f_x, f_y)$, this filter would be the conventional Wiener filter with respect to the noise term. The factor $P^*(f_x, f_y)$ takes into account the inaccuracy of motion estimation.

In general, $P(f_x, f_y)$ has low-pass characteristics. Therefore, the optimal filter $G(f_x, f_y)$ is also low-pass, in general. A careful examination of Equation (9.3.12) will

---

*Sections marked with an asterisk may be skipped or left for further exploration.

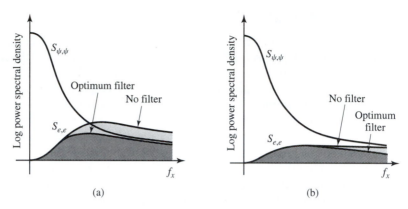

Figure 9.18 Power spectral density of the motion-compensated prediction error for (a) moderately accurate motion estimation, and (b) very accurate motion estimation. Reprinted from B. Grod, Motion compensation: Visual aspects, accuracy, and fundamental limits, in M. I. Sezan and R. L. Lagendijk, eds., *Motion Analysis and Image Sequence Processing,* Boston: Kluwer Academic Publishers, 126–52, by permission of Kluwer Academic Publishers.

show that $S_{e,e}(f_x, f_y)$ tends to be small at low spatial frequencies, and large at high frequencies. For example, if $G(f_x, f_y) = 0$ for a certain high-frequency range, the prediction error would have the same energy as the reference signal for this frequency range. This analysis shows that motion compensation works well for low-frequency components of the signal, but not for high-frequency components, even with the optimal filter applied. Figure 9.18 shows $S_{e,e}(f_x, f_y)$ for two levels of motion estimation accuracy, with and without the optimal filter applied. In both cases, it is assumed that $S_{n,n}(f_x, f_y) = 0$.

Loop filtering has been found to provide significant gains in prediction accuracy. In fact, the use of a loop filter is incorporated explicitly in the ITU-T H.261 video coding standard, which employs integer-pel accuracy motion estimation (see Section 13.2.1).

Relation among Prediction Error, Motion Estimation Accuracy, and Loop-Filtering We now analyze the relation among the prediction error, motion estimation error, and noise power, with or without the application of optimum filtering, following the approach of Girod [8, 9, 10]. From the PSD of the prediction error in Equation (9.3.12), one can determine the variance of the prediction error (i.e., the MSE of prediction) by

$$\sigma_e^2 = \int_{-1/2}^{1/2} \int_{-1/2}^{1/2} S_{e,e}(f_x, f_y) \, df_x \, df_y. \tag{9.3.14}$$

Assuming that the signal spectrum is isotropic with variance σ_s^2, the noise has a flat spectrum with variance σ_n^2, and the motion estimation error follows a Gaussian distribution with variance $\sigma_{\Delta \mathbf{d}}^2$, Girod calculated the prediction error variances for various

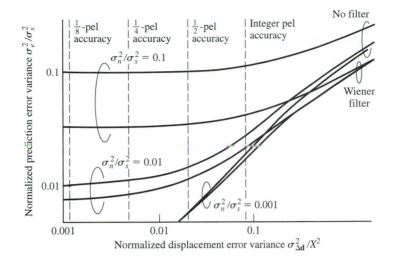

Figure 9.19   Influence of motion compensation accuracy on prediction error variance for noisy signals. X indicates the width of a pixel (the height is assumed to be equal to the width). Reprinted from B. Grod, Motion compensation: Visual aspects, accuracy, and fundamental limits, in M. I. Sezan and R. L. Lagendijk, eds., *Motion Analysis and Image Sequence Processing,* Boston: Kluwer Academic Publishers, 126–52, by permission of Kluwer Academic Publishers.

motion estimation accuracies and signal-to-noise ratios. Figure 9.19 shows the results of his analysis. The most interesting observation is that *there is a critical point for motion estimation accuracy, beyond which the possibility of further improvement in prediction accuracy is small.* This critical point increases with the noise level in the signal. That is, when the signal is very noisy, there is no point in trying to estimate the motion very accurately. Even with very accurate motion estimation, the predicted frame is still error-prone due to the presence of noise. Also revealed by this figure is that Wiener filtering can be very helpful when the noise level is high, but not very effective in the low-noise case.

In Figure 9.19, the motion estimation error variance has been related to the search accuracy in terms of pixels. This is done by assuming that the motion estimation error is due entirely to the rounding to a given fractional accuracy. The vertical dashed lines indicate the minimum displacement error variances that can be achieved with the indicated motion compensation accuracy. We can see that, in the low-noise case, no more than $\frac{1}{8}$-pel accuracy is needed, whereas for the high-noise case, $\frac{1}{2}$-pel accuracy is quite sufficient. These analytical results have been validated by actual experimental results [9]. It has been observed that for high-quality broadcast video signals, motion estimation with $\frac{1}{4}$-pel accuracy is sufficient for a practical coder, whereas for video phone–type signals, $\frac{1}{2}$-pel accuracy seems to be a desirable limit. In both cases, the gain over integer pel accuracy is significant.

9.4 SUMMARY

Transform Coding (Section 9.1)

- Transforms provide a way to represent a group of samples (e.g., an image block) as a linear combination of basic patterns. The purpose of transforms for coding is to decorrelate original samples and compact the energy into a few coefficients so that scalar quantization can be effectively applied. The optimal transform under these criteria is the KLT.

- To achieve maximum gain from transform coding, one must appropriately distribute bits among the coefficients. The optimal allocation (Equation (9.1.43)) equalizes the distortion incurred by different coefficients (Equation (9.1.44)). The final distortion depends on the geometric mean of the coefficient variances.

- The KLT depends on the second-order statistics of the signal and, thus, is image-dependent and difficult to compute. The DCT is a good approximation of the KLT for most image signals, and it can be computed using fast algorithms.

- The DCT-based coding method (Section 9.1.7) is an effective method for image coding and has been employed in international standards for image and video coding.

- Optimization of quantization and the binary encoding process is very important in a transform coder. With similar quantization and binary encoding schemes, DCT and wavelet transforms lead to similar coding efficiency.

Predictive Coding (Section 9.2)

- The purpose of prediction is also to reduce the correlation between samples to be coded, so that scalar quantization can be applied effectively. The predictor should be designed to minimize the prediction error. The LMMSE predictor can be determined by solving the Yule-Walker equation (Equation (9.2.6)).

- To avoid mismatch between the reference samples used for prediction in the encoder and those used in the decoder, closed-loop prediction is necessary, in which the encoder must repeat the same operation as the decoder.

- Theoretically, predictive coding outperforms transform coding when the predictor order is the same as the transform length. However, the bit stream resulting from predictive coding is very sensitive to transmission errors.

- For video coding, prediction can be performed in both the spatial and temporal domains. In the temporal direction, motion compensation is necessary to account for the effect of object motion. Only very low order predictors (usually using only one previous frame) are used, to maintain a reasonable complexity, and to reduce the effect of transmission error propagation.

Block-Based Hybrid Video Coding (Section 9.3)

- The block-based hybrid coder (Section 9.3.1) effectively combines motion-compensated prediction and transform coding. Because of its relatively low complexity and good coding efficiency, it has been adopted in various international standards for video coding (Chapter 13).

- Within the framework of hybrid coders, performance can be improved by appropriately performing motion estimation and compensation (e.g., with or without overlapping, fixed versus variable block size) and selecting modes of operations (intra- versus inter-, prediction mode, etc.) The RD optimized approach chooses these coding options so that the distortion is minimized under a given rate constraint (Section 9.3.3).

- Rate control (Section 9.3.4) is an important issue in video coding, especially for real-time applications, where excessive rates will lead to deterioration (as opposed to improvement) in quality due to transmission delay and loss. Rate control in hybrid coders is typically accomplished by adjusting frame rates and quantization parameters, using either heuristic or RD optimized approaches.

- Coding mode selection and rate control can affect the performance of a coder significantly. Note that all the video coding standards using the hybrid coding framework define only the bit stream syntax and, consequently, the decoder operation. This gives the encoder flexibility in optimizing its operation, including motion estimation, coding mode selection, and rate control. These are typically the factors that differentiate among video coding systems that all conform to the same standard. Beyond coding efficiency considerations, other factors that can set systems apart are how the encoder makes the bit stream more resilient to transmission errors (by judiciously selecting the coding parameters within the standard framework), and how the decoder reacts to transmission errors. These issues are discussed in Chapter 14.

- Motion compensation accuracy can be improved by applying loop-filtering (Section 9.3.5), which suppresses the propagation of quantization errors in previous frames. The achievable motion compensation accuracy depends not only on the motion estimation accuracy, but also on the loop-filter used and the noise level in the signal (Section 9.3.5).

9.5 PROBLEMS

9.1 Prove the relations in Equations (9.1.11–16).

9.2 Confirm that the 1-D DCT basis functions form an orthonormal basis.

9.3 Prove that a 2-D basis formed by the outer product of two orthonormal 1-D bases is also orthonormal.

9.4 Repeat Example 9.4 for the special case of $\rho_h = \rho_v = \rho_d = \rho$.

9.5 Repeat Example 9.5 for the special case of $\rho_h = \rho_v = \rho_d = \rho$. Compare the KLT transform basis with the DCT.

9.6 Repeat Example 9.6 for the special case of $\rho_h = \rho_v = \rho_d = \rho$.

9.7 Consider a predictive image coding system in which a pixel is predicted from two neighboring pixels: one above and one to the left. Assume that the image statistics are the same as those considered in Example 9.4. Derive the optimal predictor and the predictive coding gain. Compare the result with that in Example 9.6. Which predictor is more efficient?

9.8 Write a C or MATLAB code that examines the effect of approximating an image with a partial set of DCT coefficients. Use an 8×8 DCT, reconstruct the image with $K < 64$ coefficients, for $K = 4, 8, 16$, and 32. How many coefficients are necessary to provide a satisfactory reconstruction?

9.9 Write a C or MATLAB code that examines the effect of quantization in the DCT domain. Use an 8×8 DCT, and chose the quantization stepsizes to be multiples of the quantization table given in Figure 9.7. Try scale factors of 0.5, 1, 2, 4, 8, and 16. What is the largest scale factor that still provides a satisfactory image?

9.10 Two simple transform coding methods are: (a) only retaining the first K coefficients (called *zonal coding*), and (b) only retaining the coefficients that exceed a certain threshold (called *threshold coding*). Discuss the advantages and disadvantages of these two methods.

9.11 Consider the following three video coding methods: directly coding a frame; coding the difference between every two frames; performing block-based motion estimation between every two frames and coding the motion compensation error image. Recall that the bit rate required to achieve a given distortion is proportional to the variance of the signal to be coded. Take two adjacent video frames from a sequence, and calculate and compare the variances of (a) a frame directly; (b) the direct difference between two frames; and (c) the motion compensation error image. Based on your results, can you determine which of the three coding methods would be most efficient?

9.12 Variance is only a crude indicator of the required bit rate. If we apply DCT coding in each of the three methods in Problem 9.11, a more accurate measure of the bit rate is the number of DCT coefficients that are nonzero after quantization with a chosen quantization parameter. Write a C or MATLAB code to perform this experiment.

9.13 Show that the optimal linear estimator solution for OBMC is as given in Equation (9.3.4).

9.14 Show that the window function corresponding to the weighting coefficients given in Figure 9.16 is as given in Figure 9.17. Draw it in the style of Figure 9.15, and comment on the similarities and differences between the two window functions.

9.15 Write a C or MATLAB program that implements a basic form of the block-based hybrid coder. For simplicity, consider only the intra-mode and the P-mode. Except for the first frame, which will be coded using the intra-mode entirely, each MB in following frames will be coded using either the P-mode or intra-mode, depending on whether the original MB or the motion compensation error has the smaller variance. For motion estimation, you can choose to implement either the EBMA or some fast algorithms. For the lossless coding component, you can either use the Huffman tables given in a standard (e.g., [17]), or estimate the required bit rate by calculating the entropy of the symbols to be coded. Apply the program to several test sequences with different motion levels using different parameters: QP, frame rate, intra-frame or intra-block rate, and so on. Evaluate the trade offs between rate, distortion, and complexity (execution time). Play the decompressed video to observe the coding artifacts with different QPs.

9.6 BIBLIOGRAPHY

[1] Ahmed, N., T. Natarajan, and K. R. Rao. Discrete cosine transform. *IEEE Trans. Computer* (1974), 23:90–93.

[2] Bellman, R. *Dynamic Programming*. Princeton NJ: Princeton University Press, 1957.

[3] Chen, M. C., and A. N. Willson. Rate-distortion optimal motion estimation algorithms for motion-compensated transform coding. *IEEE Trans. Circuits Syst. for Video Technology* (April 1998), 8(2):147–58.

[4] Chen, W.-H., et al. A fast computational algorithm for the discrete cosine transform. *IEEE Trans. Commun.* (Sept. 1977), COM-25:1004–09.

[5] Chiang, T., and Y.-Q. Zhang. A new rate control scheme using quadratic rate-distortion model. *IEEE Trans. Circuits Syst. for Video Technology* (Feb. 1997), 7:246–50.

[6] Clark, R. J. *Transform Coding of Images*. London: Academic Press, 1985.

[7] Everett, H. Generalized Lagrange multiplier method for solving problems of optimum allocation of resources. *Operational Research* (1963), 11:399–417.

[8] Girod, B. The efficiency of motion-compensating prediction for hybrid coding of video sequences. *IEEE J. Select. Areas Commun.* (1987), SAC-5:1140–54.

[9] Girod, B. Motion-compensated prediction with fractional-pel accuracy. *IEEE Trans. Commun.* (1993), 41:604–12.

[10] Girod, B. Motion compensation: Visual aspects, accuracy, and fundamental limits. In M. I. Sezan and R. L. Lagendijk, eds., *Motion Analysis and Image Sequence Processing,* Boston: Kluwer Academic Publishers, 1993, 126–52.

[11] Girod, B. Rate-constrained motion estimation. In *SPIE Conf. Visual Commun. Image Processing* (Sept. 1994), 1026–34.

[12] Gonzalez, R. G., and R. E. Woods. *Digital Image Processing*. 2d ed. Reading, MA: Addison-Wesley, 1992.

[13] Hsu, C.-Y., A. Ortega, and M. Khansari. Rate control for robust video transmission over burst-error wireless channels. *IEEE J. Select. Areas Commun.* (May 1999), 17(5):756–73.

[14] Hsu, C.-Y., A. Ortega, and A. R. Reibman. Joint selection of source and channel rate for VBR video transmission under ATM policing constraints. *IEEE J. Select. Areas Commun.* (Aug. 1997), 15:1016–28.

[15] ISO/IEC. IS 10918-1: Information technology—digital compression and coding of continuous-tone still images: Requirements and guidelines, 1990. (JPEG).

[16] ISO/IEC. MPEG-2 Test Model 5 (TM5). Public document, ISO/IEC JTC1/SC29/WG11/93-225b, April 1993.

[17] ITU-T. Recommendation H.263: Video coding for low bit rate communication, 1998.

[18] Jain, A. K. *Fundamentals of Digital Image Processing*. Englewood Cliffs, NJ: Prentice Hall, 1989.

[19] Jayant, N. S., and P. Noll. *Digital Coding of Waveforms*. Englewood Cliffs, NJ: Prentice-Hall, 1984.

[20] Kossentini, F., Y.-W. Lee, M. J. T. Smith, and R. Ward. Predictive RD optimized motion estimation for very low bit rate video coding. *IEEE J. Select. Areas Commun.* (Dec. 1997), 15(9):1752–63.

[21] Lee, B. G., A new algorithm to compute discrete cosine transform. *IEEE Trans. Acoust., Speech, Signal Processing* (Dec. 1984), ASSP-32:1243–45.

[22] Li, W. Vector transform and image coding. *IEEE Trans. Circuits Syst. for Video Technology* (Dec. 1991), 1(4):297–307.

[23] Li, W. and Y.-Q. Zhang. Vector-based signal processing and quantization for image and video compression. *Proceedings of the IEEE* (Feb. 1995), 83(2):317–33.

[24] Lin, J.-L., and A. Ortega. Bit rate control using piecewise approximated rate-distortion characteristics. *IEEE Trans. Circuits Syst. for Video Technology* (Aug. 1998), 8:446–59.

[25] Nogaki, S., and M. Ohta. An overlapped block motion compensation for high quality motion picture coding. *IEEE Int. Conf. Circuits and Systems* (May 1992), 184–87.

[26] Orchard, M. T., and G. J. Sullivan. Overlapped block motion compensation: An estimation-theoretic approach. *IEEE Trans. Image Process.* (1994), 3:693–99.

[27] Ortega, A. Variable bit rate video coding. In M. T. Sun and A. R. Reibman, eds., *Compressed Video over Networks,* New York: Marcel Dekker, 2001.

[28] Ortega, A., and K. Ramchandran. Rate-distortion methods for image and video compression. *IEEE Signal Processing Magazine* (Nov. 1998), 15:23–50.

[29] Pennebaker, W. B., and J. L. Mitchell, *JPEG Still Image Data Compression Standard,* New York: Van Nostrand Reinhold, 1993.

[30] Rabiner, L. R., and R. W. Shafer. *Digital Processing of Speech Signals*. Englewood Cliffs, NJ: Prentice Hall, 1978.

[31] Ramstad, T. A. Considerations on quantization and dynamic bit allocation in subband coders. *IEEE Int. Conf. Acoust., Speech, Signal Processing* (April 1986), 841–44.

[32] Schuster, G. M., and A. K. Katsaggelos. *Rate-distortion Based Video Compression*. Boston: Kluwer Academic Publishers, 1997.

[33] Schuster, G. M., and A. K. Katsaggelos. A video compression scheme with optimal bit allocation among segmentation, motion, and residual error. *IEEE Trans. Image Process.* (Nov. 1997), 6:1487–501.

[34] Segall, A. Bit allocation and encoding for vector sources. *IEEE Trans. Inform. Theory* (March 1976), IT-22:162–69.

[35] Sullivan, G. J., and R. L. Baker. Rate-distortion optimized motion compensation for video compression using fixed vs. variable size blocks. In *Global Telecomm. Conf.* (GLOBECOM '91) Dec. 1991, 85–90.

[36] Sullivan, G. J., and T. Wiegand. Rate-distortion optimization for video compression. *IEEE Signal Processing Magazine* (Nov. 1998), 15:74–90.

[37] Vetro, A., H. Sun, and Y. Wang. MPEG-4 rate control for multiple video objects. *IEEE Trans. Circuits Syst. for Video Technology* (Feb. 1999), 9:186–99.

[38] Wiegand, T., M. Lightstone, D. Mukherjee, T. Campell, and S. K. Mitra. Rate-distortion optimized mode selection for very low bit rate video coding and the emerging H.263 standard. *IEEE Trans. Circuits Syst. for Video Technology* (April 1996), 6(2):182–90.

10

Content-Dependent Video Coding

In Section 8.1, we gave an overview of different video coding systems. One way of distinguishing coding systems is to compare their source models. Video coding based on source models that do not adapt to the shape of moving objects in a video sequence was described in Chapter 9. In this chapter, we focus on video coding algorithms that consider the shape of the objects in the video sequence. In order to efficiently transmit arbitrarily shaped video objects, we must code the object shape as well as the texture. In Section 10.1, we describe several methods for coding the 2-D shape of an object. Section 10.2 presents algorithms to efficiently code the textures of regions. Two algorithms that jointly code shape and texture are briefly described in Section 10.3. In Sections 10.4 and 10.5, we describe how these techniques can be used in region-based and object-based coding, respectively. Object-based coding may be based on 2-D and 3-D object models. Using 3-D object models, object-based coding can be extended to knowledge-based and semantic coding, as described in Sections 10.6 and 10.7. In Section 10.8, we present a concept that allows us to integrate conventional block-based hybrid coders with object-based, knowledge-based, and semantic coding.

10.1 TWO-DIMENSIONAL SHAPE CODING

The 2-D shape of an object, say the kth object, is defined by means of an *alpha-map* M_k:

$$M_k = \{m_k(x, y) \mid 0 \leq x \leq X,\ 0 \leq y \leq Y\}, \quad 0 \leq m_k \leq 255. \qquad (10.1.1)$$

The shape M_k defines for each pel $\mathbf{x} = (x, y)$ whether it belongs to the object $(m_k(\mathbf{x}) > 0)$ or not $(m_k(\mathbf{x}) = 0)$. For an opaque object, the corresponding alpha-values

are 255, for transparent objects they range from zero to 255. Usually, the alpha-map
has the same spatial and temporal resolution as the luminance signal of the video sequence. In video editing applications, the alpha-map is used to describe object shape
and object transparency. Let us assume that we have a background image $\psi_b(\mathbf{x})$, the
object represented by image $\psi_o(\mathbf{x})$, and the alpha-map $m_o(\mathbf{x})$. Overlaying the object on
the background is done according to

$$\psi(\mathbf{x}) = \left(1 - \frac{m_o(\mathbf{x})}{255}\right) \cdot \psi_b(\mathbf{x}) + \frac{m_o(\mathbf{x})}{255} \cdot \psi_o(\mathbf{x}) \qquad (10.1.2)$$

The amplitude of the alpha-map determines how visible the object becomes. In this
chapter, we will deal only with binary object shape; that is, $m_k(\mathbf{x}) \in \{0, 255\}$.

There are two classes of binary shape coders. A *bitmap*-based coder (Section
10.1.1) encodes for each pel whether it belongs to the object or not. A *contour*-based
coder (Section 10.1.2) encodes the outline of the object. In order to retrieve the bitmap
of the object shape, the contour is filled with the object label. In the case that texture is
also transmitted with the shape information, an *implicit* shape coder can be used, with
which the shape information can be derived from the texture (see Section 10.3). As with
texture coding, binary shapes can be coded in a lossless or lossy fashion. Therefore,
it is important to define a measure for the quality of lossily encoded shapes (Section
10.1.3).

10.1.1 Bitmap Coding

A bitmap-based shape coder specifies a binary alpha-map. In its simplest form, we
scan the alpha-map and transmit for each pel a zero or one depending on whether it
belongs to the object. Such an algorithm is inefficient, because it does not consider
statistical dependencies between neighboring pels. In this section, we describe three
ways of improving coding efficiency: one class of algorithms scans the bitmap in scan
line order and specifies the run-lengths of black and white pels; a second class codes
each pel of the bitmap based on the colors of the neighboring pels; and the third class
defines the shape as a tree of basic shapes arranged such that they completely fill the
object shape.

Scan-Line Coding   In this method, we scan the image line by line and transmit
the run of black and white pels using a variable-length coder, thus exploiting 1-D correlation between pels. This method is used in the fax standard G3 [40]. Next, we discuss
an extension of this coder that also considers the correlation between neighboring scan
lines.

The *modified READ (relative element address designate) code,* as used in the fax
standards G4 [23] and JBIG [24], codes each scan line with respect to the previous
scan line, thus exploiting 2-D correlation. The algorithm scans each line of the document and encodes the location of *changing pels,* where the scan line changes its color
(Figure 10.1). In this line-by-line scheme, the position of each changing pel on the
current line is coded with respect to the position of either a corresponding changing pel
in the reference line, which lies immediately above the present line, or the preceding

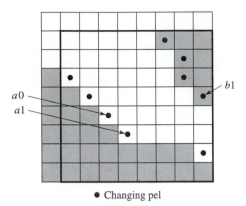

● Changing pel

Figure 10.1   Changing pels define the object boundary for a modified READ coder.

changing pel in the current line [40]. Figure 10.1 is used to explain one implementation of a modified READ coder. We assume that the top five rows of the block have already been coded, hence the coder knows the positions of the pels $a0$ and $b1$ in the current block. The unknown point $a1$ on the object boundary is encoded with reference to the two pels $a0$ and $b1$. Pel $a0$ is the last boundary pel encoded prior to $a1$. Pel $b1$ is the first changing pel on the line above $a0$ that is to the right of $a0$ and has the opposite color of $a0$, if such a point exists. If not, then $b1$ is the leftmost changing pel on the same line as $a0$. In order to encode the distance between $a1$ and $a0$, one of three modes, *vertical, horizontal,* or *vertical pass,* is selected. Assuming that all pels are numbered in raster-scan order, starting with zero in the top-left corner of the block (e.g., in Figure 10.1, num($a0$) = 34), and columns are numbered from left to right (e.g., col($a0$) = 2), a mode is selected according to

$$\text{mode} = \begin{cases} \text{vertical} & \text{if } |\text{col}(a1) - \text{col}(b1)| \leq T \\ \text{horizontal} & \text{else if num}(a1) - \text{num}(a0) < W \\ \text{vertical pass} & \text{else} \end{cases} \quad (10.1.3)$$

where T is a threshold and W the width of the block to be coded. The value of T depends on the length of the scan lines, and can be as low as $T = 5$. In the example of Figure 10.1, $a1$ will be coded in the vertical mode if $T = 5$.

In vertical mode, the distance col($a1$) $-$ col($b1$) is encoded using one of eight VLC tables that is selected according to the object boundary direction, as defined by a template positioned above pel $b1$ (Figure 10.1).

In horizontal mode, the position of $a1$ is encoded as its distance to $a0$. Just due to the fact that the horizontal and not the vertical mode is selected, the decoder can sometimes deduct the minimum distance between $a1$ and $a0$. In this case, only the difference with respect to this minimum distance is encoded.

Vertical pass mode is selected according to Equation (10.1.3) when there is at least one row of points following $a0$ without a boundary pel. In this case, we send one codeword for each line without an object boundary. One last codeword codes the remaining distance to the next point on the object boundary.

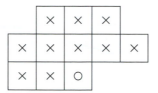

Figure 10.2   Template for defining the context of the pel to be coded (designated by "○").

Pelwise Coding   This coder exploits the spatial redundancy of the binary shape information to be coded. Pels are coded in scan-line order and row by row. The color of each pel is coded using conditional entropy coding (see Section 8.3.1). A template of n pels is used to define the causal context for predicting the color of the current pel. Figure 10.2 shows a template that uses ten pels to define the context for coding the current pel. The template extends up to two pels to the left, to the right, and above the pel to be coded [5]. This context (i.e., the colors of the pels in the template) determines the code that we use for encoding the current pel. Because each pel can only have one of two colors, there are a total of $2^{10} = 1024$ different contexts. For a given context, since each pel can only choose from two possible colors, using a Huffman code will not lead to an increase of coding efficiency, due to the minimum codeword length of one bit. However, we can code the color of the pel with less than one bit by using an arithmetic coder (Section 8.4.2). For each context, the arithmetic encoder uses a different probability table. For the template of Figure 10.2, we would require 1024 probability tables. Given that we have to store only the probability of one symbol (say, black), and can compute the probability of the other symbol (say, white), we need only store 1024 probabilities.

In the shape coding method of the MPEG-4 standard, the image is partitioned into square blocks. Three different types of blocks are distinguished: transparent and opaque blocks, as well as boundary blocks on the object boundary containing both transparent and opaque pels. The boundary blocks are coded using the preceding method, and this coding method is called *context-based arithmetic coding* [30].

Quad-Tree Coding   A quad-tree describes the shape of an object by placing nonoverlapping squares of different sizes inside the shape such that the shape is described as accurately as possible. Typically, the sizes of the squares are powers of two. We define a maximum size and a minimum size for the squares. We can describe shapes exactly if the minimum square size is one pel. In a first step, we place squares of maximum size $M \times M$ pels next to each other on the image. Then, we decide based on a homogeneity criterion whether we need to subdivide a square. If yes, then we replace the square with four squares of size $M/2 \times M/2$. We repeat this process recursively (Figure 10.3). This process creates an exact or an approximate shape approximation, depending on the homogeneity criterion and the minimum square size. The homogeneity criterion defines the number of pels n that are allowed to be misrepresented in any given square. This number can be fixed or can vary as a function of the size of the square that is currently being considered.

The quad-tree can be encoded with binary symbols. We walk through the tree in a depth-first fashion. In the example of Figure 10.3, a 1 indicates that the associated

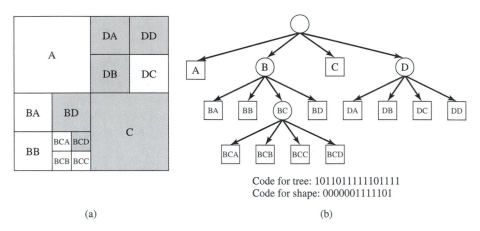

Code for tree: 1011011111101111
Code for shape: 0000001111101

(a)   (b)

Figure 10.3   (a) An image is divided into a quad-tree that describes the object shape. (b) Related quad-tree—with code that describes the tree and labels for each square, such that the shape can be recovered.

square is not further divided, a 0 indicates that the square is subdivided and that the next four symbols indicate the states of the children of this square. Again, this description is repeated recursively. After the tree is described, we must indicate for each square whether it belongs to the object. This is done by sending a binary symbol for each end node in the tree. In Figure 10.3, we use 1 for squares that belong to the object. Hence, this shape code requires as many symbols as we have 1s in the quad-tree code. Using multilevel symbols for the shape code allows us to describe several objects within an image with just one quad-tree (see Problem 10.6).

10.1.2 Contour Coding

Extensive work has been published on contour-based shape representation and coding. Different applications nurtured this research: For lossless and lossy encoding of object boundaries, chain coders [14, 11], and polygon and higher-order approximations [18, 54, 43, 15] were developed. For recognition purposes, shape representations like Fourier descriptors were developed to allow translation-, rotation-, and scale-invariant shape representations [61].

Chain Coding   A chain code follows the contour of an object. After coding the coordinates of a starting point on the object boundary, the chain code encodes the direction in which the next boundary pel is located (Figure 10.4). Chain codes can code the direction of the contour (direct chain code) or the change in direction (differential chain code) [9]. Differential chain codes are more efficient because they employ the statistical dependency between successive links. Algorithms differ by whether they consider a pel to have four or eight neighbors (for rectangular grids) or six neighbors (for hexagonal grids). In Figure 10.4 we define that the boundary pels are part of the object; however, this leads to the "dual shape" problem—for example, the white pels

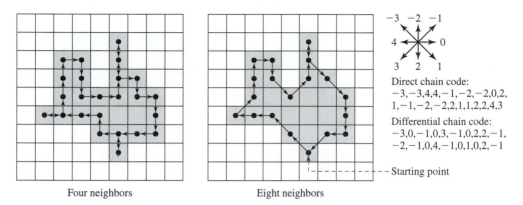

Figure 10.4   Chain code for pels with four and eight neighbors. We give examples for a direct chain code and for a differential code of the eight-connected chain. The first symbols of the two codes are identical and define the starting direction. The following symbols of the differential chain code are created by aligning direction 0 of the direction star with the direction of the last coded symbol.

in Figure 10.4 have their own set of boundary pels. This boundary pel definition causes coding of redundant information if we have two objects that touch. Several algorithms define the object boundary between pels, thus avoiding this problem [52]. Chain codes are lossless shape coders; that is, they describe a given contour exactly. However, we can preprocess (e.g., smooth or quantize) a contour prior to coding with a chain code. Sometimes, this preprocessing is not considered separately from the chain coder; in that case, chain coding may be considered as a lossy coder.

Freeman [14] originally proposed the use of chain coding for boundary quantization and encoding, which has attracted considerable attention over the last forty years [50, 33, 41, 29]. The curve is quantized using the grid intersection scheme [14], and the quantized curve is represented using a string of increments. Since the planar curve is assumed to be continuous, the increments between grid points are limited to the eight grid neighbors, and hence an increment can be represented by three bits. For lossless encoding of boundary shapes, an average rate between 1.2 and 1.4 bits/boundary pel is required [11].

There have been many extensions to this basic scheme, such as the generalized chain codes [50], where the coding efficiency has been improved by using links of different length and different angular resolution. In [29], a scheme is presented that utilizes patterns in a chain code string to increase coding efficiency. There has also been interest in the theoretical performance of chain codes. In [33], the performance of different quantization schemes is compared, whereas in [41] the rate-distortion characteristics of certain chain codes with preprocessing are studied.

Some chain codes also include simplifications of the contour in order to increase coding efficiency [42]. This is similar to filtering the object shape with morphological filters and then coding with a lossless chain code. Simplifications correspond to

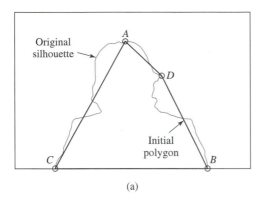

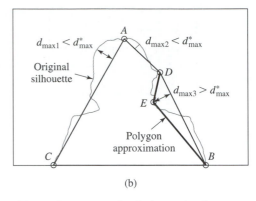

(a)   (b)

Figure 10.5   Polygon approximation of a contour. The quality measure d_{max}^* is shown. A point E is added during the refinement of the contour description at the location with the largest d_{max}^*.

amplitude quantization in image coding. Whereas the chain code is limited to a lossless shape representation, the following shape coding algorithms allow for a lossy shape representation.

Polygon Approximation   A polygon-based shape representation was developed for object-based analysis-synthesis coding [18, 19]. The object shape is approximated by a continuous chain of line segments (Figure 10.5). Intuitively, we can assume that a polygon representation is very efficient for describing geometric objects with straight edges using just a few line segments. Curved boundaries will require many line segments in order to provide a natural-looking boundary.

Since a polygon representation allows for lossy shape coding, we must define a quality measure by which we can evaluate the quality of the coded shape. The maximum of the Euclidean distances between the original and the approximated contour positions, d_{max}, is commonly used.

The efficiency of a polygon approximation depends not only on the object shape but also on the location of the starting points for the contour description. Instead of computing the representation for all starting points and choosing the most efficient ones, the polygon approximation is computed by using those two contour points with the maximum distance between them as the starting points. Then, additional points are added to the polygon where the approximation errors between the polygon and the boundary are maximum. In Figure 10.5(a), the initial approximation $\overline{AB}$ is extended by point C and then D. This is repeated until the shape approximation error is less than a threshold d_{max}^*. Figure 10.5(b) shows that the point E is added to the approximation where the largest error d_{max} is measured. Advanced algorithms select vertices that minimize the approximation error given a bit budget, by optimally positioning the vertices on the contour or within a thin stripe along the contour [30, 32].

The encoder must transmit the vertex coordinates of the line segments to the decoder. After encoding the absolute position of one vertex, coordinates of the following vertices are differentially encoded using the image size as an upper bound for the vertex coordinates.

The polygon representation requires many vertices for lossless shape coding. Chain codes usually outperform polygon representation for lossless coding, whereas polygon representations are more efficient for lossy coding.

Spline Approximation   Spline functions are defined by selected control points. We can use spline functions to compute continuous curves from discrete points. We represent the continuous boundary coordinates as a function of the contour length t as $\mathbf{c}(t) = [x(t), y(t)]$. This contour can be approximated by n node points $\mathbf{p}_n = [x_n, y_n]$, according to

$$\mathbf{c}(t) = \sum_{n=0}^{N-1} \mathbf{p}_n \cdot B_{n,q}(t). \tag{10.1.4}$$

The spline function $B_{n,q}$ is called a *B-spline*, where the letter "B" stands for "basis." $B_{n,q}$ depends on the degree q of the function and the position t_n of the nodes on the contour. It is determined by the following recursion:

$$B_{n,q}(t) = \frac{(t - t_n) \cdot B_{n,q-1}(t)}{t_{n+q-1} - t_n} + \frac{(t_{n+q} - t) \cdot B_{n+1,q-1}(t)}{t_{n+q} - t_{n-1}}; \quad q > 0 \tag{10.1.5}$$

and

$$B_{n,0}(t) = \begin{cases} 1 & t_n \leq t < t_{n+1}; \\ 0 & \text{else.} \end{cases} \tag{10.1.6}$$

Figure 10.6 shows the shape of different spline functions for equally spaced nodes. Nodes can also be placed at arbitrary intervals on the contour, and are allowed to coincide. A property of the spline function of order q is that its first $q - 1$ derivatives are continuous along the contour. As can be seen, Equation (10.1.6) is simply a hold function. A spline function of order q is created by convoluting the spline of order $q - 1$ with a spline of order 0. Since the extent of the spline function is limited, the position of a point $\mathbf{c}(t)$ is influenced by $q + 1$ points where q is the degree of the spline. Quadratic and cubic splines are used frequently. For shape coding, we must

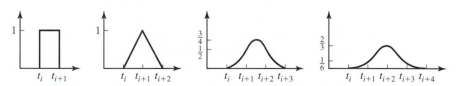

Figure 10.6   Splines of order $q = 0$, 1 (linear), 2 (quadratic), 3 (cubic), with equidistant nodes.

approximate the discrete contour $\mathbf{c}(t_i)$ with I contour points:

$$\mathbf{c}(t_i) = \sum_{n=0}^{N-1} \mathbf{p}_n \cdot B_{n,q}(t_i) \quad \text{for } 0 \leq i < I. \tag{10.1.7}$$

We can compute the node points $\mathbf{p}_n$ if we choose the number of nodes N smaller than the number of contour points I. In this case, we solve this system using a least-squares method.

In [25], B-spline curves are used to approximate a boundary. An optimization procedure is formulated for finding the optimal locations of the control points by minimizing the MSE between the boundary and the approximation. This is an appropriate objective when the smoothing of the boundary is the main problem. When the resulting control points need to be encoded, however, the tradeoff between the encoding cost and the resulting distortion must be considered. By selecting the MSE as the distortion measure and allowing the control points to be located anywhere on the plane, the resulting optimization problem is continuous and convex and can be solved easily. In order to encode the positions of the resulting control points efficiently, however, one must quantize them, and therefore the optimality of the solution is lost. It is well known that the optimal solution to a discrete optimization problem (quantized locations) need not be close to the solution of the corresponding continuous problem. In [30], a shape coder using splines is presented. Coding is optimal in the operational rate-distortion sense; that is, the control points are placed such that the required bit rate is minimum for a given distortion.

Fourier Descriptors   Fourier descriptors were developed for applications in recognition, where shape is an important key. Fourier descriptors allow a translation-, rotation-, and scale-invariant representation of closed contours [60]. There are different methods to define the Fourier descriptors of a closed contour. Using a simple descriptor, we sample the 2-D coordinates (x_n, y_n) of the contour with N pels clockwise in the image plane. We represent the coordinates as a complex number $z_n = x_n + j \cdot y_n$, and compute the Fourier series as

$$Z(m) = \sum_{n=0}^{N-1} z(n) \cdot e^{-j(2\pi nm/N)}. \tag{10.1.8}$$

Smooth contours will have their energy concentrated in the lower frequencies of the Fourier series. Setting higher-frequency coefficients to zero smoothes a contour.

It can be shown from Equation (10.1.8) that a translation of the contour will only influence coefficient $Z(0)$, the DC value. Scaling of the contour will multiply all coefficients with the scale factor. Rotation of the contour will shift the phase of the Fourier coefficients but will not effect their amplitude.

Another way of computing Fourier descriptors is to transform the list of 2-D coordinates (x_n, y_n) into an ordered list $(n, (y_{n+1} - y_n)/(x_{n+1} - x_n))$ with $(x_0, y_0) = (x_N, y_N)$ and $0 \leq n \leq N$ being the contour points. The term $(y_{n+1} - y_n)/(x_{n+1} - x_n)$

describes the change of direction of the contour. The Fourier series of this descriptor compacts more energy into the lower coefficients. This is partially due to the fact that the DC value describes a circle, which has constant change of direction, instead of a straight line as in Equation (10.1.8).

In order to preserve the main characteristics of a shape, only the large Fourier coefficients must be maintained. Fourier descriptors are not very efficient in reconstructing polygon-like shapes with only a few coefficients. This is one of the reasons that they never became very competitive in terms of coding efficiency.

10.1.3 Evaluation Criteria for Shape Coding Efficiency

Lossy shape coding can create two different types of distortions:

- The contour of an object may be changed: sharp corners smoothed, or parts of the contour displaced by a small number of pels.
- The topology of an object may be changed: holes may be added or closed, or parts of the object disconnected from the original shape.

There are two commonly used quality measures for objectively assessing the quality of coded shape parameters. The peak deviation d^*_{max} is the maximum of the Euclidean distances between each coded contour point and the closest point on the original contour. This measure allows for an easy interpretation of the shape quality due to changes of the object contours. However, if lossy shape coding results in changing the topology of an object, the peak deviation d^*_{max} is not a useful measure. A second measure, d_n, depends of the number of pels that belong to the object and its coded representation. The metric d_n is the number of erroneously represented pels of the coded shape divided by the total number of pels belonging to the original shape; that is d_n describes the relative area error. Since different objects can have very different ratios of contour pels to interior pels, a given value for d_n only allows us to compare different approximations of the same object. The quality measure d_n by itself does not provide sufficient information about subjective shape quality, because the subjective meaning of a given d_n depends not only on the object size but also on the contour length.

Subjective evaluation of several sequences indicated that a shape representation with an approximation error of $d^*_{max} > 3$ pels is not useful at all for video, independent of its resolution. It was found that allowing a peak distance of $d^*_{max} = 1.4$ pels at CIF resolution (352×288 pels) is sufficient to allow proper representations of objects in low-bit-rate video coding applications.

Subjective evaluations also showed that the preceding two objective measures truthfully reflect subjective quality when comparing different bitmap-based or different contour-based shape coders. For lossy shape coding, the bitmap-based shape coders create blocky object shapes, whereas contour-based coders create an object shape showing curved distortions, especially at object corners or polygon edges. Since the two classes of shape coders give different distortions (Figure 10.7), a comparison between algorithms belonging to different classes must be done subjectively.

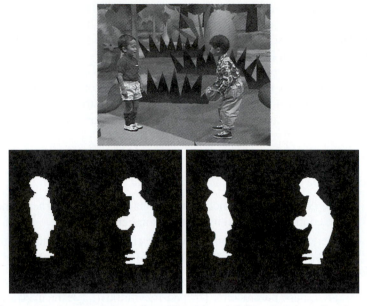

Figure 10.7   Lossy encoding using a bitmap-based (left) and a contour-based (right) shape coder. The shapes describe the two children in the top picture. Reprinted from C. L. Jordan, F. Bossen, and T. Ebrahimi. Scalable shape representation for content based visual data compression. *IEEE Int. Conf. Image Process. (ICIP '97),* Santa Barbara, CA, 1997, 1:512–15. Copyright 1997 IEEE.

10.2 TEXTURE CODING FOR ARBITRARILY SHAPED REGIONS

There are two classes of algorithms for coding the texture of arbitrarily shaped regions:

1. In the first, we *extrapolate* the texture of the region to fill a rectangle. Then we code this rectangle using an algorithm for coding the texture of rectangular regions. We give examples in Section 10.2.1. These algorithms are usually fast to compute, and will create as many coefficients as there are pels in the rectangle (this is a disadvantage, compared to the second class of algorithms).
2. In the second, we code the pels of the region *directly* using a transform that is adapted to the shape of the region. Therefore, we compute as many coefficients as there are pels in the region. Examples are given in Section 10.2.2.

10.2.1 Texture Extrapolation

Texture extrapolation or *padding* algorithms use a bounding box around the region that defines the area of texture to be coded. We call pels that belong to the region *opaque* and pels outside of the region but inside the bounding box *transparent,* since they will not be visible at the decoder. The texture of the bounding box will be coded using any algorithm suitable for coding textures of rectangular regions, such as DCT or wavelet transforms. Figure 10.8 shows the bounding box for a region. This bounding box is

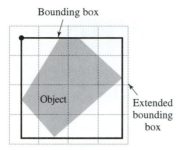

Figure 10.8 The circumscribing rectangle of the object or region of interest is extended such that the side lengths of the extended bounding box are multiples of eight for DCT coding.

extended such that its width and height are suitable for the texture coding algorithm. If we use a DCT, we might extend the rectangle such that its width and height are multiples of eight or sixteen pels.

Since the decoder will neglect the texture of all transparent pels, we define the transparent pels inside the bounding box such that they do not unnecessarily increase the data rate. Ideally, we extrapolate the texture of the region such that the signal is extended to the transparent pels without creating high-frequency components. This usually excludes setting the transparent pels to zero or 255. A first simple approach is to set the texture to the average texture value of the region or of the neighboring opaque pels. In addition, low-pass extrapolation filters can be applied to these transparent pels. Each transparent pel is set to the average of its four neighbors. This process starts at one corner of the bounding box and proceeds in scan-line order [31].

10.2.2 Direct Texture Coding

In order to limit the number of coefficients to be transmitted to the number of pels that must be coded, several algorithms were developed that compute a transform of an arbitrary image region. Starting with the DCT of the bounding box of an image region, Gilge computes a transform that is orthogonal on the image segment [16]. The resulting transform computes as many coefficients as there are pels in the image segment. Furthermore, it exploits the spatial correlation between neighboring pels. For each new shape of a segment, a new transform must be computed, which makes this approach very computationally intensive.

The shape-adaptive DCT (SA-DCT) is much faster to compute, using 1-D DCT transforms. First, we shift all pels of an image segment vertically to the block boundary (Figure 10.9(b)) [55]. Then a 1-D DCT according to the length of each column is applied. In a second step, the computed DCT coefficients are shifted horizontally to the left block boundary (Figure 10.9(e)). Again, a 1-D DCT is applied with a length according to the number of DCT coefficients in the row. Finally, the coefficients are quantized and encoded in a manner identical to the coefficients of the regular 2-D DCT. The SA-DCT is not orthonormal, due to the different lengths of the DCTs. Furthermore, the shifting of pels does not allow us to fully exploit the spatial correlation between neighboring pels. Therefore, the SA-DCT is not as efficient as the Gilge transform. Compared to padding, this more complex algorithm gains, on average, 1–3 dB in PSNR measured over an image segment for the same bit rate.

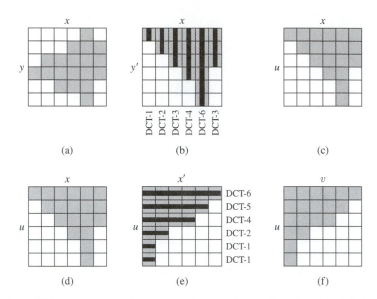

Figure 10.9   A shape-adaptive DCT requires transforms of length n: (a) original image segment, (b) pels shifted vertically, (c) location of DCT coefficients after vertical 1-D DCT, (d) location of DCT coefficients prior to horizontal 1-D DCT, (e) DCT coefficients shifted horizontally, (f) location of DCT coefficients after horizontal DCT.

Wavelet coders can also be adapted to arbitrarily shaped regions [36]. In order to compute the wavelet coefficients, the image signal at the boundary of a segment must be extended using periodic and symmetric extensions. Since no pels are shifted, the spatial correlation between neighboring pels is maintained.

10.3 JOINT SHAPE AND TEXTURE CODING

This shape coding technique was inspired by the blue-screen technique used in film and TV studios. The object to be coded is placed on a static monochrome background. The color of the background must be outside of the color space occupied by the texture of the object. Usually, highly saturated colors (e.g., pure blue) fulfill this requirement. Since the shape information is coded as part of the texture information, we sometimes call this coding method *implicit shape coding*. We present two coding algorithms that use this feature.

Today, GIF89a [17, 4] is used in Web applications to allow the description of arbitrarily shaped image objects. An image with a maximum of 256 colors is encoded based on a lossless compression scheme called LZW (after the authors Lempel, Ziv, and Welch). The particular implementation is called LZ78. In the header of a GIF file, it is possible to declare one of the 256 colors as the transparent color. All pels with that color will not be displayed, hence they appear to be transparent.

This concept of defining a transparent color was also introduced into video coding. The video signal is encoded using a frame-based coder. The chroma-key (background color) is transmitted to the decoder. The decoder decodes the images. Pels with a color similar to the chroma-key are considered to be transparent. Otherwise, pels belong to the object [6, 22, 20]. Since the shape information is typically carried by the subsampled chroma signal, this technique is not suited for lossless shape coding. Because the shape information is embedded in the texture, the shape coding is lossy as long as there is quantization of the texture. An important advantage of this method is its low computational and algorithmic complexity. As far as coding efficiency is concerned, this implicit shape coder requires a higher data rate than an explicit shape coder plus a texture coder like SA-DCT, or low-pass extrapolation with DCT.

10.4 REGION-BASED VIDEO CODING

Most image and video coders are optimized for coding efficiency. Image and video quality is measured by PSNR. It is known that the simple measure of PSNR does not capture well the properties of the human visual system. This becomes very apparent at low bit rates when blocking artifacts distort images without resulting in a low PSNR. Region-based image and video coding, also known as *second generation image and video coding* [34], tries to pay special consideration to the HVS. Properties of the HVS lead to the following basic claims that build the foundation for the choices made in designing these algorithms [59]:

- Edge and contour information are very important to the human visual system and are responsible for our perception.
- Texture information has relative importance, and influences our perception when taken together with contour information.

Based on these assumptions, region-based video coding puts more emphasis on coding contours than on coding textures. Motion (i.e., static or moving objects) is not considered when determining important contours.

A region-based video coder segments each image into regions of similar textures. Since the contours are considered to be very important, the coder transmits the contours of the regions with high precision. The textures of the regions are approximated using the mean value of the texture of the original region. Figure 10.10 shows an image segmented into regions. Depending on how stringently the similarity criterion is defined and what the minimum region size is, different segmentations result. Compared to DCT-based image coders at low data rates, this coder does not show any blocking artifacts. However, the flat texture representation creates a new kind of distortion that some people consider more disturbing. At high data rates, a DCT-based coder clearly outperforms a region-based coder, because the region-based coder requires many contours to be transmitted.

This concept can be extended to video coding [53]. In order to reduce the data rate required for shape coding, regions are tracked from one image to the next. Regions with similar or identical motion get clustered. For the new image, the encoder transmits the motion of regions in the current image, their change in shape, and newly appearing regions. Changes of the texture values are also transmitted.

Figure 10.10 The image "camera man" is segmented into several regions. Each region is represented by its contour and the mean value of the original texture. Different segmentations (requiring different amounts of data to code) are achieved by varying parameters of the segmentation algorithm.

10.5 OBJECT-BASED VIDEO CODING

Object-based analysis-synthesis coding [39] defines objects of uniform motion according to a 2-D or 3-D motion model. In contrast to region-based coding, OBASC does not further segment objects into regions of homogenous texture, thus saving on transmittal of shape parameters. OBASC subdivides each image of a sequence into uniformly moving objects, and describes each object m by three sets of parameters $A(m)$, $M(m)$, and $S(m)$, defining its motion, shape, and color, respectively. Motion parameters define position and motion of the object. Color parameters denote the luminance as well as the chrominance reflectance on the surface of the object (in computer graphics, these are

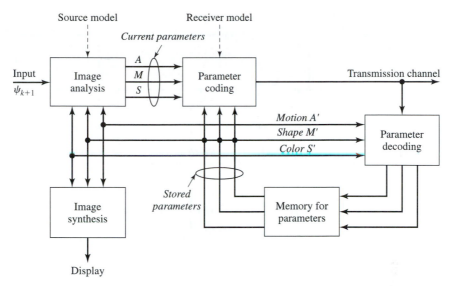

Figure 10.11   Block diagram of an object-based analysis-synthesis coder.
Reprinted from *Signal Processing: Image Communications,* 1(2), Musmann et al.,
Object-oriented analysis-synthesis coding of moving images, 117–138, copyright
1989, with permission from Elsevier Science.

sometimes called texture). Figure 10.11 explains the concept and structure of OBASC.
Instead of a frame memory as used in block-based hybrid coding, OBASC requires
a memory to store the coded and transmitted object parameters $A'(m)$, $M'(m)$, and
$S'(m)$. The parameter memories in the encoder and decoder contain the same information. Evaluating these parameter sets, image synthesis computes a model image ψ'_k that
is displayed at the decoder. The parameter sets of the memory and the current image
ψ_{k+1} are the input to image analysis.

The task of image analysis is to analyze the current image to be coded ψ_{k+1} and
to estimate the parameter sets $A_{k+1}(m)$, $M_{k+1}(m)$, and $S_{k+1}(m)$ of each object m. In
the current image, moving and static objects are detected first. For moving objects,
new motion and shape parameters are estimated in order to reuse most of the already
transmitted color parameters $S'_k(m)$. Objects for which the correct motion and shape
parameters can be estimated are denoted as MC-objects (for "model compliance").
In the final step of image analysis, image areas that cannot be described by MC-
objects using the previously transmitted color parameters and the new motion and
shape parameters $A_{k+1}(m)$, $M_{k+1}(m)$, respectively, are detected. These areas of model
failures (MF) are defined by 2-D shape and color parameters only, and are referred
to as MF-objects. The detection of MF-objects exploits the fact that small position
and shape errors of the model objects—referred to as *geometrical distortions*—do not
disturb subjective image quality. This assumption is valid for OBASC, because the
motion-compensated prediction image looks like a real image due to the object-based
image description. However, this image might be semantically incorrect. Thus, MF-
objects are reduced to those image regions with significant differences between the

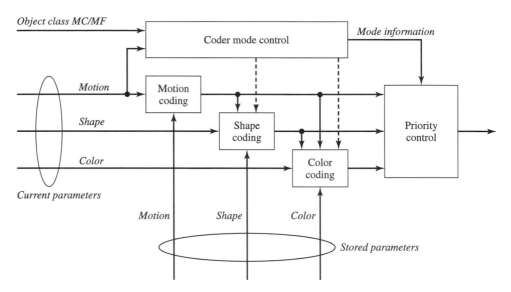

Figure 10.12   Parameter coding for OBASC. Reprinted from *Signal Processing: Image Communications,* 1(2), Musmann et al., Object-oriented analysis-synthesis coding of moving images, 117–138, copyright 1989, with permission from Elsevier Science.

motion- and shape-compensated prediction image and the current image ψ_{k+1}. They tend to be small in size. This allows us to code color parameters of MF-objects with high quality, thus avoiding subjectively annoying quantization errors. Since the transmission of color parameters is expensive in terms of data rate, the total area of MF-objects should not be more than 4% of the image area, assuming 64 kbit/s, CIF, and 10 Hz.

Depending on the object class, MC/MF, the parameter sets of each object are coded by parameter coding using predictive coding techniques (Figure 10.12). Motion and shape parameters are coded, transmitted, and decoded for MC-objects, as well as shape and color parameters for MF-objects. Since the coding of color parameters is most expensive in terms of bit rate, parameter coding and image analysis must be designed jointly. Let R_A, R_M, and R_S represent the rates for motion, shape, and texture, respectively. By minimization of the total bit rate,

$$R = R_A + R_M + R_S, \tag{10.5.1}$$

a higher coding gain can be achieved than in block-based hybrid coding techniques.

In OBASC, the suitability of source models can be judged by comparing the bit rates required for coding the same image sequence at the same image quality. Image quality is influenced mainly by the algorithm for detecting model failures, and by the bit rate available for coding the color parameters of model failures. We describe in the following several source models that have been developed in the context of OBASC.

10.5.1 Source Model F2D

The source model of *flexible 2-D objects (F2D)* assumes that the motion of a real object can be described by a smooth displacement vector field. This displacement vector field moves the projection of the real object into the image plane to its new position. The vector field defines a vector for each pel that belongs to the projections of the object into the image plane. The shape of the object is defined by the 2-D silhouette that outlines the projection of the object into the image plane.

In order to efficiently describe the displacement vector field of the object, the field is subsampled, and only one vector for every 16×16 pels is transmitted. The decoder reconstructs the displacement vector field by bilinear interpolation, resulting in a smooth displacement vector field that allows us to describe 2-D object motion as well as some flexible deformations of an object. As an example, the source model F2D is very well suited to model a sheet of rubber moving on a flat surface. Additionally, stretching of the rubber can be described. The shape of a moving object is described by a polygon approximation.

Whereas block-based hybrid coding predicts the current image $k+1$ using motion-compensated prediction from the texture of frame k, OBASC stores the texture of objects in a texture memory (Figure 10.13). Using this texture memory improves the

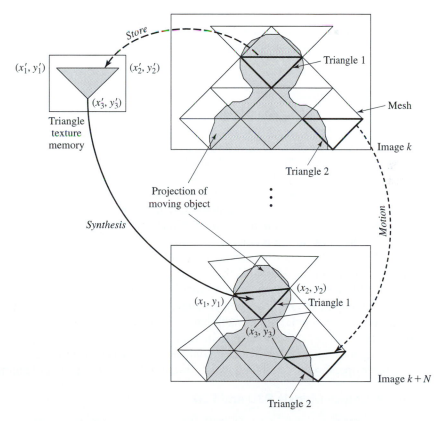

Figure 10.13 Image synthesis for MC-objects using a triangle-based mesh as texture memory.

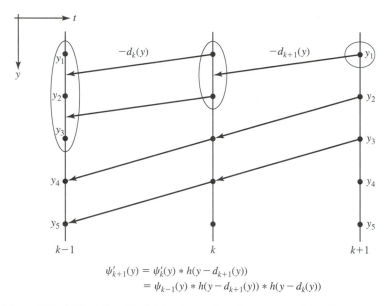

$$\psi'_{k+1}(y) = \psi'_k(y) * h(y - d_{k+1}(y))$$
$$= \psi_{k-1}(y) * h(y - d_{k+1}(y)) * h(y - d_k(y))$$

Figure 10.14   Block-based hybrid coders perform image synthesis by filter concatenation for motion compensation. Adapted from M. Hötter, Optimization and efficiency of an object-oriented analysis-synthesis coder. *IEEE Trans. Circuits and Systems for Video Technology* (April 1994), 4(2):181–94. Copyright 1994 IEEE.

performance of motion compensation. Figure 10.14 shows motion compensation by filter concatenation, as done in block-based hybrid coding. The motion-compensated prediction of image k is computed by copying a pel from image $k-1$ in the case of an integer pel motion vector, and by filtering with $h(n)$ in the case of a motion vector with fractional pel accuracy. The low-pass filter $h(n)$ is typically implemented as bilinear interpolation. If we now use frame k to predict frame $k+1$, we see that pel y_1 in frame $k+1$ in Figure 10.14 is computed by filtering previously filtered pels of frame k. Therefore, motion-compensated prediction using filter concatenation creates images that lose sharpness as we predict later frames.

This effect of filter concatenation can be avoided using a texture memory (Figure 10.15). Assuming again that this texture memory is initialized for frame k with the texture of frame k, we use the displacement vector field $D_k(y)$ from frame $k-1$ to k in order to compute the motion-compensated prediction image k. This requires one filter operation for each pel, with a fractional pel motion vector. Similarly, we can compute the motion-compensated prediction image $k+1$ by directly accessing the texture memory using the concatenated displacement vector field $D_{k+1,\text{tot}}(y)$. Again, we need only one filter operation for each pel, with a fractional pel motion vector (Figure 10.15).

10.5.2 Source Models R3D and F3D

The use of 3-D scene models with 3-D object models holds the promise that we can describe and model real-world objects more efficiently using moving *rigid 3-D model ob-*

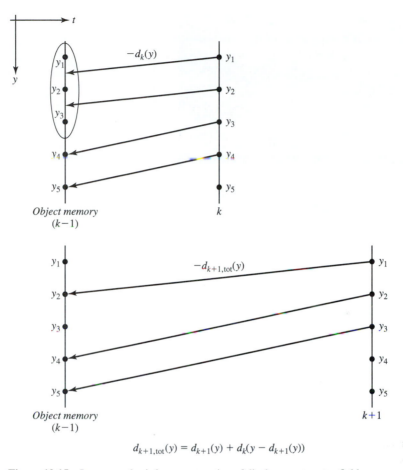

Figure 10.15 Image synthesis by concatenation of displacement vector fields (one-dimensional case) using texture memory. Adapted from M. Hötter, *Optimization and efficiency of an object-oriented analysis-synthesis coder. IEEE Trans. Circuits and Systems for Video Technology* (April 1994), 4(2):181–94. Copyright 1994 IEEE.

jects (R3D) (Section 5.3). We describe the shape of our model objects using a wireframe. The texture of the object is described using images or texture maps. Objects may move in 3-D using 3-D motion according to Equation (5.3.9). We allow an object to consist of several connected, rigid components in order to enable the modeling of joints [37]. After motion and shape estimation and compensation, areas of model failure are detected. The difference image between the prediction image and the original image (Figure 10.16(a)) is binarized with a noise-adaptive threshold (Figure 10.16(b)). The resulting mask identifies areas where there are differences between the images. Some of these differences are due to a small error in motion compensation (Figure 10.16(c)). We consider these errors as geometric distortions that are not of subjective importance. Of importance are areas of model failure where the content of the image changes. These areas exceed

(a)   (b)

(c)   (d)

Figure 10.16   Detection of model failures: (a) scaled difference image between real image ψ_{k+1} and model image after motion and shape compensation ψ_{k+1}^*; (b) synthesis error mask; (c) geometrical distortions and perceptually irrelevant regions; (d) mask MF_{R3D} indicating regions with model failures of the source model R3D.

a minimum size and have shapes that are not line structures. Figure 10.16(d) shows MF-objects due to eye-blinking, mouth motion, and specular reflections on the earring. We code the shape and texture only for the areas of model failure (MF_{R3D}-objects).

Since the surface of a model object is described as a wireframe, it is simple to enable flexible deformations of the object. The source model of moving *flexible 3-D objects* (F3D) allows control points of the wireframe to move tangentially to the object surface [45]. We describe this deformation using 2-D shift vectors. These shift vectors enable us to compensate for local deformations of the real object. Describing flexible deformations for the entire model object may require a high data rate. Therefore, we limit the use of the shift vectors to the areas of MF_{R3D}-objects.

Using the source model F3D requires in a first step the estimation of the MC_{R3D}-

objects and the MF$_{R3D}$-objects, as described in the previous paragraph. Input to the image analysis for estimating the shift vectors consists of the MC$_{R3D}$-objects, the MF$_{R3D}$-objects, and the real image $k + 1$ (Figure 10.17). Shift vectors are estimated for those vertices that are projected into the area of the current MF$_{R3D}$-objects. After compensating for the estimated shift vectors, we verify this step of image analysis by detecting model failures again, using the same algorithm as for the source model R3D. Figure 10.17 shows the estimated shift vectors and the MF$_{F3D}$-objects. These MF$_{F3D}$-objects are usually smaller than the MF$_{R3D}$-objects, because the flexible shift

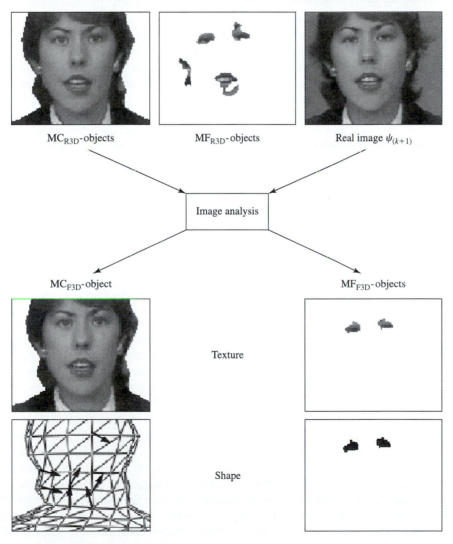

Figure 10.17 Flexible shift vectors are estimated in the areas of MF$_{R3D}$-objects. This usually results in a reduction of the MF-object size.

vectors compensate for some flexible deformations. Since it costs less in bits to code the shift vectors of the MC_{F3D}-object and the texture and shape of the MF_{F3D}-objects, instead of the texture and shape of the larger MF_{R3D}-objects, the use of F3D results in an increased coding efficiency. This is possible because the data rate for the shift vectors is only spent if they reduce the size of the MF-objects. When coding simple video telephony sequences, the use of F3D instead of R3D reduces the data rate from 64 to 56 kbps [46, 47].

OBASC codes the color parameters at a rate of 1 bit/pel, which is much higher than the 0.1–0.3 bit/pel that a block-based hybrid coder typically uses. Therefore, OBASC shows fewer quantization artifacts, and the images tend to look sharper than those of a hybrid coder. However, OBASC relies heavily on image analysis. Moving objects must be segmented correctly, and their motion must be estimated fairly precisely. If the image analysis fails, OBASC cannot use the shape parameters, which it must transmit in addition to the motion and texture parameters, to increase coding efficiency. To the contrary, it might require a data rate higher than that of a hybrid coder that transmits only motion and texture parameters. Due to the lack of robustness of image analysis, OBASC is still an ongoing area of research.

Depending on the applied source model, OBASC spends up to 80% of its data rate for coding model failures. OBASC does not require or exploit any scene knowledge. Therefore, the algorithm for the detection of model failures is not controlled by scene contents. In order to incorporate knowledge about the scene, we must extend OBASC to knowledge-based analysis-synthesis coding (KBASC), which evaluates scene knowledge for efficient coding.

10.6 KNOWLEDGE-BASED VIDEO CODING

A KBASC coder is based on OBASC. In addition, the coder tries to recognize objects (such as faces) in a video scene. As soon as the coder recognizes an object, it switches the coding mode for the object from the generic object-based mode to a knowledge-based mode. For head and shoulder scenes, the goal is to achieve a better modelling of the human face and to use the knowledge of the face location to control the coder. In order to enable automatic switching from OBASC to KBASC, an algorithm for recognizing facial features and the adaptation of a face model to the image sequences is proposed [28]. First, we must identify the eyes and mouth of the person that we want to code using KBASC (Figure 10.18(a)). In a second step, we must adapt a face model into the wireframe that currently represents the person (Figure 10.18(b)). As an example, we incorporate the face model "Candide" (Figure 10.18(c)) [49] into the object model. We can adapt the face model to the face of a person in an image by scaling the face model horizontally to match the distance between the two eyes and vertically to match the distance between mouth and eyes. Finally, we must stitch the face model into the wireframe model that we use to describe the object when coding with the OBASC layer (Figure 10.18(d)).

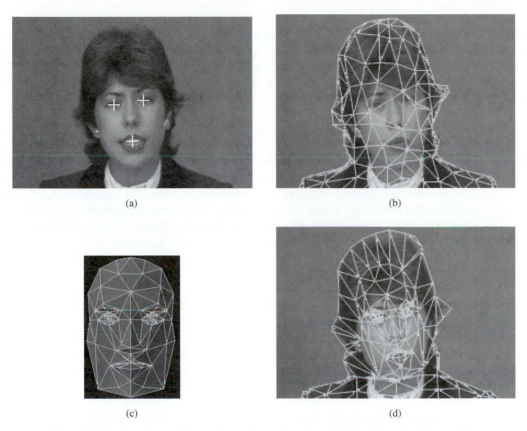

Figure 10.18 First, the positions of eye and mouth are identified (a). Using the object model of the person as a reference (b), we scale the face model (c) to the picture. Finally, we cut an opening into the face area of the wireframe of the person (b) and stitch the face model into the wireframe (d).

As soon as KBASC has detected a face, its coding efficiency increases compared to OBASC. One reason is that the use of a face model instead of a generic model improves the shape representation and allows for better motion compensation. Knowing the location of the face also allows us to define the joint between head and shoulders more precisely, which again results in improved motion estimation and compensation. Finally, we can use knowledge about the face location to compute model failures in a scene-dependent fashion. While we continue to use a sensitive detection algorithm in the face area, we can attribute most prediction errors outside of the face area to geometric distortions as well as illumination errors that we choose to neglect. In an experiment with the test sequence "Claire" at a spatial resolution of 352×288 pels and frame rate of 10 Hz, the average area of model failures was 3% for the source model F3D. The area decreased to 2.7% when we adapted the face model into the model object. Finally, exploiting the knowledge of the face location for the detection of model failures reduced their area to 2.5% without loss of subjective image quality. The corresponding data rates are 57 kbps for F3D with the OBASC coder and 47 kbps for KBASC [28].

10.7 SEMANTIC VIDEO CODING

Previously described video coding algorithms describe each frame of a video sequence using a combination of the current image signal and a predicted image signal that is derived using temporal prediction. Semantic video coding describes a video sequence using model objects with behavior that represent real objects with their behavior. We expect semantic coding to be far more efficient than the other video coders, because motion and deformation of most objects is very limited when compared to the variations possible in an array of pels required to display the object. The number of possible variations in the video representation determines the maximum entropy of the video sequence and hence the required bit rate. If we consider KBASC with a face model, changing the facial expression from neutral to joy would require a KBASC coder to transmit the mouth area as an MF-object. A semantic coder with an appropriate face model could just transmit the command "smile," and the decoder would know how to deform the face model to make it smile.

For face models, action units (AU) describing facial expressions have been proposed [7, 2, 13, 35]. Action units are derived from the facial action coding system (FACS) developed by psychologists [10]. The system was developed to "distinguish all possible visually distinguishable facial movements." The unit of measurement used is action, not muscle units—this is because FACS combines more than one muscle in its unitization of appearance changes. The term "action unit" is also used because FACS can attribute more than one action to what most anatomists describe as one muscle. For example, the frontalis muscle, which raises the brow, was separated into two action units, depending on whether the inner or outer portion of this muscle lifts the inner or outer portion of the eyebrow. Several action units can be exercised at the same time in order to create any facial expression. There are 46 AUs that account for changes in facial expression, and twelve AUs that describe changes in gaze direction and head orientation in coarser terms. Several example AUs are shown in Figure 10.19.

In order to use this system for face animation, amplitudes are assigned to action units. Temporal behavior of facial expressions can be modeled according to the onset, apex, and offset of AU measured on humans.

While the FACS system is based on human physiology, other proposals define feature points on a face that are animated by means of facial animation parameters (FAPs). The 68 FAPs used in MPEG-4 are based on the study of minimal perceptible actions, and are closely related to muscle actions [27, 58, 21]. FAPs represent a complete set of basic facial actions including head motion, tongue, eye, and mouth control. They allow representation of natural facial expressions. Figure 10.20 shows the points that can be used to define the head shape, and the feature points that are animated.

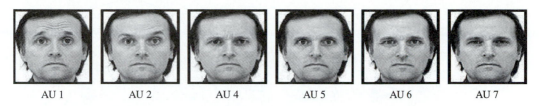

| AU 1 | AU 2 | AU 4 | AU 5 | AU 6 | AU 7 |

Figure 10.19   Examples of the six actions used in [3]. AU 1: Inner brow raiser. 2: Outer brow raiser. 4: Brow lowerer. 5: Upper lid raiser (widening the eyes). 6: Cheek raiser. 7: Lid tightener (partial squint). Reprinted from M. S. Bartlett, et al., Classifying facial action, in D. Touretzky, M. Mozer, and M. Hasselmo, eds., *Advances in Neural Information Processing Systems* 8, Cambridge, MA: MIT Press, 1996. Copyright 1996 MIT Press.

Both the FAP system and FACS animate faces with less than 100 parameters. It has been found that FAPs or AUs can be used to animate a face with less than 2 kbps [57]. However, it is not yet clear how to create a face model that allows realistic animation of faces. Shape, motion, and illumination estimation are still problems waiting for robust solutions. Furthermore, extraction of semantic parameters like FAPs or AUs from a video sequence is a difficult challenge. Semantic coding requires very reliable parameter estimation, since an error in the estimation changes the semantic meaning of the face representation at the decoder. A mix-up between a joyful and a sad facial expression can prevent effective communications. If KBASC is used to describe a traffic scene, an error might have a car turn left instead of driving straight ahead.

Since a semantic coder provides the promise of very high coding efficiency, it will be very sensitive to errors in the communications channel. A bit error will not result in a lower picture quality but in a change of semantic content.

10.8 LAYERED CODING SYSTEM

Depending on the scene content, it may not be efficient to describe each moving object with a separate object model in an object-based coder. Similarly, it is not always efficient to code an image of a sequence with a reference to previous images. Hence, we need to code a video sequence with the appropriate set of source models. In [38], a layered coding system is proposed (Figure 10.21). This system allows the application of different source models and different levels of scene understanding for coding of an image sequence, such that the best performance is guaranteed. Whereas layer I transmits only the color parameters of statistically dependent pels, layer II allows the additional transmission of motion parameters for blocks of fixed size and position. Hence layer I corresponds to an image coder or the I-frame coder of a hybrid coder, and layer II corresponds to the coding of P-frames in a hybrid coder. Layer III, the analysis-synthesis coder, allows the transmission of shape parameters in order to describe arbitrarily shaped objects. Here, layer III is represented by an OBASC coder based on the source model of F3D.

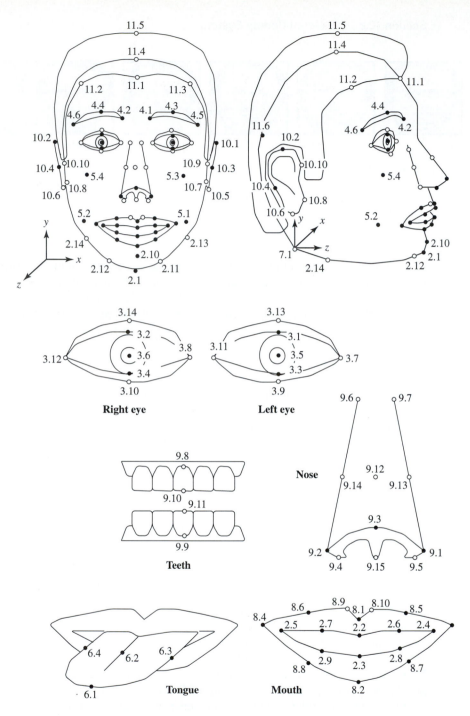

Figure 10.20 Feature points for a generic face as used in MPEG-4.

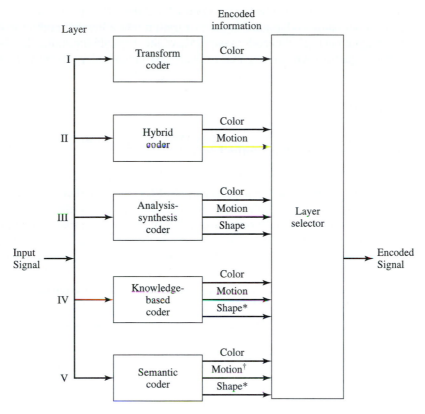

Figure 10.21   Block diagram of a layered coder. Reprinted from *Signal Processing: Image Communications* (Nov. 1995), 7(4–6), H. Mussman, A layered coding system for very low bit rate video coding, 267–68, copyright 1995, with permission from Elsevier Science.

Using a 3-D source model allows for a seamless integration with KBASC. In layer IV, knowledge of the scene contents is extracted from the video sequence and transmitted, in order to apply scene-dependent object models (such as the face model "Candide"). Layer V transmits high-level symbols for describing complex behavior of the objects.

Whereas different layers of this layered coding system have been investigated by many groups, the selection of the appropriate layer for encoding has been thoroughly investigated only for layers I and II, since these layers correspond to coding blocks in intra-mode or predictive mode in a block-based hybrid coder. Switching between layers II and III has only recently been addressed. In [8], a selection depending on the residual coding error and bit rate is proposed. In [44], the size of a moving object determines whether the object is coded in the block-based or in the object-based layer.

Switching between the OBASC of layer III and the KBASC of layer IV is addressed in [28]. A semantic coder suitable for integration into a layered coding system has been investigated in [12]; this coder saves 35% of the data rate for coding head-and-shoulder video sequences when compared to the H.263 video coder (layers I and II) [22].

10.9 SUMMARY

In this chapter, we discussed content-based video coding, which describes individual video objects and not just entire frames.

- **2-D Shape Coding (Section 10.1)**   In order to describe an object, we use an alpha-map that defines for each pel of a frame whether it belongs to the object or not. In addition, the alpha-map allows us to describe arbitrary transparency for a video object. In order to efficiently transmit the shape of a video object, we looked at 2-D shape coding algorithms. Bitmap-based shape coders describe for each pel whether it belongs to the object using a context-based arithmetic coder, whereas contour-based coders describe the outline of an object using polygons, splines, or a chain code. The human visual system is relatively sensitive to shape distortions. Therefore, we tend to code shapes with high quality.

- **Texture Coding (Section 10.2)**   As far as texture coding is concerned, we now must be able to code the texture of an arbitrarily shaped region. One possibility is to low-pass extrapolate the texture such that it fills a square block that can be coded using standard transform coders like the DCT. Alternatively, we can use the more efficient shape-adaptive DCT or wavelet transform that creates only as many coefficients as there are pels to be coded.

- **Object-based Analysis Synthesis Coding (Section 10.5)**   Object-based analysis-synthesis coding integrates arbitrarily shaped video objects into the framework of video coding. Since an object is now described as a unit, the motion-compensated prediction image looks like a natural image. It does not suffer from the block distortion of the block-based motion-compensated prediction images. We use the concept of model failures to determine those parts of an image that cannot be modeled with sufficient subjective quality. OBASC has been implemented using 2-D and 3-D source models. 3-D source models were extended to knowledge-based coding, where the coder identifies predefined models such as faces to describe the video scene more efficiently (Section 10.6). Semantic coding aims at describing video using high-level descriptions (Section 10.7). It has mainly been developed for faces, where parameter sets like AU or FAP are used to animate a face.

- **Layered Coding System (Section 10.8)**   Finally, we looked at a layered coder that integrates different coding modes into a framework. The layered coding system switches from one coder to another based on the coding efficiency, ensuring that we always code a video scene with the most efficient source model.

- **Object Segmentation and Motion Estimation**   Efficient content-dependent video coding depends on an image analysis that describes the image content accurately. Image analysis includes the problems of segmenting the objects or regions and estimating their motion. Segmentation involves partition of an initial video frame into regions of coherent texture or motion, and the tracking of these regions over time. In Section 6.8, we briefly considered the problem of segmenting a frame into regions of coherent motion. We saw there that motion estimation and region segmentation are interrelated and should be jointly optimized. A more substantial discussion of segmentation and tracking algorithms is beyond the scope of this book; interested readers can explore this topic further by beginning with the references [48, 1, 51, 56].

10.10 PROBLEMS

10.1 What is an alpha-map and how is it used in video?

10.2 What are the two main shape coding approaches? Discuss their advantages and disadvantages. How do they differ when a shape is coded lossily?

10.3 What objective measures for shape quality do you know?

10.4 Imagine a diagonal line of one pel width. How does the representation of this line change if we define an object using pels with four or eight neighbors? What is the advantage of a hexagonal grid for pels?

10.5 For the example given in Figure 10.4, what are the direct chain and differential chain codes with four neighbors?

10.6 For the shape given in Figure P-10.6, derive the quad-tree representation.

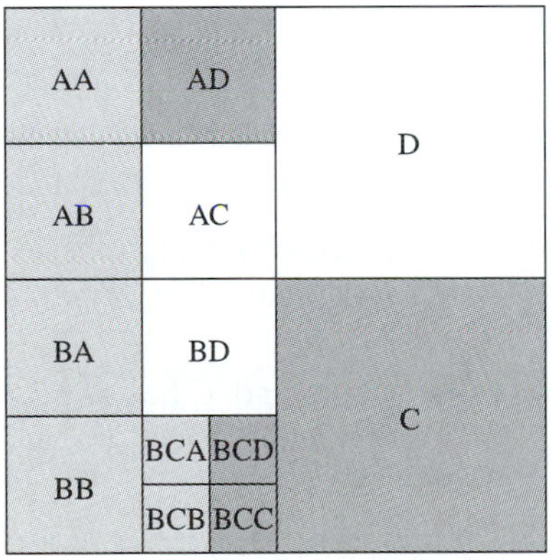

Figure P-10.6   An image with three different objects, represented by different shades.

10.7 For the Fourier descriptor representation of a contour given in Equation (10.1.8), prove that (a) a translation of the contour will only influence coefficient $Z(0)$, the DC value; (b) a scaling of the contour will multiply all coefficients with the scale factor; and (c) a rotation of the contour will shift the phase of the Fourier coefficients but it will not affect their amplitude.

10.8 Explain algorithms for joint shape and texture coding. What method is commonly used on the Internet?

10.9 How can you extend a pelwise shape coder to work efficiently not only in an image coder but also in a video coder?

10.10 What approaches to texture coding of arbitrarily shaped objects do you know? Compare their efficiency.

10.11 What are the two main assumptions that guided the design of region-based coders?

10.12 How do object-based analysis-synthesis coding and region-based coding differ? Which coder will perform better for natural video sequences and which for cartoons? Explain.

10.13 What are the most commonly used source models for OBASC?

10.14 Model failure detection ignores geometrical distortions. What are they? Will they accumulate over time?

10.15 Explain image synthesis by filter concatenation, and image synthesis by displacement vector field concatenation. Which method is more efficient?

10.16 Name at least two reasons why a knowledge-based coder can code a scene more efficiently than an OBASC coder.

10.17 Why does semantic coding hold the promise to be very efficient? Why do we not have semantic coders today?

10.18 What are the differences between action units and facial animation parameters?

10.19 Explain the layers of a layered coder. Which layers are commonly used today?

10.11 BIBLIOGRAPHY

[1] Aach, T., and A. Kaup. Bayesian algorithms for change detection in image sequences using Markov random fields. *Signal Processing: Image Communication* (Aug. 1995), 7(2):147–60.

[2] Aizawa, K., H. Harashima, and T. Saito. Model-based analysis-synthesis image coding (MBASIC) system for a person's face. *Signal Processing: Image Commun.* (Oct. 1989), 1(2):139–52.

[3] Bartlett, M. S., et al. Classifying facial action. In D. Touretzky, M. Mozer, and M. Hasselmo, eds., *Advances in Neural Information Processing Systems 8,* 823–29. Cambridge, MA: MIT Press, 1996.

[4] Born, G. *The File Formats Handbook.* Boston: International Thompson Computer Press, 1995.

[5] Brady, N., and F. Bossen. Shape compression of moving objects using context-based arithmetic encoding. *Signal Processing: Image Communications* (May 2000), 15(7–8):601–17.

[6] Chen, T., C. T. Swain, and B. G. Haskell. Coding of subregions for content-based scalable video. *IEEE Trans. on Circuits and Systems for Video Technology* (Feb. 1997), 7:256–60.

[7] Choi, C. S., K. Aizawa, H. Harashima, and T. Takebe. Analysis and synthesis of facial image sequences in model-based image coding. *IEEE Trans. on Circuits and Systems for Video Technology* (June 1994), 4(3):257–75.

[8] Chowdury, M. F., A. F. Clark, A. C. Downton, E. Morimatsu, and D. E. Pearson. A switched model-based coder for video signals. *IEEE Trans. on Circuits and Systems for Video Technology* (June 1994), 4(3):216–27.

[9] de Klerk, M., R. Prasad, J. H. Bons, and N. B. J. Weyland. Introducing high-resolution line graphics, in UK teletext using differential chain coding. *Proceedings of IEE, Part I* (1990), 137(6):325–34.

[10] Eckman, P., and V. W. Friesen. *Facial Coding System*. Palo Alto, CA: Consulting Psychologists Press, 1977.

[11] Eden, M., and M. Kocher. On the performance of a contour coding algorithm in the context of image coding. Part I: Contour segment coding. *Signal Processing* (1985), 8:381–86.

[12] Eisert, P., T. Wiegand, and B. Girod. Model-aided coding: A new approach to incorporate facial animation into motion-compensated video coding. *IEEE Trans. on Circuits and Systems for Video Technology* (April 2000), 10(3):344–58.

[13] Forchheimer, R., O. Fahlander, and T. Kronander. A semantic approach to the transmission of face images. *International Picture Coding Symposium (PCS'84)* (July 1984), p. 10.5.

[14] Freeman, H. On the encoding of arbitrary geometric configurations. *IRE Trans. Electronic Computers* (June 1961), EC-10(2):260–68.

[15] Gerken, P. Object-based analysis-synthesis coding of image sequences at very low bit rates. *IEEE Trans. Circuits and Systems for Video Technology* (June 1994), 4:228–35.

[16] Gilge, M., T. Engelhardt, and R. Mehlan. Coding of arbitrarily shaped image segments based on a generalized orthogonal transform. *Signal Processing: Image Communications* (Oct. 1989), 1(2):153–80.

[17] Graphics interchange format, version 89a. Technical report, CompuServe Incorporated, Columbus, Ohio, July 1990.

[18] Hötter, M. Object-oriented analysis-synthesis coding based on two dimensional objects. *Signal Processing: Image Communication* (Dec. 1990), 2(4):409–28.

[19] Hötter, M. Optimization and efficiency of an object-oriented analysis-synthesis coder. *IEEE Trans. on Circuits and Systems for Video Technology* (April 1994), 4(2):181–94.

[20] ISO/IEC. IS 14496-1: Information technology—coding of audio-visual objects—part 1: Systems, 1999. (MPEG-4 Systems).

[21] ISO/IEC. IS 14496-2: Information technology—coding of audio-visual objects—part 2: Visual, 1999. (MPEG-4 Video).

[22] ITU-T. Recommendation H.263: Video coding for low bit rate communication, 1998.

[23] ITU-T (International Telecommunication Union, Telecommunication Standardization Sector). Recommendation T.6: Facsimile coding schemes and coding control functions for group 4 facsimile apparatus, Geneva, 1988.

[24] ITU-T. Recommendation T.82 (03/93)—information technology—coded representation of picture and audio information—progressive bi-level image compression, Geneva, 1993.

[25] Jain, A. K. *Fundamentals of Digital Image Processing.* Englewood Cliffs, NJ: Prentice Hall, 1989.

[26] Jordan, C. L., F. Bossen, and T. Ebrahimi. Scalable shape representation for content based visual data compression. In *IEEE Int. Conf. Image Process. (ICIP'97), Special Session on Shape Coding,* Santa Barbara, CA, 1997, 1:512–15.

[27] Kalra, P., A. Mangili, N. Magnenat-Thalmann, and D. Thalmann. Simulation of facial muscle actions based on rational free form deformations. *Eurographics* (1992), 59–69.

[28] Kampmann, M., and J. Ostermann. Automatic adaptation of a face model in a layered coder with an object-based analysis-synthesis layer and a knowledge-based layer. *Signal Processing: Image Communications* (March 1997), 9(3): 201–20.

[29] Kaneko, T., and M. Okudaira. Encoding of arbitrary curves based on the chain code representation. *Commun.* (1985), 33:697–707.

[30] Katsaggelos, A. K., L. P. Kondi, F. W. Meier, J. Ostermann, and G. M. Schuster. MPEG-4 and rate distortion based shape coding techniques. *Proceedings of the IEEE, Special Issue on Multimedia Signal Processing, Part Two* (June 1998), 86(6):1126–54.

[31] Kaup, A. A new approach towards description of arbitrarily shaped image segments. *SPIE Conf. Visual Communications and Image Processing (VCIP'97)* (Feb. 1997), SPIE-3012:731–41.

[32] Kim, J. I., A. C. Bovik, and B. L. Evans. Generalized predictive binary shape coding using polygon approximation. *Signal Processing: Image Communications* (May 2000), 15(7–8):643–63.

[33] Koplowitz, J. On the performance of chain codes for quantization of line drawings. *IEEE Trans. Pattern Anal. Machine Intell.* (March 1981), 3(2):180–85.

[34] Kunt, M., A. Ikonomopoulos, and M. Kocher. Second-generation image-coding techniques. *Proceedings of the IEEE* (April 1985), 73:549–74.

[35] Li, H., P. Roivainen, and R. Forchheimer. 3-D motion estimation in model-based facial image coding. *IEEE Trans. Pattern Anal. Machine Intell.* (June 1993), 15(6):545–555.

[36] Li, S., and W. Li. Shape-adaptive discrete wavelet transforms for arbitrarily shaped visual object coding. *IEEE Trans. on Circuits, Systems, and Video Technology* (Aug. 2000), 10(5):725–43.

[37] Martinez, G. Shape estimation of articulated 3-D objects for object-based analysis-synthesis coding (OBASC). *Signal Processing: Image Communications* (March 1997), 9(3):175–99.

[38] Musmann, H. A layered coding system for very low bit rate video coding. *Signal Processing: Image Communications* (Nov. 1995), 7(4-6):267–78.

[39] Musmann, H. G., M. Hötter, and J. Ostermann. Object oriented analysis-synthesis coding of moving images. *Signal Processing: Image Communications* (Oct. 1989), 1:117–38.

[40] Netravali, A. N., and B. G. Haskell. *Digital Pictures—Representation and Compression.* New York: Plenum Press, 1988.

[41] Neuhoff, D. L., and K. G. Castor. A rate and distortion analysis of chain codes for line drawings. *IEEE Transactions on Information Theory* (1985), 31(1): 53–67.

[42] Nunes, P., F. Pereira, and F. Marques. Multi-grid chain coding of binary shapes. *IEEE Int. Conf. Image (ICIP'97),* Santa Barbara, CA, 1997, III:114–17.

[43] O'Connell, K. J. Object-adaptive vertex-based shape coding method. *IEEE Transactions on Circuits and Systems* (Feb. 1997), 7(1):251–55.

[44] Ostermann, J. The block-based coder mode in an object-based analysis-synthesis coder. *28th Asilomar Conference on Signals, Systems, and Computers* (Nov. 1994), TP 6–1.

[45] Ostermann, J. Object-based analysis-synthesis coding (OBASC) based on the source model of moving flexible 3-D objects. *IEEE Trans. on Image Processing,* (September 1994), 3(5):705–11.

[46] Ostermann, J. Object-based analysis-synthesis coding (OBASC) based on the source model of moving rigid 3-D objects. *Signal Processing: Image Communications* (1994), 6(2):143–61.

[47] Ostermann, J. Segmentation of image areas changed due to object motion considering shadows. In Y. Wang, et al., eds., *Multimedia Communications and Video Coding,* 241–246. New York: Plenum Press, 1996.

[48] Pratt, W. K. *Digital Image Processing.* 2nd ed. New York: John Wiley & Sons, 1991.

[49] Rydfalk, R. Candide, a parameterised face. Internal Report Lith–ISY–I–0866, Linköping University, Linköping, Sweden, 1987.

[50] Saghri, J. A., and H. Freeman. Analysis of the precision of generalized chain codes for the representation of planar curves. *IEEE Trans. Pattern Anal. Machine Intell.* (Sept. 1981), 3(5):533–39.

[51] Salembier, P., and F. Marques. Region-based representations of image and video: segmentation tools for multimedia services. *IEEE Trans. on Circuits and Systems for Video Technology* (Dec. 1999), 9(8):1147–69.

[52] Salembier, P., F. Marques, and A. Gasull. Coding of partition sequences. In L. Torres and M. Kunt, eds., *Video Coding: The Second Generation Approach,* 125–70. Boston: Kluwer Academic Publishers, 1996.

[53] Salembier, P., L. Torres, F. Meyer, and C. Gu. Region-based video coding using mathematical morphology. *Proceedings of the IEEE, Special Issue on Digital TV* (June 1995), 83(6):843–57.

[54] Schuster, G. M., and A. G. Katsaggelos. An optimal segmentation encoding scheme in the rate-distortion sense. *IEEE Int. Symp. Circuits and Systems (ISCAS'96),* Atlanta, 1996, 2:640–43.

[55] Sikora, T., S. Bauer, and B. Makai. Efficiency of shape-adaptive transforms for coding of arbitrarily shaped image segments. *IEEE Trans. on Circuits and Systems for Video Technology* (June 1995), 5(3):254–58.

[56] Stauder, J., R. Mech, and J. Ostermann. Detection of moving cast shadows for object segmentation. *IEEE Transactions on Multimedia* (March 1999), 1(1):65–76.

[57] Tekalp, A. M., and J. Ostermann. Face and 2-D mesh animation in MPEG-4. *Signal Processing: Image Communications, Special Issue on MPEG-4* (Jan. 2000), 15:387–421.

[58] Terzopolous, D., and K. Waters. Physically-based facial modeling, analysis and animation. *Journal of Visualization and Computer Animation* (1990), 1:73–80.

[59] Torres, L., and M. Kunt. Second Generation Video Coding Techniques. In L. Torres and M. Kunt, eds., *Video Coding: The Second Generation Approach,* 1–30. Boston, MA: Kluwer Academic Publishers, 1996.

[60] van Otterloo, P. *A Contour-Oriented Approach for Shape Analysis.* Hemel Hampstead, England: Prentice Hall International, 1991.

[61] Zahn, C. T., and R. Z. Roskies. Fourier descriptors for plane closed curves. *IEEE Transactions on Computer* (March 1972), 21(3):269–81.

11

Scalable Video Coding

The coding methods introduced in the two preceding chapters, including both waveform-based and content-dependent methods, aim to optimize the coding efficiency for a fixed bit rate. This presents a difficulty when multiple users try to access the same video through different communication links. For example, an MPEG-1 video coded at 1.5 mbps can be downloaded in real time for playback on a terminal that is connected to a server via a high-speed link (e.g., an ADSL modem). But a user with only a modem connection of 56 kbps will not be able to receive sufficient bits for playback in real time. *Scalability refers to the capability of recovering physically meaningful image or video information by decoding only partial compressed bit streams.* In the previous example, if the video stream is scalable, then the user with a high-bandwidth connection can download the entire bitstream to view the full-quality video, while the user with a 56 kbps connection will only download a subset of the stream, and see a low-quality presentation.

This example illustrates only one aspect of scalability: bandwidth scalability. A scalable stream can also offer adaptivity to varying channel error characteristics, and computing power at the receiving terminal. For wireless communications, scalability allows the adjustment of the source rate and the application of unequal error protection in response to channel error conditions. For Internet transmission, scalability enables variable-bit-rate transmission, selective bit discarding, and the adjustment of the source rate to correspond to different modem rates, changing channel bandwidth, and diverse device capabilities. As we move to the convergence of wireless, Internet, and multimedia, scalability becomes increasingly important for rich media access from anywhere, by anyone, at any time, with any device, and in any form.

Scalable coders can have *coarse granularity* (in two or three layers—these are also called *layered* coders), or *fine granularity*. In the extreme case of fine granularity, the bit stream can be truncated at any point. The more bits that are retained, the better will be the reconstructed image quality. We call such a bit stream *embedded*. Embedded coders enable precise bit rate control, which is a desirable feature in many applications. For example, network filters (described in Chapter 15) can select the number of bits to transmit from an embedded bit stream to match the available bandwidth.

Scalable coding is typically accomplished by providing multiple versions of a video, in terms of amplitude resolution (called *quality scalability* or *SNR scalability*), spatial resolution (*spatial scalability*), temporal resolution (*temporal scalability*), frequency resolution (*frequency scalability*—more often known as *data partition*), or combinations of these options. Figure 11.1 shows a set of images decoded from a scalable bit stream, which have been obtained by combining quality scalability with spatial scalability.

Scalable contents can be accessed at either the frame or object level. The latter is referred as *object-based scalability,* as defined in the MPEG-4 standard [5]. In this section, we first introduce the four basic schemes for achieving scalability, including quality, spatial, temporal, and frequency scalability (Section 11.1). We then describe how to realize scalability at the object level (Section 11.2). Although similar concepts can be applied to different types of coders, we will focus our discussion on the modification of block-based hybrid coders (Section 9.3.1) to realize different scalability modes. Finally, we will describe wavelet-based coding methods which, by the nature of wavelet transforms, naturally lead to fine granularity scalability.

Note that an alternative way to handle varying channel environments and receiver capabilities is by *simulcast,* which simply codes the same video several times, each with a different quality or resolution setting. Although simple, it is very inefficient, as a higher-quality or resolution bitstream essentially repeats the information that is already contained in a lower quality or resolution stream, plus some additional information. On the other hand, to provide scalability functionality, a coder must sacrifice a certain amount of coding efficiency, compared to the state-of-the-art nonscalable coders. *The design goal in scalable coding is to minimize the reduction in coding efficiency while realizing the requirement for scalability.*

11.1 BASIC MODES OF SCALABILITY

11.1.1 Quality Scalability

Quality scalability is defined as the representation of a video sequence with varying accuracies in the color patterns. This is typically accomplished by quantizing the color values (in the original or a transformed domain) with increasingly finer quantization stepsizes. Because different quantization accuracies lead to different PSNRs between the original and the quantized video, this type of scalability is more commonly referred to as SNR scalability.

Figure 11.2 shows a bit stream with N layers of quality scalability. Decoding the first layer (also called the *base layer*) provides a low-quality version of the reconstructed image. Further decoding the remaining layers (also called *enhancement layers*) results in quality increases of the reconstructed images, up to the highest quality. The first layer

(a)

(b)

Figure 11.1 Decoded video frames from a scalable bit stream combining quality and spatial scalability. The two images in (a) are of size 176×144, obtained by decoding at 6.5 and 133.9 kbps, respectively. The two images in (b) are of size 352×288, decoded at bit rates of 21.6 and 436.3 kbps, respectively. Courtesy of Haidi Gu.

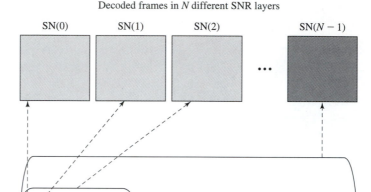

Figure 11.2 Bit stream with N layers of quality scalability. Reprinted from I. Sodagar, H.-J. Lee, P. Hatrack, and Y.-Q. Zhang, Scalable wavelet coding for synthetic/natural hybrid images, *IEEE Trans. Circuits Syst. for Video Technology* (March 1999), 9:244–54. Copyright 1999 IEEE.

is obtained by applying a coarse quantizer to the original image or in a transform (e.g., DCT) domain. The second layer contains the quantized difference between the original image and that reconstructed from the first layer, using a quantizer that is finer than that used to produce the first layer. Similarly, each of the subsequent layers contains the quantized difference between the original image and that reconstructed from the previous layers, using an increasingly finer quantizer.

An encoder with two-level quality scalability is depicted in Figure 11.3(a). For the base level, the encoder operates in the same manner as that of a typical block-based hybrid coder. For the enhanced level, the operations are performed as follows:

1. The raw video frame (or the motion compensation error frame) is DCT-transformed and quantized at the base level;
2. The base-level DCT coefficients are reconstructed by inverse quantization;
3. The base-level DCT coefficients are subtracted from the original DCT coefficient;
4. The residual is quantized by a quantization parameter that is smaller than that of the base level;
5. The quantized bits are coded by VLC.

Since the enhanced level uses a smaller quantization parameter, it achieves better quality than the base level.

The decoder operation is depicted in Figure 11.3(b). For the base level, the decoder operates exactly as a nonscalable video decoder. For the enhanced level, both levels must be received, decoded by variable length decoding (VLD), and inversely

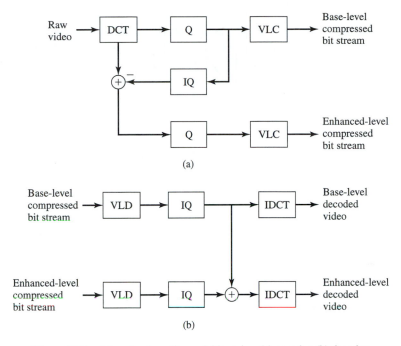

Figure 11.3   A two-level quality-scalable codec: (a) encoder, (b) decoder.

quantized. Then the base-level DCT coefficient values are added to the enhanced-level DCT coefficient refinements. After this stage, the summed DCT coefficients are inversely DCT-transformed, resulting in enhanced-level decoded video.

11.1.2 Spatial Scalability

Spatial scalability is defined as the representation of the same video in varying spatial resolutions or sizes (Figure 11.4(a) and (b)). Figure 11.5 illustrates a bit stream with M layers of spatial scalability. By decoding the first layer, the user can display a preview version of the decoded image at a lower resolution. Decoding the second layer results in a larger reconstructed image. Furthermore, by progressively decoding the additional layers, the viewer can increase the spatial resolution of the image, up to the full resolution of the original image.

To produce such a layered bit stream, a multiresolution decomposition of the original image is first obtained. The lowest-resolution image is coded directly to produce the first layer (i.e., the base layer). To produce the second layer, the decoded image from the first layer is first interpolated to the second-lowest resolution and the difference between the original and the interpolated image at that resolution is coded. The bit stream for each of the following resolutions is produced in the same way: first an estimated image at that resolution is formed, based on the previous layers, then the difference between the estimated and the original image at that resolution is coded.

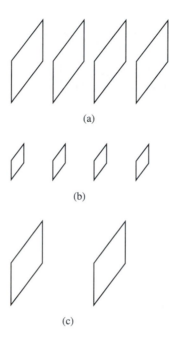

(a)

(b)

(c)

Figure 11.4 Spatial and temporal scaling of a video stream: (a) original video frames; (b) frames scaled to $\frac{1}{4}$ original size; (c) temporally scaled frames.

Decoded frames in M different spatial layers

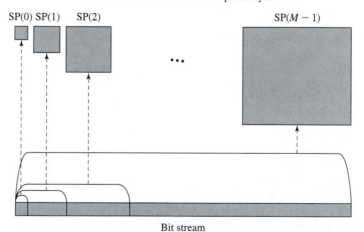

SP(0) SP(1)   SP(2)   SP($M - 1$)

Bit stream

Figure 11.5 Bit stream with M layers of spatial scalability. Reprinted from I. Sodagar, H.-J. Lee, P. Hatrack, and Y.-Q. Zhang, Scalable wavelet coding for synthetic/natural hybrid images, *IEEE Trans. Circuits Syst. for Video Technology* (March 1999), 9:244–54. Copyright 1999 IEEE.

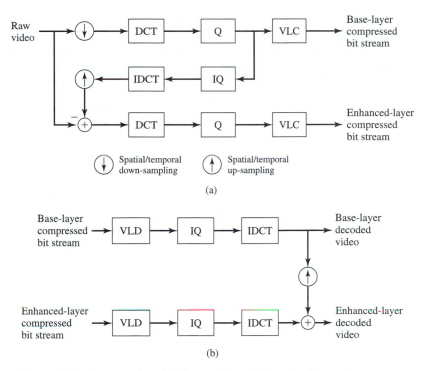

Figure 11.6 A two-level spatially/temporally scalable codec: (a) encoder, (b) decoder.

Figure 11.6(a) shows a block diagram of a two-layer spatially scalable encoder. For the base layer, the raw video is first spatially down-sampled,* then DCT-transformed, quantized, and VLC-coded. For the enhanced layer, the operations are performed as follows:

1. The raw video is spatially down-sampled, DCT-transformed, and quantized at the base layer;
2. The base-layer image is reconstructed by inverse quantization and inverse DCT;
3. The base-layer image is spatially up-sampled;†
4. The up-sampled base-layer image is subtracted from the original image;
5. The residual is DCT-transformed, and quantized by a quantization parameter, which is smaller than that of the base layer;
6. The quantized bits are coded by VLC.

---

*For example, spatially down-sampling with a ratio of 4:1 can be accomplished by replacing every 2×2 pixels by their average value. More sophisticated prefilters can be used to reduce the aliasing effect in the down-sampled image, at the expense of increased complexity.

†For example, a simple way to spatially up-sample with a ratio of 1:4 is to replicate each pixel four times.

Since the enhanced layer uses a smaller quantization parameter, it achieves higher quality than the base layer.

A spatially scalable decoder with two-layer scalability is depicted in Figure 11.6(b). For the base layer, the decoder operates exactly as a nonscalable video decoder. For the enhanced layer, both layers must be received, decoded by VLD, inversely quantized, and inversely DCT-transformed. Then the base-layer image is spatially up-sampled. The up-sampled base-layer image is combined with the enhanced-layer refinements to form the enhanced-layer decoded video.

11.1.3 Temporal Scalability

Temporal scalability is defined as the representation of the same video in varying temporal resolutions or frame rates (see Figure 11.4(a) and (c)). Temporal scalability enables different frame rates for different layers of the contents. Typically, temporally scalable video is encoded efficiently in such a way: making use of temporally up-sampled pictures from a lower layer as a prediction in a higher layer. The block diagram of a temporally scalable codec is the same as that of spatially scalable codec (see Figure 11.6). The only difference is that the spatially scalable codec uses spatial down-sampling and spatial up-sampling, whereas the temporally scalable codec uses temporal down-sampling and temporal up-sampling. The simplest way to perform temporal down-sampling is by frame skipping. For example, temporal down-sampling with a ratio of 2:1 can be achieved by discarding one frame from every two frames. Temporal up-sampling can be accomplished by frame copying. For example, temporal up-sampling with a ratio of 1:2 can be realized by making a copy of each frame and transmitting the two frames to the next stage. In this case, the base layer simply includes all the even frames, and the enhancement layer all the odd frames. For motion compensation, a base-layer frame will be predicted only from previous base layer frames, whereas an enhancement layer frame can be predicted from both base-layer and enhancement-layer frames.

11.1.4 Frequency Scalability

Another way to represent a video frame in multiple layers is by including different frequency components in each layer, with the base layer containing low-frequency components, and other layers containing increasingly higher-frequency components. This way, the base layer will provide a blurred version of the image, and the addition of enhancement layers will yield increasingly sharper images. Such decomposition can be accomplished via whole-frame transforms, such as subband-decomposition or wavelet transforms, or via block-based transforms, such as block DCT. One way to implement this idea in a block-based hybrid coder is by including the mode information, motion information, and first few DCT coefficients of each macroblock in the base layer, and including the remaining DCT coefficients in the enhancement layer. In the MPEG-2 standard, this is known as *data partitioning*. We will discuss wavelet-based scalable coders in Section 11.3.

Decoded frames in hybrid spatial/SNR layers

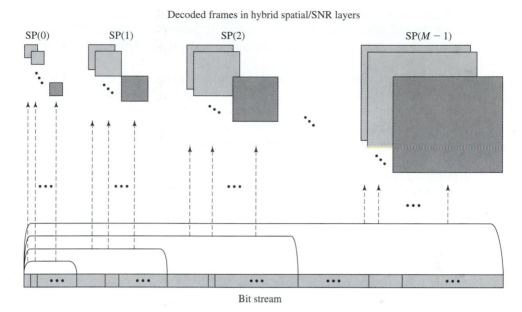

Bit stream

Figure 11.7 $N \times M$ layers of combined spatial/quality scalability. Reprinted from I. Sodagar, H.-J. Lee, P. Hatrack, and Y.-Q. Zhang, Scalable wavelet coding for synthetic/natural hybrid images, *IEEE Trans. Circuits Syst. for Video Technology* (March 1999), 9:244–54. Copyright 1999 IEEE.

11.1.5 Combination of Basic Schemes

Quality, spatial, temporal, and frequency scalability are basic scalable mechanisms. They can be combined to reach finer granularity. Figure 11.7 shows a case of combined spatial and quality scalability. In this example, the bitstream consists of M spatial layers, and each spatial layer includes N levels of quality scalability. In this case, both spatial resolution and quality of the reconstructed image can be improved by progressively transmitting and decoding the bit stream. The order is to improve image quality at a given spatial resolution until the best quality is achieved at that resolution, and then to increase the spatial resolution to a higher level and improve the quality again.

11.1.6 Fine-Granularity Scalability

The scalability methods that we have described produce bit streams that consists of several layers, a base layer followed by one or more enhancement layers. This type of coder provides only coarse granularity, in that the quality improvements are obtained with rate increases in large discrete steps. If only a partial set of bits in a particular layer is received, the reproduced video quality is determined to a large extent by the received bits in the previous layers, not benefiting from the bits received from this layer. *Fine-granularity scalability* (FGS) refers to a coding method by which the rate as well as the quality increment at a much smaller step. In the limiting case, in which a bit stream can provide continuously improving video quality with every additional bit, the underlying coding method is called embedded coding. Obviously, FGS and embedded coding can adapt to bandwidth variations in real networks more effectively than other scalable methods.

In practice, the requirement that the bit stream be completely embedded or have fine granularity is often relaxed. A base layer may be first produced to provide a low but guaranteed level of quality, and then an enhancement layer may be generated to provide improvement in fine granularity. This is the method employed, for example, in the FGS mode of MPEG-4 [14, 11]. In this case, the conventional block-based hybrid coding method is employed to produce a base-layer stream at a given frame rate, using a relatively large QP. Then, for every coded frame (intra- or inter-coded), the differences between the original DCT coefficients* and the quantized coefficients in the base layer (to be called *refinement coefficients*) are coded into a fine-granularity stream. This is accomplished by quantizing the refinement coefficients using a very small QP[†] and then representing the quantized indices through successive bit plane encoding.

Specifically, the absolute values of quantized refinement coefficients in each block are specified in binary representations. Starting from the highest bit plane that contains nonzero bits, each bit plane is successively coded using run-length coding, block by block. The run lengths can be coded using either Huffman or arithmetic coding. Typically, different codebooks are used for different bit planes, because the run-length distributions are different across planes. When only a partial set of the enhancement-layer stream is decoded, depending on how many bit planes were included in the retained bits, the reconstructed video will have a quality between that obtainable from the base layer only, to that obtainable with the QP used on the refinement DCT coefficients. The granularity of the stream is at the bit-plane level: every additional complete bit plane will yield an improvement in the accuracy of the DCT coefficients by a factor of two.[‡]

This method offers scalability in quality only at a fixed spatial-temporal resolution. To add temporal scalability, the base layer may include a subsampled set of frames coded with a higher QP. For the enhancement layer, the remaining frames are bidirectionally predicted from the base-layer frames. The DCT coefficients in the prediction error images are then quantized and coded using the bit-plane encoding method, in addition to the refinement coefficients for the base-layer frames [22]. This scheme is illustrated in Figure 11.8.

A limitation with this FGS scheme is that the base layer must be delivered completely and without error. This may or may not be possible in practical networks.

---

*For each block, the transform is applied to the original block or the motion compensation error block, depending on whether the block is coded in the intra- or inter-mode in the base layer.

[†]No quantization is performed if one wishes to obtain lossless representation at the highest rate.

[‡]In fact, the granularity can be up to the codeword level for the run-length representation, as every additional complete codeword received can improve the reconstruction accuracy of some coefficients—although, depending on the available bandwidth, it is more practical to truncate a FGS stream at the bitplane level.

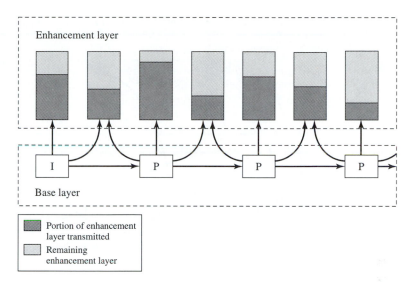

Figure 11.8   Illustration of the transmission of portions of the enhancement-layer stream for the support of joint quality-temporal scalability. Adapted from M. van der Schaar and H. Radha, A novel MPEG-4 based hybrid temporal-SNR scalability for Internet video, *IEEE Int. Conf. Image Processing* (Sept. 2000), 3:548–51. Copyright 2000 IEEE.

Another problem is that, when the base-layer bit rate is kept low (so as to increase the scalable range of the bit stream), the coding efficiency of the FGS method will be significantly reduced compared to a nonscalable coder. This is because the reconstructed base-layer frames are used as reference frames for motion-compensated prediction. Although such a choice eliminates the problem of temporal error propagation in the enhancement layer, the high quantization error in base-layer frames reduces the prediction gain. One approach to improving the coding efficiency of the FGS method is to periodically use an intermediate enhancement layer (reconstructed from some but not all bit planes) as a reference for motion-compensated prediction [24].

It is difficult to use block-based transforms to realize fully embedded coding. We will show in Section 11.3 how to achieve this goal with wavelet transforms.

11.2 OBJECT-BASED SCALABILITY

The various scalable coding methods introduced in the previous section perform the same operation over the entire video frame. In object-based temporal scalability (OTS), the frame rate of a selected object is enhanced such that it has a smoother motion than the remaining area. (In other words, the frame rate of the selected object is higher than that of the remaining area.) We will mainly introduce OTS based on MPEG-4 implementations.

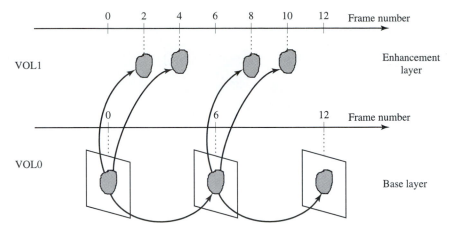

Figure 11.9   OTS enhancement structure of Type 1, with predictive VideoObjectPlanes (P-VOPs) (Courtesy of MPEG-4).

There are two types of enhancement structures in OTS. Figure 11.9 shows an example of Type 1, in which VOL0 (VideoObjectLayer 0) is an entire frame with both an object and a background, whereas VOL1 represents the particular object in VOL0. VOL0 is coded with a low frame rate, and VOL1 is coded to achieve a higher frame rate than VOL0. In this example, frames 2 and 4 are coded by predicting from frame 0 in VOL0, followed by overlapping the object of the enhancement layer onto the combined frame. Figure 11.10 shows another example of Type 1, which uses bidirectional predictions forming bidirectional VOPs (B-VOPs) in the enhancement layer. In this case, frames 2 and 4 in VOL1 are predicted from frames 0 and 6 in VOL0. In both cases, two

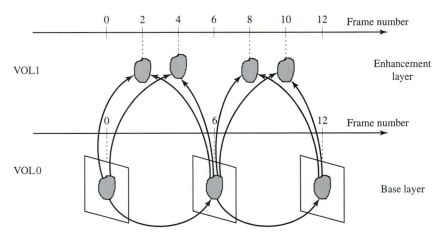

Figure 11.10   OTS enhancement structure of Type 1, with B-VOPs.

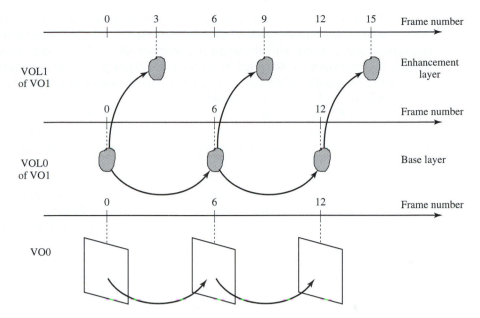

Figure 11.11   OTS enhancement structure of Type 2.

additional shape data, a forward shape and a backward shape, are encoded to perform the background composition.

Figure 11.11 shows an example of Type 2 OTS, in which VO0 (VideoObject 0) is the sequence of an entire frame that only contains a background, and has no scalability layer. VO1 is the sequence of a particular object and it has two scalability layers, VOL0 and VOL1. VOL1 represents the same object as VOL0 and is coded to achieve a higher frame rate than VOL0. In this example, VOL0 is regarded as a base layer and VOL1 is regarded as an enhancement layer of the OTS. Note that the VO0 may not have the same frame rate as other VOs.

11.3 WAVELET-TRANSFORM–BASED CODING

The discrete wavelet transform (DWT) has recently emerged as a powerful tool in image and video compression, due to its flexibility in representing nonstationary image signals and its ability to adapt to human visual characteristics [16, 15]. The wavelet representation provides a multiresolution/multifrequency expression of a signal with localization in both time and frequency. This property is very desirable in image and video coding applications. First, real-world image and video signals are nonstationary in nature. A wavelet transform decomposes a nonstationary signal into a set of multiscaled subbands in which each component becomes relatively more stationary and hence easier to code. Also, coding schemes and parameters can be adapted to the statistical properties of each subband, and hence coding each stationary component is more efficient than coding the whole nonstationary signal. In addition, the wavelet representation matches the spatially tuned frequency-modulated properties experienced in early human vision, as indicated by research results in psychophysics and physiology. Finally, the multiresolution/multifrequency decomposition offered by the wavelet transform lends itself easily to a scalable bit stream.

Like the DCT-based approach, wavelet-transform–based coding for images consists of three steps: (1) wavelet transform; (2) quantization; and, (3) entropy coding. Wavelet transform converts an image signal into a set of coefficients that correspond to scaled space representations at multiple resolutions and frequency segments. Wavelet coefficients are typically organized into a hierarchical data structure, so that bit allocation and data compaction can be employed more efficiently. Quantization allows the data rate to be reduced at the expense of some distortions. Finally, entropy coding encodes the quantized coefficients into a set of compact binary bit streams. There are many variations of wavelet-transform–based coding. Wavelet transform is typically performed globally, but blockwise wavelet transform has also been developed to boost implementation efficiency [9]. Vector wavelet transform and arbitrary-shape wavelet transform have also been recently developed [12, 10]. Quantization can be of several types: scalar, vector, or trellis-coded (TCQ). Entropy coding can be Huffman or arithmetic, using either fixed or adaptive code, implemented in terms of bit planes or entire samples. When applying wavelets to video coding, mechanisms to reduce temporal redundancy are needed. Motion compensation in both the spatial and wavelet domains, 3-D wavelets, and motion-compensated 3-D wavelet video coding algorithms have been proposed [26, 21].

For image coding, wavelet transform has been proven to outperform DCT-based coding techniques by a wide margin, in terms of compression efficiency and enhanced features such as scalability. That is the reason why both MPEG-4 and JPEG2000 have selected wavelet-based schemes as the basis for coding still texture and images [9, 20]. However, wavelet-based video coding has yet to show significant compression benefits beyond DCT-based schemes, although advanced scalability has been promised by wavelet approaches. A comparison of DCT and wavelet coding for images and video is included in [25]. Most existing video coding standards (H.261, H.263, MPEG-1 and MPEG-2) have adopted the block-based hybrid video coding framework introduced in Chapter 9 [7, 8, 3, 4]. MPEG-4 [5] also uses a DCT-based scheme for coding of natural video, although wavelet-based coding is used for still images and graphics. There has been very active research in wavelet-based video coding, aiming at both much improved compression ratios and scalabilities.

We begin this section by describing the most popular wavelet-based image coding scheme, namely, the zero-tree–based scheme. It is followed by an example motion-compensated video codec to illustrate wavelet applications to video coding. Since the basic mathematical formulation of wavelet transform has been extensively covered in the literature (see, e.g., [23, 1]), this section focuses on the compression aspect of the wavelet-based coding approach.

11.3.1 Wavelet Coding of Still Images

EZW Method   The zero-tree–based coding concept was originally developed by Shapiro, and known as embedded zero-tree wavelet (EZW) coding [19]. Besides superior compression performance, the advantages of zero-tree wavelet coding include simplicity, embedded bitstream structure, scalability, and precise bit-rate control. Zero-tree wavelet coding is based on three key ideas: (1) wavelet transform for decorrelation; (2) exploitation of the self-similarity inherent in the wavelet transform to predict the location of significant information across scales; and (3) universal lossless data compression using adaptive arithmetic coding.

A discrete wavelet transform decomposes the input image into a set of subbands of varying resolutions. The coarsest subband, known as the DC band, is a low-pass approximation of the original image, and the other subbands are finer-scale refinements. In a hierarchical subband system such as that of the wavelet transform, with the exception of the lowest frequency subband, every coefficient at a given scale can be related to a set of coefficients of similar orientation at the next finer scale. The coefficient at the coarse scale is called the *parent,* and all coefficients at the same spatial location, and of similar orientation at the next finer scale, are called *children*.

As an example, Figure 11.12 shows a wavelet tree resulted from a three-level wavelet decomposition. For the lowest frequency subband, the top left square in the example, the parent-child relationship is defined such that each parent node has three children, one in each subband, at the same scale and spatial location but different orientation. For the other subbands, each parent node has four children in the next finer scale of the same orientation.

The wavelet coefficients of the DC band are encoded independently from the other bands. As shown in Figure 11.13, the current coefficient X is adaptively predicted from three other quantized coefficients in its neighborhood (i.e. A, B, and C), and the predicted value is subtracted from the current coefficient. The predicted value is

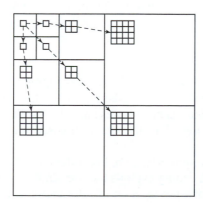

Figure 11.12 The parent-child relationship of wavelet coefficients. Reprinted from I. Sodagar, H.-J. Lee, P. Hatrack, and Y.-Q. Zhang, Scalable wavelet coding for synthetic/natural hybrid images, *IEEE Trans. Circuits Syst. for Video Technology* (March 1999), 9:244–54. Copyright 1999 IEEE.

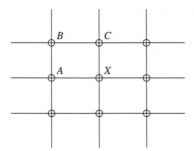

Figure 11.13 Adaptive predictive coding of the DC coefficients. Reprinted from I. Sodagar, H.-J. Lee, P. Hatrack, and Y.-Q. Zhang, Scalable wavelet coding for synthetic/natural hybrid images, *IEEE Trans. Circuits Syst. for Video Technology* (March 1999), 9:244--54. Copyright 1999 IEEE.

obtained as follows:

$$\hat{X} = \begin{cases} C & \text{if } |A - B| < |B - C|, \\ A & \text{otherwise.} \end{cases} \qquad (11.3.1)$$

If any of the neighbors, A, B, or C, is not in the image, its value is set to zero for the purpose of the prediction.

In the bit stream, the quantization stepsize is first encoded; then, the magnitude of the minimum value of the differential quantization indices, called "band_offset", and the maximum value of the differential quantization indices, called "band_max_value", are encoded into the bit stream. The parameter band_offset is a negative integer or zero, and the parameter band_max_value is a positive integer. Therefore, only the magnitudes of these parameters are coded into the bit stream. The differential quantization indices are coded using the arithmetic coder in a raster-scan order, starting from the upper-left index and ending at the lower-right. The model is updated with each encoding of a bit of the differential quantization index, to adapt to the statistics of the DC band.

EZW scans wavelet coefficients subband by subband. Parents are scanned before any of their children. Each coefficient is compared against the current threshold T. A coefficient is significant if its amplitude is greater than T. Such a coefficient is then encoded using one of two symbols, *negative significant* or *positive significant*. The *zero-tree root* symbol is used to signify a coefficient below T with all its children also below T. The *isolated zero* symbol signifies a coefficient below T but with at least one child not below T. For significant coefficients, EZW further encodes coefficient values using a successive approximation quantization scheme. Coding is done by bit planes, leading to an embedded stream.

ZTE and MZTE Methods EZW has been significantly improved over the years. Examples include SPIHT [18], predictive EZW [13], zero-tree entropy (ZTE) coding [17], and the more general case multiscale ZTE (MZTE) [20]. Shape-adaptive wavelet coding using zero-trees has also been proposed for coding of objects with arbitrary shapes [10].

Zero-tree entropy coding is based on, but differs significantly from, EZW coding. Similarly to EZW, ZTE coding exploits the self-similarity inherent in the wavelet transform of images to predict the location of information across wavelet scales. Although ZTE does not produce a fully embedded bit stream (as EZW), it gains flexibility and other advantages over EZW coding, including substantial improvement in coding efficiency, simplicity, and spatial scalability. ZTE coding is performed by assigning a zero-tree symbol to a coefficient and then coding the coefficient value with its symbol in one of the two different scanning orders. The four zero-tree symbols used in ZTE are *zero-tree root* (ZTR), *valued zero-tree root* (VZTR), *value* (VAL), and *isolated zero* (IZ). The zero-tree symbols and quantized coefficients are then losslessly encoded using an adaptive arithmetic coder with a given symbol alphabet. The arithmetic encoder adaptively tracks the statistics of the zero-tree symbols and encoded values using three models: (1) *type,* to encode the zero-tree symbols; (2) *magnitude,* to encode the values in a bit-plane fashion; and (3) *sign,* to encode the sign of the value. For each coefficient, its zero-tree symbol is encoded first; if necessary, its value is then encoded. The value is encoded in two steps. First, its absolute value is encoded in a bit-plane fashion using the appropriate probability model, and then the sign is encoded using a binary probability model.

The MZTE coding technique is based on ZTE coding, but utilizes a new framework to improve and extend ZTE to a fully scalable yet very efficient coding technique. At the first scalability layer, the zero-tree symbols are generated in the same way as in ZTE coding, and coded with the nonzero wavelet coefficients of that scalability layer. For the next scalability layer, the zero-tree map is updated along with the corresponding value refinements. In each scalability layer, a new zero-tree symbol is coded for a coefficient only if it was coded as ZTR or IZ in the previous scalability layer. If the coefficient is coded as VZTR or VAL in the previous layer, only its refinement value is coded in the current layer. An additional probability model, for the residual, is used for encoding the refinements of the coefficients that are coded with a VAL or VZTR symbol in any previous scalability layers. Furthermore, to utilize the highly correlated zero-tree symbols between scalability layers, context modeling, based on the zero-tree symbol of the coefficient in the previous scalability layer in MZTE, is used to better estimate the distribution of zero-tree symbols.

The images in Figure 11.14(a) and (b) are obtained by JPEG and MZTE compression schemes, respectively, at the same compression ratio of 45:1. The results show that the MZTE scheme generates much better image quality with good preservation of fine texture regions and absence of the blocking effect, compared with JPEG. The study in [25] showed that a wavelet-based coder can typically outperform a DCT-based coder by about 1 dB in PSNR when the bit rate is the same.

With the MZTE method, one can easily obtain spatial and quality scalabilities. By decoding increasingly more bits from the bit stream of an MZTE-compressed image, one can obtain bigger and sharper renditions of the original image.

Shape-Adaptive DWT   Wavelet-based coding can be extended to objects of arbitrary shapes [10]. With shape-adaptive DWT (SA-DWT), the object shape mask is decomposed into a pyramid of subbands, so that we know which wavelet-tree nodes have valid wavelet coefficients and which have "don't-care" values. We must pay attention to the manner of coding multiresolution arbitrarily shaped objects with such don't-care

(a)

(b)

Figure 11.14   Coded images by
(a) JPEG baseline method (PSNRs for
Y, U, and V components are 28.36, 34.74,
34.98 dB, respectively); (b) the MZTE
method (PSNRs for Y, U, and V
components are 30.98, 41.68, 40.14 dB,
respectively). Both at a compression ratio
of 45:1. Only portions of the Y
components are shown.

values (corresponding to the out-of-boundary pixels or out-nodes). Various wavelet-based coding methods including EZW and ZTE can be extended to SA-DWT. In this subsection, we discuss how to extend the ZTE method to the shape-adaptive case.

As discussed in [10], the SA-ZTE decomposes the arbitrarily shaped objects in the image domain into a hierarchical structure, with a set of subbands of varying resolutions. Each subband has a corresponding shape mask associated with it to specify the locations of the valid coefficients in that subband. There are three types of nodes in a tree: zeros, nonzeros, and out-nodes (with don't-care values). The task is to extend the zero-tree coding method to the case with out-nodes. A simple way is to set those don't-care values to zeros, and then apply the zero-tree coding method. However, this requires bits to code the out-notes, in the form of a don't-care tree (in which the parent and all of its children have don't-care values). This is a waste of bits, because out-nodes do not need to be coded as the shape mask already indicates their status. Therefore, we should treat out-nodes differently from zeros. Although we don't want to use bits to code an out-node, we must decide what to do with its children. One way is not to code any information about the status of the children of the don't-care node. This way, we always assume that it has four children to be examined further. When the decoder scans to this node, it will be informed by the shape information that this is a don't-care node, and it will continue to scan its four children. By doing so, all the don't-care nodes in a tree structure need not be coded. This approach performs well when there are only sparse valid nodes in a tree structure. One disadvantage of this approach is that, even if a don't-care node has four zero-tree root children, it still must code four zero-tree root symbols instead of one if the don't-care value is treated as a zero. Another approach is to selectively treat an out-node as a zero. This is equivalent to creating another symbol for coding some don't-care values. Through extensive experiments, it has been found that the method of not coding out-nodes performs better overall. A detailed description of the coding algorithm can be found in [10].

Extensive experiments have been conducted on the SA-DWT coding technique, and the results have been compared with that of SA-DCT coding. The object shape is coded using the MPGE-4 shape coding tool. Figure 11.15 presents PSNR versus bit rate curves obtained by SA-DCT and SA-ZTE methods. The shape bits are excluded from the bit rate since they are independent of the texture coding scheme. Only the texture bit rates are used for comparison. The bit rate (in bpp) is calculated based on the number of pixels in an object with a reconstructed shape, and the PSNR value is also calculated over the pixels in the reconstructed shape. Clearly, SA-ZTE coding achieves a better coding efficiency than SA-DCT, with 1.5–2 dB improvement in PSNR. Figure 11.16(a) and (b) show the reconstructed objects from SA-DCT and SA-ZTE coding, respectively.

11.3.2 Wavelet Coding of Video

Wavelet coding techniques for video can be classified into three categories: (1) spatial-domain motion compensation, followed by 2-D wavelet transform; (2) wavelet transform followed by frequency-domain motion compensation; (3) 3-D wavelet transforms with or without motion estimation. Different motion estimation algorithms, quantization schemes, and entropy coding methods can be applied to each of the three categories.

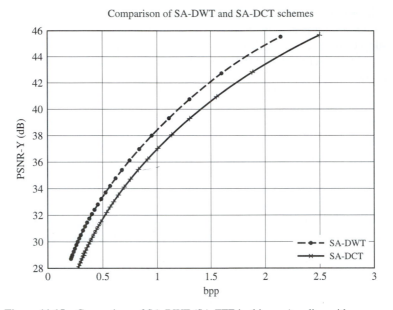

Figure 11.15   Comparison of SA-DWT (SA-ZTE in this case) coding with SA-DCT.

We use an example video codec to illustrate the wavelet-based video coding process [17]. The codec consists of five main components: (1) three-stage motion estimation and compensation; (2) adaptive wavelet transform; (3) multistage quantization; (4) zero-tree entropy coding; and (5) intelligent bit rate control. The block

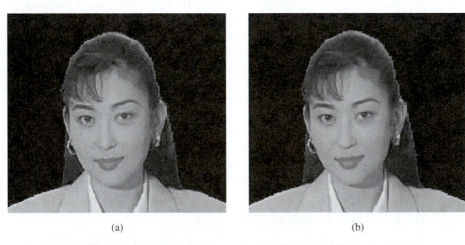

(a)   (b)

Figure 11.16   Reconstructed objects using SA-DCT and SA-ZTE: (a) SA-DCT (1.0042 bpp; PSNR-Y = 37.09 dB; PSNR-U = 42.14 dB; PSNR-V = 42.36 dB); (b) SA-ZTE (0.9538 bpp; PSNR-Y = 38.06 dB; PSNR-U = 43.43 dB; PSNR-V = 43.25 dB). Only portions of the Y components are shown.

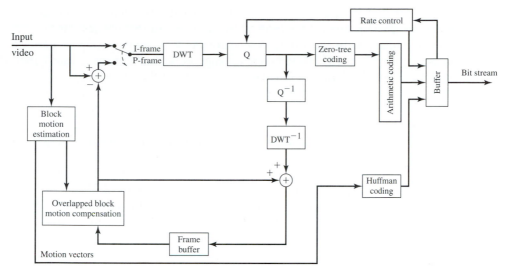

Figure 11.17   Block diagram of a wavelet-based video codec. Reprinted from S. A. Martucci, I. Sodagar, T. Chiang, and Y.-Q. Zhang, A zerotree wavelet video coder, *IEEE Trans. Circuits Syst. for Video Technology* (Feb. 1997), 7(1):109–18. Copyright 1997 IEEE.

diagram of the codec is shown in Figure 11.17. First, a three-stage motion estimation and compensation process is performed, including global motion estimation, variable block size motion estimation, and overlapped block motion compensation. The motion-compensated residual frames are then decomposed with a wavelet filter into a set of multiresolution pyramidal subbands, followed by a multistage quantizer with different stepsizes and dead zones. The quantized coefficients are finally entropy-coded using a ZTE-like encoder (but optimized for residual video wavelet coefficients.)

Global motion estimation is used to estimate effects of camera motion and to separate the image sequence into different layers. Each layer has its own global and local motion estimation. Global motion is estimated based on a 2-D affine motion model, and is applied prior to the local motion estimation.

After global motion estimation, a quad-tree structure is used to split the frames into variable-size blocks, ranging from 8×8 to 256×256 pixels. Conventional block matching is then used to find the MVs for individual blocks. OBMC (see Section 9.3.2) uses a weighted average of neighboring blocks displaced by their respective MVs as the prediction for the current block. This results in a smoother version of the motion-compensated residual frame. Due to the variable block size used for motion estimation, OBMC uses the smallest block size (in this case, 8×8) and the associated window function as the overlapping unit for prediction.

The wavelet transform is applied to the motion-compensated residual frames. Since the energy distribution across different subbands tend to be uniform for residual error frames, the normal criterion for the choice of wavelet filters—compact support—becomes less important. A more important criterion here is the time localization of a wavelet filter. A wavelet filter with good time localization tends to localize edges that are more often encountered in residual frames than original images. This suggests that short-tap filters (with better time localization but less compact support) perform better than long-tap filters (with better compact support but bad time resolution). Indeed, the Haar transform [1] has been found to perform reasonably well.

The DWT is followed by scalar quantization of the wavelet transform coefficients, which is then followed by ZTE. The bit allocation and rate control in the existing coder can be divided into three stages: (1) temporal-domain bit allocation with variable frame rates; (2) inter-frame bit allocation, to give optimal rate assignment among frames; and (3) intra-frame bit allocation, to produce optimal bit allocation within a frame.

This method has been compared to the MPEG-4 DCT-based coder [6]. The two methods produced similar performance in terms of PSNR [25]. It has been found that optimization of quantization, entropy coding, and the complex interplay among elements in the coding system are more important than optimizing the transform itself.

11.4 SUMMARY

- Scalability is a very useful feature in coding methods, since it allows flexibility and adaptivity to (1) changing network bandwidth; (2) varying channel conditions; and (3) different computing capabilities. Scalable coders are desirable for transport over networks that have time-varying bandwidth or error characteristics, or to receivers that have varying access rates or processing capabilities. Scalability is also desirable for video streaming applications, where the same video file may be accessed by users with different communication links and computing powers. (Video streaming is discussed further in Chapter 15.)

- Scalable coders can have either coarse granularity (in two or three layers; these are also called layered coders) or fine granularity. In the extreme case of fine granularity, the bit stream is fully embedded and can be truncated at any point.

- Quality, spatial, temporal, and frequency scalability are basic scalable mechanisms. These schemes can be combined to achieve finer granularity.

- One approach to obtain embedded streams at fixed spatial-temporal resolutions is by successive bit-plane encoding, as used in the fine-granularity scalability mode of the MPEG-4 standard, and in most wavelet-based coders.

- Wavelet-based coders can easily be made to produce an embedded stream, because the wavelet transform provides a multiscale representation of the signal.

11.5 PROBLEMS

11.1 What are the differences between layered and embedded coders? Give example coders in each category.

11.2 Describe the four basic scalability modes: quality, spatial, temporal, and frequency.

11.3 How does the FGS method described in Section 11.1.6 work? What is the difference between FGS and quality scalability?

11.4 Consider i.i.d. Gaussian processes. Suppose that one can design a one-layer coder that reaches the RD bound; that is, the distortion and rate are related by $D = \sigma_x^2 2^{-2R}$, where σ_x^2 is the source signal variance. Show that a quality scalable coder with two quantization layers will reach the same RD bound as the one-layer coder. Extend the proof to an arbitrary number of layers. (This result means that, for Gaussian processes, SNR scalability can achieve layered functionality without sacrificing coding efficiency; it is known as the *successive refinement property* of the Gaussian process. To explore further on this topic, see [2].)

11.5 Starting with the program that you created for the block-based hybrid video coder in Problem 9.15, generate a code for implementing two-layer quality scalability. Compare the coding efficiencies of the original and the layered coder for a test sequence.

11.6 Starting with the same program as in Problem 11.5, generate a code for implementing two-layer spatial scalability. Compare the coding efficiencies of the original code and the layered coder for a test sequence.

11.7 What are the main differences between a blockwise transform such as the DCT and the wavelet transform? What are the advantages and disadvantages of each?

11.8 How would you describe the EZW method to a fellow student who wants to know how it works?

11.9 Why is the wavelet transform more suitable for producing embedded streams? How would you categorize the type of scalability that is offered by, say, the EZW method? Does it offer quality, spatial, and/or frequency scalability?

11.10 Implement the two-stage wavelet transform for an image using the Haar wavelet filter (low-pass filter: [1, 1], high-pass filter [1, −1]). Examine the resulting image quality when coefficients in a chosen subband are set to zero. Based on these images, comment on what type of features a particular band of wavelet coefficients reveals.

11.6 BIBLIOGRAPHY

[1] Arkansu, A., and R. A. Haddad. *Multiresolution Signal Decomposition: Transforms, Subbands, Wavelets*. London: Academic Press, 1996.

[2] Equitz, W. H. R., and T. M. Cover. Successive refinement of information. *IEEE Trans. Inform. Theory* (March 1991), 37:269–75.

[3] ISO/IEC. IS 11172: Information technology—Coding of moving pictures and associated audio for digital storage media at up to about 1.5 mbit/s, 1993. (MPEG-1).

[4] ISO/IEC. IS 13818-2: Information technology—Generic coding of moving pictures and associated audio information: Video, 1995. (MPEG-2 Video).

[5] ISO/IEC. IS 14496: Information technology—coding of audio-visual objects—part 2: Visual, 1999. (MPEG-4 Video).

[6] ISO/IEC. Very low bitrate video codec. Public document, ISO/IEC MPEG96/M0637, Munich, January 1996. (MPEG-4 VM).

[7] ITU-T. Recommendation H.261: Video codec for audiovisual services at px64 kbits, 1993.

[8] ITU-T. Recommendation H.263: Video coding for low bit rate communication, 1998.

[9] ISO/IEC JTC1/SC29/WG1. FCD 15444-1: Information technology—JPEG2000 image coding system, March 2000. (Available at http://www.jpeg.org/FCD15444-1.htm.)

[10] Li, S., and W. Li. Shape-adaptive discrete wavelet transforms for arbitrarily shaped visual object coding. *IEEE Trans. Circuits Syst. for Video Technology* (Aug. 2000), 10:725–43.

[11] Li, W. Overview of fine granularity scalability in MPEG-4 video standard. *IEEE Trans. Circuits Syst. for Video Technology* (March 2001), 11(3):301–17.

[12] Li, W., and Y.-Q. Zhang. Vector-based signal processing and quantization for image and video compression. *Proceedings of the IEEE* (Feb. 1995), 671–91.

[13] Liang, J. Highly scalable image coding for multimedia applications. *Fifth ACM International Multimedia Conf.* Seattle, WA, (Nov. 1997), 11–19.

[14] Ling, F., W. Li, and H. Q. Sun. Bitplane coding of DCT coefficients for image and video compression. *SPIE Conf. Visual Commun. Image Processing* (Jan. 1999).

[15] Mallat, S. Multifrequency channel decompositions of images and wavelet models. *IEEE Trans. Acoust., Speech, Signal Processing* (Dec. 1989), 37(12):2091–2110.

[16] Mallat, S. A theory of multiresolution signal decomposition: The wavelet representation. *IEEE Trans. Pattern Analysis and Machine Intell.* (July 1989), 11(7):674–93.

[17] Martucci, S. A., I. Sodagar, T. Chiang, and Y.-Q. Zhang. A zerotree wavelet video coder. *IEEE Trans. Circuits Syst. for Video Technology* (Feb. 1997), 7(1):109–18.

[18] Said, A., and W. Pearlman. A new, fast, and efficient image codec based on set partitioning in hierarchical trees. *IEEE Trans. Circuits Syst. for Video Technology* (June 1996), 6:243–50.

[19] Shapiro, J. M. Embedded image coding using zerotrees of wavelet coefficients. *IEEE Trans. Signal Processing* (Dec. 1993), 41:3445–62.

[20] Sodagar, I., H.-J. Lee, P. Hatrack, and Y.-Q. Zhang. Scalable wavelet coding for synthetic/natural hybrid images. *IEEE Trans. Circuits Syst. for Video Technology* (March 1999), 9:244–54.

[21] Taubman, D., and A. Zakhor. Multirate 3-D subband coding of video. *IEEE Trans. Image Process.* (Sept. 1994), 3:572–88.

[22] van der Schaar, M., and H. Radha. A hybrid temporal-SNR fine-granular scalability for Internet video. *IEEE Trans. Circuits Syst. for Video Technology* (March 2001), 11(3):318–31.

[23] Vetterli, M., and J. Kovacevic. *Wavelets and Subband Coding.* Englewood Cliffs, NJ: Prentice Hall PTR, 1995.

[24] Wu, F., S. Li, and Y.-Q. Zhang. A framework for efficient progressive fine granularity scalable video coding. *IEEE Trans. Circuits Syst. for Video Technology* (March 2001), 11(3):332–44.

[25] Xiong, Z., K. Ramachandran, M. Orchard, and Y.-Q. Zhang. A comparative study of DCT- and wavelet-based image coding. *IEEE Trans. Circuits Syst. for Video Technology* (Aug. 1999), 9(5):692–95.

[26] Zhang, Y.-Q., and S. Zafar. Motion-compensated wavelet transform coding for color video compression. *IEEE Trans. Circuits Syst. for Video Technology* (Sept. 1992), 2(3):285–96.

12

Stereo and Multiview Sequence Processing

The acquisition and display of stereoscopic—and, more generally, multiview—sequences have many applications. In industries, stereo display is used for object handling and inspection by robots in hazardous environments. For medical applications, stereoscopic probes provide more useful information than monoscopic probes. For scientific exploration and visualization, stereoscopic or multiview display enables a viewer to experience a 3-D environment via a flat 2-D screen. Such applications have long existed, and stereo sequences have traditionally been acquired, stored, and displayed in an analog format. A recent boost for 3-D vision, however, comes from the advancement of digital television technology. As digital TV, including HDTV, becomes a reality, digital 3-DTV is foreseen to be the next "quantum leap" for the TV industry, promising to offer 3-D perception, in addition to sharper images. Another more recent application for stereo/3-D display lies in virtual reality and human-computer interface, where 3-D rendering and display gives viewers an illusion of physically interacting with people or objects in a remote (or nonexistent) site.

In this chapter, we start with a brief discussion of the mechanism for perceiving depth in the human visual system (Section 12.1). We then move on to the technical aspects of stereo imaging, and show how the 3-D position of an imaged point is related to its projections in a stereo image pair (Section 12.2). Next we discuss the technically most challenging problem in stereo sequence processing: disparity and, consequently, depth estimation (Section 12.3). We then describe techniques for interpolating an intermediate view from existing views (Section 12.4). Finally, we present approaches for coding of stereoscopic and multiview video (Section 12.5). These topics have been chosen because they belong to the signal processing aspects of stereo and multiview systems; for coverage of stereo and multiview capture and display systems, see, for example, [16].

12.1 DEPTH PERCEPTION

Stereoscopic imaging and display is motivated by how humans perceive depth: through two separate eyes. Although there are several monocular cues that the human brain uses to differentiate objects at different depths (e.g., through relative size and motion parallax), the most important and effective mechanism is through a binocular cue known as stereopsis. To understand how and why stereoscopic systems work, it is essential to understand this phenomenon. In this section, we first describe how stereopsis works. We then present the visual sensitivity in depth perception, which is important for designing 3-D video systems.

12.1.1 Binocular Cues—Stereopsis

To explain stereopsis, we take an example from [17]. Try this experiment: hold one finger in front of your face. When you look at the finger, your eyes are *converging* on the finger. This is accomplished by the muscle movement of your eyes, so that the images of the finger fall on the fovea of each eye—that is, the central portion of each retina. If you continue to converge your eyes on your finger, while paying attention to the background, you'll notice that the background appears to be double. On the other hand, if you try to focus on the background, your finger will appear double. This phenomenon is illustrated in Figure 12.1.

If we could take the images that are on your left and right retinas and super-impose them, we would see two almost-overlapping images. *Retinal disparity* refers to the horizontal distance between the corresponding left and right image points of the superimposed retinal images. The points for which the disparity is zero are where the eyes are converged. The mind's ability to combine two different perspective views into one image is called *fusion*, and the resultant sense of depth is called *stereopsis*.

12.1.2 Visual Sensitivity Thresholds for Depth Perception

Having learned how human beings perceive depth information, the next question is how sensitive the human visual system is to changes in depth, both in space and over time. In Section 2.4, we discussed visual thresholds in detecting spatial (in the same 2-D

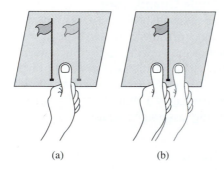

(a)   (b)

Figure 12.1   An experiment for understanding stereopsis: (a) When your eyes converge on the thumb, the background appears as a double image. (b) When your eyes converge on the background, you see two thumbs in the foreground.

plane) and temporal changes in monoscopic images. We learned that contrast sensitivity, as a function of spatial frequency, exhibits band-pass characteristics at low temporal frequencies (maximum sensitivity at approximately 4 cpd) and low-pass characteristics at high temporal frequencies. Similarly, visual sensitivity with respect to temporal changes is band-pass at low spatial frequencies (maximum sensitivity at approximately 8–15 Hz), and low-pass at high spatial frequencies. These findings have guided the design of video display systems. In order to display as much as the eye can see, a device should be able to produce a spatial frequency of at least 64 cpd (the resolution in terms of pixels depends on the picture height and width and the viewing distance), and a temporal frequency of 70 Hz.

To design a 3-D video capture and display system, an important issue is the visual threshold in depth perception. Because depth is perceived through retinal disparities, visual sensitivity to depth can be evaluated in terms of the variation in disparity. Results quoted in the following are from Pastoor [11].

Sensitivity to Depth Variation in Space and over Time   It has been found that the visual sensitivity to spatial and temporal changes in disparity lies significantly below the sensitivity to image contrast variations. Thus, a display system that meets the spatial and temporal bandwidth requirement of the contrast sensitivity function would also enable satisfactory perception of depth variation, both spatially and temporally. Visual sensitivity in terms of the spatial frequency of the disparity field has a band-pass characteristic, with a bandwidth of 3–4 cpd. The maximum sensitivity lies in the range of 0.2–0.5 cpd. The upper limit for temporal changes in disparity that are still perceptible as movements in depth lies in the region of 4–6 Hz.

Sensitivity to Disparity Quantization   It has been found that the threshold of stereoscopic acuity—that is, the ability to resolve stimuli in depth—is about 2 seconds of arc.* Such fine differences in depth can be perceived if the horizontal image resolution for both left and right images is at least 24 cpd. With digital display, the projected disparity—and, consequently, the apparent depth—is quantized. When the quantization is too coarse, the objects in the scene may appear to belong to layers of flat scenery at different depths, and object motion in depth may also appear jerky.

Sensitivity to View Angle Quantization   In natural vision, when a viewer moves, their viewpoint moves continuously. With 3-D display systems offering multiple viewpoints, only a limited number of viewpoints can be realized, which can be thought of as quantizing the direction of the viewing angle. Such quantization has the effect that, when the observer changes their position in front of the display, sudden shifts take place between perceived views. It has been demonstrated that view angle shifts must lie within a threshold of approximately 1 min of arc[†] for the subjective quality to be rated "good" on the CCIR quality assessment scale.

---

*1 second of arc is equal to 1/3600 of a degree.
[†]1 minute of arc is equal to 1/60 of a degree.

Asymmetric Spatial Resolution Requirement   Various experiments have shown that in stereo display, the resolution of one image can be considerably reduced (up to half) without causing deterioration in the subjective impression of sharpness, when the other image is displayed with sufficient resolution as dictated by the visual spatial-temporal thresholds, at least for a short period. It is not known to what degree such a display leads to visual strain when viewed over a long period of time. This asymmetrical property has been explored in stereo sequence compression (see Section 12.5.3). By reducing the sampling resolution of one view by half in both horizontal and vertical directions, a compression factor of four can be achieved immediately.

Asynchronous Fusion of Binocular Views   Within certain limits of disparity values, the visual system is capable of fusing the two images of a stereo pair presented asynchronously as long as the delay between the presentations does not exceed 50 ms [14]. This fact has been taken advantage of in field sequential display systems. The delay between the left and right views can lead to distortions of depth when the objects being displayed move. It has been found that a delay of 160 ms or longer can create visible depth distortion.

12.2 STEREO IMAGING PRINCIPLE

Stereoscopic imaging and display systems are designed to emulate human stereo perception: capturing a 3-D scene using two cameras located in slightly shifted positions, and then presenting a separate image to each eye. In this section, we describe the principle of stereoscopic imaging, and show how to deduce the depth of a 3-D point from its image positions in two different imaging planes. We will see that, in fact, to capture depth information, the two cameras do not have to be parallel (as human eyes), although such an arrangement leads to mathematically simple relationships of depth and image correspondence. We will start by considering a system with two cameras in arbitrary positions. We will then focus on two special cases: one with two cameras sharing the same imaging plane, another with two cameras oriented towards each other.

12.2.1 Arbitrary Camera Configuration

We start by considering two arbitrarily positioned cameras. Consider an arbitrary camera configuration, as shown in Figure 12.2. Let $[\mathbf{R}]_l$ and $\mathbf{t}_l$ (and $[\mathbf{R}]_r$ and $\mathbf{t}_r$) denote the rotation matrix and the translation vector required to align the left (and right) camera coordinates ($\mathbf{C}_l$ and $\mathbf{C}_r$) with a chosen world coordinate ($\mathbf{C}_w$). Then the coordinates of the left and right cameras, $\mathbf{X}_l = [X_l, Y_l, Z_l]^T$ and $\mathbf{X}_r = [X_r, Y_r, Z_r]^T$, are related to the world coordinate $\mathbf{X} = [X, Y, Z]^T$ by

$$\mathbf{X}_l = [\mathbf{R}]_l \mathbf{X} + \mathbf{T}_l; \quad \mathbf{X}_r = [\mathbf{R}]_r \mathbf{X} + \mathbf{T}_r. \qquad (12.2.1)$$

The matrices $[\mathbf{R}]_l$ and $[\mathbf{R}]_r$ are orthonormal by definition. Eliminating $\mathbf{X}$ from Equation (12.2.1) yields

$$\mathbf{X}_r = [\mathbf{R}]_r [\mathbf{R}]_l^T (\mathbf{X}_l - \mathbf{T}_{rl}) + \mathbf{T}_{rl} = [\mathbf{R}]_{rl} \mathbf{X}_l + \mathbf{T}_{rl}, \qquad (12.2.2)$$

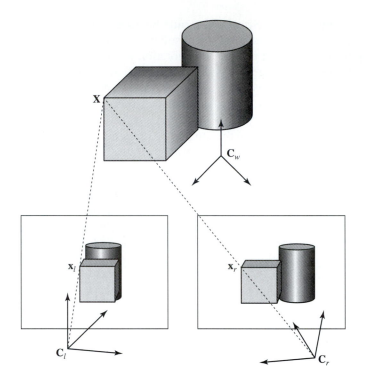

Figure 12.2   A stereo imaging system. Adapted from M. G. Strintzis and
S. Malassiotis, Object-based coding of stereoscopic and 3-D image sequences, *IEEE
Signal Processing Magazine* (May 1999), 16(3):14–28. Copyright 1999 IEEE.

with

$$[\mathbf{R}]_{rl} = [\mathbf{R}]_r[\mathbf{R}]_l{}^T, \quad \mathbf{T}_{rl} = \mathbf{T}_r - [\mathbf{R}]_r[\mathbf{R}]_l{}^T\mathbf{T}_l. \tag{12.2.3}$$

The preceding equation describes the relation between the left and right camera coordinates.

   Recall that when the world coordinate is coincident with the camera coordinate,
the world coordinate of a 3-D point $\mathbf{X} = [X, Y, Z]^T$ is related to the image coordinate
$\mathbf{x} = [x, y]^T$ by Equation (5.1.2). Applying this relation to $\mathbf{X}_l$ and $\mathbf{X}_r$, respectively,
yields the left and right image coordinates, $\mathbf{x}_l = [x_l, y_l]^T$ and $\mathbf{x}_r = [x_r, y_r]^T$:

$$x_l = F_l\frac{X_l}{Z_l}, \quad y_l = F_l\frac{Y_l}{Z_l};$$

$$x_r = F_r\frac{X_r}{Z_r}, \quad y_r = F_r\frac{Y_r}{Z_r}. \tag{12.2.4}$$

Substituting this relation into (12.2.2) yields

$$\frac{Z_r}{F_r}\begin{bmatrix} x_r \\ y_r \\ F_r \end{bmatrix} = \frac{Z_l}{F_l}[\mathbf{R}]_{rl}\begin{bmatrix} x_l \\ y_l \\ F_l \end{bmatrix} + \mathbf{T}_{rl}. \qquad (12.2.5)$$

Using the preceding relation, one can determine Z_r and Z_l of any 3-D point if its image positions $\mathbf{x}_l$ and $\mathbf{x}_r$ are known. Although there are three equations, only two are independent, from which a unique solution is obtainable. Once Z_l and Z_r are known, one can then determine X_l, X_r, Y_l, Y_r using Equation (12.2.4), and finally derive the reference coordinate (X, Y, Z) using Equation (12.2.1). This is the underlying principle for determining the structure of a 3-D object (i.e., the 3-D positions of its various points) from stereo images. This procedure is commonly known as *stereo triangulation*.

The reconstructed 3-D point (X, Y, Z) is in fact the intersection of the projection lines from the left and right image points. If the two image points are not exactly the projections of a 3-D point, their projection lines will not intersect. This will make Equation (12.2.5) inconsistent. The least-squares solution will be the midpoint in the shortest line that connects the two projection lines [22].

The displacement vector between the left and right image coordinates of the same 3-D position is known as the *disparity vector* (DV) or simply *disparity*.* The disparity vector can be defined over either the left or right image coordinate. For example, if we use the right image position as the reference, the disparity is defined as the displacement of the left coordinate with respect to a given right coordinate; that is, $\mathbf{d}(\mathbf{x}_r) = \mathbf{x}_l(\mathbf{x}_r) - \mathbf{x}_r$.[†] The main difficulty in 3-D structure estimation from stereo images lies in determining the disparity vectors; or, equivalently, establishing the correspondence between the left and right image points.

12.2.2 Parallel Camera Configuration

The most popular configuration for stereo imaging is to use two cameras with parallel imaging planes that are located on the same X-Y plane of the world coordinate. This is shown in Figure 12.3. The distance between the two cameras is called the *baseline distance,* denoted by B. When B is close to the separation between the two eyes of a person (2.5–3 in.), this configuration simulates the human binocular imaging system. If we set the origin of the world coordinate halfway between the two cameras, and assume that the two cameras have the same focal lengths, denoted by F, the relations in Equations (12.2.1) and (12.2.2) will become:

$$X_l = X + \frac{B}{2}, \quad X_r = X - \frac{B}{2}, \quad Y_l = Y_r = Y, \quad Z_l = Z_r = Z; \quad (12.2.6)$$

$$x_l = F\frac{X + B/2}{Z}, \quad x_r = F\frac{X - B/2}{Z}, \quad y_l = y_r = y = F\frac{Y}{Z}. \quad (12.2.7)$$

---

*Strictly speaking, we should call it *image disparity,* to differentiate it from retinal disparity and parallax (which refers to disparity at the display screen).

[†]Note that in previous chapters we have used $\mathbf{d}$ to represent the motion vector.

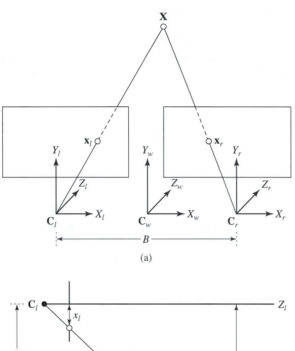

(a)

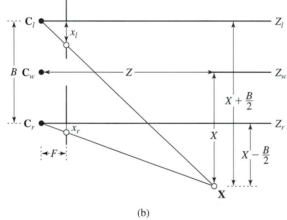

(b)

Figure 12.3 Parallel camera configuration: (a) the 3-D view; (b) the X-Z cross-section view ($Y = 0$).

One can also derive the preceding relation based directly on Figure 12.3(b). The disparity vector has only a horizontal component in this case, and is related to depth by

$$d_x = x_l - x_r = \frac{FB}{Z}. \tag{12.2.8}$$

The 3-D coordinates can be recovered from the image coordinates according to

$$X = \frac{B(x_l + x_r)}{2d_x}, \quad Y = \frac{By}{d_x}, \quad Z = \frac{FB}{d_x}. \tag{12.2.9}$$

These relations form the basis for deriving the depth and consequently 3-D structure information from the disparity information.

Equation (12.2.8) plays an important role in stereo sequence analysis. There are several interesting properties that can be derived from this relation. First, the disparity

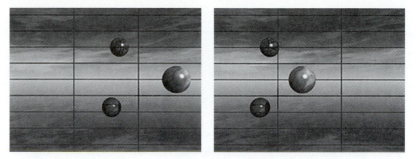

Figure 12.4   A stereo image pair of the kind captured by two parallel cameras. If you use your left and right eyes to look at the left and right images separately, you should see three balls at different depths. Courtesy of Ru-Shang Wang.

value of a 3-D point (X, Y, Z) is independent of the X and Y coordinates, and is inversely proportional to the Z value. The closer an object point is to the camera, the larger the disparity value. For an object point very far from the camera, the disparity value diminishes. Second, the range of the disparity value increases with the baseline B, the distance between the two cameras. Finally, with our definition of disparity, d_x is always positive. This is equivalent to saying that the left coordinate is always larger than the right coordinate for the same 3-D point, as is clear from Figure 12.3.

Figure 12.4 shows a stereo pair obtained using parallel cameras.* It is obvious that corresponding points in the left and right images are on the same horizontal line, and that the closer object (the larger ball) has a larger horizontal disparity.

12.2.3 Converging Camera Configuration

In a converging stereo configuration, the image planes of the two cameras are tilted with respect to each other so that their focal lines converge to a point at some distance. This is shown in Figure 12.5. The angle between the two cameras is called the *convergence angle*. In the figure, we assume the world coordinate is half-way between the left and right coordinates, and θ is half of the convergence angle. With this set-up, the left and right camera coordinates are related to the world coordinate by

$$\mathbf{R}_l = \begin{bmatrix} \cos\theta & 0 & -\sin\theta \\ 0 & 1 & 0 \\ \sin\theta & 0 & \cos\theta \end{bmatrix}, \quad \mathbf{T}_l = \begin{bmatrix} \cos\theta B/2 \\ 0 \\ \sin\theta B/2 \end{bmatrix}, \tag{12.2.10}$$

$$\mathbf{R}_r = \begin{bmatrix} \cos\theta & 0 & \sin\theta \\ 0 & 1 & 0 \\ -\sin\theta & 0 & \cos\theta \end{bmatrix}, \quad \mathbf{T}_r = \begin{bmatrix} -\cos\theta B/2 \\ 0 \\ \sin\theta B/2 \end{bmatrix}. \tag{12.2.11}$$

*In fact, the two images here and those in Figure 12.6 were artificially generated, using a ray tracing technique.

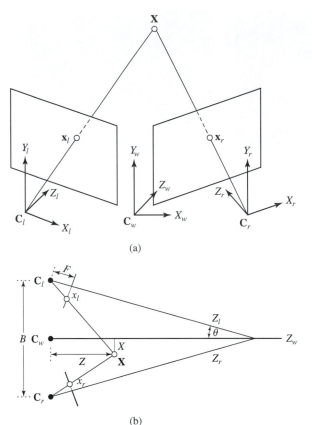

Figure 12.5   Converging camera configuration: (a) the 3-D view; (b) the X-Z cross-section view ($Y = 0$).

Substituting the preceding relations into Eqs. (12.2.2) and (12.2.4) yields

$$x_l = F\frac{\cos\theta(X + B/2) - \sin\theta Z}{\sin\theta(X + B/2) + \cos\theta Z}, \qquad y_l = F\frac{Y}{\sin\theta(X + B/2) + \cos\theta Z},$$

$$x_r = F\frac{\cos\theta(X - B/2) + \sin\theta Z}{-\sin\theta(X - B/2) + \cos\theta Z}, \qquad y_r = F\frac{Y}{-\sin\theta(X - B/2) + \cos\theta Z}.$$

$$(12.2.12)$$

One can also derive the preceding relation based on the illustration in Figure 12.5(b). From these equations, one can derive the relation between the horizontal (and vertical) disparity and the 3-D world coordinate (X, Y, Z). The relation is not as straightforward as in the parallel camera case. When θ is small (less than $1°$), the vertical disparity can be neglected.

With a converging configuration, the imaging system can get a better depth perception of objects close to the cameras than with a parallel set-up. However, when a stereo pair acquired by a converging stereo camera is directly projected on a screen

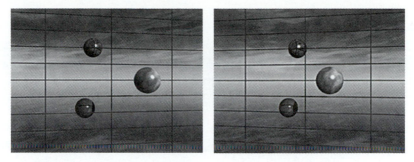

Figure 12.6   An example stereo pair of the kind obtained by converging cameras—notice the keystone effect. Courtesy of Ru-Shang Wang.

and viewed, the perceived depth is distorted. One well known artifact is the *keystone distortion*. This refers to the fact that when imaging a rectangular grid with equally spaced lines, the left camera shows the vertical distance between two adjacent lines as larger on the left side than on the right (and the image produced by the right camera has the reverse effect). An example stereo pair such as those obtained by converging cameras is given in Figure 12.6. When displayed using a parallel projection configuration and viewed by the human eye, such vertical disparity will cause difficulty in depth perception. To display the images acquired by a converging camera system, geometric correction (also known as image rectification) is needed so that the corrected images appear as if they were acquired using a parallel set-up.

12.2.4 Epipolar Geometry

Figure 12.7 illustrates the relation between the left and right images of a 3-D point in an arbitrary stereo camera set-up. For an arbitrarily chosen point in 3-D, designated $\mathbf{X}$, the plane Π—defined by this point and the optical centers of the left and right cameras, $\mathbf{C}_l$ and $\mathbf{C}_r$—is called the *epipolar plane*. The intersections of this plane with the left and right image planes Π_l and Π_r are called the left (the line connecting $\mathbf{x}_l$ and $\mathbf{e}_l$) and right (the line connecting $\mathbf{x}_r$ and $\mathbf{e}_r$) *epipolar lines*, denoted by ep_l and ep_r respectively. The image of $\mathbf{C}_l$ in the right image, $\mathbf{e}_r$, is called the *right epipole*. Similarly, the image of $\mathbf{C}_r$ in the left image, $\mathbf{e}_l$, is the *left epipole*. We see that, for any imaged point that falls on the left epipolar line, its corresponding pixel in the right image must be on the right epipolar line. This is known as the *epipolar constraint*. This property can be used to constrain the search range for disparity estimation, as discussed in Section 12.3.

Note that there is an epipolar plane and two corresponding epipolar lines associated with each selected image point. For example, given $\mathbf{x}_l$, a point in the left image, its left epipolar line is the one connecting this point with the left epipole, $\mathbf{e}_l$. Its corresponding right epipolar line is the line connecting the image of this point in the right view, $\mathbf{x}_r$ and the right epipole $\mathbf{e}_r$. All the left epipolar lines pass through the left epipole, and all the right epipolar lines pass through the right epipole.

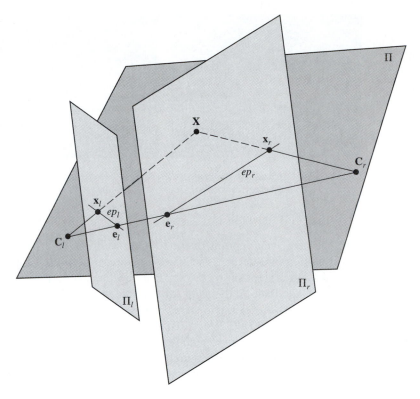

Figure 12.7 Epipolar geometry for an arbitrary camera configuration.

The relation between an image point and its epipolar line depends on the camera set-up. A very important result in stereo imaging is that this relation can be completely characterized by a 3×3 matrix known as the *fundamental matrix*, $[\mathbf{F}]$. Let $\tilde{\mathbf{x}}_i^T = [\mathbf{x}_i^T, 1]$, $i = l, r$, where $\mathbf{x}_l$ and $\mathbf{x}_r$ represent the left and right images of the same 3-D point.* They are related by:

$$\tilde{\mathbf{x}}_r^T [\mathbf{F}] \tilde{\mathbf{x}}_l = 0, \quad \tilde{\mathbf{x}}_l^T [\mathbf{F}]^T \tilde{\mathbf{x}}_r = 0. \tag{12.2.13}$$

In the simplified configuration shown in Figure 12.2, the fundamental matrix depends only on the relative relations of the two camera coordinates, $[\mathbf{R}]_{rl}$, $\mathbf{t}_{rl}$.† In general, it will also depend on intrinsic camera parameters. For the derivation of the preceding

---

*$\tilde{\mathbf{x}}^T = [\mathbf{x}^T, 1]$ is the representation of $\mathbf{x}$ in the homogeneous coordinate (also known as the projective coordinate). Use of the homogeneous coordinate can convert most nonlinear relations encountered in perspective imaging into linear relations.

†The fundamental matrix relates the corresponding points in the left and right images via Equation (12.2.13). In Chapter 7, we learned that the images of a 3-D object before and after a rotation and a translation are related by the essential matrix via Equation (7.1.11). Obviously, the fundamental matrix is equivalent to the essential matrix if we think of $[\mathbf{R}]_{rl}$ and $\mathbf{T}_{rl}$ as the motion parameter of the imaged object.

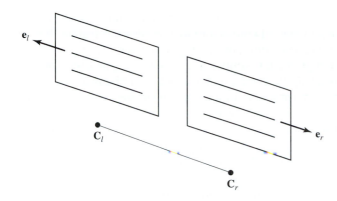

Figure 12.8 Epipolar geometry for a parallel camera configuration: epipoles are at infinity, and epipolar lines are parallel. Adapted from O. Faugeras, *Three-Dimensional Computer Vision—A Geometric Viewpoint*, Cambridge, MA: MIT Press, 1993. Copyright 1993 MIT Press.

result and the relation between [**F**] and camera parameters, the reader is referred to the excellent book by Faugeras [3]. When the camera parameters are unknown, the elements in [**F**] can be estimated from a set of detected corresponding points in the left and right images, based on Equation (12.2.13). A good reference on this subject is [23].

With a parallel camera set-up, the epipoles are at infinity and the epipolar lines are all horizontal, as illustrated in Figure 12.8. For any given point in, say, the right image, $\mathbf{x}_r$, the left and right epipolar lines associated with this point are simply the horizontal lines with the same y coordinate as this point. On the other hand, with a converging configuration, the left and right epipolar lines are tilted with respect to each other. This can easily be observed in the example image given in Figure 12.6. However, when the angle θ is sufficiently small, the epipolar lines can be treated as horizontal.

Given a pair of images obtained by two converging cameras, one can rectify them so that the rectified images appear as if they were captured using two parallel cameras. Visually, this has the effect of mapping a pair of images so that the originally tilted set of epipolar lines become horizontal and parallel. Such rectification, for example, greatly simplifies the disparity estimation problem. It is also needed for displaying such images by a parallel projection system. Recall from the preceding discussion that the epipolar constraint is completely characterized by the fundamental matrix relating the two cameras. Therefore, the key to the rectification process is to determine the fundamental matrix. In practice, the camera configuration parameters may not be known exactly, so that the fundamental matrix must be determined from the corresponding feature points in the two given images. A discussion of camera calibration and rectification techniques is beyond the scope of this chapter; the reader is referred to [19, 3] for a good treatment of this subject.

12.3 DISPARITY ESTIMATION

As introduced in the previous section, disparity refers to the displacement between the 2-D coordinates in the left and right image planes of a 3-D point in an imaged scene. As shown there, from the disparity associated with two corresponding points in a stereo image pair, one can derive the depth associated with the underlying 3-D point and, consequently, its complete 3-D coordinate in the real world. For this reason, disparity estimation is an essential step in any 3-D image or video processing system. It is also important for generating intermediate views between given views.

The disparity estimation problem is similar to the motion estimation problem in that both require establishment of the correspondence between pixels in two images. In a non–feature-based approach, for every point in one image, the task is to determine its corresponding point in the other image, or mark it as being occluded in the other image. For a feature-based approach, the task is to first detect certain feature points in one image, and then find their corresponding positions in the other. In motion estimation, the two images differ in time, whereas in disparity estimation, the two images are taken at the same time but from different camera viewpoints.

Disparity estimation tends to be more complex and challenging than motion estimation. First, only a limited number of objects usually move in a scene, so that the MVs are zero for most pixels. In stereo imaging, however, almost all pixels have nonzero disparities, except those pixels that are very far away from the image plane. Second, the range of MVs is usually quite limited, whereas the DV can be very large for objects close to the camera. For example, with BT.601 video, the MVs usually do not exceed a few pixels, whereas the DVs can reach 40–50 pixels in the horizontal direction and up to five pixels in the vertical direction (assuming that a slightly convergent camera configuration is used) [6]. Obviously, if one were to use a search technique similar to the block-matching algorithm for motion estimation, the search region must be greatly increased. The blocking artifacts also become more visible, because the blockwise constant model does not approximate the disparity variation as well.

The general approach to disparity estimation is very similar to motion estimation, except that there are more physically induced constraints in disparity fields, as described in the following. Basically, one must set up an objective function that measures the error between a given view and its predicted version from the other view, using *disparity-compensated prediction* (DCP). If we assume that the disparity field is smooth, then some smoothness measures can be added to the objective function. These smoothness constraints should be relaxed at object boundaries or occluded regions. The estimated parameters are further subjected to some physical constraints. The differences between approaches lie in the parameterization of the disparity field as well as the search method. In the following, we first describe some important constraints that are useful for disparity estimation. We then present several estimation methods.

12.3.1 Constraints on Disparity Distribution

The DVs are constrained by the camera geometry and the continuity of the object surface. These constraints, if used judiciously, can be very helpful in disparity estimation.

Epipolar Constraint   Given a stereo imaging configuration, the epipolar constraint refers to the fact that the corresponding pixels in a stereo pair always lie on respective epipolar lines. Recall that if the fundamental matrix is known, then given any point in, say, the right image, one can determine the epipolar line in the left image using Equation (12.2.13), which defines all possible locations of this point in the left image. Therefore, one only needs to search along this line.

For the parallel camera configuration, the epipolar lines are parallel to horizontal scan lines of the images, so that one can constrain the search within the horizontal line on which the right image point is located. When the camera configuration is nonparallel, although one can confine the search to the epipolar lines corresponding to the actual camera set-up, one simpler approach is to first map the left and right images into a parallel configuration, and then apply the procedure for the parallel configuration [21]. In either case, one must first estimate the camera geometry parameters or the fundamental matrix, which is required for determining the epipolar lines and for reprojection.

Unidirectionality with Parallel Cameras   As shown in Section 12.2.2, with the parallel camera configuration, the DV has only horizontal components and is always positive (i.e., the left horizontal coordinate is always larger than the right).

Ordering Constraint   Assuming opacity of the imaged objects, the relative positions of points in an object are the same in both views of the scene: a feature point to the left of another in the left view lies to the left in the right view as well. Let $x_{r,1}$ and $x_{r,2}$ represent two points in the right image on the same horizontal line, and assume that $x_{r,1} < x_{r,2}$. Then, by requiring $x_{l,1} < x_{l,2}$, we have $d_{x,2} > d_{x,1} + x_{r,1} - x_{r,2}$, which constrains the search range for $d_{x,2}$ given the previously estimated $d_{x,1}$. Specifically, for neighboring pixels in the right image (i.e., $x_{r,2} = x_{r,1} + 1$), their disparities are related by $d_{x,2} > d_{x,1} - 1$.

12.3.2 Models for the Disparity Function

Disparity is a deterministic function of depth, for a given camera configuration. Given a model for the actual surface function, one can derive the corresponding model for the disparity function. Such a model is very helpful for disparity estimation.

Consider the simplest case, in which the surface of the imaged scene can be approximated by a plane, so that

$$Z(X, Y) = aX + bY + c. \tag{12.3.1}$$

Using Equations (12.2.6) and (12.2.7), we can rewrite Z in terms of the right image coordinates:

$$Z(x_r, y_r) = a \left(\frac{Zx_r}{F} + \frac{B}{2} \right) + b \frac{Zy_r}{F} + c \tag{12.3.2}$$

or

$$Z(x_r, y_r) = \frac{c + aB/2}{1 - ax_r/F - by_r/F}. \tag{12.3.3}$$

We can see that the surface function varies nonlinearly in terms of image coordinates. Substituting the preceding result into Equation (12.2.8) yields:

$$d_x(x_r, y_r) = \frac{1}{c/B + a/2}(F - ax_r - by_r). \tag{12.3.4}$$

Therefore, *the disparity function is affine in the image coordinate when the surface is a plane.*

In reality, the entire surface (i.e., the depth distribution) of the imaged scene cannot be approximated well by a plane. However, it usually can be divided into small patches such that each patch is approximately planar. Using the preceding result, we can model the disparity function over each patch as affine. The disparity estimation problem can then be turned into the estimation of three affine parameters for each patch, which is equivalent to estimating the disparity (d_x only) at three corner points (nodes) of each patch. If we represent the entire surface by a triangular mesh, then the disparity estimation problem can be reduced to estimating the disparity at nodal points. The disparity function within each patch can then be interpolated from the nodal points, using the affine model.

Using similar approaches, one can also derive the disparity model for higher-order (curved) surfaces. Such parametric models can be very helpful for disparity estimation.

12.3.3 Block-Based Approach

This approach assumes that the disparity function over each small block of the reference view can be described by a constant or a low-order polynomial. The constant disparity or the polynomial coefficients are determined by minimizing the error between the two views after warping, based on the estimated disparity function (i.e., disparity compensation). Either exhaustive or gradient-descent search methods used for block-based motion estimation can be applied. But the search should be constrained properly, using the constraints described in Section 12.3.1. In the special case that the disparity is assumed to be constant, only a single horizontal shift must be searched, and only in one direction, positive or negative, depending on which view is used as reference. However, compared to motion estimation, the search range should be much larger, if the depth variation in the imaged scene is relatively large or the baseline separation between cameras is large. To overcome the difficulty associated with the potentially very large disparity value, a hierarchical search strategy, such as that described in Section 6.9, should be used. A good review of block-based estimation methods can be found in [20].

Unlike in motion estimation, where the blockwise constant model is fairly accurate for a large number of blocks, this model is only appropriate here when the surface patch corresponding to the block is a flat surface that is parallel with the imaging plane, which is rarely the case. On the other hand, the affine model is quite good when the block size is sufficiently small. In this case, the surface patch corresponding to each block is approximately flat, and (as shown in Section 12.3.2) the disparity function over a planar patch is well described by an affine function.

12.3.4 Two-Dimensional Mesh–Based Approach

Instead of using a blockwise constant model, the mesh-based approach described for motion estimation can also be applied to disparity estimation. In this case, one first applies a mesh in a reference view, say the left view, and tries to find the corresponding nodal positions in the right view. Note that each pair of corresponding 2-D mesh elements can be considered as the projections of a 3-D surface patch on the left and right

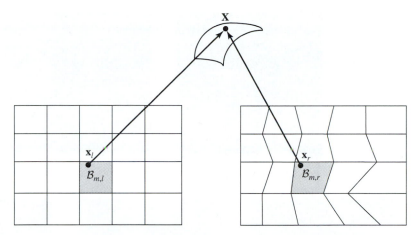

Figure 12.9   Correspondence between 3-D and 2-D meshes. Reprinted from R. Wang and Y. Wang, Multiview video sequence analysis, compression, and virtual viewpoint synthesis, *IEEE Trans. Circuits Syst. for Video Technology* (April 2000), 10(3):397–410. Copyright 2000 IEEE.

images, as shown in Figure 12.9. As with motion estimation, it is best to parameterize the disparity function by the nodal displacement between the left and right views, rather than polynomial coefficients in affine or bilinear models. One can estimate the nodal disparities by minimizing the disparity-compensated prediction error between corresponding elements, similar to the mesh-based motion estimation method (Section 6.6). When the underlying images are obtained with a parallel set-up, only horizontal disparities must be searched.

One such approach is described in [21]. As shown in Figure 12.9, the left view is described by a rectangular mesh, and the nodes in the right view are shifted horizontally with respect to the corresponding nodes in the left view. The disparity function within each element is modeled as bilinear, so that the 3-D patch corresponding to this element is a curved surface. To determine the horizontal disparity at each node, the disparity compensation error summed over the four elements attached to this node is minimized. Instead of performing exhaustive search, a gradient-based fast algorithm is used.

Figure 12.10 compares the disparity estimation results for a sample stereo image pair ("Man"), using the block-based approach and the preceding mesh-based approach [21]. The image pair was obtained with two parallel cameras with a baseline distance of 50 cm. This large baseline distance leads to very large disparity values between the two images. A global disparity is first estimated, and used to compensate the right image before local disparity estimation. The block-matching algorithm (assuming constant disparity within each block) uses 16×16 blocks and a search range of ± 100 pixels. The mesh-based scheme applies a rectangular mesh on the left image with element size 32×16, and moves the nodal position horizontally in the right image to minimize the disparity compensation error between corresponding elements. Although the

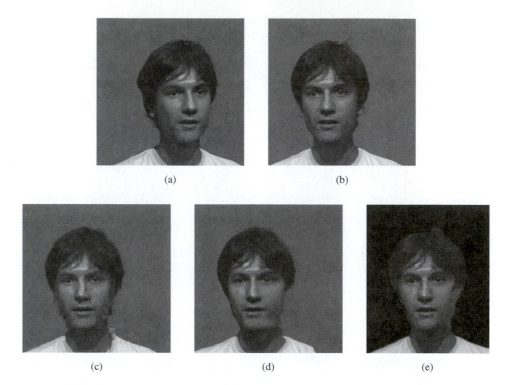

(a)   (b)

(c)   (d)   (e)

Figure 12.10   (a) Original left image; (b) original right image; (c) predicted right image obtained by a block-matching algorithm (PSNR of 32.03 dB over the foreground); (d) predicted right image obtained by a mesh-based algorithm (PSNR of 27.28 dB over the foreground); (e) predicted right image obtained by the dynamic programming method of [2]. Parts (a)–(d) reprinted from R. Wang and Y. Wang, Multiview video sequence analysis, compression, and virtual viewpoint synthesis, *IEEE Trans. Circuits Syst. for Video Technology* (April 2000), 10(3):397–410. Copyright 2000 IEEE. Part (e) courtesy of Fatih Porikli.

block-matching algorithm leads to higher PSNR for the predicted image, the mesh-based scheme yields a visually more accurate prediction.

In this approach, a regular mesh is overlaid over the entire reference image. This will lead to estimation errors near object boundaries and places with depth discontinuity. A better approach, although more complex, is to generate a mesh that follows the object contour, and allow discontinuity at object boundaries. For example, if the mesh is generated so that each mesh element corresponds to a surface patch that is approximately flat, then the disparity function over this element can be modeled accurately by an affine function, as shown in Section 12.3.2. In this case, triangular meshes would be more appropriate.

A benefit of using mesh-based disparity estimation is that it is easy to construct intermediate views using the resulting disparity information. This topic is discussed further in Section 12.4.

12.3.5 Intra-Line Edge Matching Using Dynamic Programming

With parallel imaging geometry (or after image rectification), the epipolar lines are horizontal scanlines. Therefore, given a horizontal scan line, a pair of corresponding points in the right and left images should be searched for within this line only. Instead of searching for matches between all the pixels, one can instead find all the edge points on this line, and try to match each edge pixel on the right image with one edge pixel on the left image. The searching of matching pairs along this line can be converted to a problem of finding a path on a 2-D search plane whose vertical and horizontal axes are the right and left edge positions, as illustrated in Figure 12.11. The figure assumes that there are M and N edge points in the left and right images, respectively. Each node corresponds to the association of a right edge point with a left edge point. In general, not all edge pixels seen in the right image will also appear in the left image. Also, a single edge pixel in the right image may correspond to several or none of the edge points in the left image. These situations happen when there are parts of the objects that are observable in one view but are occluded in the other view. Therefore, the matching could be one-to-many, many-to-one, or one-to-none. However, the legitimate matches should satisfy the ordering constraint. Starting from the leftmost edge pixel in the right image, for each new edge pixel in the right image, its matching edge point in the left image must be among the remaining unmatched edges or the one matched in the previous step. Therefore, the legitimate paths are those that do not go back, either horizontally or vertically. If we assign a cost at each possible node in the graph, the problem is to find the best path from the beginning pair (m_0) to the last pair (m_e), so that the total cost is minimized. The cost for a given node can be determined, for example, as the sum of squared or absolute errors between corresponding pixels in a small window surrounding the two edge pixels defining the node. It can also include

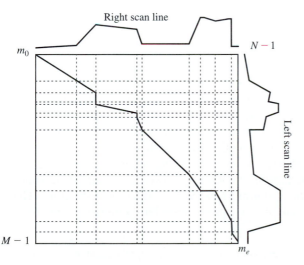

Figure 12.11 Path-finding analogy of the stereo matching problem. Adapted from O. Faugeras, *Three-Dimensional Computer Vision—A Geometric Viewpoint*, Cambridge, MA: MIT Press, 1993. Copyright 1993 MIT Press.

the errors between the corresponding pixels in the intervals between the current node and the previous node.

The preceding method does not take into account the mutual dependency of edge pixels between scan lines. If there is a vertical edge extending across scan lines, the correspondences in one scan line have strong dependency on the correspondences in the neighboring scan lines. Specifically, if two points are on a vertically connected edge in the right image, with a high probability, their corresponding points would also lie on a vertically connected edge in the left image. One way to promote such consistency among the scan lines is by including a penalty term in the cost definition, so that a higher cost is assigned for matches that are not consistent with matches found in the previous line. With this modification, the stereo matching problem can be cast as that of finding a surface that best satisfies the intra-scan line matches and inter-scan line consistency. Here the matching surface is defined by stacking 2-D matching paths.

Ohta and Kanade [10] first proposed to cast the disparity estimation problem as a surface matching problem as outlined here. Furthermore, they developed a fast search algorithm using the dynamic programming method. The original work of Ohta and Kanade considered only the constraint imposed by edges along the vertical direction. Falkenhagen extended this scheme to consider more complicated 2-D constraints [2]. Figure 12.10(e) shows the predicted right image for "Man" using this method.*

12.3.6 Joint Structure and Motion Estimation

The preceding discussions have focused on the estimation of the disparity function from a pair of stereoscopic views obtained at the same time. From the disparity function, one can derive the structure (i.e., spatial distribution) of the imaged scene, as described in Section 12.2. This allows one to reconstruct the 3-D surface of the imaged object. With stereo sequences, one is interested in the estimation not only of structure at a given frame, but also 3-D motion between frames. The most challenging problem is to estimate structure and motion jointly.

One common approach to this problem is to model the surface of the imaged object with a 3-D mesh (also known as a wireframe). Then the structure and motion estimation problem is to find the 3-D coordinates of all the nodes in a starting frame, and the 3-D nodal displacements between successive frames. Note that the 3-D mesh will project to 2-D meshes in the left and right images, as illustrated in Figure 12.12. To determine the initial 3-D mesh, one can first construct a 2-D mesh in one view, and then find the corresponding nodal positions in the other view. This can be accomplished using the mesh-based disparity estimation method described in Section 12.3.4. From the disparities between corresponding nodes, one can then determine the 3-D positions of all the nodes. For the produced 3-D mesh to fit the actual object surface, the initial 2-D mesh must be adapted to the object boundary, and each element should correspond to a flat patch. This, however, requires knowledge of the object structure. Therefore, it is important to obtain a good object segmentation based on the stereo views. Also,

---

*Courtesy of Fatih Porikli, who implemented the method of [2] with slight modifications.

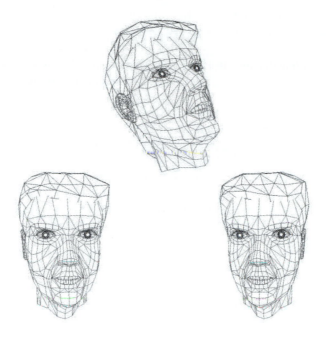

Figure 12.12   Projection of 3-D surface mesh to 2-D meshes in the left and right images. Courtesy of Khalid Salim Goudeaux.

an iterative procedure may be necessary, each time starting with a new 2-D mesh adapted based on the previous estimate of the object structure. To determine the 3-D nodal displacements, one can determine 2-D nodal displacements in both views, by minimizing the motion-compensated prediction error between two adjacent frames of the same view, as well as the disparity-compensated prediction error between two views at the same frame time. The main difficulty in both structure and motion estimation lies in how to handle occluded regions, including regions seen only in one view at the same frame time, and regions disappearing or newly appearing over time. For research papers on this topic, see [5, 7, 18].

12.4 INTERMEDIATE VIEW SYNTHESIS

One interesting stereo processing task is to interpolate or synthesize an intermediate view from given views—for example, generating a central view from the left and right views. This is often required in virtual reality displays, in which the displayed view must be constantly updated based on users view points. View synthesis is also required in advanced stereo or multiview video coding systems, where the views to be coded are first synthesized from a reference view, and then the synthesis error images are coded. A naive approach to view synthesis that does not consider disparity is linear interpolation, which generates the interpolated central view $\psi_c(\mathbf{x})$ from the left and right views, $\psi_l(\mathbf{x})$ and $\psi_r(\mathbf{x})$, by

$$\psi_c(\mathbf{x}) = w_l(\mathbf{x})\psi_l(\mathbf{x}) + w_r(\mathbf{x})\psi_r(\mathbf{x}). \tag{12.4.1}$$

If the baseline distance from the central to the left view is D_{cl} and that to the right view is D_{cr}, then the weighting factors are determined according to

$$w_l(\mathbf{x}) = \frac{D_{cr}}{D_{cl} + D_{cr}}, \quad w_r(\mathbf{x}) = 1 - w_l(\mathbf{x}). \tag{12.4.2}$$

Although simple, this approach will not yield satisfactory results, because pixels with the same image coordinate in different views correspond to different object points, and averaging among them will yield blurred images.

Figure 12.13 illustrates disparity-compensated interpolation, which is more appropriate. Let $\mathbf{d}_{cl}(\mathbf{x})$ and $\mathbf{d}_{cr}(\mathbf{x})$ represent the disparity field from the central to the left and to the right view, respectively. The interpolated central view is determined according to

$$\psi_c(\mathbf{x}) = w_l(\mathbf{x})\psi_l(\mathbf{x} + \mathbf{d}_{cl}(\mathbf{x})) + w_r(\mathbf{x})\psi_r(\mathbf{x} + \mathbf{d}_{cr}(\mathbf{x})). \tag{12.4.3}$$

For pixels that are visible in both views, the weighting factors can be chosen as in Equation (12.4.2). If a pixel is only visible in one view, then the weighting factor for the other view should be zero. Specifically,

$$w_l(\mathbf{x}) = \begin{cases} \frac{D_{cr}}{D_{cl} + D_{cr}}, & \text{if } \mathbf{x} \text{ is visible in both views,} \\ 1, & \text{if } \mathbf{x} \text{ is visible only in the left view,} \\ 0, & \text{if } \mathbf{x} \text{ is visible only in the right view.} \end{cases} \tag{12.4.4}$$

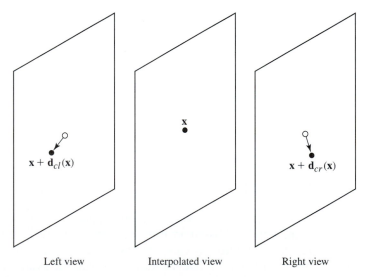

Left view   Interpolated view   Right view

Figure 12.13 Disparity-compensated interpolation: $\mathbf{x}$ is interpolated from $\mathbf{x} + \mathbf{d}_{cl}(\mathbf{x})$ in the left view and $\mathbf{x} + \mathbf{d}_{cr}(\mathbf{x})$ in the right view.

For example, if the imaged object is a face, then for the central part of the face, both views will be used for the interpolation. For the left (resp. right) part of the face, only the left (resp. right) view should be used.

The above described method assumes that $\mathbf{d}_{cl}(\mathbf{x})$ and $\mathbf{d}_{cr}(\mathbf{x})$ are known. In reality, one can only estimate the disparity, say from the left to the right view, $\mathbf{d}_{lr}(\mathbf{x})$, based on the given left and right images. It is not a straightforward task to generate $\mathbf{d}_{cl}(\mathbf{x})$ and $\mathbf{d}_{cr}(\mathbf{x})$ from $\mathbf{d}_{lr}(\mathbf{x})$. Note that it is easy to interpolate the disparity field from the left to the central view, by

$$\mathbf{d}_{lc}(\mathbf{x}) = \frac{D_{cr}}{D_{cl} + D_{cr}}\mathbf{d}_{lr}(\mathbf{x}). \tag{12.4.5}$$

But this does not allow one to find the corresponding point in the left view for every pixel in the central view. Rather, it determines the corresponding point in the central view for every pixel in the left view. In general, there will be pixels not covered, or pixels that correspond to more than one pixel in the left view.

The problem is feasible if $\mathbf{d}_{lr}(\mathbf{x})$ is estimated using the mesh-based approach. In this case, one can easily generate the mesh for the intermediate view from those in the left and right views, as shown in Figure 12.14. Specifically, the nodal positions in the central view, $\mathbf{x}_{c,n}$, are generated from the nodal positions in the left and right views, $\mathbf{x}_{l,n}$ and $\mathbf{x}_{r,n}$, using

$$\mathbf{x}_{c,n} = \frac{D_{cr}}{D_{cl} + D_{cr}}\mathbf{x}_{l,n} + \frac{D_{cl}}{D_{cl} + D_{cr}}\mathbf{x}_{r,n}. \tag{12.4.6}$$

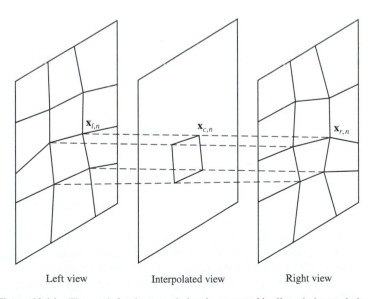

| Left view | Interpolated view | Right view |

Figure 12.14   The mesh for the central view is generated by linearly interpolating the nodal positions in the left and right views.

Figure 12.15 An example of intermediate view synthesis: the center image is interpolated from the left and right images. The result is obtained using mesh-based disparity-compensated interpolation, with the disparity map between the left and right views estimated using a mesh-based approach. Reprinted from R. Wang and Y. Wang, Multiview video sequence analysis, compression, and virtual viewpoint synthesis, *IEEE Trans. Circuits Syst. for Video Technology* (April 2000), 10(3):397–410. Copyright 2000 IEEE.

Then, for each pixel in an element in the central view, one can determine its corresponding point in the left (or right) view, by interpolating from the nodal positions, using the shape functions of the element (see Section 6.6). In the case of parallel imaging geometry, only horizontal disparities exist, and the problem is even simplified. For an explicit description of this case, see [21]. Figure 12.15 shows the interpolation result for the stereo pair "Man" using this approach.

12.5 STEREO SEQUENCE CODING

One major obstacle for the application of stereo and multiview sequences, including 3DTV, is the extremely large amount of data associated with a stereo or multiview sequence. To enable the storage or transmission of a stereo sequence at a reasonable cost, substantial compression of the data must be accomplished. In this section, we present several compression schemes that have been developed for stereo sequences.

12.5.1 Block-Based Coding and MPEG-2 Multiview Profile

With this approach, the coder first compresses, say, the left view sequence with a monoscopic video coding algorithm. For the right view sequence, each frame is predicted from the corresponding frame in the left view sequence, based on an estimated disparity field, and the disparity field and the prediction error image are coded. The disparity estimation and compensation processes both use a block-based approach; that is, for each image block in the right view, they find a most-closely resembling block in the left view. A better approach is to switch between disparity-compensated prediction between different views at the same time and motion-compensated prediction between different frames in the same view, depending on which gives a smaller prediction error.

This approach has been adopted in the *multiview profile* of the MPEG-2 standard [4], which can be implemented using the standard's temporal scalability mode. As will

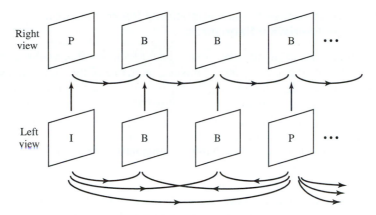

Figure 12.16   MPEG-2 multiview profile.

be discussed in Section 13.5.3 (and shown in Figure 13.24), with MPEG-2 temporal scalability, a sequence is divided into two sets of subsampled frames. The base layer—consisting of, say, the even frames—is coded using unidirectional MCP, where the reference frames are from the previously coded even frames. The enhancement layer, consisting of the remaining odd frames, is coded using bidirectional MCP, with one reference frame coming from the previously coded odd frame, and another from a nearby even frame in the base layer. Figure 12.16 illustrates how the temporal scalability is applied to a pair of stereo views. In this case, the left view is coded as the base layer, and the right view as the enhancement layer. For the left view, an image is coded in either I-, B-, or P-mode. For B- and P-images, the reference frames used for prediction come from the left view only. For the right view, an image is coded in either the P- or B-mode. In the P-mode, the image is predictively coded, using the corresponding frame in the left view as the reference. In the B-mode, one reference comes from a previous frame in the right view, and another reference frame is the corresponding frame in the left view. The motion vector used in the latter case is actually a disparity vector, and the prediction process is disparity-compensated prediction.

   Remember that with bidirectional prediction, the predicted block from either reference frame or the average of both can be used, depending on which provides the best prediction. Usually, the similarity between successive frames within the same view is higher than that between corresponding frames in the left and right views. The deficiency of the block-based disparity model further limits the accuracy obtainable with DCP. Because of these factors, MCP is usually preferred over DCP, so that the overall performance of the multiview profile is only slightly better than simulcast, where each view is coded independently using the MPEG-2 main profile [8]. Simulation results comparing the multiview profile with simulcast can be found in [13]. Note that with both simulcast and the MPEG-2 multiview profile, typically one view is coded at much lower quality than the other, reference view. This is motivated by the fact that the perceived quality of a stereo sequence is largely determined by the view with a higher quality (see Section 12.1.2).

12.5.2 Incomplete Three-Dimensional Representation of Multiview Sequences

In a multiview sequence, the same object region is often visible in multiple views. If we code only the texture map of this region in one of the views (the view that provides the highest resolution representation), and the disparity information between this view and other views, then we can reconstruct its images in other views reasonably well using DCP. For applications not requiring very high-quality reproduction of each view, the error between the predicted and the original images need not be coded. Instead of coding the texture map of each region separately, we can put the texture maps of all the different regions in an augmented image and treat this image as a single texture image. This augmented texture map, with the associated region segmentation and disparity information for each region, constitute the incomplete 3-D (I3D) representation, proposed by Ohm and Müller [9, 8]. An example of the I3D representation of the sequence "Man" is shown in Figure 12.17. In this case, the left and right sides of the face texture are obtained from the left and right views, respectively. Coding the I3D representation instead of the original two views can lead to significant savings in bit rates. With the coded disparity information, one can also interpolate intermediate views.

12.5.3 Mixed-Resolution Coding

As described in Section 12.1.2, the resolution of one of the two images can be considerably reduced (to approximately half the cut-off frequency of the original, using today's television standards) without causing irritation or deterioration in the subjective impression of sharpness when the image is presented for a short period of time [11]. Based on this asymmetric property of the HVS, it is not necessary to present two sequences in a stereo pair with the same spatial and temporal resolution. In mixed-resolution coding [12, 1, 15], one sequence, say the left, is coded at the highest affordable spatial and temporal resolution, while the other is first down-sampled spatially and

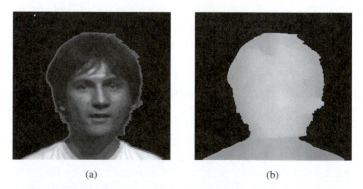

(a)   (b)

Figure 12.17   Incomplete 3-D representation for the sequence "Man": (a) augmented texture surface, and (b) disparity map. The original left and right images are shown in Figure 12.15. Reprinted from J.-R. Ohm, Encoding and reconstruction of multiview objects, *IEEE Signal Processing Magazine* (May 1999), 16(3):47–54. Copyright 1999 IEEE.

temporally, and then coded. It has been found [15] that one can subsample one view by a factor of two in both horizontal and vertical directions, without incurring degradation in an observer's ability to perceive depth.

12.5.4 Three-Dimensional Object-Based Coding

These approaches go one step further than coders based on DCP and MCP. Instead of deriving 2-D motion and disparity for performing MCP and DCP, 3-D structure and motion parameters are estimated from the stereo or multiple views. The structure, motion, and surface texture (color) of each object are coded, instead of individual image frames. At the decoder, desired views are synthesized based on the structure, motion, and texture information. Such an approach has more potential than a 2-D MCP/DCP combined approach. First, 3-D motion and structure parameters are physically constrained in a more direct and simpler form than 2-D motion and disparity, and proper use of these constraints should lead to more accurate estimation of 3-D motion and structure. In fact, accurate 3-D motion estimation should yield more accurate 2-D motion, which could in turn reduce the number of bits needed to represent each single video. Second, with the 3-D information derived from the stereo pair, one can generate any intermediate view. This feature is desired in many applications—for example, virtual reality, in which continuous views in the 3-D world are required but are too expensive to acquire or transmit. Finally, the coded 3-D information enables manipulation (change of view angle, annotation, animation, etc.) of the imaged object or scene, which is an important feature of interactive multimedia communications.

A general framework for such a coding scheme is shown in Figure 12.18. Based on the input stereo views, the coder first performs object segmentation and object-level motion and structure estimation. As described in Section 12.3.6, these three tasks are

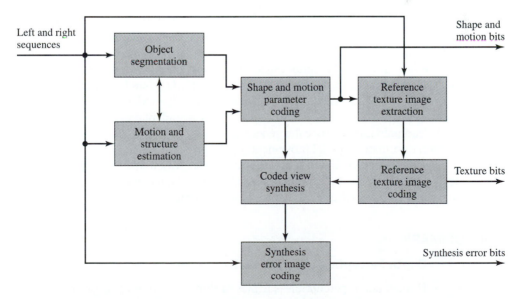

Figure 12.18   A general framework for object-based coding of stereoscopic sequences.

interrelated, and an iterative procedure is usually employed to yield the final solution. Each object is described by a wireframe, so that the motion and structure information can be described by initial nodal positions and nodal displacement vectors. For each object, a reference texture map is also extracted from both views, as in the I3D representation described in Section 12.5.2. Based on the segmentation map and the structure and motion parameters, one can then synthesize the left and right views from the reference texture map and code the synthesis error image if necessary. Instead of using a reference texture map that is constructed from the given views, one can also choose one of the input views as the reference texture map, and synthesize the other view. The synthesis can be accomplished using disparity-compensated texture warping from the reference texture map (see Section 12.4). The parameters to be coded include the segmentation map, the structure and motion information, the reference texture map, and the synthesis error image. Similar methodology can be applied for coding of multiview sequences. Coders that follow this general framework can be found in [18, 21].

The success of a 3-D object-based coder depends critically on its accuracy in 3-D structure and motion estimation. Without knowing what and how many objects may exist in an imaged scene, trying to recover the entire 3-D information from a generic stereo sequence is a very difficult, if not impossible task. Although substantial research efforts have been devoted to this problem and many promising approaches have been developed, existing solutions are still limited in their robustness and accuracy, and require very intense computations. Because of these problems, the performance of 3-D object-based coders has yet to compete favorably with non-object–based coders.

12.5.5 Three-Dimensional Model-Based Coding

The preceding discussion attempts to derive the 3-D structure of the objects in a scene automatically from the observed stereo sequences. Because of the very complex object composition that may exist in the imaged scene, this is a very difficult problem. A less difficult problem obtains when there are only a few objects in the scene, and the 3-D models of the objects follow some known structure. In this case, a generic model can be built for each potential object. For example, in teleconferencing applications, the imaged scene usually consists of one or a few head-and-shoulder type of objects. The coder can use predesigned generic face and body models. The structure estimation problem reduces to that of estimating the necessary adaptations of the generic model to fit the captured image. Once the generic model is adapted to the imaged object, only the motion information (nodal movements) must be estimated and coded. This approach is similar to the knowledge-based coding previously discussed in Chapter 10. Compared to the method developed for monoscopic sequences, the availability of stereo or more views helps to improve the model adaption accuracy.

12.6 SUMMARY

Depth Perception (Section 12.1)

- Human depth perception is achieved through fusion of displaced views by two eyes, an ability known as stereopsis.

- The eye is less sensitive to spatial and temporal variation in depth than in contrast. Therefore, a stereo system in which each view is captured and displayed with the standard resolution for monoscopic sequences can render depth properly.
- In fact, the two views need not be rendered with the same high spatial and temporal resolution. The perceived quality of a stereo sequence is dominated by the better view. This property has been exploited in stereo sequence compression.

Stereo Imaging Principle (Section 12.2)

- With stereo imaging, the images of a 3-D point in two different views are displaced with respect to each other, and it is this disparity between the two positions that enables the recovery of the depth information of the 3-D point.
- The most general relation between depth and disparity is given by Equation (12.2.5). For the most popular parallel camera configuration, this is reduced to the famous inverse relationship between disparity and depth (Equation (12.2.8)). For an arbitrary camera configuration, the relation between the left and right image positions of the same object point is described by the epipolar geometry (Figure 12.7; Equation (12.2.13)).

Disparity Estimation (Section 12.3)

- Disparity estimation is similar to motion estimation (in that one must find corresponding points in two different images) and hence can be accomplished using similar approaches, including block-based and mesh-based methods.
- Compared to motion estimation, disparity estimation can be aided by adding physical constraints. The most important is the epipolar constraint, which says that for any point in a right image, its corresponding point in the left image must fall on the epipolar line. For a parallel camera set-up, this means that corresponding points in two images are on the same horizontal line.
- Disparity estimation is a fundamental step for recovering 3-D structure and motion of imaged objects from a stereo sequence. To achieve this goal, disparity estimation must be performed in conjunction with object segmentation, 3-D modeling using wireframes, and detection and treatment of occluded regions.

Intermediate View Synthesis (Section 12.4)

- View synthesis is required in virtual reality display. It is also an important component in advanced multiview sequence coding systems.
- Disparity compensated interpolation is necessary to obtain satisfactory results.

Stereo and Multiview Sequence Coding (Section 12.5)

- There are generally two types of approaches. The first is waveform-based: it combines disparity-compensated and motion-compensated prediction, and codes the prediction error image and the disparity and motion vectors.
- The second is object-based: it tries to recover the 3-D structure and motion of the imaged object, and codes this information directly. The second approach has the

potential of achieving very high compression ratios, for simply structured scenes. It also facilitates the synthesis of any intermediate view at the decoder.

- Object-based and model-based approach described in Sections 12.5.4 and 12.5.5 can be considered as the extensions of the object-based and knowledge-based coders discussed in Chapter 10, which are developed for coding of monoscopic sequences.

12.7 PROBLEMS

12.1 What are the advantages and disadvantages of parallel versus converging camera configurations for stereo imaging?

12.2 Derive the horizontal and vertical disparity values, $d_x(x_r, y_r), d_y(x_r, y_r)$, in the converging camera configuration. Show that the vertical disparity is zero for $X = 0$ or $Y = 0$. Also derive the method for determining the 3-D position from $d_x(x_r, y_r), d_y(x_r, y_r)$.

12.3 Derive the fundamental matrix $[\mathbf{F}]$ for (a) parallel, and (b) converging camera configurations.

12.4 For a curved surface, $Z = aX + bY + cXY + d$, what is the parameterized form of the disparity function, assuming a parallel camera configuration?

12.5 When the baseline between two parallel cameras is large, the horizontal disparities are very large. In this case, it is better to first estimate a global horizontal disparity value, before applying block-based or mesh-based disparity estimation. This global disparity can be found by finding a global horizontal shift between the two images that will yield the least error after shifting. Write a program to estimate the global disparity between a stereo pair.

12.6 Implement (in C or MATLAB) a hierarchical block-matching algorithm for disparity estimation for the case of parallel imaging geometry (only horizontal disparity must be searched). Evaluate its performance over a stereo image pair. You should first estimate the global disparity and then apply the block-matching algorithm to the images after global disparity compensation.

12.7 Implement the mesh-based approach for disparity estimation for the case of parallel imaging geometry. For simplicity, use a regular square mesh over the reference view. Compare its performance for the same stereo pair to the block-matching algorithm.

12.8 Implement the dynamic programming scheme for disparity estimation described in [10]. Evaluate its performance over a stereo image pair. Compare its performance to the block-matching algorithm and the mesh-based approach.

12.9 Generate an intermediate view between the views of a stereo pair acquired by parallel cameras. Compare two approaches: linear interpolation, and disparity-compensated interpolation using the mesh-based approach (Section 12.4). Ideally, you should estimate the disparity values using the mesh-based approach. If you choose to use the simpler block-based approach, then you can assume that the disparity found for each block is the disparity at the center of the block, and that all the block centers are the nodes in the mesh for the reference view.

12.10 Implement a stereo image coder using block-based disparity compensation. Code one image (reference view) using the DCT method (Section 9.1.7). For the other image, for each block, find its prediction using block-based disparity, and then code the error block using the DCT approach.

12.11 Implement a stereo video coder using block-based disparity compensation and motion compensation. Code one view (reference view) using the hybrid coding method (Section 9.3.1). For simplicity use unidirectional prediction only. For the other view, for each block in a frame, estimate it using both motion compensation from the previous frame in this view, and disparity compensation from the same frame in the reference view. Use either the average of the two predictions or one of the predictions, depending on the prediction errors. Then code the prediction error using the DCT method.

12.8 BIBLIOGRAPHY

[1] Dinstein, I., et al. Compression of stereo images and the evaluation of its effects on 3-D perception. *SPIE Conf. Applications of Digital Image Processing* XII (1989), SPIE-1153:522–30.

[2] Falkenhagen, L. Depth estimation from stereoscopic image pairs assuming piecewise continuous surfaces. In *European Workshop on Combined Real and Synthetic Image Processing for Broadcast and Video Production,* 115–27. Hamburg, November 1994.

[3] Faugeras, O. *Three-Dimensional Computer Vision—A Geometric Viewpoint.* Cambridge, MA: MIT Press, 1993.

[4] ISO/IEC. IS 13818-2 AMD3: MPEG-2 multiview profile, Sept. 1996.

[5] Koch, R. Dynamic 3-D scene analysis through synthesis feedback control. *IEEE Trans. Pattern Anal. Machine Intell.* (1993), 15:556–68.

[6] Kopernik, A., R. Sand, and B. Choquet. The future of three-dimensional TV. In E. Dubois and L. Chiariglione, eds., *Signal Processing of HDTV.* Vol. IV, 17–29. Amsterdam: Elsevier Science, 1993.

[7] Malassiotis, S., and M. G. Strintzis. Model-based joint motion and structure estimation from stereo images. *Computer Vision, Graphics, and Image Understanding* (Jan. 1997), 65(1):79–94.

[8] Ohm, J.-R. Encoding and reconstruction of multiview video objects. *IEEE Signal Processing Magazine* (May 1999), 16(3):47–54.

[9] Ohm, J.-R., and K. Müller. Incomplete 3-D—multiview representation of video objects. *IEEE Trans. Circuits Syst. for Video Technology* (March 1999), 9(2):389–400.

[10] Ohta, Y., and T. Kanade. Stereo by intra- and inter-scanline search using dynamic programming. *IEEE Trans. Pattern Anal. Machine Intell.* (1985), 7:139–45.

[11] Pastoor, S. 3-D-television: A survey of recent research results on subjective requirements. *Signal Processing: Image Commun.* (1991), 4:21–32.

[12] Perkins, M. Data compression of stereopairs. *IEEE Trans. Comm.* (1992), 4:684–96.

[13] Puri, A., R. V. Kollarits, and B. G. Haskell. Stereoscopic video compression using temporal scalability. *SPIE Conf. Visual Commun. Image Processing (VCIP95)*, SPIE-2501:745–56.

[14] Ross, J., and J. H. Hogeben. Short-term memory in stereopsis. *Vision Res.* (1974), 14:1195–201.

[15] Sethuraman, S., M. W. Siegel, and A. G. Jordan. A multiresolution framework for stereoscopic image sequence compression. *IEEE Int. Conf. Image Processing (ICIP94)*, 2:361–65.

[16] Sexton, I., and P. Surman. Stereoscopic and autostereoscopic display systems—an in-depth review of past, present, and future technologies. *IEEE Signal Processing Magazine* (May 1999), 16(3):85–99.

[17] Stereographics, Inc. *Perceiving Stereoscopic Images*. (http://www.stereographics.com/html/chapter_2.html).

[18] Strintzis, M. G., and S. Malassiotis. Object-based coding of stereoscopic and 3-D image sequences. *IEEE Signal Processing Magazine* (May 1999), 16(3):14–28.

[19] Tsai, R. Y. A versatile camera calibration technique for high-accuracy 3-D machine vision metrology using off-the-shelf TV cameras and lenses. *IEEE Journal of Robotics and Automation* (Aug. 1987), 3:323–44.

[20] Tzovaras, D., M. S. Strintzis, and H. Sahinoglou. Evaluation of multiresolution techniques for motion and disparity estimation. *Signal Processing: Image Commun.* (June 1994), 6(1):59–67.

[21] Wang, R., and Y. Wang. Multiview video sequence analysis, compression, and virtual viewpoint synthesis. *IEEE Trans. Circuits Syst. for Video Technology* (April 2000), 10(3):397–410.

[22] Weng, J., T. S. Huang, and N. Ahuja. *Motion and Structure from Image Sequences*. New York: Springer-Verlag, 1993.

[23] Zhang, Z. Determining the epipolar geometry and its uncertainty: A review. Technical report No. 2927, INRIA, Sophia-Antipolis, France, July 1996. (Available from http://www-sop.inria.fr/robotvis/personnel/zzhang/publications.html)

13

Video Compression Standards

Digital video communication is a complex and computationally intensive process that requires many people to receive video signals from different sources. There are mainly three classes of devices that are currently used for digital video communications:

- Digital television sets or settop boxes, which are mainly designed to receive video signals from various content providers. These devices rely on a single fixed video decoding algorithm that is implemented in hardware or in a combination of hardware and a programmable reduced instruction set computer (RISC) processor or digital signal processor (DSP). Currently, there are no means provided for uploading a new algorithm once the hardware is deployed to the customer.

- Video phones, which are usually implemented on DSPs, with hardware acceleration for some computationally intensive parts of the video coding and decoding algorithm (such as DCT and motion estimation). Usually, the set of algorithms used in a particular video phone cannot be replaced.

- Personal computers, which are the most flexible and most expensive platform for digital video communication. While a PC with a high-end Pentium III processor is able to decode DVDs, the software is usually preinstalled with the operating system in order to avoid hardware and driver problems. Video decoders for streaming video may be updated using automatic software download and installation, as done for commercial software such as Real Player, Windows Media Player, Apple QuickTime, and Microsoft Netmeeting, among other.

Digital video communications standards were mainly developed for digital television and video phones, in order to enable industry to provide consumers with bandwidth-efficient terminals at an affordable price. We describe the standards organizations, the meaning of compatibility, and applications for video coding standards in Section 13.1. We begin the description of actual standards with the ITU video coding standards H.261 and H.263 (Section 13.2) for interactive video communications. In Section 13.3, we present the standards H.323 and H.324, which define multimedia terminals for audio-visual communications. Within the ISO, the Moving Picture Expert Group (MPEG) defined MPEG-1 (Section 13.4) and MPEG-2 (Section 13.5) for entertainment and digital TV. MPEG-4 (Section 13.6) is the first international standard that standardizes not only audio and video communications but also graphics for use in entertainment and interactive multimedia services. All standards describe the syntax and semantics of a bit stream. Section 13.7 presents an overview of the organization of a bit stream as used by H.261, H.263, and MPEG-1,-2, and -4. Finally, we give a brief description of the ongoing MPEG-7 standardization activity (Section 13.8), which intends to standardize the interface for describing the content of an audiovisual document.

13.1 STANDARDIZATION

Developing an International Standard requires collaboration among many partners from different countries, and an organization that is able to support the standardization process as well as enforce the standards. In Section 13.1.1 we describe organizations such as the ITU and ISO. In Section 13.1.2, the meaning of compatibility is defined. Section 13.1.3 briefly describes the workings of a standardization body. In Section 13.1.4, applications for video communications are listed.

13.1.1 Standards Organizations

Standards are required if we want multiple terminals from different vendors to exchange or receive information from a common source like a TV broadcast station. Standardization organizations have their roots in the telecom industry (ITU) and international trade (ISO).

ITU   The telecom industry has established a long record of setting international standards [7]. At the beginning of electric telegraphy in the nineteenth century, telegraph lines did not cross national frontiers, because each country used a different system and each had its own telegraph code to safeguard the secrecy of its military and political messages. Messages had to be transcribed, translated, and handed over at frontiers before being retransmitted over the telegraph network of the neighboring country. The first International Telegraph Convention was signed in May 1865, and it harmonized the different systems used. This event marked the birth of the International Telecommunication Union (ITU; www.itu.int).

Following the invention of the telephone and the subsequent expansion of telephony, the Telegraph Union began, in 1885, to draw up international rules for telephony. In 1906, the first International Radiotelegraph Convention was signed. Subsequently, several committees were set up for establishing international standards, including the International Telephone Consultative Committee (CCIF) in 1924, the International

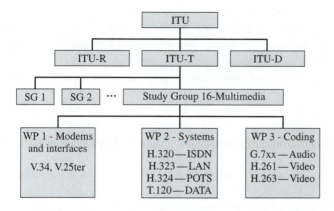

Figure 13.1 Organization of the ITU, with its subgroups relevant to digital video communications. Working Parties are organized into Questions that define the standards.

Telegraph Consultative Committee (CCIT) in 1925, and the International Radio Consultative Committee (CCIR) in 1927. In 1927, the Telegraph Union allocated frequency bands to the various radio services existing at the time. In 1934, the International Telegraph Convention of 1865 and the International Radiotelegraph Convention of 1906 were merged to become the ITU. In 1956, the CCIT and the CCIF were amalgamated to create the International Telephone and Telegraph Consultative Committee (CCITT). In 1989, CCITT published the first digital video coding standard, the CCITT Recommendation H.261 [41], which is still relevant today. In 1992, ITU reformed itself which resulted in the renaming of the CCIR as the ITU-R and the CCITT as the ITU-T. Consequently, the standards of the CCITT are now referred to as ITU-T Recommendations (for example, CCITT H.261 is now known as ITU-T H.261).

Figure 13.1 shows the structural organization of the ITU, detailing the parts that are relevant to digital video communications. The ITU-T is organized in study groups, with Study Group 16 (SG 16) being responsible for multimedia. SG 16 divided its work into different Working Parties (WPs), each dealing with several Questions. Questions that SG 16 worked on in 2001: Question 15 (advanced video coding), which developed the video coding standards ITU-T Recommendation H.261 and H.263 [50], and Question 19 (extension to existing ITU-T speech coding standards at bit rates below 16 kbps), which developed speech coding standards such as ITU-T Recommendation G.711 [36], G.722 [37], and G.728 [39]. Question numbers tend to change every four years.

ITU is an international organization, created by a treaty signed by its member countries. Countries consider dealings of the ITU relevant to their sovereign power. Accordingly, any recommendation (i.e., standard) of the ITU must be agreed upon unanimously by the member states. As a result, the standardization process in the ITU is often unable to keep up with the progress in modern technology. Sometimes, unanimous decisions cannot be reached and ITU recommends regional standards (such as seven-bit or eight-bit representation of digital speech in the United States of America and Europe, respectively). As far as mobile telephony is concerned, ITU did not play a leading role. In the United States of America, there is not even a national mobile telephony standard, as every operator is free to choose a standard of their own liking. This contrasts with the approach adopted in Europe, where the GSM standard has been so successful that it is expanding all over the world, the United States of America included. With the UMTS (the so-called third generation mobile standard), the ITU-T is retaking its original role as developer of global mobile telecommunication standards.

ISO   The need to establish international standards developed with the growth of trade [7]. The International Electrotechnical Commission (IEC) was founded in 1906 to prepare and publish international standards for all electrical, electronic, and related technologies. The IEC is currently responsible for standards of such communication means as "receivers," audio and video recording systems, and audiovisual equipment, currently all grouped in TC 100 (Audio, Video and Multimedia Systems and Equipment). International standardization in other fields (particularly mechanical engineering) was the concern of the International Federation of the National Standardizing Associations (ISA), set up in 1926. The ISA's activities ceased in 1942, but a new international organization called the International Organization for Standardization (ISO; www.iso.ch) began to operate in 1947 with the stated objective "to facilitate the international coordination and unification of industrial standards." All computer-related activities are currently in the Joint ISO/IEC Technical Committee 1 (JTC 1) on Information Technology. This committee has become very large; about 30 percent of all ISO and IEC standards work is done in JTC 1.

The subcommitees SC 24 (Computer Graphics and Image Processing) and SC 29 (Coding of Audio, Picture, Multimedia and Hypermedia Information) are relevant to multimedia communications. Whereas SC 24 defines computer graphics standards such as VRML, SC 29 developed the well known audiovisual communication standards MPEG-1, -2, and -4 (Figure 13.2). The standards were developed at meetings attended by 200–400 delegates from industry, research institutes, and universities.

ISO has been an agency of the United Nations since 1947, and both ISO and IEC have the status of private not-for-profit companies established according to the Swiss Civil Code. Similarly to the ITU, the ISO requires consensus in order to publish a standard. The ISO also fails sometimes to establish truly international standards, as can be seen in the case of digital TV. While the same video decoder (MPEG-2 Video) is used worldwide, the audio representation is different in the United States and in Europe.

Both the ISO and the ITU are in constant competition with industry. While the ISO and the ITU have been very successful in defining widely used audio and video coding standards, they were less successful in defining transport of multimedia signals over the Internet. This is currently handled by the Internet Engineering Task Force (IETF; www.ietf.org), which is a large, open, international community of network designers, operators, vendors, and researchers concerned with the evolution of the Internet architecture and the smooth operation of the Internet. The IETF is open to any interested individual. Other de facto standards (such as JAVA) are defined by one or a few companies, thus limiting the access of outsiders and newcomers to the technology.

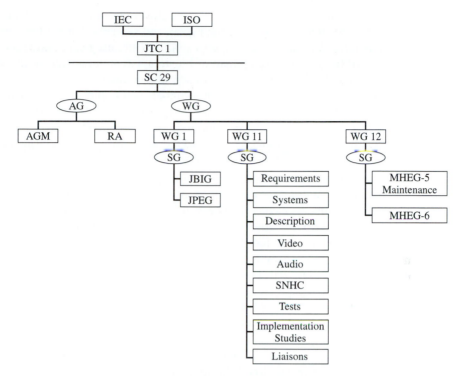

Figure 13.2   Audiovisual communications standards such as MPEG-1, MPEG-2, and MPEG-4 are developed by Working Group 11 of Subcommittee 29 under ISO/IEC JTC 1.

13.1.2 Requirements for a Successful Standard

International standards are developed in order to allow interoperation of communications equipment provided by different vendors. Consequently, the following requirements enable a successful deployment of audiovisual communications equipment in the marketplace:

1. Innovation: In order for a standard to distinguish itself from other adopted or widely accepted industry standards it must provide a significant amount of innovation. Innovation in the context of video coding means that the standard provides new functionalities—such as broadcast-quality interlaced digital video, video on CD-ROM, or improved compression. If the only distinction of a new standard is better compression, the standard should provide at least an improvement that is visible to the consumer and nonexpert, before its introduction makes sense commercially. This usually translates to a gain of 3 dB in PSNR of the compressed video at a generally acceptable level of picture quality.

2. Competition: Standards should not prevent competition between manufacturers. Therefore, standards specifications must be open and available to everyone. Free

software for encoders and decoders also helps to promote a standard. Furthermore, a standard should only define the syntax and semantics of a bit stream—that is, *a standard defines how a decoder works*. Bit stream generation is not standardized. Although the development of bit stream syntax and semantics requires an encoder and a decoder, the standard does not define the encoder. Therefore, manufacturers of standards-compliant terminals can compete not only on price but also on additional features, such as postprocessing of the decoded media—and, more importantly, encoder performance. In video encoding, major differences in performance result from particulars of motion estimation, scene change handling, rate control and optimal bit allocation.

3. Independence from transmission and storage media: A content provider should be able to transmit or store digitally encoded content independent of the network or storage media. As a consequence of this requirement, audio and video standards are used to encode audiovisual information, then a systems standard is used to format the audio and video bit streams into a format that is suitable for the selected network or storage system. The systems standard specifies the packetization, multiplexing, and packet header syntax for delivering the audio and video bit streams. The separation of transmission media and media coding usually creates overhead for certain applications. For example, we cannot exploit the advantages of joint source and channel coding.

4. Forward compatibility: A new standard should be able to understand the bit streams of prior standards; for example, a new video coding standard like H.263 [50] should be able to decode bit streams encoded according to the previous video coding standard H.261 [41]. Forward compatibility ensures that new products can be gradually introduced into the market. The new features of the latest standard are used only when terminals conforming to the latest standard can communicate. Otherwise, terminals interoperate according to the previous standard.

5. Backward compatibility: A new standard is backward compatible with an older standard if the old standard can decode bit streams of the new standard. A very important backward-compatible standard was the introduction of analog color TV. Black-and-white receivers were able to receive the color TV signal and display a slightly degraded black-and-white version of the signal. Backward compatibility in today's digital video standards can be achieved by defining reserved bits in the bit stream that a decoder can ignore. A new standard would transmit extra information using these reserved bits of the old standard. Thus old terminals will be able to decode bit streams according to the new standard. Furthermore, they will understand those parts of the bit stream that comply to the old standard. Backward compatibility can put severe restrictions on the improvements that a new standard may achieve over its predecessor. Therefore, backward compatibility is not always implemented in a new standard.

6. Upward compatibility: A new receiver should be able to decode bit streams that were made for similar receivers of a previous or cheaper generation. Upward compatibility is important if an existing standard is extended. A new HDTV set

should be able to receive standard-definition TV, since both receivers use the same MPEG-2 standard [19].

7. Downward compatibility: An old receiver should be able to receive and decode bit streams for the newer generation receivers. Downward compatibility is important if an existing standard is extended. This may be achieved by decoding only parts of the bit stream, which is easily possible if the new bit stream is scalable (Chapter 11).

Obviously, not all of the above requirements are essential for the wide adoption of a standard. We believe that the most important requirements are innovation, competition, and forward compatibility, in that order. Compatibility is most important for devices like TV settop boxes or mobile phones, which cannot be upgraded easily. On the other hand, any multimedia PC today comes with more than ten software video codecs installed, relaxing the importance of compatible audio and video coding standards for this kind of terminal.

13.1.3 Standard Development Process

Video coding standards are developed in three phases, *competition, convergence,* and *verification.* Figure 13.3 shows the process for the video coding standard H.261 [41]. The competition phase started in 1984. The standard was published in December 1990 and revised in 1993.

During the competition phase, the application areas and requirements for the standard are defined. Furthermore, experts gather and demonstrate their best algorithms. Usually, the standardization body issues a *Call for Proposals* as soon as the requirements are defined, in order to solicit input from the entire community. This phase can be characterized by independently working competing laboratories.

The goal of the convergence phase is to collaboratively reach an agreement on the coding method. This process starts with a thorough evaluation of the proposals for the standard. Issues like coding efficiency, subjective quality, implementation complexity

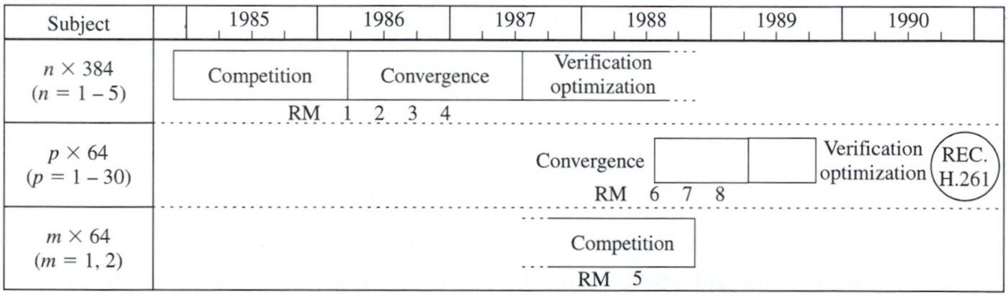

Subject	1985	1986	1987	1988	1989	1990	
$n \times 384$ ($n = 1-5$)	Competition	Convergence		Verification optimization			
	RM 1 2 3 4						
$p \times 64$ ($p = 1-30$)				Convergence		Verification optimization	REC. H.261
				RM 6 7 8			
$m \times 64$ ($m = 1, 2$)				Competition			
				RM 5			

Figure 13.3   Overview of the H.261 standardization process. Initially, the target was to develop two standards for video coding at rates of $n \times 384$ and $m \times 64$ kbps. Eventually, ITU settled on one standard at a rate of $p \times 64$ kbps. Reprinted from S. Okubo, Reference model methodology—a tool for the collaborative creation of video coding standards, *IEEE* (Feb. 1995), 83(2):139–50. Copyright 1995 IEEE.

and compatibility are considered when agreeing on a first common framework for the standard. This framework is implemented at different laboratories and the description is refined until the different implementations achieve identical results. This framework has different names in different standards, such as *Reference Model* (RM) for H.261, *Test Model Near term* (TMN) for H.263, *Simulation Model* (SM) for MPEG-1, *Test Model* (TM) for MPEG-2, *Verification Model* (VM) for MPEG-4, and *Test Model Long term* (TML) for H.26L. After the first version of the framework is implemented, researchers suggest improvements, such as new elements for the algorithm, or better parameters for existing elements. These are evaluated against the current framework. Proposals that achieve significant improvements are included in the next version of the framework, which serves as the new reference for further improvements. This process is repeated until the desired level of performance is achieved.

During the verification phase, the specification is checked for errors and ambiguities. *Conformance bit streams* and correctly decoded video sequences are generated. A standard-compliant decoder must decode every conformance bit stream into the correct video sequence.

The standardization process of H.261 can serve as a typical example (Figure 13.3). In 1985, the initial goal was to develop a video coding standard for bit rates between 384–1920 kbps. Due to the deployment of ISDN telephone lines, another standardization process for video coding at 64–128 kbps began two years later. In 1988, the two standardization groups realized that one algorithm could be used for coding video at rates between 64–1920 kbps. RM 6 was the first reference model that covered the entire bit-rate range. Technical work was finished in 1989, and one year later, H.261 was formally adopted by the ITU.

13.1.4 Applications for Modern Video Coding Standards

As mentioned, there have been several major initiatives in video coding, which have led to a range of video standards for different applications [12].

- Video coding for video teleconferencing led to the ITU standards H.261 for ISDN video conferencing [41]; H.263 for video conferencing over analog telephone lines, and desktop and mobile terminals connected to the Internet [50]; and H.262/MPEG-2 video [43, 17] for ATM/broadband video conferencing.

- Video coding for storing movies on CD-ROM and other consumer video applications, with about 1.2 mbps allocated to video coding and 256 kbps allocated to audio coding, led to the initial ISO MPEG-1 standard [16]. Today, MPEG-1 is used for consumer video on CD, Karoke machines, in some digital camcorders, and on the Internet. Some digital satellites used MPEG-1 to broadcast TV signals prior to the release of MPEG-2.

- Video coding for broadcast and for storing of digital video on DVD, with on the order of 2–15 mbps allocated to video and audio coding, led to the ISO MPEG-2 standard [19], and specifications for DVD operations by the Digital Audio Visual Council (DAVIC; www.davic.org) and the DVD consortium [25]. This work was extended to video coding for HDTV with bit rates ranging from 15–400 mbps allocated to video coding. Applications include satellite TV, cable TV, terrestrial broadcast, video editing, and storage. Today, MPEG-2 video is used in every digital settop box. It has also been selected as the video decoder for the American HDTV broadcast system.

- Coding of separate audiovisual objects, both natural and synthetic, is standardized in ISO MPEG-4 [22]. Target applications are Internet video, interactive video, content manipulation, professional video, 2-D and 3-D computer graphics, and mobile video communications.

In the following sections, we will first describe H.261; then, for H.263, we will highlight the differences from H.261 and compare their coding efficiency. Then, we will discuss MPEG-1, -2, and -4, again focusing on their differences.

13.2 VIDEO TELEPHONY WITH H.261 AND H.263

Video coding at 64 kbps was first demonstrated at a conference in 1979 [57]. However, it took more than ten years to define a commercially viable video coding standard at that rate. The standard H.261 was published in 1990, in order to enable video conferencing using 1–30 ISDN channels. At that time, video conferencing hardware became available from different vendors. Companies such as PictureTel, who sold video conferencing equipment with a proprietary algorithm, quickly offered H.261 as an option. Later, the ITU developed the similar standard H.263 that enables video communications over analog telephone lines. Today, H.263 video encoder and decoder software is installed on every PC with the Windows operating system.

13.2.1 H.261 Overview

Figure 13.4 shows the block diagram of an H.261 encoder that processes video in the 4:2:0 sampling format. It is a block-based hybrid coder with motion compensation (Section 9.3.1). It subdivides the image into macroblocks of size 16×16 pels. An MB consists of four luminance and two chrominance blocks (one for the Cr and another for the Cb component). H.261 uses an 8×8 DCT for each block to reduce spatial redundancy, a DPCM loop to exploit temporal redundancy, and unidirectional integer-pel forward motion compensation for MBs (box P in Figure 13.4) to improve the performance of the DPCM loop.

A simple two-dimensional loop filter (see Section 9.3.5) may be used to low-pass filter the motion-compensated prediction signal (box F in Figure 13.4). This usually decreases the prediction error and reduces the blockiness of the prediction image. The loop filter is separable into 1-D horizontal and vertical functions with the coefficients $[\frac{1}{4}, \frac{1}{2}, \frac{1}{4}]$. H.261 uses two quantizers for DCT coefficients. A uniform quantizer with stepsize eight is used in intra-mode for DC coefficients, and a nearly uniform midtread quantizer with the stepsize between two and 62 is used for AC coefficients in intra-mode and in inter-mode (Figure 13.5). The input between $-T$ and T, which is called the

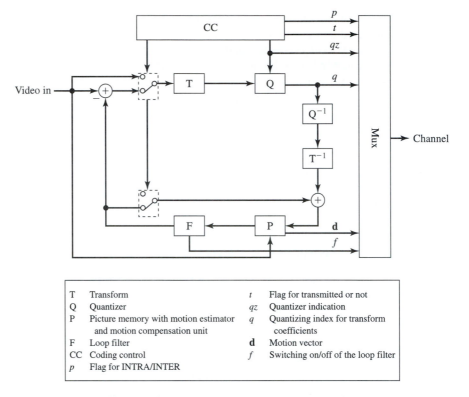

Figure 13.4   Block diagram of an H.261 encoder [41].

dead zone, is quantized to zero. Except for the dead zone, the stepsize is uniform. This dead zone avoids coding many small DCT coefficients that would mainly contribute to coding noise.

The encoder transmits mainly two classes of information for each MB that is coded: DCT coefficients resulting from the transform of the prediction error signal (q in Figure 13.4) and motion vectors that are estimated by the motion estimator (v and box P in Figure 13.4). The motion vector range is limited to ± 16 pels. The control data that tells the decoder whether and how an MB and its blocks are coded are named *macroblock type* (MTYPE) and *coded block pattern* (CBP). Table 13.1 shows the different MB types and their corresponding VLC codewords. In intra-mode, the bit stream contains transform coefficients for each block. Optionally, a change in the quantizer stepsize (MQUANT) of ± 2 levels can be signaled. In inter-mode, the encoder has a choice of just sending a differentially coded motion vector (MVD) with or without the loop filter on. Alternatively, a CBP may be transmitted in order to specify the blocks for which transform coefficients will be transmitted. Since the standard does not specify an encoder, it is up to the encoder vendor to decide on an efficient coding control (CC in Figure 13.4) to optimally select the MTYPE, CBP, MQUANT, loop filter, and motion vectors [70]. As a rough guideline, we can select MTYPE, CBP, and MVD such that the

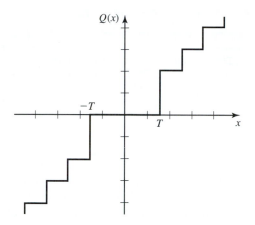

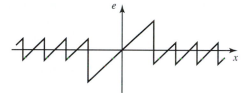

Figure 13.5 A midtread quantizer with a dead zone is used in H.261 for all DCT coefficients but the DC coefficient in the intra-mode. The bottom part shows the quantization error $e = x - Q(x)$ between the input amplitude x and the output amplitude $Q(x)$.

TABLE 13.1   VLC TABLE FOR MACROBLOCK TYPE*

Prediction	MQUANT	MVD	CBP	TCOEFF	VLC
Intra				x	0001
Intra	x			x	0000 001
Inter			x	x	1
Inter	x		x	x	0000 1
Inter + MC		x			0000 0000 1
Inter + MC		x	x	x	0000 0001
Inter + MC	x	x	x	x	0000 0000 01
Inter + MC + FIL		x			001
Inter + MC + FIL		x	x	x	01
Inter + MC + FIL	x	x	x	x	0000 01

*Two MTYPEs are used for intra-coded MBs and eight are used for inter-coded MBs. An "x" indicates that the syntactic element is transmitted for the MB [41].

prediction error is minimized. However, since the transmission of MVs costs extra bits, we do this only if the prediction error using the MV is significantly lower than without it. The quantizer stepsize is varied while coding the picture, such that the picture does not require more bits than the coder can transmit during the time between two coded frames. (Coding mode and parameter selection was discussed in Section 9.3.3.)

Most information within an MB is coded using a variable-length code that was derived from statistics of test sequences. Coefficients of the 2-D DCT are coded using

TABLE 13.2   PICTURE FORMATS SUPPORTED BY H.261 AND H.263

	Sub-QCIF	QCIF	CIF	4CIF	16CIF	Custom sizes
Lum Width (pels)	128	176	352	704	1408	<2048
Lum Height (pels)	96	144	288	576	1152	<1152
H.261		√	Opt.	Still		
H.263	√	√	Opt.		Opt.	Opt.

"Opt." = optional; "Still" = still picture.

the run-length coding method discussed in Section 9.1.7. Specifically, the quantized DCT coefficients are scanned using a zigzag scan (Figure 9.8) and converted into symbols (Figure 9.9). Each symbol includes the number of coefficients that were quantized to zero since the last nonzero coefficient, together with the amplitude of the current nonzero coefficient. Note that in the example depicted in Figure 9.9, the DC coefficient is coded separately, following the JPEG image coding method. With H.261 and other video coding standards, all the DCT coefficients are converted into (run-length, value) symbols. If the DC coefficient is nonzero, then the first run-length is zero. Thus, the first coding symbol in Figure 9.9 should be (0, 5). Each symbol is coded using VLC. The encoder sends an End Of Block (EOB) symbol after the last nonzero coefficient of a block.

H.261 does not specify the video encoder capabilities. However, the picture formats that an H.261 decoder must support are listed in Table 13.2. Several standards that set up video conferencing calls also exchange video capabilities between terminals. At a minimum level as defined in H.320, a decoder must be capable of decoding QCIF frames at a rate of 7.5 Hz [46]. An optional level of capability is defined as decoding CIF frames at 15 Hz [46]. The maximum level requires the decoding of CIF frames at 30 Hz (30,000/1001 Hz, to be precise) [46].

13.2.2 H.263 Highlights

The H.263 standard is based on the framework of H.261. Due to progress in video compression technology and the availability of high-performance desktop computers at reasonable cost, ITU decided to include more computationally intensive and efficient algorithms in the H.263 standard. The development of H.263 had three phases. The technical work for the initial standard was finished in November 1995. An extension of H.263, nicknamed H.263+, was incorporated into the standard in September 1997. The results of the third phase, nicknamed H.263++, were folded into the standard in 1999 and formally approved in November 2000. In this section, we focus on the differences between H.263 and H.261 as of 1995. We also briefly describe H.263 as of 2000.

H.263 Baseline (as of 1995) versus H.261   H.263 consists of a baseline decoder with features that any H.263 decoder must implement. In addition, optional features are defined. The following mandatory features distinguish H.263 as defined in November 1995, from H.261 [6, 12]:

1. Half-pixel motion compensation: This feature significantly improves the prediction capability of the motion compensation algorithm in cases where there is

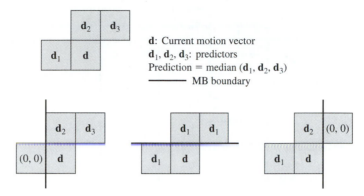

Figure 13.6 Prediction of motion vectors uses the median vector of d_1, d_2, and d_3. We assume the motion vector is zero if one MB is outside the picture or group of blocks. If two MBs are outside, we use the remaining motion vector as prediction.

object motion that needs fine spatial resolution for accurate modeling. Bilinear interpolation (simple averaging) is used to compute the predicted pels in case of noninteger motion vectors. The coding of motion vectors uses the median motion vector of the three neighboring MBs as prediction for each component of the vector (Figure 13.6).

2. Improved variable-length coding, including a 3-D VLC for improved efficiency in coding DCT coefficients. Whereas H.261 codes the symbols (run, level) and sends an EOB word at the end of each block, H.263 integrates the EOB word into the VLC. The events to be coded are (last, run, level), where last indicates whether the coefficient is the last nonzero coefficient in the block.

3. Reduced overhead at the Group of Block level as well as coding of MTYPE and CBP.

4. Support for more picture formats (Table 13.2).

In addition to these improvements, H.263 offers a list of *optional* features that are defined in annexes to the standard.

1. Unrestricted motion vectors (Annex D), which are allowed to point outside the picture, improve coding efficiency in case of camera motion or motion at the picture boundary. The prediction signal for a motion vector that points outside of the image is generated by repeating the boundary pels of the image. The motion vector range is extended to $[-31.5, 31]$.

2. Syntax-based arithmetic coding (Annex E) may be used in place of variable-length (Huffman) coding, resulting in the same decoded pictures at an average bit-rate saving of 4 percent for P frames and 10 percent for I frames. However, decoder computational requirements increase by more than 50 percent [10]. This will limit the number of manufacturers implementing this annex.*

---

*This Annex uses a suboptimal implementation of arithmetic coding.

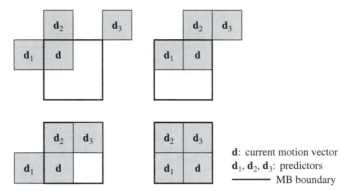

Figure 13.7   Prediction of motion vectors in advanced prediction mode, with four motion vectors in an MB. The predicted value for a component of the current motion vector *d* is the median of its predictors.

3. Advanced prediction mode (Annex F) includes unrestricted motion vector mode. Advanced prediction mode allows for two additional improvements: Overlapped block motion compensation (OBMC) may be used to predict the luminance component of a picture, which improves prediction performance and reduces significantly blocking artifacts (see Section 9.3.2) [59]. Each pixel in an 8×8 luminance prediction block is a weighted sum of three predicted values computed from three motion vectors: the vector of the current MB and the vectors of the two MBs that are closest to the current 8×8 block. The weighting coefficients and the equivalent window function for motion estimation were given in Figures 9.16–17.

   The second improvement of advanced motion prediction is the optional use of four motion vectors for an MB, one for each luminance block. This enables better modeling of motion in real images. However, it is up to the encoder to decide in which MB the benefit of four motion vectors is sufficient to justify the extra bits required for coding them. Again, the motion vectors are coded predictively (Figure 13.7).

4. *PB pictures* (Annex G) is a mode that codes a bidirectionally predicted picture with a normal forward-predicted picture. The B-picture temporally precedes the P-picture of the PB picture. In contrast to bidirectional prediction (Section 9.2.4, Figure 9.12), which is computed on a frame by frame basis, PB pictures use bidirectional prediction on an MB level. In a PB frame, the number of blocks per MB is twelve rather than six. Within each MB, the six blocks belonging to the P-picture are transmitted first, followed by the blocks of the B-picture (Figure 13.8). Bidirectional prediction is derived from the previous decoded frame and the P-blocks of the current MB. As seen in Figure 13.8, this limits backward predictions to those pels of the B-blocks that are aligned to pels inside the current P-macroblock in the case of motion between the B-picture and the P-picture (light grey area in Figure 13.8). For the light grey area of the B-block, the prediction is

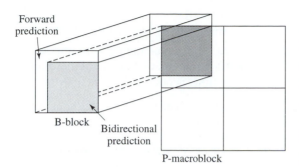

Figure 13.8   In a PB frame, forward prediction can be used for all B-blocks; backward prediction is only used for those pels that the backward motion vector aligns with pels of the current MB. Reprinted from B. Girod, E. Steinback, and N. Färber, Comparison of the H.263 and H.261 video compression standards, in *Standards and Common Interfaces for Video,* SPIE vol. CR60, 233–51, by permission of SPIE.

computed by averaging the results of forward and backward prediction. Pels in the white area of the B-block are predicted using forward motion compensation only. An *improved PB-frame mode* (Annex M) was later adopted, which removes this restriction, enabling the efficiency of regular B-frames (Section 9.2.4).

PB-pictures are efficient for coding image sequences with moderate motion. They tend not to work very well for scenes with fast or complex motion or when coding at low frame rates. Since the picture quality of a B-picture has no effect on the coding of subsequent frames, H.263 defines that the B-picture of a PB-picture set is coded at a lower quality than the P-picture by using a smaller quantizer stepsize for P-blocks than for the associated B-blocks. PB-pictures increase the delay of a coding system, since they allow the encoder to send bits for the B-frame only after the following P-frame has been captured and processed. This limits their usefulness for interactive real-time applications.

Due to the larger number of coding modes, the encoder decisions are more complex than in H.261. An RD-optimized H.263 encoder with the options Unrestricted Motion Vector Mode and Advanced Prediction was compared to TMN5, the test model encoder used during standards development [70]. The optimized encoder increases the PSNR 0.5–1.2 dB over TMN5 at bitrates between 20–70 kbps.

H.263 as of 2000   After the initial version of H.263 was approved, work continued, with further optional modes being added. However, since these more than fifteen modes are optional, it is questionable that any manufacturer will implement all of them. The ITU recognized this, and added to the standard recommendations for preferred modes. The most important preferred modes not previously are listed here.

1. Advanced Intra Coding (Annex I): Intra-blocks are coded using the block to the left or above as a prediction, provided that block is also coded in intra-mode. This mode increases coding efficiency by 10–15 percent for I pictures.

2. Deblocking filter (Annex J): An adaptive filter is applied across the boundaries of decoded 8×8 blocks to reduce blocking artifacts. This filter affects the predicted image also, and is implemented inside the prediction loop of coder and decoder.

3. Supplemental Enhancement Information (Annex L): This information may be used to provide tagging information for external use, as defined by an application

using H.263. Furthermore, it can be used to signal enhanced display capabilities such as frame freeze, zoom, or chroma-keying (see Section 10.3).

4. Improved PB-frame Mode (Annex M): As mentioned, this mode removes the restrictions placed on the backward prediction of Annex G. Therefore, this mode enables regular bidirectional prediction (Section 9.2.4).

These tools have been developed for enhancing coding efficiency. In order to enable transport of H.263 video over unreliable networks (such as wireless networks and the Internet), a set of tools also have been developed for the purpose of error resilience. These are included in Annex H (Forward Error Correction Using BCH Code), Annex K (Flexible Synchronization Marker Insertion Using the Slice Structured Mode), Annexes N and U (Reference Picture Selection), Annex O (Scalability), Annex R (Independent Segment Decoding), Annex V (Data Partitioning and RVLC), and Annex W (Header Repetition). These tools are described in Chapter 14. Further discussion of H.263 can be found in [6] and in the standard itself [50].

13.2.3 Comparison

Figure 13.9 compares the performance of H.261 and H.263 [10]. H.261 is shown with and without using the filter in the loop (Curves 3 and 5). Since H.261 was designed for data rates of 64 kbps and up, we discuss Figure 13.9 at this rate. Without options, H.263 outperforms H.261 by 2 dB (Curves 2 and 3). Another 1 dB is gained if we

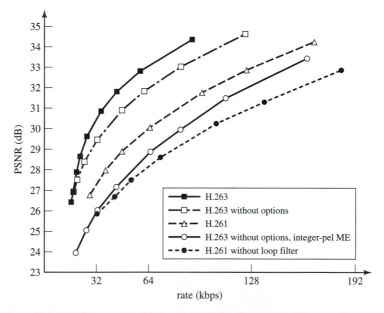

Figure 13.9   Performance of H.261 and H.263 for the sequence "Foreman" at QCIF and 12.5 Hz. Reprinted from B. Girod, E. Steinback, and N. Färber, Comparison of the H.263 and H.261 video compression standards, in *Standards and Common Interfaces for Video,* SPIE vol. CR60, 233–51, by permission of SPIE.

use the options of advanced prediction, syntax-based arithmetic coding, and PB frames (Curve 1). Curve 4 shows that restricting motion vectors in H.263 to integer pel reduces coding efficiency by 3 dB. This is due to the reduced motion compensation accuracy and the lack of the low-pass filter that bilinear interpolation brings for half-pel motion vectors. Comparing curves 3 and 5 shows the effect of this low-pass filter on coding efficiency. The differences between curves 4 and 5 are mainly due to the 3-D VLC for coding of transform coefficients, as well as improvements in coding MTYPE and CBP.

13.3 STANDARDS FOR VISUAL COMMUNICATION SYSTEMS

In order to enable useful audiovisual communications, terminals must establish a common communication channel, exchange their capabilities, and agree on the standards for exchanging audiovisual information. In other words, we need much more than just an audio and video codec to enable audiovisual communication. The setup of communication between a server and a client over a network is handled by a *systems* standard. The ITU-T developed several system standards (including H.323 and H.324) to enable bidirectional multimedia communications over different networks, several audio coding standards for audio communications, and the two important video coding standards H.261 and H.263. Table 13.3 provides an overview of the standards for audio, video, multiplexing, and call control that these system standards use [5]. In the following, we briefly describe the functionality of the recent standards H.323 [51] and H.324 [44].

13.3.1 H.323 Multimedia Terminals

Recommendation H.323 [51] provides the technical requirements for multimedia communication systems that operate over packet-based networks like the Internet, where guaranteed quality of service is usually not available.

Figure 13.10 shows the different protocols and standards that H.323 requires for video conferencing over packet networks. An H.323 call scenario optionally starts with a gatekeeper admission request(H.225.0 RAS, [48]). Then, call signaling establishes the connection between the communicating terminals (H.225.0, [48]). Next, a communication channel is established for call control and capability exchange (H.245, [49]).

TABLE 13.3   ITU-T MULTIMEDIA COMMUNICATION STANDARDS

Network	System	Video	Audio	Mux	Control
PSTN	H.324	H.261/3	G.723.1	H.223	H.245
N-ISDN	H.320	H.261	G.7xx	H.221	H.242
B-ISDN/ATM	H.321	H.261	G.7xx	H.221	Q.2931
	H.310	H.261/2	G.7xx/MPEG	H.222.0/1	H.245
QoS LAN	H.322	H.261/3	G.7xx	H.221	H.242
Non-QoS LAN	H.323	H.261	G.7xx	H.225.0	H.245

PSTN: Public Switched Telephone Network; N-ISDN: Narrow band ISDN (2 × 64 kbps); B-ISDN: Broadband ISDN; ATM: Asynchronous Transfer Mode; QoS: guaranteed quality of service; LAN: local area network; H.262 is identical to MPEG-2 Video [43, 17]; G.7xx represents G.711, G.722, and G.728.

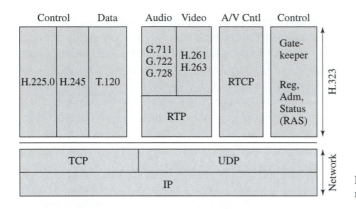

Figure 13.10   H.323 protocols for multimedia communications over TCP/IP.

Finally, the media flow is established using RTP and its associated control protocol RTCP [65]. A terminal may support several audio and video codecs, however, the support of G.711 audio (64 kbps) [36] is mandatory. G.711 is the standard currently used in the Public Switched Telephone Network (PSTN) for digital transmission of telephone calls. If a terminal claims to have video capabilities, it must include at least an H.261 video codec [41] with a spatial resolution of QCIF. Modern H.323 video terminals usually use H.263 [50] for video communications.

13.3.2 H.324 Multimedia Terminals

H.324 [44] differs from H.323 in that it enables the same communication over networks with guaranteed quality of service as is available when using V.34 [42] modems over

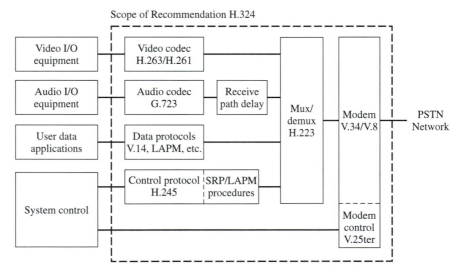

Figure 13.11   Block diagram of an H.324 multimedia communication system over the PSTN.

the PSTN. H.324 may optionally support the media types voice, data, and video. If a terminal supports one or more of these media, it uses the same audiovisual codecs as H.323. However, it also supports H.263 for video and G.723.1 [38] for audio at 5.3 and 6.3 kbps. Audio quality of a G.723.1 codec at 6.3 kbps is very close to that of a regular phone call. Call control is handled using H.245. Transmission of these different media types over the PSTN requires the media to be multiplexed (Figure 13.11) following the multiplexing standard H.223 [45]. The multiplexed data is sent to the PSTN via a V.34 modem, using the V.8 or V.8bis procedure [53, 52] to start and stop transmission. The modem control V.25ter [47] is used, if the H.324 terminal uses an external modem.

13.4 CONSUMER VIDEO COMMUNICATIONS WITH MPEG-1

The MPEG standards were developed by ISO/IEC JTC1 SC29/WG11, which is chaired by Leonardo Chiariglione. MPEG-1 was designed for progressively scanned video used in multimedia applications, and the target was to produce near-VHS quality video at a bit rate of around 1.2 mbps (1.5 mbps including audio and data). It was foreseen that much multimedia content would be distributed on CD-ROM. At the time of MPEG-1 development, 1.5 mbps was the access rate of CD-ROM players. The video format is SIF. The final standard supports higher rates and larger image sizes. Another important consideration when developing MPEG-1 was the implementation of functions that support basic VCR-like interactivity—such as fast forward, fast reverse, and random access into the stored bit stream at every half-second [55].

13.4.1 Overview

The MPEG-1 standard, formally known as ISO 11172 [16], consists of five parts: Systems, Video, Audio, Conformance, and Software.

MPEG-1 Systems provides a packet structure for combining coded audio and video data. It enables the system to multiplex several audio and video streams into one stream that allows synchronous playback of the individual streams. This requires all streams to refer to a common system time clock (STC). From this STC, presentation time stamps (PTS), which define the instant when a particular audio or video frame should be presented on the terminal, are derived. Since video coded with B-frames requires a reordering of decoded images, decoding time stamps (DTS) are used to indicate by when a certain image must be decoded.

MPEG-1 Audio is a generic standard that does not make any assumptions about the nature of the audio source. However, audio coding exploits perceptual limitations of the human auditory system for irrelevancy reduction. MPEG-1 audio is defined in three layers, I, II, and III. Higher layers have higher coding efficiency and require increased resources for decoding. Layer III especially was controversial due to its computational complexity at the time of standardization in the early 1990s. However, it is precisely this Layer III MPEG-1 Audio codec that is now known to every music fan as MP3. The reasons for its popularity are sound quality, coding efficiency—and above all—the fact

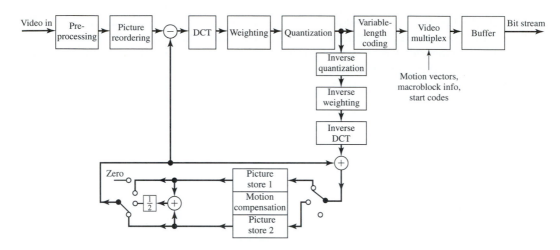

Figure 13.12 Block diagram of an MPEG-1 encoder.

that for a limited time the proprietary high-quality encoder source code was available for download on a company's web site. This started the revolution within the music industry (see Section 13.1.2 for criteria of a successful standard).

13.4.2 MPEG-1 Video

The MPEG-1 Video convergence phase started after subjective tests in October 1989, and resulted in the standard published in 1993. Since H.261 was published in 1990, there are many similarities between H.261 and MPEG-1. Figure 13.12 shows the conceptual block diagram of an MPEG-1 coder. Compared to H.261 (Figure 13.4), we notice the following differences:

1. The loop filter is gone. Since MPEG-1 uses motion vectors with half-pel accuracy, there is no need for the filter (see Section 13.2.3). The motion vector range is extended to ±64 pels.

2. MPEG-1 uses I-, P-, and B-frames. The use of the B-frames requires a more complex motion estimator and motion compensation unit. Motion vectors for B-frames are estimated with respect to two reference frames, the preceding I- or P-frame and the next I- or P-frame. Hence, we can associate two motion vectors with each MB of a B-frame. For motion-compensated prediction, we now need two frame stores for these two reference pictures. The prediction mode for B-frames is decided for each MB. Furthermore, the coding order is different from the scan order (see Figure 9.12) and, therefore, we need a picture reordering unit at the input to the encoder and at the decoder.

3. For I-frames, quantization of DCT coefficients is adapted to the human visual system by dividing the coefficients with a weight matrix. Figure 13.13 shows the default table. Larger weights result in a coarser quantization of the coefficient. It

$w_{u,v}$	0	1	2	3	4	5	6	8
0	8	16	19	22	26	27	29	34
1	16	16	22	24	27	29	34	37
2	19	22	26	27	29	34	34	38
3	22	22	26	27	29	34	37	40
4	22	26	27	29	32	35	40	48
5	26	27	29	32	35	40	48	58
6	26	27	29	34	38	46	56	69
7	27	29	35	38	46	56	69	83

Figure 13.13   Default weights for quantization of I-blocks in MPEG-1. Weights for horizontal and vertical frequencies differ.

can be seen that the weights increase with the frequency that a coefficient represents. When comparing a coder with and without the weight matrix at identical bit rates, we notice that the weight matrix reduces the PSNR of decoded pictures but increases the subjective quality.

Another difference from H.261 is that the DC coefficient of an I-block may be predicted from the DCT coefficient of its neighbor to the left. This concept was later extended in JPEG [15, 40], H.263, and MPEG-4.

MPEG-1 uses a group of pictures (GOP) structure (Figure 9.12). Each GOP starts with an I-frame followed by a number of P- and B-frames. This enables random access to the video stream as well as the VCR-like functionalities of fast forward and reverse.

Because of the large range of bit-stream characteristics supported by the standard, a special subset of the coding parameters, known as the *constrained parameters set* (CPS), has been defined (Table 13.4). CPS is a limited set of sampling and bit rate parameters designed to limit computational decoder complexity, buffer size, and memory bandwidth, while still addressing the widest possible range of applications. A decoder implemented with the CPS in mind needs only 4 Mb of DRAM while supporting SIF and CIF. A flag in the bit stream indicates whether it is CPS.

Compared to an analog consumer-quality VCR, MPEG-1 codes video with only half the number of scan lines. At a video bit rate of 1.8 mbps, however, it is possible for a good encoder to deliver video quality that exceeds that of a video recorded by an analog consumer VCR onto used video tape.

TABLE 13.4   CONSTRAINED PARAMETER SET FOR MPEG-1 VIDEO

Parameter	Maximum value
pixels/line	768 pels
lines/picture	576 lines
number of MBs per picture	396 MBs
number of MBs per second	$396 \cdot 25 = 330 \cdot 30 = 9900$
input buffer size	327,680 bytes
motion vector component	± 64 pels
bit rate	1.856 mbps

13.5 DIGITAL TV WITH MPEG-2

Towards the end of the MPEG-1 standardization process, it became obvious that MPEG-1 would not be able to efficiently compress interlaced digital video at broadcast quality. Therefore, the MPEG group issued a call for proposals to submit technology for digital coding of audio and video for TV broadcast applications. The best-performing algorithms were extensions of MPEG-1 to deal with interlaced video formats. During the collaborative phase of the algorithm development, a great deal of similarity with MPEG-1 was maintained.

The main purpose of MPEG-2 is to enable MPEG-1–like functionality for inter-laced pictures, primarily using the ITU-R BT.601 (formerly CCIR601) 4:2:0 format [35]. The target was to produce TV-quality pictures at data rates of 4–8 mbps and high quality pictures at 10–15 mbps. MPEG-2 deals with high-quality coding of possibly interlaced video, of either SDTV or HDTV. A wide range of applications, bit rates, resolutions, signal qualities, and services are addressed, including all forms of digital storage media, television (including HDTV) broadcasting, and communications [13].

The MPEG-2 standard [19] consists of nine parts: Systems, Audio, Video, Conformance, Software, Digital Storage Media—Command and Control (DSM-CC), Advanced Audio Coder (AAC), Real Time Interface, and DSM-CC Conformance. In this section, we provide a brief overview MPEG-2 systems, audio, and video, and the MPEG-2 concept of *profiles*.

13.5.1 Systems

MPEG-2 systems must be somewhat compatible with MPEG-1 systems, be error resilient, support transport over ATM networks, and transport more than one TV program in a stream without requiring a common time base for the programs. An MPEG-2 *program stream* (PS) is forward compatible with MPEG-1 system stream decoders. A PS contains compressed data from a single program, in packets of variable length, usually between 1, 2 Kb, and up to 64 Kb. The MPEG-2 *transport stream* (TS) is not compatible with MPEG-1. A TS offers error resilience as required for cable TV networks or satellite TV, uses packets of 188 bytes, and may carry several programs with independent time bases that can be easily accessed for channel hopping.

13.5.2 Audio

MPEG-2 audio comes in two parts: In part 3 of the standard, MPEG defines a forward- and backward-compatible audio format that supports five-channel surround sound. The syntax is designed such that a MPEG-1 audio decoder is able to reproduce a meaningful downmix out of the five channels of an MPEG-2 audio bit stream [18]. In part 7, the more efficient multichannel audio decoder, MPEG-2 AAC, with sound effects and many other features, is defined [20]. MPEG-2 AAC requires 30 percent fewer bits than MPEG-1 Layer III Audio for the same stereo sound quality. AAC has been adopted by the Japanese broadcasting industry; it is not popular as a format for the Internet, however, because no "free" encoder is available.

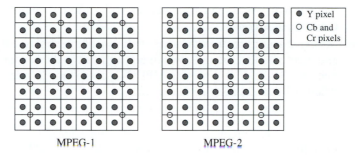

Figure 13.14   Luminance and chrominance samples in a 4:2:0 progressive frame.

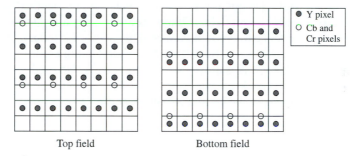

Figure 13.15   Luminance and chrominance samples in a 4:2:0 interlaced frame, in which the top field is temporally first.

13.5.3 Video

MPEG-2 is targeted at TV studios and TV broadcasting for standard TV and HDTV. As a consequence, it must support efficiently the coding of interlaced video at bit rates adequate for the applications. The major differences between MPEG-1 and MPEG-2 are the following:

1. Chroma samples in the 4:2:0 format are located horizontally shifted by 0.5 pels compared to MPEG-1, H.261, and H.263 (Figure 13.14).
2. MPEG-2 is able to code interlaced sequences in the 4:2:0 format (Figure 13.15).
3. As a consequence of the preceding, MPEG-2 allows additional scan patterns for DCT coefficients and motion compensation with blocks of size 16×8 pels.
4. Several differences—for example, ten-bit quantization of the DC coefficient of the DCT, nonlinear quantization, and better VLC tables—improve coding efficiency also for progressive video sequences.
5. MPEG-2 supports varied modes of scalability. Spatial scalability enables different decoders to extract videos of different picture sizes from the same bit stream. MPEG-2 supports temporal scalability, such that a bit stream can be decoded into video sequences of different frame rates. Furthermore, SNR scalability provides

the ability to extract video sequences with different amplitude resolutions from the same bit stream.

6. MPEG-2 defines profiles and levels that constitute subsets of the MPEG-2 features and their parameter ranges, which are signaled in the header of a bit stream (see Section 13.5.4). In this way, an MPEG-2–compliant decoder knows immediately whether it can decode the bit stream.

7. MPEG-2 allows for much higher bit rates (see Section 13.5.4).

In the following, we will discuss the extensions introduced to support interlaced video and scalability.

Coding of Interlaced Video   Interlaced video is a sequence of alternating top and bottom fields (see Section 1.3.1). Two fields are of identical *parity* if they are both top fields or both bottom fields. Otherwise, two fields are said to have opposite parity. MPEG-2 considers two types of *picture structures* for interlaced video (Figure 13.16). A *frame picture* consists of lines from the top and bottom fields of an interlaced picture in an interlaced order. This frame-picture structure is also used when coding progressive video. A *field picture* keeps the top and the bottom fields of the picture separate. For each of these pictures, I-, P-, and B-picture coding modes are available.

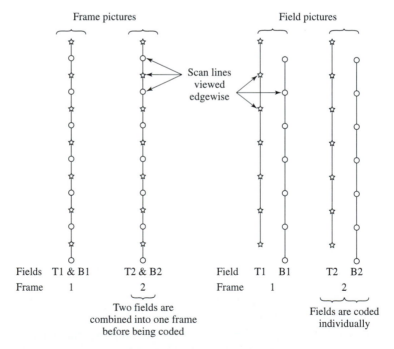

Figure 13.16   Frame and field picture structures (side views of the individual fields): each frame consists of a top and a bottom field; either one may be temporally first.

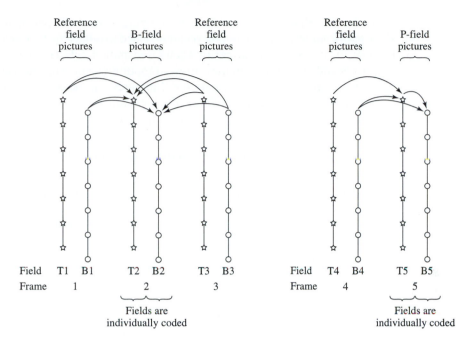

Figure 13.17   Every MB relevant for field prediction for field pictures is located within one field of the reference picture. Pictures may have different parity.

MPEG-2 adds new prediction modes for motion compensation, all related to interlaced video:

1. *Field prediction for field pictures* is used to predict an MB in a field picture. For P-fields, the prediction may come from either field of the two most recently coded fields. For B-fields, we use the two fields of the two reference pictures (Figure 13.17).

2. *Field prediction for frame pictures* splits an MB of the frame into the pels of the top field and those of the bottom field, resulting in two 16×8 field blocks (Figure 13.18). Each field block is predicted independent of the other, similarly to the method just described for field pictures. This prediction method is especially useful for rapid motion.

3. *Dual prime for P-pictures* transmits one motion vector per MB that can be used for predicting field and frame pictures from the preceding P- or I-picture. The target MB is represented as two field blocks. The coder computes two predictions for each field block and averages them. The first prediction of each field block is computed by performing motion compensation, using the transmitted motion vector and the field with the same parity as the reference. The second prediction of each field block is computed using a *corrected motion vector* and the field with the opposite parity as reference. The corrected motion vector is computed

assuming linear motion. Considering the temporal distance between the fields of same parity, the transmitted motion vector is scaled to reflect the temporal distance between the fields of opposite parity. Then a transmitted *differential motion vector* (DMV) is added, resulting in the corrected motion vector. For interlaced video, this prediction mode of dual prime for P-pictures can be as efficient as using B-pictures—without adding the delay of a B-picture.

4. *16 × 8 MC for field pictures* corresponds to field prediction for frame pictures. Within an MB, the pels belonging to different fields have their own motion vectors for motion compensation; that is, two motion vectors are transmitted for P-pictures and four for B-pictures.

These many choices for prediction makes the design of an optimal encoder obviously very challenging.

In interlaced video, neighboring rows in an MB come from different fields, thus the vertical correlation between lines is reduced when the underlying scene contains motion with a vertical component. MPEG-2 provides two new coding modes to increase the efficiency of prediction error coding:

1. *Field DCT* reorganizes the pels of an MB into two blocks for the top field and two blocks for the bottom field (Figure 13.18). This increases the correlation within a block in case of motion, and thus increases the coding efficiency.

2. MPEG-2 provides an *alternate scan* that the encoder may select on a picture-by-picture basis. This scan puts coefficients with high vertical frequencies earlier than the zigzag scan. Figure 13.19 compares the new scan to the conventional zigzag scan.

Scalability in MPEG-2   The MPEG-2 functionality described so far is achieved with the nonscalable syntax of MPEG-2, which is a superset of MPEG-1. The scalable syntax structures the bit stream in layers. The base layer can use the nonscalable syntax

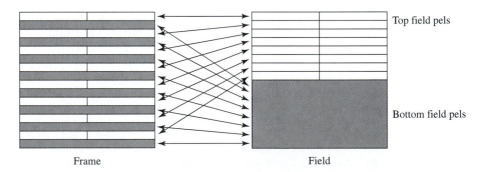

Frame   Field

Figure 13.18   Field prediction for frame pictures: the MB to be predicted is split into top field pels and bottom field pels. Each 16 × 8 field block is predicted separately with its own motion vector (P-frame) or two motion vectors (B-frame).

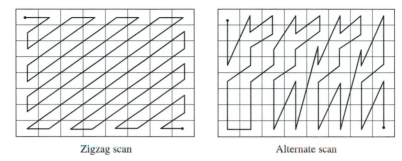

Zigzag scan	Alternate scan

Figure 13.19   The zigzag scan as known from H.261, H.263, and MPEG-1 is augmented by the alternate scan in MPEG-2, in order to code interlaced blocks that have more correlation in the horizontal than in the vertical direction.

and thus be decoded by an MPEG-2 terminal that does not understand the scalable syntax. The basic MPEG-2 scalability tools are *data partitioning, SNR scalability, spatial scalability* and *temporal scalability* (see Section 11.1). Combinations of these basic scalability tools are also supported.

When using scalable codecs, *drift* may occur in a decoder that decodes the base layer only. Drift is created if the reference pictures used for motion compensation at the encoder and the base-layer decoder differ. This happens if the encoder uses information of the enhancement layer when computing the reference picture for the base layer. The drift is automatically set to zero at every I-frame. Drift does not occur in scalable codecs if the encoder does not use any information of the enhancement layer for coding the base layer. Furthermore, a decoder decoding layers in addition to the base layer may not introduce data from the upper layers into the decoding of the lower layers.

Data Partitioning: Data partitioning splits the video bit stream into two or more layers. The encoder decides which syntactic elements are placed into the high-priority base layer and which into the low-priority enhancement layers. Typically, high-frequency DCT coefficients are transmitted in the low-priority enhancement layer, whereas all headers, side information, motion vectors, and the first few DCT coefficients are transmitted in the high-priority base layer. Data partitioning is appropriate when two transmission channels are available. Due to data partitioning, the decoder can decode the base layer only if the decoder implements a bit stream loss concealer for the higher layers. This concealer can be as simple as setting to zero the missing higher-order DCT coefficients in the enhancement layer. Figure 13.20 shows a high-level view of the encoder and decoder. The data partitioning functionality may be implemented independent of the encoder and decoder. Data partitioning does not incur any noticeable overhead. However, its performance in an error-prone environment may be poor compared to other methods of scalability [13]. Obviously, we will encounter the drift problem if we decode only the base layer.

Data partitioning encoder   Data partitioning decoder

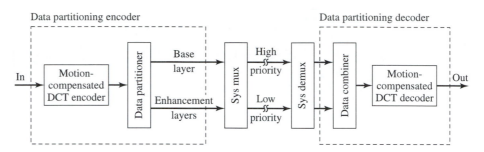

Figure 13.20   A data partitioning codec suited for ATM networks that support two degrees of quality of service.

SNR Scalability: SNR scalability is a frequency-domain method in which all layers are coded with the same spatial resolution, but with differing picture quality achieved through different MB quantization stepsizes. The lower layer provides the basic video quality, whereas the enhancement layer carries the information that, when added to the lower layer, generates a higher-quality reproduction of the input video. Figure 13.21 shows an SNR scalable coder, which includes a nonscalable base encoder. The base encoder feeds the DCT coefficients, after transform and quantization, into the SNR enhancement coder. The enhancement coder requantizes the quantization error of the base encoder, and feeds the coefficients that it sends to the SNR enhancement decoder back into the base encoder, which adds them to its dequantized coefficients and to the encoder feedback loop. Due to the feedback of the enhancement layer at the encoder, drift occurs for any decoder that decodes only the base layer.

At a total bit rate between 4–9 mbps, the combined picture quality of base and enhancement layers is 0.5–1.1 dB less than that obtained with nonscalable coding. Obviously, SNR scalability outperforms data partitioning in terms of picture quality for the base layer [61, 13].

Spatial Scalability: In MPEG-2, spatial scalability is achieved by combining two complete encoders at the transmitter and two complete decoders at the receiver. The base layer is coded at low spatial resolution using a motion-compensated DCT encoder such as H.261, MPEG-1, or MPEG-2 (Figure 13.22). The image in the frame store of the feedback loop of this base encoder is made available to the spatial enhancement encoder. This enhancement coder is also a motion-compensated DCT encoder that codes the input sequence at the high resolution. It uses the upsampled input from the lower layer to enhance its temporal prediction. The prediction image in the enhancement layer coder is the weighted sum of the temporal prediction image of the enhancement coder with the spatial prediction image from the base encoder. Weights may be adapted on an MB level. There are no drift problems with this coder, since neither the encoder nor the decoder introduces information of the enhancement layer into the base layer. At a total bit rate of 4 mbps, the combined picture quality of base and enhancement layers is 0.75–1.5 dB less than that obtained with nonscalable coding [13].

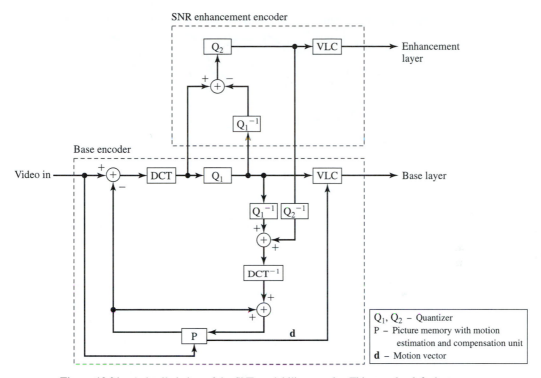

Figure 13.21   A detailed view of the SNR scalability encoder. This encoder defaults to a standard encoder if the enhancement encoder is removed.

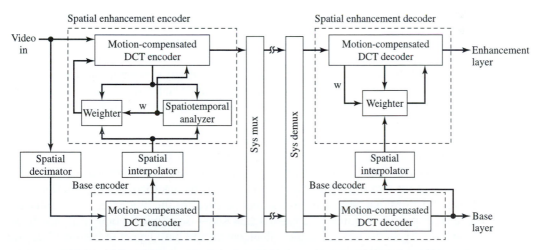

Figure 13.22   An encoder with spatial scalability consists of two complete encoders that are connected using a spatial interpolation filter.

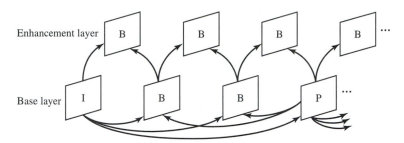

Figure 13.23   A configuration in which temporal scalability may use only the base layer to predict images in the enhancement layer. Obviously, errors in the enhancement layers to do not propagate over time.

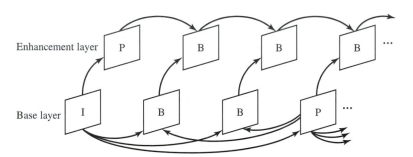

Figure 13.24   A configuration in which temporal scalability enhancement layer may use the base layer and the enhancement layer for prediction. This arrangement is especially useful for coding of stereoscopic video.

Compared to simulcast (i.e., sending two independent bit streams, one having the base-layer and the other the enhancement-layer resolution), spatial scalability is more efficient by 0.5–1.25 dB [13, 62]. Spatial scalability is the appropriate tool to be used in applications in which interoperability of video standards is necessary, and the increased coding efficiency (compared to simulcasting) is able to offset the extra cost for complexity of encoders and decoders.

Temporal Scalability: In temporal scalability, the base layer is coded at a lower frame rate using a nonscalable codec, and the intermediate frames can be coded in a second bit stream, using the first bit stream reconstruction as prediction [63]. MPEG-2 defines that only two frames may be used for the prediction of an enhancement layer picture. Figures 13.23 and 13.24 show two typical configurations. If we mentally collapse the images of the enhancement and base layers in Figure 13.23, we notice that the resulting sequence of images and the prediction arrangement are similar to a nonscalable coder—and identical to a nonscalable coder if the base layer uses only I- and P-frames. Accordingly, the picture quality of temporal scalability is only

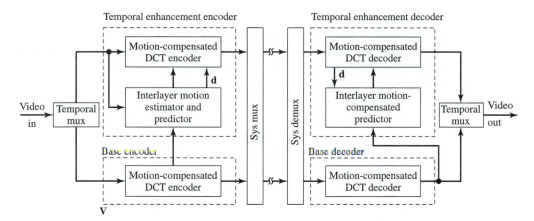

Figure 13.25   A temporal scalability encoder consists of two complete encoders, with the enhancement encoder using the base-layer video as an additional reference for prediction. The temporal demux sends pictures alternately to the base encoder and the enhancement encoder.

0.2–0.3 dB lower than a nonscalable coder [13]. In Figure 13.25, we see that enhancement- and base-layer encoders are two complete codecs that both operate at half the rate of the video sequence. Therefore, the computational complexity of temporal scalability is similar to a nonscalable coder operating at the full frequency of the input sequence; there are no drift problems.

Temporal scalability is an efficient means of distributing video to terminals with different computational capabilities (for example, a mobile terminal and a desktop PC). Another application is stereoscopic video transmission, in which right and left channels are transmitted as the enhancement and base layers, respectively. This was discussed in Section 12.5.1.

13.5.4 Profiles

The full MPEG-2 syntax covers a wide range of features and parameters. Extending the MPEG-1 concept of a constrained parameter set (Table 13.4), MPEG-2 defines *profiles* that describe the tools required for decoding a bit stream and *levels* that describe the parameter ranges for these tools. MPEG-2 initially defined five profiles for video, each adding new tools in a hierarchical fashion. Later, two profiles that do not fit the hierarchical scheme were added.

1. The *Simple* profile supports I and P frames, the 4:2:0 format, and no scalability. It is currently not used in the market.
2. The *Main* profile adds support for B-frames. The Main profile at Main level (MP@ML) is used for TV broadcasting. This profile is the most widely used.
3. The *SNR* profile supports SNR scalability in addition to the functionality of the Main profile. It is currently not used in the market.

4. The *Spatial* profile supports the functionality of the SNR profile and adds spatial scalability. It is currently not used in the market.

5. The *High* profile supports the functionality of the Spatial profile and adds support for the 4:2:2 format. This profile is far too complex to be useful.

6. The *4:2:2* profile supports studio postproduction and high-quality video for storage and distribution. It basically extends the main profile to higher bit rates and quality. The preferred frame order in a group of frames is IBIBIBIBI. ... Equipment with this profile is used in digital studios.

7. The *Multiview* profile enables the transmission of several video streams in parallel, thus enabling stereo presentations. This functionality is implemented using temporal scalability, thus enabling Main profile decoders to receive one of the video streams. Prototypes exist.

For each profile, MPEG-2, defined levels. Levels essentially define the size of the video frames, the frame rate, and picture types, thus providing an upper limit for the processing power of a decoder. Table 13.5 shows the levels defined for most profiles. The fact that only two fields in Table 13.5 are used in the market (MP@ML and 4:2:2@ML) is a strong indication that standardization is a consensus process—MPEG

TABLE 13.5 PROFILES AND LEVELS IN MPEG-2

						Profile			
			Simple (I, P) (4:2:0)	Main (I, P, B) (4:2:0)	SNR (I, P, B) (4:2:0)	Spatial (I, P, B) (4:2:0)	High (I, P, B) (4:2:0; 4:2:2)	Multiview (I, P, B) (4:2:0)	4:2:2 (I, P, B) (4:2:0; 4:2:2)
	Low	Pels/line			352	352		352	
		Lines/frame			288	288		288	
		fps			30	30		30	
		mbps			4	4		8	
	Main	Pels/line	720	720	720		720	720	720
		Lines/frame	576	576	576		576	576	512/608
		fps	30	30	30		30	30	30
		mbps	15	15	15		20	25	50
Level	High-1440	Pels/line	1440		1440	1440	1440		
		Lines/frame	1152		1152	1152	1152		
		fps	60		60	60	60		
		mbps	60		60	80	100		
	High	Pels/line	1920			1920	1920	1920	
		Lines/frame	1152			1152	1152	1152	
		fps	60			60	60	60	
		mbps	80			100	130	300	

I, P, B: allowable picture types. Maximum bit rates include all layers in case of scalable bit streams.

had to accommodate many individual desires to get patented technology required into a profile, without burdening the main applications (i.e., TV production and broadcasting).

13.6 CODING OF AUDIOVISUAL OBJECTS WITH MPEG-4

The MPEG-4 standard is designed to address the requirements of a new generation of highly interactive multimedia applications, while simultaneously supporting traditional applications. Such multimedia applications, in addition to efficient coding, also require advanced functionalities, such as interactivity with individual objects, scalability of contents, and a high degree of error resilience. MPEG-4 provides tools for object-based coding of natural and synthetic audio and video, as well as graphics. The MPEG-4 standard, similarly to its predecessors, consists of a number of parts, the primary parts being Systems, Visual, and Audio. The visual and audio parts of MPEG-4 include coding of both natural and synthetic video and audio, respectively.

13.6.1 Systems

MPEG-4 Systems enables the multiplexing of audiovisual objects and their composition into a scene. Figure 13.26 shows a scene that is composed in the receiver and then presented on the display and speakers. A mouse and keyboard may be provided to enable user input. If we neglect the user input, presentation is as on a regular MPEG-1 or MPEG-2 terminal. However, the audiovisual objects are composed into a scene at the receiving terminal, whereas all other standards discussed in this chapter require scene composition to be done prior to encoding. The scene in Figure 13.26 is composed in a local 3-D coordinate system. It consists of a 2-D background; a video playing on the screen in the scene; a presenter, coded as a 2-D sprite object, with audio; and 3-D objects like the desk and the globe. (A sprite object is a video object that is mapped onto the image plane using an affine mapping.) MPEG-4 enables user interactivity by providing the tools to interact with such a scene. Obviously, this object-based content description gives tremendous flexibility in creating interactive content and in creating presentations that are customized to an individual viewer, in language, text, advertisements, logos, and so on.

Figure 13.27 shows the different functional components of an MPEG-4 terminal [1]:

1. Media or compression layer: This is the component of the system performing the decoding of the media (such as audio, video, graphics, and other suitable media). Media are extracted from the sync layer through the *elementary stream* interface. Specific MPEG-4 media include a Binary Format for Scenes (BIFS) for specifying scene compositions and graphics contents. Another specific MPEG-4 media type is the Object Descriptor (OD). An OD contains pointers to elementary streams, similar to URLs. Elementary streams are used to convey individual MPEG-4 media. ODs also contain additional information such as quality of service parameters. This layer is media aware, but delivery unaware; that is, it does not consider transmission [67].

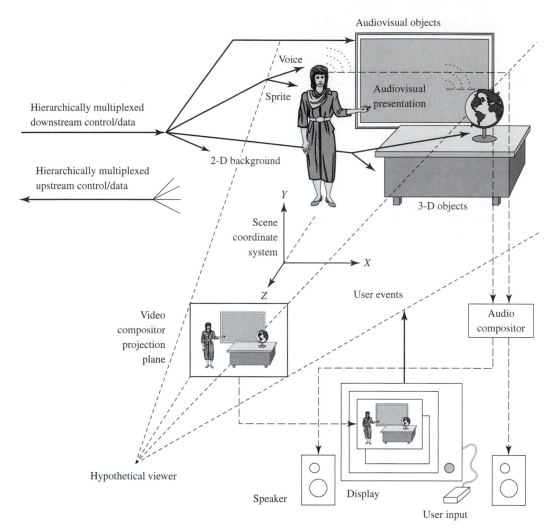

Figure 13.26   Audiovisual objects are composed into a scene within the receiver of an MPEG-4 presentation [Courtesy of MPEG-4].

2. Sync or elementary stream layer: This component of the system is in charge of the synchronization and buffering of individual compressed media. It receives sync layer (SL) packets from the delivery layer, unpacks the elementary streams according to their time stamps, and forwards them to the compression layer. A complete MPEG-4 presentation transports each media type in different elementary streams. Some media may be transported in several elementary streams, for instance if scalability is involved. This layer is media unaware and delivery

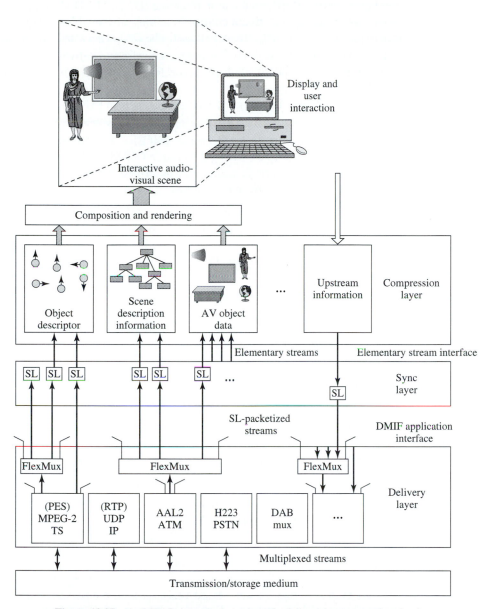

Figure 13.27   An MPEG-4 terminal consists of a delivery layer, a synchronization layer, and a compression layer. MPEG-4 does not standardize the actual composition and rendering. [Courtesy of MPEG-4].

unaware, and talks to the delivery layer through the Delivery Multimedia Integration Framework (DMIF) application interface (DAI). The DAI, in addition to the usual session set-up and stream control functions, also enables the quality of service requirements for each stream to be set. The DAI is network independent [14].

3. Delivery layer: The delivery layer is media unaware and delivery aware. MPEG-4 does not define any specific delivery layer. Rather, MPEG-4 media can be transported on existing delivery layers (such as, for instance, RTP, MPEG-2 transport stream, H.223 or ATM), using the DAI as specified in [31, 2].

MPEG-4's Binary Format for Scenes (BIFS)   The BIFS scene model is a superset of the *Virtual Reality Modeling Language* (VRML) [21, 11]. VRML allows synthetic 3-D objects in a synthetic scene to be described, and rendered using a virtual camera. MPEG-4 extends VRML in three areas:

1. 2-D scene description is defined for placement of 2-D audiovisual objects onto a screen. This is important if the coded media are only video streams that do not require the overhead of 3-D rendering. 2-D and 3-D scenes may be mixed. Figure 13.28 shows a scenegraph that places several 2-D objects on a screen. The

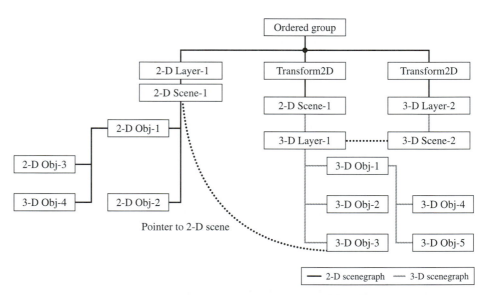

Figure 13.28   A scenegraph with 2-D and 3-D components. The 2-D scenegraph requires only simple placement of 2-D objects on the image using Transform2D nodes. 3-D objects are rendered and then placed on the screen as defined in the 3-D Layer nodes. Interaction between objects can be defined using pointers from one node to another. Reprinted from *Signal Processing: Image Communications,* 15(4–5), J. Signes, Y. Fisher, and A. Eleftheriadis, MPEG-4's binary format for scene description, 321–45, copyright 2000, with permission from Elsevier Science.

object position is defined using transform nodes. Some of the objects are 3-D objects that require 3-D rendering. After rendering, these are used as 2-D objects and placed into the 2-D scene.

2. BIFS enables the description and animation of scenes and graphics objects using its new compression tools based on arithmetic coders.
3. MPEG-4 recognizes the special importance of human faces and bodies. It introduces special tools for very efficient description and animation of virtual humans.

13.6.2 Audio

The tools defined by MPEG-4 audio [30, 3] can be combined into different audio coding algorithms. Since no single coding paradigm was found to span the complete range from very low bit rate coding of speech signals up to high-quality multichannel audio coding, a set of different algorithms has been defined to establish optimum coding efficiency for the broad range of anticipated applications (Figure 13.29). The scalable audio coder can be separated into several components.

- At its lowest rate, a Text-to-Speech (TTS) synthesizer is supported using the MPEG-4 TTS Interface (TTSI)
- Low-rate speech coding (3.1 kHz bandwidth) is based on a Harmonic Vector eXcitation Coding (HVXC) coder at 2–4 kbps.
- Telephone speech (8 kHz bandwidth) and wideband speech (16 kHz bandwidth) are coded using a Code Excited Linear Predictive (CELP) coder at rates between 3850–23,800 bps. This CELP coder can create scalable bit streams with five layers.

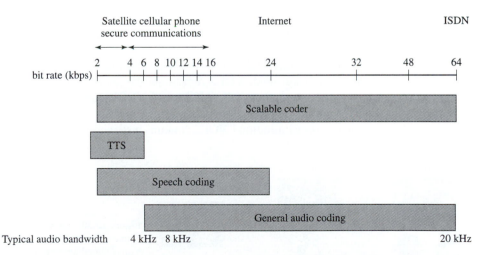

Figure 13.29   MPEG-4 audio supports coding of speech and audio starting at rates below 2 kbps, up to more than 64 kbps/channel for multichannel audio coding.

- General audio is coded from 16 up to more than 64 kbps per channel, using a more efficient development of the MPEG-2 AAC coder. Transparent audio quality can be achieved.

In addition to audio coding, MPEG-4 audio defines music synthesis at the receiver using a *structured audio* toolset, which provides a single standard to unify the world of algorithmic music synthesis, and to implement scalability and the notion of audio objects [9].

13.6.3 Basic Video Coding

Many of the MPEG-4 functionalities require access not only to an entire sequence of pictures, but to an entire object—furthermore, not only to individual pictures, but also to temporal instances of these objects within a picture. A temporal instance of a video object can be thought of as a snapshot of an arbitrarily shaped object that occurs within a picture. Like a picture, an object is intended to be an access unit; unlike a picture, it is expected to have a semantic meaning. MPEG-4 enables content-based interactivity with video objects by coding objects independently, using motion, texture, and shape. At the decoder, different objects are composed into a scene and displayed. In order to enable this functionality, higher syntactic structures had to be developed. A scene consists of several VideoObjects (VOs). The VO has three dimensions (2-D plus time). A VO can be composed of several VideoObjectLayers (VOLs). Each VOL (2-D plus time) represents various instantiations of a VO. A VOL can represent different layers of a scalable bit stream or different parts of a VO. A time instant of a VOL is called a VideoObjectPlane (VOP). A VOP is a rectangular video frame or a part thereof. It can be fully described by its texture variations (a set of luminance and chrominance values) and its shape. The video encoder applies motion, texture, and shape coding tools to the VOP using I, P, and B modes similar to the modes of MPEG-2. For editing and random access purposes, consecutive VOPs can be grouped into a GroupOfVideoObjectPlanes (GVOP). A video session, the highest syntactic structure, may consist of several VOs.

The example in Figure 13.30 shows one VO composed of 2 VOLs. VOL1 consists of the tree and the background; VOL2 represents the person. In the example, VOL1 is represented by two separate VOPs, VOP1 and VOP3. Hence, VOL1 may provide content-based scalability in the sense that a decoder may choose not to decode one VOP of VOL1 due to resource limitations. VOL2 contains just one VOP, namely VOP2. VOP2

Figure 13.30 Object-based coding requires the decoder to compose different VideoObjectPlanes into a scene. VideoObjectLayers enable content-based scalability.

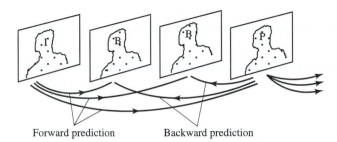

Forward prediction   Backward prediction

Figure 13.31   An example of a prediction structure using I-, P-, and B-VOPs. Reprinted from J. Ostermann and A. Puri, Natural and synthetic video in MPEG-4, *IEEE International Conference on Acoustics, Speech and Signal Processing (ICASSP'98)*, 3805–09, November 1998. Copyright 1998 IEEE.

may be represented using a temporal, spatial, or quality scalable bitstream. In this case, a decoder might again decide to decode only the lower layers of VOL 2. The example in Figure 13.30 shows the complex structures of content-based access and scalability that MPEG-4 supports. However, the given example could also be represented in a straightforward fashion with three VOs. The background, the tree and the person could be coded as separate VOs with one layer each, and each layer represented by one VOP. The VOPs are encoded separately and composed in a scene at the decoder.

To see how MPEG-4 video coding works, consider a sequence of VOPs. Extending the concept of intra-, predictive, and bidirectionally predictive pictures of MPEG-1/2 to VOPs, we get I-VOP, P-VOPs and B-VOP. If two consecutive B-VOPs are used between a pair of reference VOPs (I- or P-VOPs), the resulting coding structure is as shown in Figure 13.31.

Coding Efficiency Tools   In addition to the obvious changes due to the object-based nature of MPEG-4, the following tools were introduced in order to increase coding efficiency compared to MPEG-1 and MPEG-2:

- DC Prediction: This is improved compared to MPEG-1/2. Either the previous block or the block above the current block can be chosen as predictors to predict the current DC value.

- AC Prediction: AC prediction of DCT coefficients is new in MPEG-4. The block that was chosen to predict the DC coefficient is also used for predicting one line of AC coefficients. If the predictor is the previous block, the AC coefficients of its first column are used to predict the colocated AC coefficients of the current block. If the predictor is the block from the previous row, it is used to predict the first row of AC coefficients. AC prediction does not work well for blocks with coarse texture, diagonal edges, or horizontal as well as vertical edges. Switching AC prediction on and off on a block level is desirable but too costly. Consequently, the decision is made on the MB level.

- Alternate Horizontal Scan: This scan is added to the two scans of MPEG-2 (Figure 13.19). The alternate scan of MPEG-2 is referred to as alternate vertical scan in MPEG-4. The alternate horizontal scan is created by mirroring the vertical scan. The scan is selected at the same time as the AC prediction is decided. In case of AC prediction from the previous block, alternate vertical scan is selected. In case of AC prediction from the block above, alternate horizontal scan is used. No AC prediction is coupled to the zigzag scan.

- 3-D VLC: Coding of DCT coefficients is achieved similarly to H.263.
- Four Motion Vectors: Four motion vectors for an MB are allowed. This is done similarly to H.263.
- Unrestricted Motion Vectors: This mode is enabled. Compared to H.263, a much wider motion vector range of ± 2048 pels may be used.
- Sprite: A Sprite is basically a large background image that is transmitted to the decoder. For display, the encoder transmits affine mapping parameters that map a part of the image onto the screen. By changing the mapping, the decoder can zoom in and out of the sprite, and pan to the left or right [8].
- Global Motion Compensation: In order to compensate for global motion due to camera motion, camera zoom, or large moving objects, global motion is compensated according to the eight-parameter motion model of Equation 5.5.14 (see Section 5.5.3):

$$
\begin{aligned}
x' &= \frac{ax + by + c}{gx + hy + 1} \\
y' &= \frac{dx + ey + f}{gx + hy + 1}
\end{aligned}.
\tag{13.6.1}
$$

Global motion compensation is an important tool to improve picture quality for scenes with large global motion. These scenes are difficult to code using block-based motion. In contrast to scenes with arbitrary motion, the human eye is able to track detail in case of global motion. Thus global motion compensation helps to improve the picture quality in the most critical scenes.

- Quarter-Pel Motion Compensation: The main target of quarter-pel motion compensation is to enhance the resolution of the motion compensation scheme with only small syntactical and computational overhead, leading to more accurate motion description and less prediction error to be coded. Quarter-pel motion compensation will only be applied to the luminance pels; chrominance pels are compensated in half-pel accuracy.

As pointed out, some tools are similar to those developed in H.263. As in H.263, the MPEG-4 standard describes overlapped block motion compensation. However, this tool is not included in any MPEG-4 profile, due to its computational complexity for large picture sizes and its limited improvements for high-quality video; in other words, there is no MPEG-4–compliant decoder that needs to implement overlapped block motion compensation.

Error Resilience Tools   Besides the tools developed to enhance coding efficiency, a set of tools are also defined in MPEG-4 for enhancing the resilience of the compressed bit streams to transmission errors. These will be described in Chapter 14.

13.6.4 Object-Based Video Coding

In order to enable object-based functionalities for coded video, MPEG-4 allows the transmission of shapes for video objects. Though MPEG-4 does not standardize the method of defining or segmenting the video objects, it defines the decoding algorithm and, implicitly, an encoding algorithm for describing the shape. Shape is described using alpha-maps that have the same resolution as the luminance signal. An alpha map is colocated with the luminance picture. MPEG-4 defines the alpha-map as having two parts. The binary alpha-map defines pels that belong to the object. In the case of grey-scale alpha-maps, we have an additional alpha-map that defines the transparency using 8 bpp. Alpha-maps extend a macroblock. The 16×16 binary alpha-map of an MB is called a binary alpha block (BAB). In the following, we describe the individual tools that MPEG-4 uses for object-based video coding.

- Binary Shape: A context-based arithmetic coder (as described in Section 10.1.1) is used to code boundary blocks of an object. A boundary block contains pels of the object and of the background. It is colocated with an MB. For nonboundary blocks, the encoder just signals whether the MB is part of the object or not. A sequence of alpha-maps may be coded and transmitted without texture. Alternatively, MPEG-4 uses tools like padding and DCT or SA-DCT to code the texture that goes with the object. BABs are coded in intra-mode and inter-mode. Motion compensation may be used in inter-mode. Shape motion vector coding uses the motion vectors associated with the texture coding as a predictor.

- Padding: In order to code the texture of BABs using block-based DCT, the texture of the background may be set to any color. In intra-mode, this background color has no effect on the decoded pictures and can be choosen by the encoder. However, for motion compensation, the motion vector of the current block may refer to a boundary block in the previous reference picture. Part of the background pels of the reference picture might be located in the area of the current object—hence the value of these background pictures influences the prediction loop. MPEG-4 uses padding (as described in Section 10.2.1) to define the background pels used in prediction.

- Shape-Adaptive DCT: The encoder may choose to use SA-DCT for coding the texture of BABs (Section 10.2.2). However, padding of the motion-compensated prediction image is still required.

- Greyscale Shape Coding: MPEG-4 allows the transmission of arbitrary alpha-maps. Since the alpha-maps are defined with eight bits, they are coded the same way as the luminance signal.

Figure 13.32(a) shows the block diagram of an object-based MPEG-4 video coder. MPEG-4 uses two types of motion vectors: In Figure 13.32, we designate the conventional motion vectors used to compensate the motion of texture *texture motion;* motion vectors describing the shift of the object shape are called *shape motion.* A shape motion vector may be associated with a BAB. Image analysis estimates texture and shape

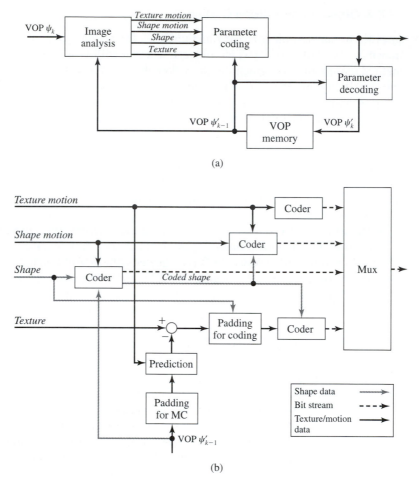

Figure 13.32 An object-based MPEG-4 video encoder for coding of arbitrarily shaped video objects: (a) the overall block diagram, (b) the block diagram for parameter coding.

motion of the current VOP ψ_k with respect to the reference VOP ψ'_{k-1}. Parameter coding encodes the parameters predictively. The parameters are transmitted and decoded, and the new reference VOP is stored in the VOP memory. The increased complexity due to the coding of arbitrarily shaped video objects becomes evident in Figure 13.32(b). First, shape motion vectors and shape pels are encoded. The shape motion coder knows which motion vectors to code by analyzing the potentially lossily encoded shape parameters. For texture prediction, the reference VOP is padded as previously described. The prediction error is padded using the original shape parameters to determine the area to be padded. Then, each MB is encoded using DCT.

13.6.5 Still Texture Coding

One of the functionalities supported by MPEG-4 is the mapping of static textures onto 2-D or 3-D surfaces. MPEG-4 Visual supports this functionality by providing a separate mode for encoding static texture information. It is envisioned that applications involving interactivity with texture-mapped synthetic scenes require continuous scalability.

For coding static texture maps, DWT coding was selected for the flexibility it offers in spatial and quality scalability while maintaining good coding performance (Section 11.3.1). In DWT coding, a texture-map image is first decomposed via a 2-D separable decomposition (using Daubechies (9,3)-tap biorthogonal filters). Next, the coefficients of the lowest band are quantized, and coded predictively using implicit prediction (similar to that used in intra-DCT coding) and arithmetic coding. This is followed by coding of coefficients of higher bands by use of multilevel quantization, zero-tree scanning, and arithmetic coding. The resulting bit stream is flexibly arranged, allowing many layers of spatial and quality scalability to be easily derived.

This algorithm was extended to code arbitrarily shaped texture maps. In order to adapt a scan line of the shape to the coding with DWT, MPEG-4 uses leading and trailing boundary extensions that mirror the image signal (Section 11.3.1).

13.6.6 Mesh Animation

Mesh-based representation of an object is useful for a number of functionalities, such as animation, content manipulation, content overlay, merging natural and synthetic video, and others [68].

Figure 13.33 shows a mesh coder and its integration with a texture coder. The mesh encoder generates a 2-D mesh-based representation of a natural or synthetic video object at its first appearance in the scene. The object is tesselated with triangular patches, resulting in an initial 2-D mesh (Figure 13.34). The node points of this initial mesh are then animated in 2-D as the VOP moves in the scene. Alternatively, the motion of the node point can by animated from another source. The 2-D motion of a video object can

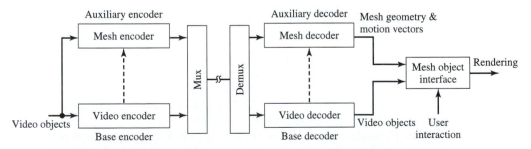

Figure 13.33 Simplified architecture of an encoder/decoder supporting the 2-D mesh object. The video encoder provides the texture map for the mesh object. Reprinted from *Signal Processing: Image Communications, Special Issue on MPEG-4,* 15, A. M. Tekalp and J. Ostermann, Face and 2-D mesh animation in MPEG-4, 387–421, copyright 2000, by permission of Elsevier Science.

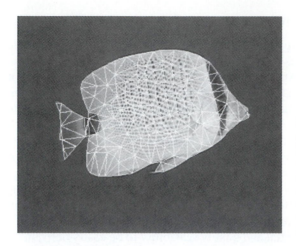

Figure 13.34 A content-based mesh designed for the "Bream" video object. Reprinted from *Signal Processing: Image Communications, Special Issue on MPEG-4,* 15, A. M. Tekalp and J. Ostermann, Face and 2-D mesh animation in MPEG-4, 387–421, copyright 2000, by permission of Elsevier Science.

thus be compactly represented by the motion vectors of the node points of the mesh. Motion compensation can be achieved by warping of the texture map corresponding to patches via affine transform from one VOP to the next. Textures used for mapping onto object mesh models or facial wireframe models are derived from either video or still images. Though mesh analysis is not part of the standard, MPEG-4 defines the encoding of 2-D meshes and the motion of their node points. Furthermore, the mapping of a texture onto the mesh may be described using MPEG-4.

13.6.7 Face and Body Animation

An MPEG-4 terminal supporting face and body animation is expected to include a default face and body model. The systems part of MPEG-4 provides means to customize this face or body model by means of face and body definition parameters (FDP, BDP) or to replace it with one downloaded from the encoder. The definition of a scene (including 3-D geometry) and of a face/body model can be sent to the receiver using BIFS [23]. Figure 13.35 shows a scenegraph that a decoder built according to the BIFS stream. The *Body* node defines the location of the Body. Its child BDP describes the look of the body using a skeleton with joints, surfaces, and surface properties. The *bodyDefTable* node describes how the model is deformed as a function of the body animation parameters. The *Face* node is a descendent of the body node. It contains the face geometry, as well as the geometry for defining the face deformation as a function of the facial animation parameters (FAPs). The visual part of MPEG-4 defines how to animate these models using FAPs and body animation parameters (BAPs) [24].

Figure 13.36 shows two phases of a left eye blink (plus the neutral phase) which have been generated using a simple animation architecture [68]. The dotted half-circle in the figure shows the ideal motion of a vertex in the eyelid as it moves down according to the amplitude of FAP 19. In this example, the faceDefTable for FAP 19 approximates the target trajectory with two linear segments, on which the vertex actually moves as FAP 19 increases.

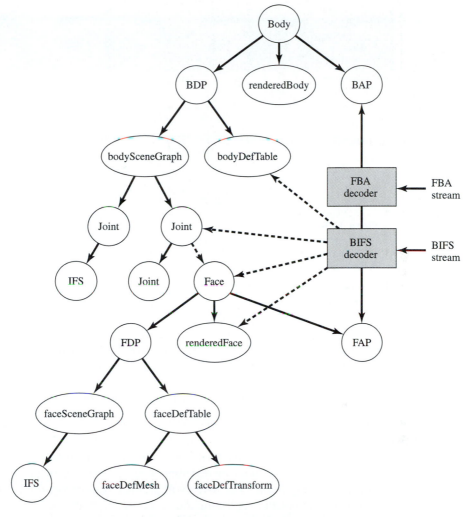

Figure 13.35   A scenegraph describing a human body, transmitted in a BIFS stream. The nodes Body and Face are animated using the FAPs and BAPs of the FBA stream. The BDP and FDP nodes and their children describe the virtual human. Reprinted from T. K. Capin, E. Petajan, and J. Ostermann, Efficient modeling of virtual humans in MPEG-4, *IEEE International Conference on Multimedia and Expo (ICME2000)*, 2:1103–6, New York, 2000. Copyright 2000 IEEE.

Facial Animation   Three groups of facial animation parameters are defined [68]. First, for low-level facial animation, a set of 66 FAPs is defined. These include head and eye rotations as well as motion of feature points on mouth, ear, nose, and eyebrow deformation (Figure 10.20). Since these parameters are model independent, their amplitudes are scaled according to the proportions of the actual animated model.

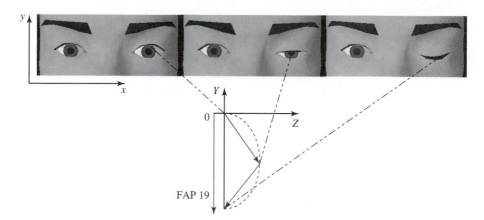

Figure 13.36 Neutral state of the left eye (left) and two deformed animation phases for the eye blink (FAP 19). The FAP defines the motion of the eyelid in the negative y direction; the faceDefTable defines the motion in one of the vertices of the eyelid in the x and z directions.

Second, for high-level animation, a set of primary facial expressions such as joy, sadness, surprise, and disgust are defined. Third, for speech animation, fourteen visemes define mouth shapes that correspond to phonemes. Visemes are transmitted to the decoder or are derived from the phonemes of the TTS synthesizer of the terminal.

The FAPs are linearly quantized and entropy coded using arithmetic coding. Alternatively, a time sequence of sixteen FAPs can also be DCT coded. Due to efficient coding, it takes only about 2 kbps to achieve lively facial expressions.

Body Animation   BAPs manipulate independent degrees of freedom in the skeleton model of the body to produce animation of body parts [4]. As with the face, the remote manipulation of a body model in a terminal with BAPs can accomplish lifelike visual scenes of the body in real time without sending pictorial and video details of the body every frame. The BAPs will produce reasonably similar high-level results in terms of body posture and animation on different body models, also without the need to transmit a model to the decoder. There are a total of 186 predefined BAPs in the BAP set, with an additional set of 110 user-defined extension BAPs. Each predefined BAP corresponds to a degree of freedom in a joint connecting two body parts. These joints include toe, ankle, knee, hip, spine, shoulder, clavicle, elbow, wrist, and fingers. Extension BAPs are provided to animate additional features in connection with body deformation tables [1]—for example, for clothing animation, or body parts that are not part of the human skeleton.

The BAPs are categorized into groups with respect to their effect on the body posture. This grouping scheme offers a number of advantages. First, it allows us to adjust the complexity of the animation by choosing a subset of BAPs. For example, the total number of BAPs in the spine is 72, but significantly simpler models can be obtained by choosing only a predefined subset. Second, assuming that not all the motions

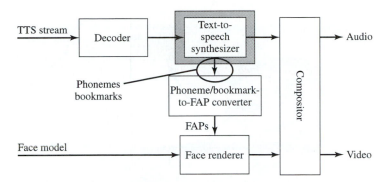

Figure 13.37   MPEG-4 architecture for facial animation, allowing synchronization of facial expressions and speech generated by a proprietary text-to-speech synthesizer.

contain all the BAPs, only the active BAPs can be transmitted to decrease required bit rate significantly. BAPs are coded similar to FAPs, using arithmetic coding.

Integration of Speech Synthesis   MPEG-4 acknowledges the importance of TTS for multimedia applications, by providing a text-to-speech synthesizer interface to a proprietary TTS. A TTS stream contains text in ASCII and optional prosody in binary form. The decoder decodes the text and prosody information according to the interface defined for the TTS synthesizer. The synthesizer creates speech samples that are handed to the compositor. The compositor presents audio and (if required) video to the user.

Figure 13.37 shows the architecture for speech-driven facial animation that allows synchronized presentation of synthetic speech and talking heads. A second output interface of the TTS sends the phonemes of the synthesized speech, as well as start time and duration information for each phoneme, to a phoneme/bookmark-to-FAP-converter. The converter translates the phonemes and timing information into FAPs that the face renderer uses in order to animate the face model. In addition to the phonemes, the synthesizer identifies bookmarks in the text that convey nonspeech related FAPs such as joy to the face renderer. The timing information of the bookmarks is derived from their position in the synthesized speech. Since the facial animation is driven completely from the text input to the TTS, transmitting an FAP stream to the decoder is optional. Furthermore, synchronization is achieved because the talking head is driven by the asynchronous proprietary TTS synthesizer.

13.6.8 Profiles

MPEG-4 has developed an elaborate structure of profiles (Figure 13.38). Whereas an application may choose to implement one or more MPEG-4 profiles, a MPEG-4 terminal must implement several profiles. An object descriptor profile is required to enable the transport of MPEG-4 streams and identify these streams in the terminal. A scene description profile provides the tools to compose the audio, video, or graphics

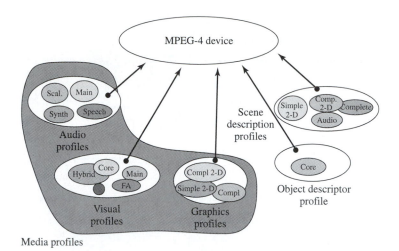

Figure 13.38   An MPEG-4 terminal must implement at least one profile each of the object descriptor, scene description, and media profiles. Not all profiles within a group are listed. Reprinted from *Signal Processing: Image Communications,* 15(4–5), R. Koenen, Profiles and levels in MPEG-4. Approach and overview, 463–78, copyright 2000, with permission from Elsevier Science.

objects into a scene. A 2-D scene description profile enables just the placement of 2-D video objects; higher profiles provide more functionality. A media profile must be implemented in order to present actual content on the terminal. MPEG-4 supports audio, video, and graphics as media. Several video profiles are defined; we list here only a subset of them, and mention their main functionalities:

- Simple Profile: The Simple profile was created with low-complexity applications in mind. The first usage is mobile use of (audio) visual services, and the second is putting very low-complexity video on the Internet. It supports up to four rectangular objects in a scene with, at the lowest level, a maximum total size of a QCIF picture. There are three levels for the Simple profile with bitrates from 64–384 kbps. It provides the following tools: I- and P-VOPs, AC/DC prediction, four motion vectors, unrestricted motion vectors, slice resynchronization, data partitioning, and reversible VLC. This profile is able to decode H.263 video streams that do not use any of the optional annexes of H.263.

- Simple Scalable Profile: This profile adds support for B-frames, temporal scalability, and spatial scalability to the Simple profile. It is useful for applications that provide services at more than one level of quality due to bit-rate or decoder resource limitations, such as Internet use and software decoding.

- Advanced Real-Time Simple (ARTS) Profile: This profile extends the capabilities of the Simple profile and provides more sophisticated error protection of rectangular video objects using a back-channel, which signals transmission errors from the decoder to the encoder such that the encoder can transmit video information in intra-mode for the affected parts of the newly coded images. It is suitable for real-time coding applications such as video phone, teleconferencing, and remote observation.

- Advanced Simple Profile: This profile enhances coding efficiency compared to the Simple profile by adding support for quarter-pel motion compensation, global motion compensation, and B-VOPS.

- Fine Granular Scalability Profile: This profile allows up to eight scalable layers, so that delivery quality can easily adapt to decoding resources. In case a network supports different levels of QoS, higher layers may be sent with a lower priority than lower layers. The enhancement layers are coded using a bit-plane coder for DCT coefficients. It can be used with the Simple or Advanced Simple profile as a base layer.

- Core Profile: In addition to the tools of the Simple profile, this profile enables scalable still textures, B-frames, binary shape coding, and temporal scalability for rectangular as well as arbitrarily shaped objects. It is useful for higher-quality interactive services, combining good quality with limited complexity and supporting arbitrarily shaped objects. Also, mobile broadcast services could be supported by this profile. The maximum bitrate is 384 kbps at Level 1 and 2 mbps at Level 2.

- Core Scalable Visual Profile: This adds to the Core profile object-based SNR as well as spatial and temporal scalability.

- Main Profile: The Main profile adds support for interlaced video, greyscale alpha-maps, and sprites. The Main profile was created with broadcast services in mind, addressing progressive as well as interlaced material. It combines the highest quality with the versatility of arbitrarily shaped objects using grey-scale coding. The highest level accepts up to 32 objects for a maximum total bit rate of 38 mbps.

- Advanced Coding Efficiency (ACE): This profile targets transmission of entertainment videos at bit rates less than 1 mbps. However, in terms of specification, it adds to the Main profile by extending the range of bit rates and adding the tools of quarter-pel motion compensation, global motion compensation, and shape-adaptive DCT. This profile does not support sprites.

- Simple Studio Profile: This profile supports I-VOPs only for very high quality, and bit rates up to 1200 mbps. Each VOP may be of arbitrary shape and have several alpha planes associated with it. The formats 4:0:0, 4:2:2, and 4:4:4 are supported at resolutions suitable for HDTV and digital cinema.

- Core Studio Profile: This profile adds P-VOPs to the Simple Studio profile, which makes this profile more complex but also more efficient.

More profiles are defined for face, body, and mesh animation. At the time of this writing, it is still too early to know what profiles will eventually be implemented in products. First-generation prototypes implement only the Simple and Advanced Simple profile, and they target applications in the area of mobile video communications.

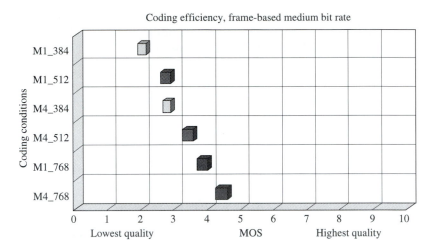

Figure 13.39 Subjective quality of MPEG-4 Main profile versus MPEG-1. M4_x is an MPEG-4 coder operating at the rate of x kbps; M1_x is an MPEG-1 encoder operating at the given rate [27].

13.6.9 Evaluation of Subjective Video Quality

MPEG-4 introduces new functionalities (such as object-based coding) and claims to improve coding efficiency. These claims were verified by means of subjective tests. User-perceived video quality is measured using the mean opinion score (MOS) [70]. Figure 13.39 shows the results of subjective coding efficiency tests that compared MPEG-4 and MPEG-1 video at bit rates between 384–768 kbps. These results indicated that MPEG-4 outperforms MPEG-1 significantly in these bit rates. The MPEG-4 video was coded using the tools of the Main profile (Section 13.6.8). In Figure 13.40, we see the improvements in coding efficiency due to the additional tools of the ACE profile (Section 13.6.8). The quality provided by the ACE profile at 768 kbps equals the quality provided by the Main profile at 1024 kbps. This makes the ACE profile very attractive for delivering movies to the home over cable modems or digital subscriber lines (DSL). Further subjective tests showed that the object-based functionality of MPEG-4 does not decrease the subjective quality of the coded video object when compared to frame-based video—that is, the bits spent on shape coding are compensated for in bits saved by not coding pels outside of the video object. Hence, the advanced tools of MPEG-4 enable content-based video representation without increasing the bit rate for video coding.

13.7 VIDEO BIT STREAM SYNTAX

As mentioned, video coding standards define the syntax and semantics of the video bit stream, instead of the actual encoding scheme. They also specify how the bit stream must be parsed and decoded to produce the decompressed video signal. In order to support different applications, the syntax must be flexible. This is achieved by having a hierarchy

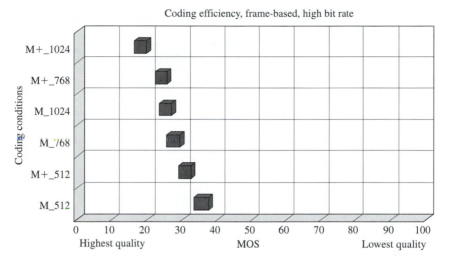

Figure 13.40 Subjective quality of MPEG-4 ACE versus MPEG-4 Main profile. M_x is an MPEG-4 coder according to the Main profile operating at the rate of x kbps; M+_x is an MPEG-4 encoder according to the ACE profile operating at the given rate [26].

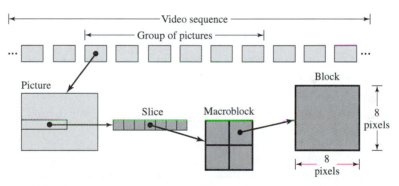

Figure 13.41 Visualization of the hierarchical structure of an MPEG-2 bit stream from a video sequence layer down to the block level, shown for the luminance component. Each layer has also two chrominance components associated with it.

of different layers that each start with a *header*. Each layer performs a different logical function (Table 13.6). Most headers can be uniquely identified in the bit stream because they begin with a *start code* that is a long sequence of zeros (23 for MPEG-2) followed by a one and a start code identifier. Figure 13.41 visualizes the hierarchy for MPEG-2. In the following list, we describe the elements of the hierarchical bit stream structure:

1. Sequence: A video sequence commences with a sequence header, and may contain additional sequence headers. It includes one or more groups of pictures and ends with an end-of-sequence code. The sequence header and its extensions

TABLE 13.6   SYNTAX HIERARCHY AS USED IN DIFFERENT VIDEO CODING STANDARDS*

Syntax layer	Functionality	Standard
Sequence (SC)	Definition of entire video sequence	H.261/3, MPEG-1/2
VOL (SC)	Definition of entire video object	MPEG-4
GOP (SC)	Enables random access in video stream	MPEG-1/2
GVOP (SC)	Enables random access in video stream	MPEG-4
Picture (SC)	Primary coding unit	H.261/3, MPEG-1/2
VOP (SC)	Primary coding unit	MPEG-4
GOB (SC)	Resynchronization, refresh, and error recovery in a picture	H.261/3
Slice (SC)	Resynchronization, refresh, and error recovery in a picture	MPEG-1/2
Video Packet (SC)	Resynchronization and error recovery in a picture	MPEG-4
MB	Motion compensation and shape coding unit	H.261/3, MPEG-1/2/4
Block	Transform and compensation unit	H.261/3, MPEG-1/2/4

*Each layer starts with a header. An SC in a syntax layer indicates that the header of that layer starts with a start code.

contain basic parameters such as picture size, image aspect ratio, picture rate, and other global parameters. The VideoObjectLayer header has the same functionality, however it carries additional information that an MPEG-4 decoder needs in order to compose sequences of arbitrarily shaped video objects into one video sequence to be displayed.

2. Group of Pictures: A GOP is a header followed by a series of one or more pictures intended to allow random access into the sequence, fast search, and editing. Therefore, the first picture in a GOP is an intra-coded picture (I-picture). This is followed by an arrangement of forward-predictive coded pictures (P-pictures) and optional bidirectionally predicted pictures (B-pictures). This GOP header also contains a time code for synchronization and editing. A GOP is the base unit for editing and random access, since it is coded independent of previous and consecutive GOPs. In MPEG-4, the function of the GOP is served by a GroupOfVideoObjectPlanes. Since H.261 and H.263 were designed mainly for interactive applications, they do not use the concept of GOP. However, the encoder may choose at any time to send an I-picture, thus enabling random access and simple editing.

3. Picture: A picture is the primary coding unit of a video sequence. A picture consists of three rectangular matrices representing luminance (Y) and two chrominance (Cb and Cr) values. The picture header indicates the picture type (I, P, B), picture structure (field/frame), and perhaps other parameters (such as motion-vector ranges). A VOP is the primary coding unit in MPEG-4. It has the size of the bounding box of the video object.

Each standard divides a picture into groups of MBs. Whereas H.261 and H.263 use a fixed arrangement of MBs, MPEG-1/2 allow for a flexible arrangement. MPEG-4 arranges a variable number of MBs into one group.

4. GOB: H.261 and H.263 divide the image into GOBs of three lines of MBs, with eleven MBs in one GOB line. The GOB headers define the position of the GOB within the picture. For each GOB, a new quantizer stepsize may be defined. GOBs are important in the handling of errors. If the bit stream contains an error, the decoder can skip to the start of the next GOB, thus limiting the extent of bit errors to within one GOB of the current frame. However, error propagation may occur when predicting the following frame.

5. Slice: MPEG-1, MPEG-2, and H.263 Annex K extend the concept of GOBs to a variable configuration. A slice groups several consecutive MBs into one unit. Slices may vary in size. In MPEG-1, a slice may be as big as one picture. In MPEG-2 however, at least each row of MBs in a picture starts a new slice. Having more slices in the bit stream allows better error concealment, but uses bits that could otherwise be used to improve picture quality.

6. Video Packet Header: The video packet approach adopted by MPEG-4 is based on providing periodic resynchronization markers throughout the bit stream. In other words, the length of the video packets are not based on the number of MBs, but instead on the number of bits contained in that packet. If the number of bits contained in the current video packet exceeds a threshold as defined by the encoder, then a new video packet is created at the start of the next MB. This way, a transmission error causes less damage to regions with higher activity than to regions that are stationary, as opposed to the more rigid slice and GOB structures. The video packet header carries position information and repeats information of the picture header that is necessary to decode the video packet.

7. Macroblock: An MB is a 16×16 pixel block in a picture. Using the 4:2:0 format, each chrominance component has one-half the vertical and horizontal resolution of the luminance component. Therefore an MB consists of four Y, one Cr, and one Cb block. Its header carries relative position information, quantizer scale information, MTYPE information (I, P, B), and a CBP indicating which of the six blocks of an MB are coded (and how). As with other headers, other parameters may or may not be present in the header, depending on the MTYPE. Since MPEG-4 also needs to code the shape of video objects, it extends the MB by a binary alpha block that defines for each pel in the MB whether it belongs to the VO. In the case of greyscale alpha-maps, the MB also contains four blocks for the coded alpha-maps.

8. Block: A block is the smallest coding unit in standardized video coding algorithms. It consists of 8×8 pixels and can be one of three types: Y, Cr, or Cb. The pixels of a block are represented by their DCT coefficients, via a Huffman code that records the number of zeros before the next nonzero coefficient, and the amplitude of this coefficient.

The different headers in the bit stream allow a decoder to recover from errors in the bit stream and to start decoding as soon as it receives a start code. The design of the bit stream syntax assures that no valid combination of code words can emulate a start code. Often a reserved bit in the syntax is used to avoid this startcode emulation. The behavior of a decoder when receiving an erroneous bit stream is not defined in the standard. Different decoders may behave very differently: some decoders crash and require rebooting of the terminal, others recover within a picture, yet others wait until the next I-frame before they start decoding again.

13.8 MULTIMEDIA CONTENT DESCRIPTION USING MPEG-7

With the ubiquitous use of video, the ability to index and search for speech, audio, images, and video sequences becomes very important. MPEG-7 is an ongoing standardization effort for content description of audiovisual documents [32, 64]. In principle, MPEG-1/2/4 are designed to represent the information itself, whereas MPEG-7 is meant to represent information about the information. Looking at it from another perspective: *MPEG-1/2/4 make content available, whereas MPEG-7 allows you to find the content you need* [64]. MPEG-7 is intended to provide complementary functionality to other MPEG standards: representing information about the content, not the content itself ("the bits about the bits"). Whereas MPEG-4 enables the attachment of limited textual meta-information to its streams, the MPEG-7 standard provides a full set of indexing and search capabilities, such that we can search for a video sequence not only with text keys but also with keys like color histograms, motion trajectory, and so on. MPEG-7 became an international standard in 2001.

In this section, we first provide an overview of the elements standardized by MPEG-7. We then describe multimedia description schemes, focusing on content description. We explain how MPEG-7 decomposes an AV document to arrive at both structural and semantic descriptions. Finally, we describe the visual descriptors used in these descriptions. The descriptors and description schemes presented in the following assume that semantically meaningful regions and objects can be segmented, and that the shape and motion parameters, and even semantic labels of these regions and objects, can be accurately extracted. We note, however, that generation of such information remains an unsolved problem and may require manual assistance. The MPEG-7 standard defines only the syntax that can be used to specify such information, not the algorithms that can be used to extract them.

13.8.1 Overview

The main elements of the MPEG-7 standard [32] are:

- Descriptor (D): The MPEG-7 descriptors are designed to represent features, including low-level audiovisual features; high-level features of semantic objects, events, and abstract concepts; information about storage media; and so on. Descriptors define the syntax and the semantics of each feature representation.

- Description Scheme (DS): The MPEG-7 DSs expand on the MPEG-7 descriptors by combining individual descriptors as well as other DSs within more complex structures, and by defining the relationships among the constituent descriptors and DSs.
- Description Definition Language (DDL): The DDL is a language that allows the creation of new DSs and, possibly, new descriptors. It also allows the extension and modification of existing DSs. The XML schema language has been selected to provide the basis for the DDL.
- System tools: These are tools that are needed to prepare MPEG-7 descriptions for efficient transport and storage, to allow synchronization between content and descriptions, and to manage and protect intellectual property.

13.8.2 Multimedia Description Schemes

In MPEG-7, the DSs are categorized as pertaining specifically to the audio or visual domain, or pertaining generically to the description of multimedia. The multimedia DSs are grouped into the following categories according to their functionality (see Figure 13.42):

- Basic elements: These deal with basic datatypes, mathematical structures, schema tools, linking and media localization tools, as well as basic DSs, which are elementary components of more complex DSs.
- Content description: These DSs describe the structural and conceptual aspects of an AV document.

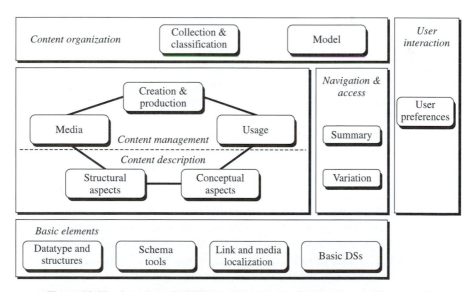

Figure 13.42 Overview of MPEG-7 multimedia description schemes. [Courtesy of MPEG-7].

- Content management: These tools specify information about the storage media, the creation, and the usage of an AV document.
- Content organization: These tools address the organization of the content by classification, by the definition of collections of AV documents, and by modeling.
- Navigation and access: These include summaries for browsing and variations of the same AV content for adaptation to capabilities of client terminals, network conditions, or user preferences.
- User interaction: These DSs specify user preferences pertaining to the consumption of the multimedia material.

Content Description   In the following, we briefly describe the DSs for content description. More detailed information can be found in [29]. The DSs developed for content description fall into two categories: those describing the structural aspects of an AV document, and those describing the conceptual aspects.

Structural Aspects: These DSs describe the syntactic structure of an AV document in terms of segments and regions. An AV document (e.g., a video program with audio tracks) is divided into a hierarchy of segments, known as a *segment tree*. For example, the entire document is segmented into several story units, each story unit is then divided into different scenes, and finally each scene is split into many camera shots. A segment at each level of the tree can be further divided into video and audio segments, corresponding to the video frames and the audio waveform, respectively. In addition to using a video segment that contains a set of complete video frames (which may not be contiguous in time), a still or moving region can also be extracted. A region can be recursively divided into subregions, to form a *region tree*. The concept of the segment tree is illustrated on the left-hand side of Figure 13.43.

Conceptual Aspects: These DSs describe the semantic content of an AV document in terms of events, objects, and other abstract notions. The semantic DS describes events and objects that occur in a document, and attaches corresponding "semantic labels" to them. For example, the event type could be a news broadcast, a sports game, and so on. The object type could be a person, a car, and so on. As with the structure description, MPEG-7 also uses hierarchical decomposition to describe the semantic content of an AV document. An event can be further broken up into many subevents to form an event tree (right-hand side of Figure 13.43). Similarly, an object tree can be formed. An event-object relation graph describes the relation between events and objects.

Relation between Structure and Semantic DSs: An event is usually associated with a segment, and an object with a region. Each event or object may occur multiple times in a document, and their actual locations (which segments or regions) are described by a set of links, as shown in Figure 13.43. In this sense, the syntactic structure, represented by the segment tree and the region tree, is like the table of contents in the beginning of a book, whereas the semantic structure, the event tree and the object tree, is like the index at the end of the book.

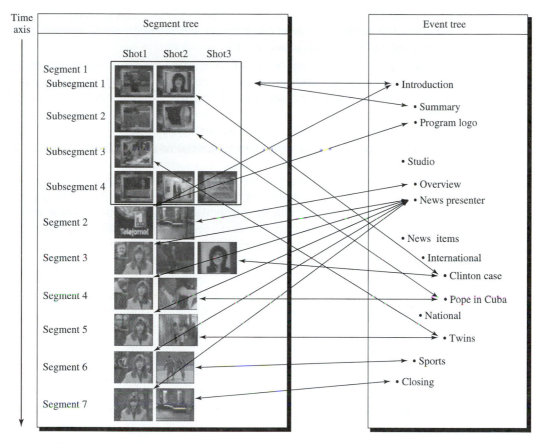

Figure 13.43   Description of an AV document (a news program in this case) based on segment and event trees. The segment tree is like the table of contents in the beginning of a book, whereas the event tree is like the index at the end of the book. [Courtesy of MPEG-7].

13.8.3 Visual Descriptors and Description Schemes

For each segment or region at any level of the segment or region tree, a set of audio and visual descriptors and DSs are used to characterize this segment or region. In this section, we briefly describe the visual descriptors and DSs that have been developed to describe the color, texture, shape, motion, and location of a video segment or object. More complete descriptions can be found in [28, 33].

   Color   These descriptors describe the color distributions in a video segment, a moving region or a still region.

- Color space: Five color spaces are defined, RGB, YCrCb, HSV, HMMD, and monochrome. Alternatively, one can specify an arbitrary linear transformation matrix from the RGB coordinate.

- Color quantization: This descriptor is used to specify the quantization parameters, including the number of quantization levels and starting values for each color component. Only uniform quantization is considered.
- Dominant color: This descriptor specifies the dominant colors in the underlying segment, including the number of dominant colors, a value indicating the spatial coherence of the dominant color (i.e., whether the dominant color is scattered over the segment or forms a cluster), and for each dominant color, the percentage of pixels taking that color, the color value, and its variance.
- Color histogram: The color histogram is defined in the HSV space. Instead of using the color histogram itself, the Haar transform is applied to the histogram and the Haar coefficients are specified using variable precision, depending on the available bit rate. Several types of histograms can be specified. The common color histogram, which includes the percentage of each quantized color among all pixels in a segment or region, is called ScalableColor. GoF/GoP Color refers to the average, median, or intersection (minimum percentage for each color) of conventional histograms over a group of frames or pictures.
- Color layout: This descriptor is used to describe on a coarse level the color pattern of an image. An image is reduced to 8×8 blocks, with each block represented by its dominant color. Each color component (Y/Cb/Cr) in the reduced image is then transformed using DCT, and the first few coefficients are specified.
- Color structure: This descriptor is intended to capture the spatial coherence of pixels with the same color. The counter for a color is incremented as long as there is at least one pixel with this color in a small neighborhood around each pixel, called the structuring element. Unlike the color histogram, this descriptor can distinguish between two images in which a given color is present in identical amounts, but where the structure of the groups of pixels having that color is different in the two images.

Texture   This category is used to describe the texture pattern of an image.

- Homogeneous texture: This descriptor is used to specify the energy distribution in different orientations and frequency bands (scales). The first two components are the mean value and the standard deviation of the pixel intensities. The following 30 components are obtained through a Gabor transform with six orientation zones and five scale bands.
- Texture browsing: This descriptor specifies the texture appearances in terms of regularity, coarseness, and directionality, which are in line with the type of descriptions that a person may use in browsing or retrieving a texture pattern. In addition to regularity, up to two dominant directions and the coarseness along each direction can be specified.
- Edge histogram: This descriptor is used to describe the edge orientation distribution in an image. Three types of edge histograms can be specified, each with five

entries, describing the percentages of directional edges in four possible orientations, and nondirectional edges. The global edge histogram is accumulated over every pixel in an image; the local histogram consists of sixteen subhistograms, one for each block in an image divided into 4×4 blocks; the semiglobal histogram consists of thirteen subhistograms, one for each subregion in an image.

Shape   These descriptors are used to describe the spatial geometry of still and moving regions.

- Contour-based descriptor: This descriptor is applicable to a 2-D region with a closed boundary. MPEG-7 has chosen to describe boundaries using the peaks in the curvature scale space (CSS) representation [56], which has been found to reflect human perception of shapes—similar shapes have similar parameters in this representation. The CSS representation of a boundary is obtained by recursively blurring the original boundary using a smoothing filter, computing the curvature along each filtered curve, and finally determining zero-crossing locations of the curvature after successive blurring. The descriptor specifies the number of curvature peaks in the CSS, the global eccentricity and circularity of the boundary, the eccentricity and circularity of the prototype curve (which is the curve leading to the highest peak in the CSS), the prototype filter, and the positions of the remaining peaks.

- Region-based shape descriptor: This descriptor makes use of all pixels constituting the shape, and thus can describe any shape—not only a simple shape with a single connected region, but also a complex shape that consists of several disjoint regions. Specifically, the original shape represented by an alpha-map is projected onto angular radial transform (ART) basis functions, and the descriptor includes 35 normalized and quantized magnitudes of the ART coefficients.

- Shape 3-D: This descriptor provides an intrinsic description of 3-D mesh models. It exploits some local attributes of the 3-D surface. To derive this descriptor, the so-called shape index is calculated at every point on the mesh surface, which depends on the principal curvature at the point. The descriptor specifies the shape spectrum, which is the histogram of the shape indices calculated over the entire mesh. Each entry in the histogram essentially specifies the relative area of all the 3-D mesh surface regions with a shape index lying in a particular interval. In addition, the descriptor includes the relative area of planar surface regions of the mesh for which the shape index is not defined, and the relative area of all the singular polygonal components, which are regions for which reliable estimation of the shape index is not possible.

Motion   These descriptors describe the motion characteristics of a video segment or a moving region, as well as global camera motion.

- Camera motion: Seven possible camera motions are considered: panning, tracking (horizontal translation), tilting, booming (vertical translation), zooming, dollying (translation along the optical axis), and rolling (rotation around the optical axis) (see Figure 5.4). For each motion, two moving directions are possible. For each motion type and direction, the presence (i.e., duration), speed, and amount of motion are specified. The last term measures the area that is covered or uncovered due to a particular motion.

- Motion trajectory: This descriptor is used to specify the trajectory of a nonrigid moving object, in terms of the 2-D or 3-D coordinates of certain key points at selected sampling times. For each key point, the trajectory between two adjacent sampling times is interpolated by a specified interpolation function (either linear or parabolic).

- Parametric object motion: This descriptor is used to specify the 2-D motion of a rigid moving object. Five types of motion models are included: translation, rotation/scaling, affine, planar perspective, and parabolic. The planar perspective and parabolic motions refer to the projective mapping defined in Equation 5.5.14, and the biquadratic mapping defined in Equation 5.5.19, respectively. In addition to the model type and model parameters, the coordinate origin and time duration must be specified.

- Motion activity: This descriptor is used to describe the intensity and spread of activity over a video segment (typically at the shot level). Five attributes can be specified: (1) intensity of activity, measured by the standard deviation of the motion vector magnitudes; (2) direction of activity, which specifies the dominant or average direction of all motion vectors; (3) spatial distribution of activity, derived from the run lengths of blocks with motion magnitudes lower than the average magnitude, (4) spatial localization of activity; and (5) temporal distribution of activity, described by a histogram of quantized activity levels over individual frames in the shot.

Localization　　These descriptors and DSs are used to describe the location of a still or moving region.

- Region locator: This descriptor specifies the location of a region by a brief and scalable representation of a bounding box or polygon.

- Spatial-temporal locator: This DS describes a moving region. It decomposes the entire duration of the region into a few subsegments, with each segment being specified by the shape of the region in the beginning of the segment (known as a reference region), and the motion between this region and the reference region of the next segment. For a nonrigid object, a FigureTrajectory DS is developed, which defines the reference region by a bounding rectangle, ellipse, or polygon, and specifies the motion between reference regions using the MotionTrajectory descriptor, which specifies the coordinates of selected key points over successive sampling times. For a rigid region, a ParameterTrajectory DS is used, which uses the RegionLocator descriptor to specify a reference region, and the parametric object motion descriptor to describe the motion.

13.9 SUMMARY

Video communications require standardization in order to build reasonably priced equipment that interoperates and caters to a large market. Personal video telephony was the first application that was targeted by a digital video compression standard. H.261 was published in 1990—101 years after Jules Verne wrote down the idea of a video telephone, and 899 years earlier than he predicted [69]. The subsequent important video compression standards H.263, MPEG-1, MPEG-2, and MPEG-4 were established in 1993, 1995, 1995, and 1999, respectively.

Whereas the H.261 and H.263 standards describe only video compression, MPEG-1/2/4 standards describe also the representation of audio, as well as a system that enables the joint transmission of audiovisual signals. H.261 is a block-based hybrid coder with integer-pel motion compensation. The main application for H.261 is video coding for video conferencing over ISDN lines at rates between 64 kbps and 2 mbps. H.263 extends H.261, and adds many features (including half-pel motion compensation), thus enabling video coding for transmission over analog telephone lines at rates below 56 kbps.

MPEG-1 was also derived from H.261. It added half-pel motion compensation, bidirectional prediction for B-pictures, and other improvements in order to meet the requirements for coding video at rates around 1.2 mbps, for consumer video on CD-ROM at CIF resolution. MPEG-2 was the first standard that was able to code interlaced video at full TV and HDTV resolution. It extended MPEG-1 to include new prediction modes for interlaced video. Its main applications are in TV broadcasting at rates around 4 mbps, and 15 mbps for high-quality video. MPEG-4 video, based on MPEG-2 and H.263, is the latest video coding standard, which introduces object-based functionalities that describe video objects not only by motion and texture but also by shape. Shape information is colocated with the luminance signal and coded using a context-based arithmetic coder.

MPEG-2 and MPEG-4 define profiles that require a decoder to implement a subset of the tools that the standard defines. This enables the building of standard decoders that are to some extent tailored towards certain application areas.

Whereas MPEG-1/2/4 standards were developed to enable the exchange of audiovisual data, MPEG-7 aims to enable the searching and browsing of such data. MPEG-7 can be used independently of the other MPEG standards—an MPEG-7 description might even be attached to an analog movie. MPEG-7 descriptions could be used to improve the functionalities of previous MPEG standards, but will not replace MPEG-1, 2, or 4.

Since the computational power in terminals increases every year, standardization bodies try to improve on their standards. The ITU is currently working on the video coding standard H.26L, which promises to outperform H.263 and MPEG-4 by more than 1 dB for the same bit rate, or to reduce the bit rate by more than 20 percent for the same picture quality when coding video at rates above 128 kbps.

13.10 PROBLEMS

1. In the content of standards, what kinds of compatibility do you know about?
2. What are the most computationally intensive parts of an H.261 video encoder? What are the most computationally intensive parts of a decoder?
3. What is a loop filter? Why is H.261 the only standard that implements it?
4. What are the tools that improve the coding efficiency of H.263 over H.261?
5. What is the main difference between MPEG-1 B-frames and H.263 PB-frames, according to the improved PB-frame mode?
6. What is the purpose of the standards H.323 and H.324?
7. Why does MPEG-2 have more than one scan mode?
8. What effect does the perceptual quantization of I-frames have on the PSNR of a coded picture? How does perceptual quantization affect picture quality? What is a good guideline for choosing the coefficients of the weight matrix?
9. Explain the concept of profiles and levels in MPEG-2.
10. Which of the MPEG-2 profiles are used in commercial products? Why do the others exist?
11. What kind of scalability is supported by MPEG-2?
12. What is drift? When does it occur?
13. Discuss the error resilience tools that H.261, H.263, and MPEG-1/2/4 provide. Why are MPEG-4 error resilience tools best suited for lossy transmission channels?
14. What are the differences between MPEG-1 Layer III audio coding and MPEG-2 AAC in terms of functionality and coding efficiency?
15. MPEG-4 allows the encoding of shape signals. In the case of binary shape, how many blocks are associated with a macroblock? What is their size? What about greyscale shape coding?
16. Why does MPEG-4 video according to the ACE profile outperform MPEG-1 video?
17. What part of an MPEG-4 terminal is not standardized?
18. Why do video bit streams contain start codes?
19. What is meta-information?
20. Which standard uses a wavelet coder, and for what purpose?
21. Why is the definition of FAPs as done in MPEG-4 important for content creation?
22. How is synchronization achieved between a speech synthesizer and a talking face?
23. What is the functionality and purpose of MPEG-4 mesh animation?

24. What is the difficulty with video indexing and retrieval? How can a standardized content description interface such as MPEG-7 simplify video retrieval?

25. How does the segment tree in MPEG-7 describe the syntactic structure of a video sequence? How does the event tree in MPEG-7 describe the semantic structure of a video sequence? What are their relations?

26. What are the visual descriptors developed by MPEG-7? Assume that these descriptors are attached to every video sequence in a large video database. Describe ways that you may use them to retrieve certain type of sequences.

13.11 BIBLIOGRAPHY

[1] Avaro, O., A. Elftheriadis, C. Herpel, G. Rajan, and L. Ward. MPEG-4 systems: Overview. In A. Puri and T. Chen, eds., *Multimedia Systems, Standards, and Networks*. New York: Marcel Dekker, 2000, 331–65.

[2] Basso, A., M. R. Civanlar, and V. Balabanian. Delivery and control of MPEG-4 content over IP networks. In A. Puri and T. Chen, eds., *Multimedia Systems, Standards, and Networks*. New York: Marcel Dekker, 2000, 501–23.

[3] Brandenburg, K., O. Kunz, and A. Sugiyama. MPEG-4 natural audio coding. *Signal Processing: Image Communication* (2000), 15(4-5):423–44.

[4] Capin, T. K., E. Petajan, and J. Ostermann. Efficient modeling of virtual humans in MPEG-4. In *IEEE International Conference on Multimedia and Expo (ICME2000)*, 2:1103–6, New York, 2000.

[5] Chen, T., Emerging standards for multimedia applications. In L. Guan, S. Y. Kung, and J. Larsen, eds., *Multimedia Image and Video Processing*. CRC Press, 2000, 1–18.

[6] Chen, T., G. J. Sullivan, and A. Puri. H.263 (including H.263++) and other ITU-T video coding standards. In A. Puri and T. Chen, eds., *Multimedia Systems, Standards, and Networks*. New York: Marcel Dekker, 2000, 55–85.

[7] Chiariglione, L., Communication standards: Götterdämmerung? In A. Puri and T. Chen, eds., *Multimedia Systems, Standards, and Networks*. New York: Marcel Dekker, 2000, 1–22.

[8] Dufaux, F., and F. Moscheni. Motion estimation techniques for digital TV: A review and a new contribution. *IEEE* (June 1995), 83(6):858–76.

[9] Lee, Y., E. D. Scheirer, and J.-W. Yang. Synthetic and SNHC audio in MPEG-4. *Signal Processing: Image Communication* (2000), 15(4-5):445–61.

[10] Girod, B., E. Steinbach, and N. Färber. Comparison of the H.263 and H.261 video compression standards. In *Standards and Common Interfaces for Video*, SPIE (Oct. 1995), CR60:233–51.

[11] Hartman, J., and J. Wernecke. *The VRML Handbook*. Reading, MA: Addison Wesley, 1996.

[12] Haskell, B. G., et al. Image and video coding: Emerging standards and beyond. *IEEE Trans. Circuits Syst. for Video Technology* (Nov. 1998), 8(7):814–37.

[13] Haskell, B. G., A. Puri, and A. N. Netravali. *Digital Video: An Introduction to MPEG-2.* New York: Chapman & Hall, 1997.

[14] Herpel, C., A. Elftheriadis, and G. Franceschini. MPEG-4 systems: Elementary stream management and delivery. In A. Puri and T. Chen, eds., *Multimedia Systems, Standards, and Networks.* New York: Marcel Dekker, 2000, 367–405.

[15] ISO/IEC. IS 10918-1: Information technology—digital compression and coding of continuous-tone still images: Requirements and guidelines, 1990. (JPEG).

[16] ISO/IEC. IS 11172: Information technology—coding of moving pictures and associated audio for digital storage media at up to about 1.5 mbit/s, 1993. (MPEG-1).

[17] ISO/IEC. IS 13818-2: Information technology—generic coding of moving pictures and associated audio information: Video, 1995. (MPEG-2 Video).

[18] ISO/IEC. IS 13818-3: Information technology—generic coding of moving pictures and associated audio information—part 3: Audio, 1995. (MPEG-2 Audio).

[19] ISO/IEC. IS 13818-1: Information technology—generic coding of moving pictures and associated audio information: Systems, 1995. (MPEG-2 Systems).

[20] ISO/IEC. IS 13818-3: Information technology—generic coding of moving pictures and associated audio information—part 7: Advanced audio coding (AAC), 1997. (MPEG-2 AAC).

[21] ISO/IEC. IS 14772-1: Information technology—computer graphics and image processing—the virtual reality modeling language—part 1: Functional specification and UTF-8 encoding, 1997. (VRML).

[22] ISO/IEC. IS 14496-X: Information technology—coding of audio-visual objects, 1999. (MPEG-4).

[23] ISO/IEC. IS 14496-1: Information technology—coding of audio-visual objects—part 1: Systems, 1999. (MPEG-4 Systems).

[24] ISO/IEC. IS 14496-2: Information technology—coding of audio-visual objects—part 2: Visual, 1999. (MPEG-4 Video).

[25] ISO/IEC. IS 16500: Information technology—generic digital audio-visual systems, 1999. (DAVIC).

[26] ISO/IEC. Report of the formal verification tests on advanced coding efficiency ACE (formerly Main Plus) profile in version 2. Public document, ISO/IEC JTC 1/SC 29/WG 11 N2824, July 1999.

[27] ISO/IEC. Report of the formal verification tests on MPEG-4 coding efficiency for low and medium bit rates. Public document, ISO/IEC JTC 1/SC 29/WG 11 N2826, July 1999.

[28] ISO/IEC. CD 15938-3: MPEG-7 multimedia content description interface—part 3: Visual. Public document, ISO/IEC JTC1/SC29/WG11 W3703, La Baule, France, October 2000.

[29] ISO/IEC. CD 15938-5: Information technology—multimedia content description interface—part 5: Multimedia description schemes. Public document, ISO/IEC JTC1/SC29/WG11 N3705, La Baule, France, October 2000.

[30] ISO/IEC. IS 14496-3: Information technology—coding of audio-visual objects—part 3: Audio, 2000. (MPEG-4 Audio).

[31] ISO/IEC. IS 14496-6: Information technology—coding of audio-visual objects—part 6: Delivery multimedia integration framework (DMIF), 2000. (MPEG-4 DMIF).

[32] ISO/IEC. Overview of the MPEG-7 standard (version 4.0). Public document, ISO/IEC JTC1/SC29/WG11 N3752, La Baule, France, October 2000.

[33] ISO/IEC. MPEG-7 visual part of experimentation model version 9.0. Public document, ISO/IEC JTC1/SC29/WG11 N3914, Pisa, January 2001.

[34] ITU-R. BT. 500-10: Methodology for the subjective assessment of the quality of television pictures, 2000.

[35] ITU-R. BT.601-5: Studio encoding parameters of digital television for standard 4:3 and wide-screen 16:9 aspect ratios, 1998. (Formerly CCIR601).

[36] ITU-T. Recommendation G.711: Pulse code modulation (PCM) of voice frequencies, 1988.

[37] ITU-T. Recommendation G.722: 7 kHz audio-coding within 64 kbit/s, 1988.

[38] ITU-T. Recommendation G.723.1: Dual rate speech coder for multimedia communications transmitting at 5.3 and 6.3 kbit/s, 1988.

[39] ITU-T. Recommendation G.728: Coding of speech at 16 kbit/s using low-delay code excited linear prediction, 1992.

[40] ITU-T. Recommendation T.81—Information technology—Digital compression and coding of continuous-tone still images—Requirements and guidelines, 1992. (JPEG).

[41] ITU-T. Recommendation H.261: Video codec for audiovisual services at $p \times 64$ kbits, 1993.

[42] ITU-T. Recommendation V.34: A modem operating at data signaling rates of up to 28,800 bit/s for use on the general switched telephone network and on leased point-to-point 2-wire telephone-type circuits, 1994.

[43] ITU-T. Recommendation H.262: Information technology—generic coding of moving pictures and associated audio information: Video, 1995.

[44] ITU-T. Recommendation H.324: Terminal for low bit rate multimedia communication, 1995.

[45] ITU-T. Recommendation H.223: Multiplexing protocol for low bit rate multimedia communication, 1996.

[46] ITU-T. Recommendation H.320: Narrow-band visual telephone systems and terminal equipment, 1997.

[47] ITU-T. Recommendation V.25ter: Serial asynchronous automatic dialling and control, 1997.

[48] ITU-T. Recommendation H.225.0: Call signaling protocols and media stream packetization for packet based multimedia communications systems, 1998.

[49] ITU-T. Recommendation H.245: Control protocol for multimedia communication, 1998.

[50] ITU-T. Recommendation H.263: Video coding for low bit rate communication, 1998.

[51] ITU-T. Recommendation H.323: Packet-based multimedia communications systems, 1998.

[52] ITU-T. Recommendation V.8 bis: Procedures for the identification and selection of common modes of operation between data circuit-terminating equipments (DCEs) and between data terminal equipments (DTEs) over the public switched telephone network and on leased point-to-point telephone-type circuits, 1998.

[53] ITU-T. Recommendation V.8: Procedures for starting sessions of data transmission over the public switched telephone network, 1998.

[54] Koenen, R., Profiles and levels in MPEG-4: Approach and overview. *Signal Processing: Image Communications* (2000), 15(4-5):463–78.

[55] Mitchell, J. L., W. B. Pennebaker, C. E. Fogg, and D. J. LeGall. *MPEG video compression standard*. Digital Multimedia Standards Series. Bonn; Germany: Chapman and Hall, 1996.

[56] Mokhtarian, F. S., S. Abbasi, and J. Kittler. Robust and efficient shape indexing through curvature scale space. In *British Machine Vision Conference*, 53–62, Edinburgh, England, 1996.

[57] Musmann, H. G., and J. Klie. TV transmission using a 64 kbit/s transmission rate. In *International Conference on Communications*, Boston, MA, IEEE 23.3.1–5, 1979.

[58] Okubo, S., Reference model methodology—a tool for the collaborative creation of video coding standards. *IEEE* (Feb. 1995), 83(2):139–50.

[59] Orchard, M. T., and G. J. Sullivan. Overlapped block motion compensation: An estimation-theoretic approach. *IEEE Trans. Image Process* (1994), 3:693–99.

[60] Ostermann, J., and A. Puri. Natural and synthetic video in MPEG-4. In *IEEE International Conference on Acoustics, Speech and Signal Processing (ICASSP'98)*, IEEE, Seattle, WA, USA (Nov. 1998), 3805–09.

[61] Puri, A. Video coding using the MPEG-2 compression standard. *SPIE Visual Communications and Image Processing* (Nov. 1993), 1199:1701–13.

[62] Puri, A., and A. Wong. Spatial domain resolution scalable video coding. *SPIE Visual Communications and Image Processing* (Nov. 1993), 1199:718–29.

[63] Puri, A., L. Yan, and B. G. Haskell. Temporal resolution scalable video coding. *International Conference on Image Processing (ICIP 94)*, IEEE, Austin, Texas, USA (Nov. 1994), 2:947–51.

[64] Salembier, P., and J. R. Smith. MPEG-7 multimedia description schemes. *IEEE Trans. Circuits Syst. for Video Technology* (June 2001), 6:748–759.

[65] Schulzrinne, H., S. Casner, R. Frederick, and V. Jacobson. RTP: A transport protocol for real-time applications. IETF RFC 1889, Jan. 1996 (Available via ftp://ftp.isi.edu/in-notes/rfc1889.txt).

[66] Signés, J., Y. Fisher, and A. Eleftheriadis. MPEG-4's binary format for scene description. *Signal Processing: Image Communications* (2000), 15(4-5):321–45.

[67] Signés, J., Y. Fisher, and A. Elftheriadis. MPEG-4: Scene representation and interactivity. In A. Puri and T. Chen, eds., *Multimedia Systems, Standards, and Networks*. New York: Marcel Dekker, 2000:407–47.

[68] Tekalp, A. M., and J. Ostermann. Face and 2-D mesh animation in MPEG-4. *Signal Processing: Image Communications, Special Issue on MPEG-4* (Jan. 2000), 15:387–421.

[69] Verne, J. In the twenty-ninth century. The day of an American journalist in 2889. In *Yesterday and Tomorrow*, London: Arco, 1965. (Translation by I. O. Evans of *La Journée dún journaliste américain ex 2890,* Amiens: Atelier deGué, 1979, a short story written in 1888 and first published in English in 1889 as *In the Year 2089.*)

[70] Wiegand, T., M. Lightstone, D. Mukherjee, T. Campell, and S. K. Mitra. Rate-distortion optimized mode selection for very low bit rate video coding and the emerging H.263 standard. *IEEE Trans. Circuits Syst. for Video Technology* (April 1996), 6(2):182–90.

14

Error Control in Video Communications

In previous chapters, we have discussed various video coding techniques and standards. For effective video communications, reduction of raw video data rates is only one of the necessary steps. Another equally important task is handling errors and losses in a communication network. In contrast with data communications, which are not usually subject to strict delay constraints and can therefore be handled using network protocols that use retransmission to ensure error-free delivery, real-time video is delay sensitive and cannot easily make use of retransmission. The extensive use of predictive and variable-length coding in video coding renders compressed video especially vulnerable to transmission errors, and successful video communication in the presence of errors requires careful designs of the encoder, decoder, and other system layers.

In this chapter, we present approaches that have been developed for error control in video communications. We start by describing the necessity and the challenges of error control for video communications, and provide an overview of various approaches that have been developed (Section 14.1). To help understand the issues in error control in video communications, we describe in Section 14.2 the QoS requirements of various video services and the characteristics of different networks. Section 14.3 presents error control mechanisms at the transport level. Sections 14.4 and 14.5 review techniques for error-resilient encoding and error concealment. Section 14.6 describes techniques that rely on encoder and decoder interactions. Finally, Section 14.7 summarizes error resilience tools adopted by the H.263 and MPEG-4 standards.

14.1 MOTIVATION AND OVERVIEW OF APPROACHES

A video communication system typically involves five steps, as shown in Figure 14.1. The video is first compressed by a video encoder to reduce the data rate, and the compressed bit stream is then segmented into fixed or variable-length packets and multiplexed with other data types, such as audio. The packets might be sent directly over the network, if the network guarantees bit-error–free transmission. Otherwise, they usually undergo a channel encoding stage, typically using forward error correction (FEC) and interleaving, to protect them from transmission errors. At the receiver end, the received packets are FEC decoded and unpacked, and the resulting bit stream is then input to the video decoder to reconstruct the original video. In practice, many applications embed packetization and channel encoding in the source coder as an adaptation layer to the network.

Transmission errors can be roughly classified into two categories: *random bit* errors and *erasure* errors. Random bit errors are caused by the imperfections of physical channels, which result in bit inversion, insertion, and deletion. Depending on the coding methods and the affected information content, the impact of random bit errors can range from negligible to objectionable. When fixed-length coding is used, a random bit error will affect only one codeword, and the damage caused is generally acceptable. But if VLC (e.g., Huffman coding) is used, random bit errors can desynchronize the coded information, such that many following bits are undecodable until the next synchronization codeword appears. Erasure errors, on the other hand, can be caused by packet loss in packet networks such as the Internet, burst errors in storage media due to physical defects, or system failures for a short time. Random bit errors in VLC coded streams can also cause effective erasure errors, since a single bit error can lead to many following bits becoming undecodable, hence useless. The effect of erasure errors (including those due to random bit errors) is much more destructive than random bit errors, due to the loss of or damage to a contiguous segment of bits.

Error control in video communications is very challenging for several reasons. First, compressed video streams are extremely vulnerable to transmission errors because of the use of temporal predictive coding and VLC by the source coder. Due to the use of temporal prediction, a single erroneously recovered sample can lead to errors in the following samples in the same and following frames, as illustrated in Figure 14.2. Note that the error not only propagates temporally, but also spreads spatially due to motion-compensated prediction. Second, with VLC, the effect of a bit error is equivalent

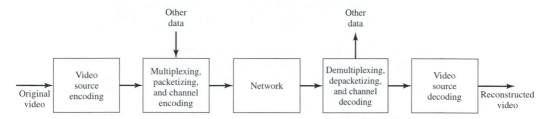

Figure 14.1   A typical video communication system.

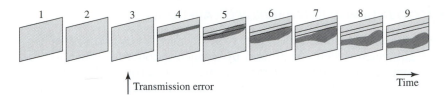

Transmission error   Time

Figure 14.2   Illustration of spatiotemporal error propagation. Reprinted from
B. Girod and N. Färber, Feedback-based error control for mobile video transmission,
Proceedings of the IEEE (Oct. 1999), 87:1707–23. Copyright 1999 IEEE.

to that of an erasure error, causing damage over a large portion of a video frame.
Typically, in a block-based hybrid coder (Section 9.3.1), each group of blocks (typically
a row of macroblocks) is led by a resynchronization marker, which helps the decoder to
regain synchronization. If a transmission error is detected in the middle of a GOB, the
entire GOB is discarded. The decoder resumes decoding in the next GOB. To illustrate
the visual artifacts caused by transmission errors, Figure 14.3 shows a typical example
of reconstructed frames in the presence of packet loss. In this example, a damaged GOB

Figure 14.3   Effect of transmission errors on a compressed video stream using the
H.263 standard for a selected frame. Upper left: no transmission errors were present
and the picture quality is as high as the bit rate allows; upper right: 3% packet loss;
lower left: 5% packet loss; lower right: 10% packet loss. Reprinted from Y. Wang,
S. Wenger, J. Wen, and A. G. Katsaggelos, Error resilient video coding techniques,
IEEE Signal Processing Magazine (July 2000), 17(4):61–82. Copyright 2000 IEEE.

is recovered using the so-called motion-compensated temporal interpolation method (Section 14.5.1). This method can conceal errors somewhat when the packet loss rate is not too high (see the figure for the cases with 3% and 5% packet loss rate), but breaks down when too many packets are lost (see the 10% loss case). Note that the artifacts in the frame shown are due to packet losses in data not only for this frame, but also for previous frames.

To make the compressed bit stream resilient to transmission errors, one must add redundancy to the stream, so that it is possible to detect and correct errors. Typically, this is done at the channel by using FEC, which operates over coded bit streams generated by a source coder. Classical Shannon information theory states that one can separately design the source and channel coders, to achieve error-free delivery of a compressed bit stream, as long as the source is represented by a rate below the channel capacity. Therefore, the source coder should compress a source as much as possible (to below the channel capacity) for a specified distortion, and then the channel coder can add redundancy through FEC to the compressed stream to enable the correction of transmission errors. However, such ideal error-free delivery can be achieved only with infinite delays in implementing FEC, which are not acceptable in practice. Therefore, joint source and channel coding is often a more viable scheme, which allocates a total amount of redundancy between the source and channel coding. All *error-resilient encoding* methods essentially work under this premise, and intentionally make the source coder less efficient than it could be, so that the erroneous or missing bits in a compressed stream will not have a disastrous effect on reconstructed video quality. This is usually accomplished by carefully designing both the predictive coding loop and the variable-length coder, to limit the extent of error propagation.

When an image sample or a block of samples is missing due to transmission errors, the decoder can estimate them based on surrounding received samples, by making use of inherent correlation among spatially and temporally adjacent samples. Such techniques are known as *error concealment*. This is possible because real source coders do not completely eliminate the redundancy in a signal in the encoding process. Error concealment, in contrast to error-resilient source coding, has the advantage of not employing any additional bit rate,* but adds computational complexity at the decoder.

Finally, for the embedded redundancy in the source coder to be useful, and to facilitate error concealment in the decoder, the codec and the network transmission protocol must cooperate. For example, if the bit stream is such that some bits are more important than others, then the important parts should be assigned a more stringent set of QoS parameters for delivery over a network. To suppress error propagation, the network may also provide a feedback channel, so that the encoder knows which part of the reconstructed signal at the decoder is damaged, and does not use this part for prediction of future samples.

To summarize, error control mechanisms devised for video transport can be categorized into four groups: (1) those exercised at the transport level, including channel

---

*To facilitate error concealment, a small amount of redundancy is usually inserted in the encoder, in the form of restricted prediction and data interleaving.

coders, packetizer/multiplexers, and transport protocols; (2) those introduced at the source encoder, to make the bit stream more resilient to potential errors; (3) those invoked at the decoder upon detection of errors, to conceal the effect of errors; and (4) those that require interactions between the source encoder and decoder, so that the sender can adapt its operations based on the loss conditions detected at the decoder. We will describe techniques in these four categories separately in Sections 14.3–6.

14.2 TYPICAL VIDEO APPLICATIONS AND COMMUNICATION NETWORKS

In this section, we describe different types of video applications and characteristics of practical networks. These are important factors to consider, because the necessity for error control and the effectiveness of a technique depend on the type of application as well as the underlying channel characteristics and network protocols.

14.2.1 Categorization of Video Applications

When considering error control in video transport, it is important to know whether the underlying application requires real-time delivery, and what the maximally allowed average end-to-end delay (known as *latency*) and delay variation (known as *jitter*) are. "Real-time" delivery means that the compressed data are transferred at a speed that matches the coded video source rate. For example, if the source video is coded at 10 mbps, then the throughput* of the communication channel devoted this source should be at least 10 mbps. Note that the source can be generated in real time (as in video conferencing applications) or off-line (as in video streaming applications). For all video applications except simple downloading, real-time delivery is required. But some applications (e.g., streaming) can afford a relatively large play-out delay. In this case, the receiver can use a large buffer to smooth out the jitter, so that the decoded video can be played out at a constant frame rate, after an initial play-out delay. For such applications, a limited number of retransmissions can be used to handle transmission errors. In the following, we classify video applications based on their requirements in terms of real-time delivery, latency, and jitter.

Note that the end-to-end delay from a video source to a destination can be contributed by several factors (see Figure 14.4): encoder processing delay ΔT_e (including acquiring the data and encoding); encoder buffer delay ΔT_{eb} (to smooth the rate variation in the compressed bit stream); transmission delay ΔT_c (delay caused by the transmission itself, which is usually very small, and the delay due to queuing and perhaps retransmission in a packet-based network); decoder buffer delay ΔT_{db} (to smooth out transmission jitter); and decoder processing delay ΔT_d (including both decoding and display buffer for constant frame play-out). Of course, when the source is precompressed, as in video broadcast and streaming applications, the encoding delay is not of concern. In this section, we focus on transmission delay, assuming that the delay caused

---

*"Throughput" means the effective end-to-end delivery rate. For example, a network may have a raw bandwidth of 10 mbps, but some data have to be retransmitted due to packet loss or timeout, so that the actual throughput is lower than 10 mbps.

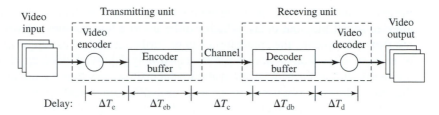

Figure 14.4   Factors contributing to the end-to-end delay in a video communication system. Adapted from A. Ortega and K. Ramchandran, Rate-distortion methods for image and video compression, *IEEE Signal Processing Magazine* (Nov. 1998), 15:23–50. Copyright 1998 IEEE.

by video encoding and decoding and by encoder and decoder buffers are acceptable to the underlying application, and that these delays are relatively fixed.

Interactive Two-Way Visual Communications   Examples in this category include teleconferencing, video telephony, virtual classrooms, and so on. Such applications have very stringent delay requirements. For effective communications, the latency and jitter must be kept within a certain limit. For example, for intercontinental telephone conversations, the ITU-T G.114 standard recommends a maximum one-way delay of 150 ms for acceptable quality. Between 150–400 ms is acceptable provided the network administrations are aware of the impact of the delay on the user applications [37]. The same numbers apply for video conferencing applications. Delay of more than 400 ms is intolerable for effective communications. Also, to maintain lip synchronization when playing audio and video, the video delay relative to the audio should be limited. Usually, audio is intentionally delayed slightly in the sender and receiver, to maintain lip synchronization.

For such applications, encoding and decoding must be accomplished in real time. Otherwise, incoming frames may build up at the encoder, or received bits at the decoder may have to be thrown away. To satisfy the stringent delay requirement, encoder and decoder buffers are typically very small. Retransmission is usually unacceptable. Jitter in the network also must be limited. Because of these constraints, interactive applications are considered the most demanding among the various types of video communications.

Fortunately, only low-to-intermediate video quality, in terms of both spatial and temporal resolution, is expected. For example, QCIF at 5–10 fps is considered acceptable for video phone applications,* and CIF at 10–20 fps is satisfactory for most video conferencing scenarios. Furthermore, a moderate amount of compression and transmission artifacts is often tolerable.

One-Way Video Streaming   In this type of application, a live or precompressed video source is delivered to one or multiple destinations. Instead of waiting for

---

*At such low frame rates, the round-trip delay for video often exceeds the 400-ms limit previously cited, causing problems in lip synchronization.

the entire video file to arrive, the receiving terminal starts to decode and play the video after an initial play-out delay, which can be up to a few seconds. Depending on the number of recipients, the application can be further classified as *broadcast, multicast,* or *unicast.* Television broadcast is the most widely deployed video application. Video multicast, such as the Mbone application, which allows any user on the Internet to view the video of a live conference or seminar, is another important application. An application that is becoming increasingly popular is streaming video from a precompressed video database over the Internet, typically received by a single user.

Obviously, the bandwidth requirements depend on the type of video material. For digital TV broadcast over terrestrial, cable, or direct broadcast satellite (DBS) channels, video programs typically are in the BT.601 resolution, compressed using the MPEG-2 format with bit rates between 3–10 mbps. For HDTV, the bit rates could be up to 20 mbps. For multicasting and unicasting over the Internet or a wireless network—because of the limited bandwidth that can be devoted to each user, and the variability of the bandwidth and delay—lower quality video (CIF or lower, compressed in MPEG-1 or MPEG-4, or other, proprietary formats) is typically used.

Because there is no interactive communication between the sender and receiver, latency can be fairly large, but jitter must be limited, so that the received stream can be decoded and played out smoothly at a constant frame rate, after an initial delay. TV broadcast uses dedicated channels, which have guaranteed bandwidth and very low jitter. For video multicast or unicast over the Internet, where jitter could be large due to variability of network conditions, a large smoothing buffer is typically installed in the receiver to absorb the jitter. The larger the buffer, the longer the play-out delay, but the smoother the displayed video.

In the case of multicast or unicast over the Internet, the potential recipients may be connected to the network with different access links (from a 100 mbps fast Ethernet, to a slow wireless modem link). Also, the receiving terminals may have very different computing powers (from a powerful workstation, to a handheld battery-powered device). The video server must take these variables into account. For example, with scalable coding (Chapter 11), users with different bandwidths and computing powers may choose to extract different portions of a compressed video stream. Video streaming over the Internet and wireless IP networks is discussed in more detail in Chapter 15.

The major difference between live and precompressed sources is, obviously, that the live source must be compressed in real time, which significantly increases the complexity of the system at the transmission end. The encoding process and encoder buffer will also contribute to additional delays.

One-Way Video Downloading   In this case, a prerecorded video (and audio) is downloaded to a destination. The receiving terminal does not start to play the material until the entire video has arrived. This application is the least demanding in terms of delay requirement, although excessive delay (because of very low transmission bandwidth) may cause the receiver to abort the downloading in the middle of a communication session. Scalable coding and progressive transmission are desirable, so that a low-resolution version of the video can be delivered relatively quickly. Because of the relaxed delay requirement, retransmission can be used to handle any lost or erroneously delivered part of the data. No special error control mechanisms are necessary, beyond those typically used for data transfer.

14.2.2 Communication Networks

In this subsection, we describe different types of networks and associated protocols that may be used for video transport. For each network, we describe its characteristics in terms of bandwidth, delay, jitter, and loss rate (including bit error and packet loss), as well as typical video applications carried over such a network.

Public Switched Telephone Network   The PSTN refers to the plain old telephone system (POTS). The best thing about the PSTN is that it is accessible by almost anyone in the United States and, to a large extent, in the world. The fastest transmission rate by the most advanced modem technology has now reached 56 kbps, but this is still too low for carrying video with pleasing quality. The ITU-T H.324 standard series (Section 13.3.2) has been developed for multimedia communications over the PSTN [36]. In H.324-based systems, the trade-off between bit error rate and bit rate of the modem link is adjusted via modem control mechanisms. Most systems employ strong error control coding to achieve very low error rates, at some expense of bit rate [81].

Even at 56 kbps (typically, the achievable transmission rate for the payload on a 56K modem is much lower), only a small-window video (QCIF or lower) can be provided. Although this may be adequate for visual communication purposes, it is not nearly as pleasing as most people would like. Higher rates, up to 6 mbps in the downlink direction (from a central office to the home), are possible with the ADSL (Asymmetric Digital Subscriber Loop) technology [54]. ADSL is one of the main transport media for streaming MPEG-1 and MPEG-2 movies through video-on-demand services. A good discussion of issues involved in multimedia transport over ADSL can be found in [96].

Integrated Service Digital Network   ISDN is the first public network using digital transmission [66]. The data rate can be multiples of basic channels (known as B channels) of 64 kbps each, with the multiplying factor p ranging from 1–24, corresponding to an aggregated bandwidth of 64–1,536 kbps. The basic subscription of ISDN, which is known as basic rate interface, comes with 2B + D channels, where the D channel is 16 kbps and is used for return signaling. At 128 kbps ($p = 2$) for the signal delivery, only very low-quality video (e.g., QCIF at 10 fps) can be achieved. For better video quality (i.e., CIF at 15–30 fps), at least 384 kbps ($p = 6$) is required.

Like the phone line, communication over ISDN is based on *circuit switching;* that is, an ISDN connection is dedicated to a particular conversation once the connection is established. For this reason, the connection is very reliable. The first ITU-T recommendation series for audiovisual conferencing, H.320, was developed for ISDN [40]. H.320 is currently employed by the vast majority of video conferencing and video telephony equipment. The multiplex protocol H.221 [39] used by H.320 systems offers a bit-oriented, practically error-free video transmission channel with a fixed video bit rate.

Beyond the mandatory intra-MB refresh mechanisms of the video coding standard, no other error control tools are necessary [81].

Broadband ISDN (B-ISDN)   The ISDN service just described, with the basic channel rate being 64 kbps, is more accurately known as narrowband ISDN or N-ISDN. This can be offered over existing twisted-pair local-loop wiring. The term broadband ISDN or B-ISDN [66] describes ISDN services that offer high bandwidth channels, including the H0 channel with a rate of 384 kbps, the H11 channel with a rate of 1,536 mbps, and the H12 channel with a rate of 1,920 mbps. This requires the use of higher-bandwidth coaxial cables or optical fibers. The bandwidth of B-ISDN is also specified in multiples of 64 kbps, with the multiplying factor ranging from 1–65,535.

For efficient usage of the bandwidth, the B-ISDN uses ATM packet-switching technology with fixed-size packets (known as cells). No dedicated links are created for an established connection, which makes it less reliable (i.e., with more delay variation and potential cell loss) than circuit-switched ISDN. The very short cell size (53 bytes, with 48 bytes payload) makes it suitable for real-time applications with low delay requirements. In such networks, cell loss can occur due to traffic congestion, although the loss ratio is quite low; typically the cell loss rate (CLR) is in the range of 10^{-6}–10^{-4} for video services [3]. Generally, cell loss can be considered as a sub-form of packet loss, whereby cells are extremely small packets. The mechanisms to cope with cell losses, however, are different, because it is not effective to add synchronization markers at the beginning of each cell, from a resilience versus overhead trade-off perspective.

The ATM network provides four types of services: constant bit rate (CBR), variable bit rate (VBR), available bit rate (ABR), and unspecified bit rate (UBR). Typically, video is carried in the VBR service, where the compressed bit stream must satisfy preset constraints in terms of average and peak rates. The allowance of rate variation makes it possible to choose the video coding parameters to maintain a relatively constant video quality, which is a desirable feature. Good reviews of ATM networks and issues related to video transport over ATM can be found in [3, 5].

Internet   The Internet is composed of multiple physical networks interconnected by computers known as routers. The most basic building block is known as a *local area network* (LAN), which typically has a bandwidth from 10–100 mbps. Interconnected LANs in a local region form a *metropolitan area network* (MAN), and finally several interconnected MANs form a *wide area network* (WAN). Communications on the Internet are based on *packet-switching*. That is, the data to be delivered are divided into packets, with each packet given a header that specifies the source and destination addresses. Each packet is delivered independently. Packets may be discarded due to buffer overflow at intermediate network nodes (such as switches or routers), or considered lost due to excessively long queuing delay. For real-time video applications, any packet arriving after the allowed delay time is also considered lost. The transport of the packets is governed by the TCP/IP protocol suite, which includes three important protocols: the Internet protocol (IP), the user datagram protocol (UDP) for unreliable connectionless packet-delivery service, and the transmission control protocol (TCP) for reliable stream services. IP is the lower layer protocol on top of which UDP and TCP operate. TCP guarantees delivery by using automatic retransmission request (ARQ) [16].

The Internet and its associated TCP/IP protocol were originally designed for data communications, which are not delay sensitive. For real-time services such as video conferencing and video streaming, the retransmission mechanism in TCP is not appropriate. To support such applications, real-time transport protocol (RTP) and its companion real-time control protocol (RTCP) have been developed. RTP is built on top of UDP, but has added information in the header, including sequence number and timing and synchronization information. RTP allows the detection of lost packets based on sequence numbers as well as the discarding of late packets at the receiver end. RTCP is designed to provide QoS feedback (such as packet loss rate) to the participants of an RTP session, so that the sender can adjust its operation correspondingly (e.g., adjust the sending rate or change the error control mechanism). A good discussion of the Internet protocols for supporting continuous media services, including RTP/RTCP, real-time streaming protocol (RTSP) for medium-on-demand, session initiation protocol (SIP) for internet telephony, session announcement protocol (SAP) for broadcast applications, and session description protocol (SDP), can be found in [62].

Typically, headers added at different network layers will contribute about 40 bytes to a packet. To gain a reasonable payload versus overhead tradeoff, large data packets of around 1500 bytes are usually employed. By using a checksum, certain bit errors in a packet can be detected and a packet is discarded completely if any bit error is detected. Thus any received packets can be considered bit-error free. The packet loss rate depends on the network conditions and can be as low as 0% in the case of a highly overprovisioned private IP network, or as high as 30% and more for long-distance connections during peak times on the Internet [6, 81]. With RTP, lost packets can be easily identified by the use of the sequence number. This information can be conveyed to the video decoder, to enable decoder-based error concealment.

The most widely used physical network for supporting local Internet traffic is Ethernet, which has a bandwidth of 10 mbps. More advanced networks include fast-Ethernet and fiber distributed data interface (FDDI) networks; both support data rates up to 100 mbps. Such faster networks are typically used in corporate Intranet environments.

Because of its ubiquity, the Internet has been envisioned as the future platform for carrying various video services. However, because the Internet is a best-effort network that does not guarantee lossless or timely delivery, error control in video communication over the Internet is a challenging research field. More substantial discussion of the Internet and the RTP/RTCP protocol can be found in Chapter 15, which focuses on video streaming applications over the Internet.

Wireless Networks   There are a variety of wireless networks targeting different environments and applications. These systems have very different capacities, coverages, and error characteristics. In the following paragraphs, we briefly summarize different wireless services. We would like to note that wireless communication is a quickly evolving technology and that newer, more reliable and faster services are likely to become available by the time this book reaches the reader.

Cellular networks [27] are traditionally developed for mobile voice communication services, but have now progressed to allow data transfer through wireless modems. The data rate is limited to under 20 kbps (the payload rate is much lower). Alternatively, the cellular network can be connected to a packet-based *wireless data network*, which enables one to transfer data between a mobile user and a wired terminal over the Internet (such as in web browsing applications). For example, the basic GPRS (general packet radio service) wireless access network used in the global system for mobile communications (GSM), which employs time division multiple access (TDMA), offers a payload bit rate ranging from 9–21.4 kbps using a single time slot, whereas enhanced GRPS wireless access technology (known as enhanced data rates for GSM evolution or EDGE) can provide bit rates ranging from 8.8–59.2 kbps [9]. By using multiple time slots, the raw data rate can be as high as 170 kbps. The available data rate is location dependent, with higher rates when the user is closer to the base station. The upcoming *third generation (3G) wireless systems* promise to provide higher data rates and ubiquitous coverage by using higher carrier frequencies, wider bandwidth, and more sophisticated techniques for multiple access, error control, and signal detection. The outdoor data rates will range from 144–384 kbps, depending on the moving speed of the user and distances between terminals, and the indoor data rates will be at least 2 mbps. The bit-error rate (BER) for data transmission will be below 10^{-6}. Real-time audiovisual communication will be an integral part of the 3G services. In general, any wireless channel is quite noisy, with high BERs. But the use of rate adaptation, FEC, and ARQ yields an almost error-free environment for data transmission. For video transmission, where ARQ must be limited, one must cope with quite high BERs and packet loss rates. This is discussed further in Section 14.3.2 when we consider wireless video transmission using the H.223 multiplex protocol.

A *wireless LAN* refers to a set of information devices (typically indoor, and stationary or slowly moving) connected into a LAN using wireless connections. Either IP or ATM protocols can be used, which are referred to as mobile IP and wireless ATM, respectively. The main challenges in adapting conventional IP and ATM protocols to wireless environments are providing continuous network connectivity to mobile nodes, and handling hand-offs, when the mobile terminal is switching from the coverage area of one access point to another. Achievable bit rates depend on the carrier frequencies. For example, LANs conforming to the IEEE 802.11 standard can have bit rates up to 11 mbps (with lower payload rate). The connection in a wireless LAN is either very good (e.g., BER lower than 10^{-5}), when the user is close to the access point, or very bad (e.g., BER higher than 10^{-2}), when the user is far from the access point. The packet loss rates depend on the packet size and the error detection and correction methods incorporated within the packet.

Broadband wireless IP networks refer to the access to the Internet by home or business users through small rooftop antennas communicating to a base station using

TABLE 14.1   DIFFERENT TYPES OF WIRELESS NETWORKS

Network	Data rate	Mobility	Range	Channel quality
Cellular network	Low ($\leq$ 20 kbps)	High (vehicular)	Long (2500 meters)	Poor
Wireless data network	Intermediate (64–384 kbps)	High (vehicular)	Long (2500 meters)	Poor
Wireless LAN	High (2–25 mbps)	Low (indoor)	Short (50 meters)	Location dependent
Wireless IP network	High (1–600 mbps, downlink)	Low (indoor)	Long 3–30 miles	Good

microwave radio. This includes *multichannel multipoint distribution service* (MMDS) and *local multipoint distribution service* (LMDS). LMDS uses higher carrier frequencies, and thus enables a higher data rate (typically up to 100 mbps, but can be as high as 600 mbps) than MMDS (up to 1 mbps). However, LMDS is effective within a much shorter range between the base station and the client site (three versus 30 miles).

Table 14.1 summarizes the characteristics of the networks that we have discussed. A good review of various wireless networks can be found in [9].

Difficulties for video transport over wireless networks lie in the low bandwidth and high error rate (random bit error, as well as long bursts because of the multipath fading effect)—and, most importantly, the *fluctuation* of available bandwidth and error characteristics. Such variability is particularly acute when mobile stations are involved. A wireless video communication system therefore must be adaptive both in rate and error resilience. To meet such requirements, most proposals for wireless video transport use layered coding with unequal error protection. It is worth noting that, for real-world applications, quite sophisticated FEC codes are used to reduce the BERs significantly. On top of FEC, real-world systems generally employ certain forms of channel multiplexers that often include transport protocol functionality as well, thereby reducing the error rates further.

Most wireless interactive multimedia communication systems employ H.223 [38] and its "mobile extensions" as the transport/multiplex protocol on top of the bit-oriented channel [81]. The mobile extensions form a hierarchy with five different levels, which allows for a scalable trade-off between robustness against bit errors of the multiplexer itself and the overhead incurred by that multiplexer. Most research for video coding on top of the H.223 transport is performed with average burst lengths in the neighborhood of sixteen bits, and BERs between 10^{-5} and 10^{-3}. H.223 conveys media data, including compressed video, in the form of packets of variable size. Typical packet sizes are around 100 bytes to ensure good delay characteristics. If bit errors damage the protocol structures of H.223 beyond the repair facilities of the employed level (a condition known as multiplex errors), then whole packets can get lost as well. Therefore, video transmission on top of H.223 must cope with packet losses.

For a good discussion of the issues involved in video transport over wireless systems, see [76, 25]. Video streaming over wireless IP networks is further considered in Chapter 15.

Broadcast Channels: Terrestrial, Cable, Satellite   This transport environment is used for broadcasting of digital TV (including HDTV programs) using the MPEG-2 video coding and transport streams. The compressed data are carried via fixed-size transport packets of 188 bytes each. Regardless of the underlying wireline or wireless physical layer, the channel coders and the MPEG-2 transport layer ensure an almost error-free environment under good weather conditions. For SDTV, the available bandwidth is divided into program channels that can carry between 3–10 mbps each. For HDTV, 20 mbps is typically allocated. A good reference on digital TV services using DBS is [55].

Table 14.2 summarizes the characteristics of major video communications applications, including the standards used, the target network, and transmission characteristics (such as packet sizes and typical error rates).

TABLE 14.2   CHARACTERISTICS OF MAJOR VIDEO COMMUNICATIONS APPLICATIONS

Application and standard family	Multiplex protocol	Video coding standard	Typical video bit rate	Packet size	Error characteristics
ISDN video phone (H.320)	H.221	H.261 and H.263	64–384 kbps	N/A	Practically error free (BER = 10^{-10}–10^{-8})
PSTN video phone (H.324)	H.223	H.263	20 kbps	100 bytes	Very few bit errors and packet losses
Mobile video phone (H.324 wireless)	H.223 with mobile extensions	H.263	10–300 kbps	100 bytes	BER = 10^{-5}–10^{-3}, occasional packet loss
Video phone over packet network (H.323)	H.225/RTP/ UDP/IP	H.261, H.263, H.262	10–1000 kbps	≤1500 bytes	BER = 0, 0–30% packet losses
Terrestrial/cable/ satellite TV	MPEG-2 system	MPEG-2 video	6–12 mbps	188 bytes	Almost error free, depending on weather
Video conferencing over "Native" ATM (H.310, H.321)	H.222.0	H.262	1–12 mbps	53 bytes (ATM cell)	Almost error free (CLR = 10^{-6}–10^{-4})

N/A: not applicable; H.262 is identical to MPEG-2 video, and H.222.0 is identical to MPEG-2 systems. Adapted from Y. Wang, S. Wenger, J. Wen, and A. G. Katsaggelos, Error resilient video coding techniques, *IEEE Signal Processing Magazine* (July 2000), 17(4):61–82. Copyright 2000 IEEE.

14.3 TRANSPORT-LEVEL ERROR CONTROL

Beginning with this section, we describe various error control mechanism that have been developed for video communications. We start with transport-level error control, because this is typically the most important part of error control. It provides a basic QoS level that can be further improved upon by additional error control mechanisms at the encoder and decoder. Transport-level error control can be exercised at the channel coder, packetizer/multiplexer, and transport protocol levels. We discuss these separately in this section.

14.3.1 Forward Error Correction

FEC is well known for both error detection and correction in data communications [52]. However, since FEC has the effect of increasing transmission overhead and therefore reducing usable bandwidth for the payload data, it must be used judiciously in video services, which are very demanding in bandwidth but can tolerate a certain degree of loss.

When applied directly to the compressed bits, FEC is effective only for channels where bit errors dominate. For example, in the H.261 standard for video conferencing over ISDN [35], an eighteen-bit error correction code, BCH (511,493), is computed and appended to 493 video bits (one fill bit, 492 coded bits). Together with an additional framing bit, the resulting data are grouped into a 512-bit frame. The FEC code is able to correct single bit errors and to detect double bit errors in each frame. The same FEC mechanism can also be used in H.263 [42, Annex H] (Section 14.7.1). However, when H.263 is applied for transporting video over wireless networks or the Internet, where the error burst is usually longer than two bits, this method is not very useful and hence seldom used. Other transport-level error control techniques discussed in this section are more effective.

For packet-based transmission, it is much more difficult to apply error correction, because several hundred bits must be recovered when a packet loss occurs. Usually, FEC is applied across data packets, so that a packet loss will lead to the loss of only one byte in an FEC block. For example, in the approach of Lee [51], Reed-Solomon (RS) codes are combined with block interleaving to recover lost ATM cells. As shown in Figure 14.5, a RS (32,28,5) code is applied to every block of 28 bytes of data to form a block of 32 bytes. After applying the RS code row by row in the memory up to the 47th row, the payload of 32 ATM cells is formed by reading column by column from the memory, with the attachment of one byte indicating the sequence number. In this way, a cell loss detected at the decoder corresponds to one byte erasure in each row of 32 bytes after deinterleaving. Up to two lost cells out of 32 cells can be recovered. The Grand-Alliance HDTV broadcast system has adopted a similar technique for combating transmission errors [10]. In [2], Ayanoglu et al. explored the use of FEC for MPEG-2 video in a wireless ATM. FEC is employed at the byte level for random bit-error correction, and at the ATM cell level for cell loss recovery. These FEC techniques are applied to both single-layer and two-layer MPEG data.

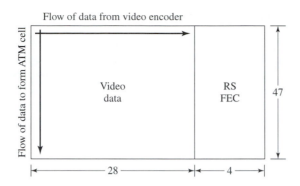

Figure 14.5   Illustration of packet-level FEC for ATM cell loss recovery. The numbers in the figure are in bytes. Reprinted from Y. Wang and Q.- F. Zhu, Error control and concealment for video communication: A review, *Proceedings of the IEEE* (May 1998), 86:974–97. Copyright 1998 IEEE.

14.3.2 Error-Resilient Packetization and Multiplexing

Depending on how compressed data are packetized, a packet loss can have different effects on the reconstructed video. It is important to perform packetization in such a way that the error can be isolated within a small region. This means that compressed bits cannot be packetized blindly into equal-sized packets; rather, the packets should be constructed according to the underlying encoding algorithm, so that a packet contains one or several independently coded data blocks. For example, for MPEG-2 coded video, a packet may contain one or several slices, possibly with repetition of the picture header at the beginning of each packet. Similarly, for H.263 coded video, a packet may include one or several GOBs. Such application-aware packetization is supported in the Internet, and known as *application layer framing* (ALF) [13]. In networks supporting variably sized packets, it is desirable to use the largest possible packets to minimize the reduction in effective data rates by packet headers and the packet processing overheads. For example, with the Internet, the maximum transmission unit (MTU) is around 1500 bytes. This size is sufficiently large to pack data from several slices or GOBs. For low–bit-rate interactive applications, the acceptable delay may limit the largest packet size to a value much smaller than the MTU. In such cases, header compression may be used to reduce the overhead [13].

As described previously, for video transport over the Internet, the RTP/UDP/IP protocol stack is typically used. RTP provides support for different video coding formats through the definition of RTP payload formats [73, 32, 15, 7]. A detailed description of the payload formats for H.261, H.263, and MPEG-1 and MPEG-2 can be found in [13].

In order to prevent the loss of contiguous blocks because of a single packet loss, *interleaved packetization* can be used, by which data from adjacent blocks or rows are put into separate packets, as illustrated in Figure 14.6. This way, a damaged block will usually be surrounded by undamaged blocks, which will ease the error concealment task at the decoder. Note that the use of interleaved packetization in the transport layer requires the source coder to perform block-level prediction only within blocks that are to be packetized sequentially. This will reduce the prediction gain slightly. Also, certain headers may have to be repeated so that each received packet is self-decodable. For

Packet 1	Packet 2	Packet 3	Packet 4	···
Frame 1 Even row data	Frame 1 Odd row data	Frame 2 Even row data	Frame 2 Odd row data	···

Figure 14.6   An example of interleaved packetization.

example, in the packetization format shown in Figure 14.6, the header for each frame should be repeated in both of the packets (containing even and odd row data).

Another factor that influences the error characteristics of video communication is how the video and other data (audio, control) are multiplexed and how the header is protected. An error in the multiplex header can cause the video data be misdelivered, causing large chunks of data to disappear at the receiver [76]. To prevent such events from happening, each packet is typically led by a long synchronization flag, and the header is heavily protected using FEC. A good example is the multiplexing standard H.223 [38], developed under the umbrella of the H.324 system for low–bit-rate multimedia communications. It is designed to support multiplexing of data from multiple sources on a circuit-switched network (e.g., wireline and wireless modems). H.223 offers a hierarchical, multilevel multiplexing structure, allowing for a scalable trade-off between the robustness against bit errors of the multiplexer itself and the overhead incurred by that multiplexer. Level 0, which is targeted for wireline environments, uses a short sync flag and no FEC on the header. Higher levels, which are targeted for more error-prone environments (such as mobile wireless modems), use increasingly longer sync flags and stronger protection for the header.

14.3.3 Delay-Constrained Retransmission

Beyond the low-level error control mechanisms such as FEC and error-resilient packetization and multiplexing, the upper-layer transport protocol can exercise error control, typically in the form of ARQ, which requests retransmission upon the detection of lost or overly delayed packets, as is done in TCP. Retransmission has been used very successfully for non–real-time data transmission, but it has been generally considered unacceptable for real-time video applications because of the delay incurred. In fact, this is not always the case. For example, for a coast-to-coast interactive service, one retransmission adds only about 70 ms delay, which can be acceptable [59]. For one-way real-time video applications such as Internet video streaming and broadcast, the delay allowance can be further relaxed to a few seconds, so that several retransmissions are possible. Retransmission has also been considered inappropriate for multipoint video conferencing because the retransmission requests from a large number of decoders can overwhelm the encoder. However, when a multipoint control unit (MCU) is used in a multipoint conference, the paths between the encoder and the MCU, and between the MCU and the decoders, are simply point-to-point. Retransmission can be applied in these paths separately. Another concern about using retransmission is that it may worsen the problem, because it will add more traffic on the network and thus further increase the packet loss rate. However, if retransmission is controlled appropriately, the end-to-end quality can be improved. For example, the encoder can reduce its current output rate so that the sum of the encoder output and the retransmitted data is kept below a given total data rate.

Obviously, for real-time applications, retransmission must be constrained so that the incurred delay is within the acceptable range. Instead of trying retransmission indefinitely to recover a lost packet, as done in TCP, the number of retransmission trials can be determined by the desired delay [53]. One can also combine layered coding (Section 14.4.4) with prioritized retransmission, by placing the base-layer packets of a layered coder in the front of the transmission queue to increase the number of retransmission trials for the base layer [64]. Finally, one can send multiple copies of a lost packet in each single retransmission trial, hoping that at least one gets through in time [98]. The number of retransmission trials, and the number of retransmission copies for a lost packet, can be determined according to the packet's importance. For example, a base-layer packet can be retransmitted several times, each time with multiple copies, whereas an enhancement-layer packet can be simply dropped. Another constrained retransmission scheme specifically designed for video streaming applications is described in Chapter 15.

14.3.4 Unequal Error Protection

The binary bits in a compressed video bit stream are not equally important. For example, the picture header and other side information are much more important than the block data, in a block-based hybrid coder. These important bits should be protected so that they can be delivered with a much lower error rate. When layered coding is used at the source coder (Section 14.4.4), the transport controller must assign appropriate priority to different layers, which is a form of transport-level control.

Different networks may implement transport prioritization using different means. In ATM networks, there is one bit in the ATM cell header that signals its priority. When traffic congestion occurs, a network node can choose to discard first the cells having low priority. Transport prioritization can also be implemented by using different levels of power to transmit the substreams in a wireless network. In the Internet, the latest RTP specification defines generic mechanisms for transporting layered video bit streams [63]. There is also network support for prioritized delivery through differentiated services [4]. In either wireless networks or the Internet, prioritization can be realized with different error-control treatments to various layers. For example, retransmission and strong FEC can be applied for the base layer, whereas no retransmission and weaker FEC may be applied to the enhancement layers. In the extreme case, the most important information can be duplicated. For example, dual transmission for picture header information and quantization matrix has been proposed for MPEG video [70]. Another way to provide different levels of protection is by using different transport protocols. For example, in the video-on-demand system described in [14], TCP is used for transmission of a very small amount of high-priority data (session control data) before a service session, and UDP is employed for the remaining low-priority data (video data) during the session.

14.4 ERROR-RESILIENT ENCODING

The previous section described various error control mechanisms that can be exercised at the transport level. These methods are applied to coded video streams, attempting to detect, correct, and if necessary and feasible, retransmit damaged data. Such methods, even when applicable, may not completely correct the error, so that the received bit stream at the decoder may still contain errors (bit errors as well as packet loss). In this section, we describe source coding methods that can produce a bit stream that is robust to transmission errors, so that an error will not adversely affect the decoder operation and lead to unacceptable distortions in the reconstructed video quality.

Compared to source coders that are optimized for coding efficiency, such coders typically are less efficient, in that they use more bits to obtain the same video quality in the absence of any transmission errors. These extra bits are called *redundancy bits,*[*] and they are introduced to enhance the video quality when the bit stream is corrupted by transmission errors. *The design goal in error-resilient coding is to achieve the best decoded video quality for a given amount of redundancy, or minimize the incurred redundancy while maintaining a prescribed quality level, both under an assumed channel environment.* There are many ways to introduce redundancy into the bit stream. Some of the techniques help to isolate the error (Section 14.4.1), whereas others enable the decoder to perform better error concealment upon detection of errors (Sections 14.4.2 and 14.4.3). Yet another group of techniques are aimed at guaranteeing a basic level of quality and providing a graceful degradation upon the occurrence of transmission errors (Sections 14.4.4 and 14.4.5).

14.4.1 Error Isolation

One main cause for the sensitivity of a compressed video stream to transmission errors is that a video coder uses VLC to represent various symbols. Any bit errors or lost bits in the middle of a codeword can not only make this codeword undecodable, but also make the following codewords undecodable or decoded to wrong symbols, even if they are received correctly. Error-isolation techniques, as the name implies, try to isolate the effect of a transmission error within a limited region. This is often achieved by placing "resynchronization markers" in the compressed bit stream, and by a technique known as "data partitioning." Both have been incorporated into MPEG-4 and H.263 standards.

Inserting Resynchronization Markers   One simple and effective approach for enhancing encoder error resilience is by inserting resynchronization markers periodically. These markers are designed such that they can be easily distinguished from all other codewords and small perturbation of these codewords. Usually some header information (regarding spatial and temporal locations, or other information that is needed

---

[*]Such bits are also referred to as "overhead."

to decode subsequent bits) is attached immediately after the resynchronization information. This way, the decoder can resume proper decoding upon the detection of a resynchronization marker. Obviously, insertion of resynchronization markers will reduce coding efficiency. First, the longer and more frequent such markers are, the more bits that will be used for them. Second, the use of synchronization markers typically interrupts in-picture prediction mechanisms, such as MV or DC coefficient prediction, which adds even more bits. But longer and frequently inserted markers would also enable the decoder to regain synchronization more quickly, so that a transmission error affects a smaller region in the reconstructed frame. Hence, in practical video coding systems, relatively long synchronization codewords are used.

Data Partitioning*   In the absence of any other error-resilience tools, the data between the error location and the first following resynchronization marker would have to be discarded. To achieve better error isolation, data between two synchronization points can be further partitioned into finer logic units, with secondary markers in between. This way, logic units before an error location can still be decoded. Such markers can be shorter than the primary markers, because they need only be free from emulation by data in the logic units that immediately precede them. This approach is used in the error-resilience modes of the MPEG-4 and H.263 standards, by which MB headers, motion vectors, and DCT coefficients of all MBs in a slice or GOB are put into separate logic units. This way, if an error happens, say, in the logic unit containing DCT coefficients, the header and motion information of the MBs contained in the previous logic units are still decodable.

14.4.2 Robust Binary Encoding

In addition to error isolation by inserting synchronization codewords or partitioning data into independent segments, one can also directly modify binary encoding methods so that the resulting bit stream is more robust to transmission errors. We present two such techniques.

Reversible Variable-Length Coding (RVLC)   In the preceding discussion, we assumed that once an error occurs, the decoder discards all the bits until a resynchronization codeword is identified. With RVLC [72], the decoder can not only decode bits after a resynchronization codeword, but also decode the bits before the next resynchronization codeword, from the backward direction, as shown in Figure 14.7. Thus, with RVLC, fewer correctly received bits will be discarded, and the area affected by a transmission error will be reduced. By providing the capability of cross-checking between the output of the forward and backward decoder, at a modest increase in complexity, RVLC

---

*Note that the data partitioning described here is different from that described in Section 11.1.4, which generates a scalable bit stream. There, bits representing headers, MVs, and perhaps the first few DCT coefficients from all MBs are put into a base-layer stream, whereas remaining bits from all MBs are put into one or more enhancement streams. Here, data from different logic units in the MBs included between two resynchronization markers are arranged sequentially, generating a single bit stream.

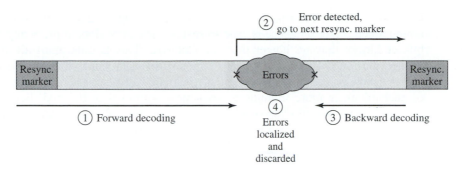

Figure 14.7   RVLC codewords can be parsed in both the forward and backward direction, making it possible to recover more data from a corrupted data stream.

can also help the decoder to detect errors that are not detectable when nonreversible VLC is used; or to provide more information on the position of the errors, and thus decrease the amount of data unnecessarily discarded. RVLC has been adopted in both MPEG-4 and H.263, in conjunction with the insertion of synchronization markers and data partitioning.

In spite of the "reversibility" constraint in the designing of the variable-length code, the application of RVLC does not necessarily lead to reduction in coding efficiency. In fact, it has been shown that compressed video data (e.g., quantized and run-length coded DCT coefficients, MV differences), which can be very well modeled by the so-called "generalized Gaussian distribution" (GGD), can be entropy coded with the Golomb-Rice (GR) and exp-Golomb (EG) codes at near-perfect efficiency. The EG code is more attractive than the GR, because its performance is very stable even when there is a mismatch between model distribution and actual data. Therefore, it is possible to achieve near-optimal efficiency with a fixed EG code table for most video sequences. Both the GR and the EG codes are very well structured. This makes it possible to perform entropy encoding and decoding without look-up tables, and to design reversible versions of GR and EG codes with exactly the same coding efficiency. More detailed descriptions of how to design RVLC tables with identical coding efficiency as GR and EG codes can be found in [86]. In addition to providing cross-checking capability of forward and backward decoded results, RVLC can provide better error detection (and in some cases even better error correction) than nonreversible VLC. More detailed analysis and comparison of the decoder operation for reversible and nonreversible VLC can be found in [85].

Error-Resilient Entropy Coding (EREC)   Instead of using synchronization codewords to enable the decoder to resume decoding, the EREC method employs a way of reordering coded bits so that the decoder can regain synchronization at the start of each block [60]. Specifically, variable-length bit streams from groups of blocks (or MBs) are distributed into slots of equal sizes. Initially, the coded data for each image block are either fully or partially placed into the designated slot for the block. Then, a predefined offset sequence is used to search for empty slots to place any remaining bits of blocks that are bigger than the slot size. This is done until all the bits are packed into one of the slots. Because the size of each slot is fixed, the decoder can regain synchronization at the start of each block. This also ensures that the beginning of each block is more immune to error propagation than the end. Because EREC does not insert any bits for synchronization purposes, the redundancy introduced is negligible. Applications of EREC in MPEG-2 and H.263 coders have been considered in [71, 48].

14.4.3 Error-Resilient Prediction

As we have seen, the use of VLC makes a compressed bit stream very sensitive to transmission errors. Another major cause for sensitivity is the use of temporal prediction. Once an error occurs such that a reconstructed frame at the decoder differs from that assumed at the encoder, the reference frames used in the decoder from that point onwards will differ from those used at the encoder, and consequently all subsequent reconstructed frames will be erroneous. This error propagation effect was shown in Figure 14.2. The use of spatial prediction for the DC coefficients and MVs will also cause error propagation, although it is confined within the same frame. Error-resilient prediction refer to techniques that constrain the prediction loop so as to confine error propagation within a short time interval.

Insertion of Intra-Blocks or Frames   One way to stop temporal error propagation is by periodically inserting intra-coded pictures or MBs. Using an I-frame can cause a surge in the output bit rate, and thus is typically not acceptable for interactive applications, due to delay constraints. For such applications, using a sufficiently high number of intra-coded MBs is an efficient and scalable tool for error resilience.

When employing intra-MBs for error resilience purposes, both the number of such MBs and their spatial placement must be determined. The number of necessary intra-MBs is dependent on the quality of the channel and the error-control mechanism applied at the transport level. Many practical systems provide information about network quality, or heuristic means to gain such information. Examples include antenna signal strength in wireless environments, or RTCP receiver reports on Internet connections. The study in [67] analyzed the dependence of the overall distortion at the decoder on the intra-MB rate at the encoder, the channel coding rate (assuming Reed-Solumn coding), and the channel error parameters (random loss rate and burst length). Based on this relation, one can find the optimal intra-MB rate and/or channel coding rate, for given channel error characteristics.

For the spatial placement of intra-MBs, the approaches proposed can be categorized as either heuristic or rate-distortion optimized. Heuristic methods include random placement and placement in the area of high activity. Hybrid schemes that additionally consider the time of the last intra update of a given MB were also considered. These methods are simple and in general perform well. The RD optimization approach can further improve performance, at the expense of increased encoder complexity.

Recall that in the absence of transmission errors, the coding mode for each MB can be determined based on the RD trade-off obtained by different modes (Section 9.3.3). Ideally, the mode that leads to the maximum reduction in the distortion per bit spent should be used. For error-resilience purposes, the same RD optimization approach can be taken, but the encoder must take into account the fact that the current MB and previous MBs may be lost, when calculating the distortion associated with each coding mode. Several approaches have been developed based on this framework, and they differ in the way that the expected distortion in the decoder is calculated, and the way that the channel error is modeled [17, 94, 89].

For example, the algorithm in [17] takes as input a mid-term prediction of the packet loss rate, p. Each MB is coded in the intra-, inter-, and skip modes, and for each mode, the resulting rate $R(\text{mode}_i)$ and distortion $D_1(\text{mode}_i)$, assuming lossless transmission of the MB, are measured. Then, for the same set of coding modes, another set of distortions, $D_2(\text{mode}_i)$, is calculated under the assumption that the coded MB got lost during transmission. The error propagation effect in an inter-coded MB, as well as error concealment in the decoder for a lost MB, are taken into account in the preceding distortion measurement. For each coding mode, the expected distortion $D(\text{mode}_i) = (1 - p)D_1(\text{mode}_i) + pD_2(\text{mode}_i)$ is determined. The coding mode that leads to the best rate-distortion trade-off is determined. This is accomplished by determining the Lagrangian $D(\text{mode}_i) + \lambda R(\text{mode}_i)$ for different modes, for a fixed λ, and finding the mode that has the minimal Lagrangian. The Lagrangian multiplier λ is chosen based on the target bit rate.

Finally, when a feedback channel is available, it can be used to convey information about missing or damaged MB data to trigger intra-coding at the sender. These schemes are discussed in Section 14.6.

Independent Segment Prediction   Another approach to limit the extent of error propagation is to split the data domain into several segments, and perform temporal and spatial prediction only within the same segment. This way, the error in one segment will not affect another segment. For example, a frame can be divided into multiple regions (e.g., a region can be a GOB or slice), and region 1 can be predicted only from region 1 in the previous frame. This is known as independent segment decoding (ISD) in H.263.

Another approach in this category is to include even-indexed frames in one segment and odd-indexed frames in another segment. This way, even frames are predicted only from even frames. This approach is called *video redundancy coding* and has been incorporated in H.263 [87, 88]. It can also be considered an approach for accomplishing multiple description coding, to be described in Section 14.4.5.

14.4.4 Layered Coding with Unequal Error Protection

Layered coding (LC) refers to coding a video into a base layer and one or several enhancement layers. The base layer provides a low but acceptable level of quality, and each additional enhancement layer incrementally improves the quality. As described in Chapter 11, layered coding is a special case of scalable coding, which enables receivers with

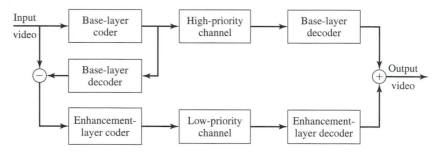

Figure 14.8   Block diagram of a system using layered coding and prioritized transport.

different bandwidth capacities or decoding powers to access the same video at different quality levels. Typical implementations include SNR scalability, spatial scalability, and temporal scalability (see Sections 11.1 and 13.5.3). To serve as an error-resilience tool, layered coding must be paired with unequal error protection (Section 14.3.4) in the transport system, so that the base layer is protected more strongly. Figure 14.8 shows the block diagram of a generic two-layer coding and transport system.

The redundancy in layered coding comes mainly from two sources. First, in order to avoid error propagation, the enhancement layer may choose to use only the base-layer frames as reference frames for temporal prediction. In this case, the prediction gain and consequently the coding efficiency will be reduced. Second, similar side information (e.g., header information, coding modes, and MVs) must be transmitted in each layer. A good study of the trade-off between the redundancy and error resilience of different scalability modes in MPEG-2 is described in [1]. Layered coding plus unequal error protection has been explored extensively for video transport over wireless networks; see, for example, [95, 46, 31, 24].

14.4.5 Multiple-Description Coding

Layered coding as described in the previous subsection can offer error resilience when the base layer is transmitted in an essentially error-free channel, realized via strong FEC and retransmission. However, in certain applications, it may not be feasible or cost effective to guarantee lossless transmission of any data. In this case, a loss in the base layer can lead to a disastrous effect in the decoded visual quality. An alternative approach to combat transmission errors is by using *multiple-description coding* (MDC). With this coding scheme, several bit streams (referred to as descriptions) of the same source signal are generated and transmitted over separate channels. Each channel may temporarily go down or suffer from long burst errors. At the destination, depending on which descriptions are received correctly, different reconstruction schemes (or decoders) will be invoked. The MDC coder and decoder are designed such that the quality of the reconstructed signal is acceptable with any one description, and that incremental improvement is achievable with additional descriptions. A conceptual schematic for a

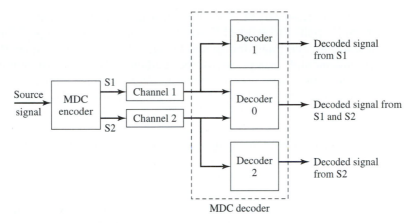

Figure 14.9   Illustration of multiple-description coding and decoding.

two-description coder is shown in Figure 14.9. In this case, there are three decoders at the destination, and only one operates at any given time.

With MDC, the channels carrying different descriptions could be physically distinct paths between the source and destination in, for example, an adhoc wireless network or a packet-switched network such as the Internet. Even when only a single physical path exists between source and destination, the path can be divided into several virtual channels by using time interleaving, frequency division multiplexing, and so on. For example, in the Internet, if the packet size is relatively large and two descriptions are put into alternating packets, then the loss characteristics for the two descriptions will be close to independent.

For each description to provide an acceptable level of quality, all the descriptions must share some fundamental information about the source, and thus must be correlated. This correlation is what enables the decoder to estimate a missing description from a received one, and thus provide an acceptable quality level from any description. On the other hand, this correlation is also the source of redundancy in MDC. An advantage of MDC over LC is that it does not require special provisions in the network to provide a reliable subchannel. For example, in a very lossy network, many retransmissions must be invoked or much redundancy must be added in FEC to realize error-free transmission. In this case, it may be more effective to use MDC.

To accomplish their respective goals, LC uses a hierarchical, decorrelating decomposition, whereas MDC uses a nonhierarchical, correlating decomposition. Some approaches that have been proposed for yielding such decompositions include overlapping quantization [74, 44, 21], correlated predictors [34], correlating linear transforms [80, 28], blockwise correlating lapped orthogonal transforms [11, 12], correlating filterbanks [91, 65, 45], and interleaved spatial-temporal sampling [79, 87]. The interleaved temporal sampling approach is known as video redundancy coding in the H.263 standard. In the following, we describe the method using correlating linear transform in more detail.

Multiple-Description Transform Coding (MDTC)   In this approach, a linear transform is applied to original signal samples to produce groups of correlated coefficients. Ideally, the transform should be such that the transform coefficients can be divided into multiple groups, so that the coefficients between different groups are correlated. This way, if some coefficient groups are lost during transmission, they can be estimated from the received groups. To minimize the loss of coding efficiency, the coefficients within the same group should be uncorrelated. To simplify the design process for a source signal with memory, one can assume the presence of a prewhitening transform, so that the correlation-inducing transform can operate on uncorrelated samples.

To simplify the design of the transform, in the approach of [80], a pairwise correlating transform (PCT) is applied to each pair of uncorrelated coefficients. The two coefficients resulting from the PCT are split into two streams that are then coded independently. If both streams are received, then an inverse PCT is applied to each pair of transformed coefficients, and the original variables can be recovered exactly, up to the quantization error. If only one stream is received, the coefficients in the missing stream are estimated from those in the received stream, based on the correlation between the two sets of coefficients. Figure 14.10 shows the block-diagram of this coding scheme for a single pair of variables. The overhead introduced by this approach can be controlled by the number of coefficients that are paired, the pairing scheme, and the transform parameters for the paired coefficients. The optimal transform, which minimizes the distortion from a single description at a fixed redundancy, has the form

$$
T = \begin{bmatrix} \sqrt{\dfrac{\cot\theta}{2}} & \sqrt{\dfrac{\tan\theta}{2}} \\[2ex] -\sqrt{\dfrac{\cot\theta}{2}} & \sqrt{\dfrac{\tan\theta}{2}} \end{bmatrix}.
$$

The parameter θ controls the amount of redundancy introduced by coding each pair of variables. Given $N \geq 2$ variables to code, there exists an optimal pairing strategy that,

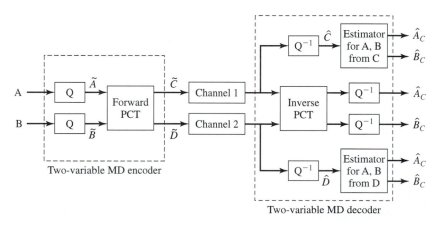

Figure 14.10   Multiple-description transform coding using a pairwise correlating transform.

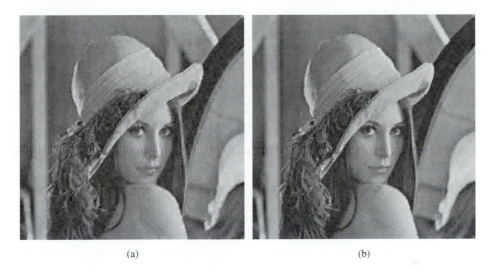

(a)   (b)

Figure 14.11   Reconstructed images of "Lena" from a single description, using the
multiple-description transform coding method: (a) redundancy = 0.088 bpp (15%),
PSNR = 28.81 dB; and (b) redundancy = 0.133 bpp (22%), PSNR = 29.63. The
reconstruction PSNR = 35.78 dB when both descriptions are received. The reference
bit rate required by the single-description coder (JPEG) is 0.60 bpp. The original
uncompressed version of this image was shown in Figure 9.10.

together with optimal redundancy allocation among the chosen pairs, can minimize the
total single-description distortion for a given total redundancy.

The MDTC method has been incorporated into a JPEG-like image coder, where
the DCT (serving as a decorrelating transform) coefficients are split into two streams
using PCTs. Figure 14.11 shows the reconstructed images from a single description
(i.e., half of the bits are lost) at two different redundancies, defined as the extra number
of bits per pixel required over the reference single-description coder (in this case, the
JPEG coder) to obtain the same distortion in the absence of transmission loss. We can
see that satisfactory images can be obtained from a single description at relatively low
redundancies.

The MDTC method has also been integrated into a block-based hybrid video
coder, in which the motion-compensated prediction error is coded using MDTC to
produce two descriptions. The challenging problem in this case is how to control the
mismatch between the reference frames used in the source coder and those at the
decoder, which may have received only one description. For specific implementation
options, see [61].

Given the relatively large overhead associated with MDC, this approach is appropriate only for channels that have relatively high loss rates. When the channel loss rate
is small, the reconstruction performance in the error-free case dominates and, therefore,
single-description coding is more appropriate. On the other hand, when the loss rate
is very high, the reconstruction quality in the presence of loss is more critical, so that

the MDC approach becomes more suitable. A challenging task is designing the MDC coder so that it can automatically adapt the amount of added redundancy according to underlying channel error characteristics.

14.4.6 Joint Source and Channel Coding

In layered coding and MDC, the interaction between the source and channel coders is exercised at a high level. In layered coding, the source coder produces a layered stream, assuming that the channel coder can guarantee the delivery of the base layer. On the other hand, with MDC, the source coder assumes that all coded bits will be treated equally and all are subject to loss. The source-channel interaction can also happen at a lower level; for example, quantizer and entropy coder design at the source coder, and the design of FEC and modulation schemes at the channel coder. This type of approach is traditionally referred to as joint source and channel coding although, in a broader sense, layered coding and MDC can also be considered as belonging to this category.

A popular FEC scheme for compressed video is the rate-compatible punctured convolutional (RCPC) code [29], with which one can easily and precisely control the incurred channel redundancy rate. By jointly designing RCPC codes and bit-rate assignments in the source coder, a desired trade-off between the source and channel coding rates can be achieved. It is also quite easy to realize unequal error protection with RCPC codes [30].

Another approach to joint source and channel coding is to jointly design the quantizer and the binary encoder for given channel error characteristics, to minimize the effect of transmission errors [49, 19, 56, 57, 75, 20, 78]. One can also carefully design the mapping between quantized source codewords and their modulation symbols. By assigning proximate points in the modulation constellation space to proximate source codewords, the effect of channel error can be reduced [93, 22].

Note that techniques in this category are mainly targeted at bit errors, and are not very effective in communication systems where compressed data are packetized and packet loss is the dominant contributor to transmission errors.

14.5 DECODER ERROR CONCEALMENT

As mentioned in the previous sections, due to various channel or network errors, compressed video data can be damaged or lost during transmission or storage. In Section 14.3, we described the mechanisms that can be invoked at the transport layer, to minimize transmission loss. Unless retransmission can be used indefinitely, which is unacceptable for real-time applications, there will be remaining errors and losses in the received data. In Section 14.4, we presented various means that the encoder can use to suppress the effect of such errors. Still, a transmission error may lead to objectionable visual distortion in the reconstructed signal at the decoder. Depending on many factors (such as source coding, transport protocol, the amount and type of information loss, and so on), the introduced distortion can range from momentary degradation to a completely unusable image or video signal. In this section, we describe techniques that can be employed at the decoder to conceal the effect of transmission errors, so as to make the decoded signal much less visually objectionable.

Error concealment is possible because, due to various constraints (such as coding delay, implementation complexity, and availability of a good source model), a compressed video bit stream still possesses a certain degree of statistical redundancy, despite the tremendous research effort that has been devoted towards achieving the highest possible compression gain. Also, the source coder may incorporate error-resilient mechanisms, intentionally introducing redundancy into the coded bits, so as to facilitate the estimation of lost data. In addition, the human perception system can tolerate a limited degree of signal distortion. All these factors can be exploited for error concealment at the decoder.

Error concealment belongs to the general problem of image recovery or restoration. However, because errors are incurred at the compressed bit level, the resulting error patterns in the pixel domain are very peculiar, and special measures are usually needed to handle such errors. As described, because of the use of predictive coding and VLC, a bit error alone can cause damage to a large region. To contain the error effect, various measures can be taken in the encoder to make the compressed stream more error-resilient, at the expense of a certain degree of coding efficiency or complexity. Here, we assume that synchronization codewords are inserted periodically within a picture (see Section 14.4.1) and that the prediction loop is periodically reset (see Section 14.4.3), so that a bit error or packet loss will only cause damage to a limited region in a picture.

Depending on the frequency of synchronization codewords, transport packet size, and bit rate, a damaged region can range from part of a macroblock to an entire picture. For high–bit-rate transmission over small packets as in ATM networks, a lost packet may only damage part of an MB or several adjacent MBs. In this case, a damaged MB is typically surrounded by multiple undamaged MBs in the same frame. When this is the case, we can use spatial interpolation to recover the damaged MB. On the other hand, for low–bit-rate application over relatively large packets (for example, 128 kbps over IP networks), a lost packet is likely to damage a large portion of a frame or an entire frame. In such situations, we must rely on the previous and possibly the following video frames for concealment of the damaged frame. Obviously, information from adjacent frames can be incorporated even if neighboring MBs are available in the same frame.

Given the block-based hybrid coding paradigm, there are three types of information that may need to be estimated in a damaged MB: the texture information, including the pixel or DCT coefficient values for either an original image block or a prediction error block; the motion information, consisting of MV(s) for an MB coded in either P- or B-mode; and, finally, the coding mode of the MB. The recovery techniques for these information types are somewhat different.

It is well-known that images of natural scenes have predominantly low-frequency components; that is, the color values of spatially and temporally adjacent pixels vary smoothly, except in regions with edges. *All the techniques that have been developed for recovering texture information make use of this smoothness property of image and video signals, and essentially they all perform some kind of spatial/temporal interpolation.*

The MV field, to a lesser extent, also shares the smoothness property, and can also be recovered by using spatial/temporal interpolation. For the coding mode information, the methods developed tend to be driven more by heuristics. In the following subsections, we review some representative techniques for each category. More extensive coverage of methods discussed here (as well as other methods) can be found in [82, 99, 47]. We will assume that error locations have been detected via other means, such as lost packet detection based on sequence numbers in services using RTP. For coverage of various error-detection methods, see [82].

Note that in a typical (non–error-resilient) encoder implementation, bits for these three types of data are stored sequentially for each MB, so that they are all lost in a damaged block. With the use of data partitioning (Section 14.4.1), it is possible that the coding mode, MV(s), and possibly the first few DCT coefficients are still available in a damaged block. As will be seen, knowledge of the coding mode and MV can greatly help in the recovery of the texture information.

14.5.1 Recovery of Texture Information

Motion-Compensated Temporal Interpolation   The simplest approach to recovering a damaged MB is by copying the corresponding MB in the previously decoded frame. This, however, will give unsatisfactory results if there is a large motion in the scene. A more effective solution is to use the MB in the previous frame pointed to by the MV for the damaged MB. The recovery performance by this approach is critically dependent on the availability of the MV. When the MV is also missing, it must first be estimated (Section 14.5.2). To reduce the impact of the error in the estimated MVs, temporal prediction may be combined with spatial interpolation.

Spatial Interpolation   Another simple approach is to interpolate pixels in a damaged block from pixels in adjacent correctly received blocks in the same frame. Usually, all blocks (or MBs) in the same row are put into the same packet, so that a packet loss typically leads to the loss of all blocks in a row. In this situation, the only available neighboring blocks for a damaged block are those above and below. Because most pixels in these blocks are too far away from the missing samples, usually only the boundary pixels in neighboring blocks are used for interpolation. Instead of interpolating individual pixels, a simpler approach is to estimate the DC coefficient (i.e., the mean value) of a damaged block and replace the damaged block by a constant equal to the estimated DC value. The DC value can be estimated by averaging the DC values of surrounding blocks. One way to facilitate such spatial interpolation is by using interleaved packetization (Section 14.3.2), so that the loss of one packet will damage only alternating blocks or MBs.

Maximally Smooth Recovery (MSR)   A problem with spatial interpolation approaches is the determination of appropriate interpolation filters. Another shortcoming is that they ignore received DCT coefficients, if any. These problems are resolved in the MSR approach of [100, 83], by requiring the recovered pixels in a damaged block to be smoothly connected with neighboring pixels, both spatially in the same frame and temporally in the previous/following frames. If some but not all DCT coefficients are received for this block, then the estimation should be such that the recovered block be as smooth as possible, subject to the constraint that the DCT on the recovered block would produce the same values for the received coefficients. These objectives can be formulated as an optimization problem, and the solutions under different loss patterns will correspond to different interpolation filters in the spatial, temporal, and frequency domains.

Spatial Interpolation Using Projection onto Convex Sets (POCS)   Another way of accomplishing spatial interpolation is by using the POCS method [69, 92]. The general idea behind POCS-based estimation methods is to formulate each constraint about the unknowns as a convex set. The optimal solution is the intersection of all the convex sets, which can be obtained by recursively projecting a previous solution onto individual convex sets. When applying POCS for recovering an image block, the spatial smoothness criterion is formulated in the frequency domain, by requiring the DFT of the recovered block to have energy only in several low-frequency coefficients. If the damaged block is believed to contain an edge in a particular direction, then one can require the DFT coefficients to be distributed along a narrow stripe orthogonal to the edge direction, that is, low-pass along the edge direction, and all-pass in the orthogonal direction. The requirement on the range of each DFT coefficient magnitude can also be converted into a convex set. Similarly, the constraint imposed by any received DCT coefficient also forms a separate convex set. Because the solution can only be obtained through an iterative procedure, this approach may not be suitable for real-time applications.

14.5.2 Recovery of Coding Modes and Motion Vectors

As indicated, some of the preceding algorithms are contingent upon knowledge of the coding mode and MVs for P- or B-mode MBs. To facilitate decoder error concealment, the encoder may perform data partition, to pack the mode and MV information in a separate partition and transmit them with stronger error protection (see Section 14.4.1). For example, this is an error-resilience mode in both H.263 and MPEG-4. Still, the mode and MV information can be damaged.

One way to estimate the coding mode for a damaged MB is by collecting the statistics of the coding-mode pattern of adjacent MBs, and find a most-likely mode given the modes of surrounding MBs [68]. A simple and conservative approach, when the coding mode is lost, is to assume that the MB is coded in the intra-mode, and use only spatial interpolation for recovering the underlying blocks.

For estimating lost MVs, there are several simple operations [50]: (1) assuming the lost MVs to be zeros, which works well for video sequences with relatively small motion; (2) using the MVs of the corresponding MB in the previous frame; (3) using the average of the MVs from spatially adjacent MBs; (4) using the median of the MVs from the spatially adjacent MB's; and, (5) reestimating the MVs [33]. Typically, when an MB is damaged, its horizontally adjacent MBs are also damaged, and hence the average or mean is taken over the MVs above and below. A simpler approach is to use the MV for the MB above the damaged MB. Research has shown that no significant improvement in picture quality can be achieved when using mean or median values of more than one MV

[81]. To facilitate the estimation of lost MVs, interleaved packetization (Section 14.3.2) can be used. For example, if even and odd rows of MBs are put into alternating packets, then the loss of a packet will damage only alternating rows.

14.5.3 Syntax-Based Repair

The preceding error-concealment techniques work in the signal domain, assuming that the error in the transmitted stream has been detected, "questionable" data units have been discarded, and the remaining bits have been decoded into signal variables (e.g., side information, MVs, DCT coefficients) by the VLC decoder. Syntax-based repair, on the other hand, tries to directly detect and repair errors in the bit stream. This is possible because, for any finite length data packet, there are only relatively few valid combinations of codewords. In addition, a compressed bit stream is never an i.i.d. white binary sequence. The distribution of ones and zeros in the bit stream is often location dependent [81]. Based on such information, the decoder can repair the bit stream so that the repaired stream conforms to the syntax dictated by the encoding scheme. The use of data partitioning and RVLC in H.263 and MPEG-4 helps to enable syntax-based repair, as they allow more "check points" and syntax constraints, which makes the total number of valid combinations of codewords smaller than the number of all possible binary strings of the same length. One such repair scheme, known as "soft-decoding," is described in [84]. Syntax-based repair is an effective error concealment method for a channel in which bit errors dominate. For packet lossy networks, in which many complete packets may be lost, syntax-based repair is only used to detect errors in the received packets, but cannot be used to reconstruct the mission data.

14.6 ENCODER–DECODER INTERACTIVE ERROR CONTROL

In the techniques presented thus far, the encoder and decoder operate independently, as far as combating transmission errors is concerned. Conceivably, if a backward channel from the decoder to the encoder is available, better performance can be achieved if the sender and the receiver cooperate in the process of error control. This cooperation can be realized at either the source coding or transport level. At the source coder, coding parameters can be adapted based on the feedback information from the decoder. At the transport level, the feedback information can be employed to change the percentage of the total bandwidth used for FEC or retransmission. We described transport-level adaptations in Section 14.3. In this section, we describe several techniques that adapt source coding strategy based on feedback information from the decoder. *These techniques are developed under the assumption that it is often acceptable to have errors as long as they do not last too long. Therefore, even if one cannot afford to correct the errors in every occurrence, it is important to limit the propagation scope of such errors.*

Before we proceed to describe these techniques, it is worthwhile to note how the feedback information is delivered. Typically, the feedback message is not part of the video syntax but is transmitted in a different layer of the protocol stack, in which control information is exchanged. For example, in the H.323 standards for multimedia communications over unreliable channels [43], the control protocol H.245 [41] allows reporting of the temporal and spatial location of damaged MBs. Usually, such messages are delivered error-free, by the use of retransmission when necessary.

14.6.1 Coding-Parameter Adaptation Based on Channel Conditions

In a channel with varying bandwidth and error characteristics, it is important to match the encoding rate with the available channel bandwidth, and to embed appropriate error resilience in the coded bit stream. When a source is coded at a rate higher than the underlying channel can deliver in time, some data will be thrown away at the discretion of the network layer, typically leading to more disturbing artifacts than the source coding distortion incurred by encoding the source at a lower rate. Furthermore, when a channel is very noisy, it is better to represent a source with lower quality, leaving more bits for error protection, either in the form of FEC or error resilience in the encoded stream.

There are two problems that must be solved: First, the transport controller must be able to periodically estimate and update the QoS parameters (e.g., bandwidth, delay, packet loss rates) of an established connection based on the feedback information or other transport level interactions; second, given the available bandwidth and error characteristics, the encoder must appropriately determine the coding parameters (such as intra-block rate, frequency of synchronization markers, range of prediction, and so on, in a block-based hybrid coder) so as to meet the target bit rate and desired amount of error resilience. We discussed the rate-control issue in Section 9.3.4, without considering the error-resilience factors. In Section 14.4.3, we described several approaches to coding-mode selection (intra- versus inter-) that take into account loss characteristics of the network. The paper by Wu et al. considers both rate estimation and rate adaptation for video delivery over the Internet [90]. Adjusting the target encoding rate based on channel feedback, for the application of video streaming over the Internet, is further discussed in Chapter 15.

14.6.2 Reference Picture Selection Based on Feedback Information

One way of taking advantage of an available feedback channel is to employ reference picture selection (RPS). If the encoder learns through a feedback channel about damaged parts of a previously coded frame, it can decide to code the next P-frame relative not to the most recent, but to an older reference picture, which is known to be available in the decoder. This requires that the encoder and decoder both store multiple previously decoded frames. Information about the reference picture to be used is conveyed in the bit stream. Compared to coding the current picture as an I-frame, the reduction in coding efficiency due to use of an older reference picture is significantly lower, if the reference picture is not too distant. A study on the delay and overhead incurred by using RPS can be found in [26].

Note that using RPS does not necessarily mean extra delay in the encoder. The encoder need not wait for the arrival of the feedback information about the previous frame to code a current frame. Rather, it can choose to use as reference a frame before the damaged frame whenever it receives the feedback information. For example, if the information about the damage of frame n does not arrive at the encoder until the time to code frame $n + d$, the decoded frames between $n + 1$ and $n + d - 1$ would all have errors, because the decoder uses different reference frames than the encoder for these frames. By selecting frame $n - 1$ as the reference frame to code frame $n + d$, error propagation will be stopped from frame $n + d$ onwards. Of course, the longer it takes to generate and deliver the feedback information, the greater will be the loss in coding efficiency.

14.6.3 Error Tracking Based on Feedback Information

Instead of using an earlier, undamaged frame as the reference frame, the encoder can track how the damaged areas in frame n would have affected decoded pixels in frames $n + 1$ to $n + d - 1$, and then perform one of the following when coding frame $n + d$. The encoder can (1) code the MBs in frame $n + d$ that would have been predicted from damaged pixels in frame $n + d - 1$ using the intra-mode; (2) avoid using the affected area in frame $n + d - 1$ for prediction in coding frame $n + d$; and (3) perform the same type of error concealment as the decoder for frames $n + 1$ to $n + d - 1$, so that the encoder's reference picture matches that at the decoder, when coding frame $n + d$. The first two approaches require only that the encoder track the locations of damaged pixels or blocks, whereas the last approach requires the duplication of the decoder operation for frames $n + 1$ to $n + d - 1$, which is more complicated. With either approach, the decoder will recover from errors completely at frame $n + d$. A problem with option 1 is that, if many errors occurred in frame n, too many MBs would need to be coded in the intra-mode. To maintain a constant bit rate, all the MBs would need to be quantized more coarsely. An alternative is to invoke the intra-mode only when the effect of damaged pixels in frame $n + d - 1$ on an MB in frame $n + d$ is severe. This approach would not stop error propagation immediately at frame $n + d$. Consequently the same remedy must be applied in the following few frames [26]. This approach is illustrated in Figure 14.12. More information on error tracking, correction, and fast algorithms can be found in [77, 18, 26]. Selectively coding MBs in the intra-mode based on the NACKs is supported by the MPEG-4 ARTS profile (Section 13.6.8).

14.6.4 Retransmission without Waiting

In order to make use of retransmitted data, a typical implementation of the decoder will have to wait for the arrival of the requested retransmission data before processing subsequently received data. In fact, this is not necessary. It is possible to use retransmission for recovering lost information without introducing delay. In the approach of Zhu [97, 82] and Ghanbari [23], when a video data unit is damaged, say in frame n, a retransmission request is sent to the encoder for recovering the damaged data. Instead of waiting for the arrival of retransmitted data, the damaged video part is concealed by a chosen error-concealment method. Then, normal decoding resumes, while a trace of the affected pixels and their associated coding information (coding mode and motion vectors) is

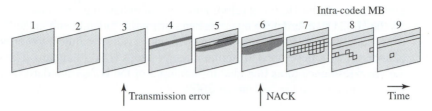

Figure 14.12   Illustration of error propagation when error tracking is used and damaged blocks are intra-coded upon the receipt of the NACK (negative acknowledgment) information. Reprinted from B. Girod and N. Färber, Feedback-based error control for mobile video transmission, *Proceedings of the IEEE* (Oct. 1999), 87:1707–23. Copyright 1999 IEEE.

recorded (similarly to the error-tracking method described in Section 14.6.3). Upon the arrival of the retransmitted data, say at frame $n + d$, the affected pixels are corrected, so that they are reproduced as if no transmission loss had occurred. The correction signal is obtained from the retransmitted data and the recorded trace.

This method can achieve lossless recovery, except during the time between the information loss and the arrival of the retransmission data. During that interval, any error-concealment technique (Section 14.5) can be applied to the damaged regions. This scheme eliminates the delay associated with conventional retransmission schemes without compromising video quality. The price paid is the relatively high implementation complexity. In contrast to the error-tracking approach in Section 14.6.3, which adapts encoding operation to stop error propagation, this approach makes use of retransmitted data to correct and stop error at the decoder.

14.7 ERROR-RESILIENCE TOOLS IN H.263 AND MPEG-4

In this section, we summarize the error-resilience tools defined in the H.263 and MPEG-4 standards. These tools fall in the error-resilient encoding category, even though the standards only define the bit-stream syntax. It is worth noting that, as with any error-resilient encoding methods, these tools tend to reduce coding efficiency slightly. However, they give a good decoder the tools to localize errors in the bit stream, to limit the image area that is affected by these errors, and to apply appropriate error-concealment methods (Section 14.5). How these tools are used to help the decoder operation is outside the scope of the standard, and that fact enables competition between different decoder vendors.

14.7.1 Error-Resilience Tools in H.263

The initial target of the H.263 standard was to serve as a video coding standard for the H.324 system, which aims to enable video telephony over wireline and wireless modems. As described in Section 14.2, by applying appropriate FEC and multiplexing, both wireline and wireless modem channels can be considered error-free. Therefore the initial focus of the H.263 development was on enhancing the coding efficiency only. As the H.323 standard develops, targeting primarily video telephony and conferencing over the best-effort Internet, the error-resilience issues become a major concern. A set of annexes were developed to augment the previous H.263 standard. The following are the error-resilience tools included in Version 3 of the H.263 standard (i.e., H.263++), which was approved in November 2000.

Forward Error Correction Using BCH Code (Annex H)   This tool allows the inclusion, for blocks of 492 coded video bits, of nineteen-bit BCH (511, 492) FEC parity information in the bit stream. Together with a single additional bit to allow for resynchronization to the resulting 512-bit block structure, Annex H introduces an overhead of roughly 4% of the bit rate. The BCH (511, 492) FEC code is able to correct single, and to reliably detect double bit errors in a 512-bit block. This tool is a leftover from the H.261 standard, which is targeted for video delivery over ISDN channels, to handle the occasional bit errors. For highly bit-error–prone mobile channels, Annex H is not efficient, since an error burst longer than two bits is neither correctable, nor reliably detectable. Furthermore, the fixed structure of 492 video bit blocks does not allow the precise alignment of the block boundaries to synchronization markers. For these reasons, Annex H is not very useful for combating errors in the Internet or wireless networks [81].

Flexible Synchronization Marker Insertion Using the Slice Structured Mode (Annex K)   The slice structure, when used, replaces the GOB structure. Slice headers serve as synchronization markers and interrupt in-picture prediction of MVs and DC coefficients. Whereas the GOB structure contains a fixed number of MBs, each scan-order slice has about the same number of bits. If the number of bits contained in the current slice exceeds a predetermined threshold, then a new slice is created at the start of the next MB. As illustrated in Figure 14.13, the active area in a video (in which each MB requires more bits) will have more slices and consequently more synchronization markers than the stationary area. This facilitates error concealment at the decoder in the active region.

A slice in H.263 can also correspond to a rectangular area in a frame, aligned to MB boundaries. This enables independent segment decoding, discussed in a subsequent paragraph.

Reference Picture Selection (Annex N and Annex U)   As introduced in Section 14.6.2, RPS enables the encoder to use, for motion compensation, an older reference picture that is known to be correctly received at the decoder's site. In H.263, RPS can be used on whole pictures, picture segments (slices or GOBs), or on individual MBs. The former two mechanisms are defined in Annex N and were introduced as an error-resilience tool only, whereas the latter, defined in Annex U, was designed with both error resilience and coding efficiency in mind.* In H.263, RPS can be used with or without feedback information. Without feedback information, the video redundancy

---

*By allowing an MB to choose the best matching MB among a set of past frames, coding efficiency can be increased at the expense of motion estimation complexity.

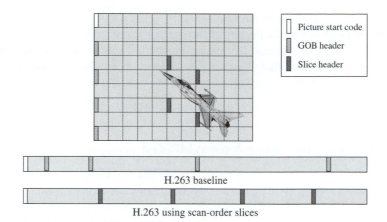

Figure 14.13   Scan-order slice versus GOB in H.263. With the GOB structure, the GOBs containing the moving airplane (the third and fourth GOB) are more likely to be hit by errors because they contain more bits. A damaged GOB will lead to visually unacceptable distortion. With the slice structure, the same region is covered by more slices. Error resilience is improved, because a larger number of synchronization markers appear in this region and a lost slice affects a smaller region. Reprinted from Y. Wang, S. Wenger, J. Wen, and A. G. Katsaggelos, Error resilient video coding techniques, *IEEE Signal Processing Magazine* (July 2000), 17(4):61–82. Copyright, 2000 IEEE.

coding method described in Section 14.4.3 can be used, which employs a prefixed interleaved RPS scheme. This technique is, however, much less efficient than feedback-based mechanisms [87, 88].

Scalability (Annex O)   Annex O defines temporal, spatial, and SNR scalability similar to those used in MPEG-2. As described in Section 14.4.4, these tools can be used to enhance error resilience if multiple transport paths with different QoS characteristics are available between the source and the destination. Today's networks usually do not support different QoS, hence scalability is rarely used in practice for the purpose of error resilience.

Independent Segment Decoding (Annex R)   ISD forces encoder and decoder to treat segment (slice or GOB) boundaries like picture boundaries, thereby preventing the propagation of corrupted data from one segment to another. This method is a special case of error-resilient prediction as discussed in Section 14.4.3. ISD in conjunction with rectangular slices has been shown to improve error resilience [88]. The overhead of ISD is roughly reversely proportional to the picture size, and is impractically high for picture sizes smaller than CIF [81].

Data Partitioning and RVLC (Annex V)   As mentioned, with data partitioning, MB header, MV, and DCT information are no longer interleaved on an MB by MB basis; rather, they are grouped into partitions separated by specially designed markers. Header and motion information is coded with RVLC, whereas the DCT coefficients are coded using the nonreversible VLC table used in the baseline H.263.

Header Repetition (Annex W)   This allows the repetition of the header of the previous frame in the header of the current frame, which enables, with a frame delay, the decoding of a frame for which the frame header is lost. Note that H.263 contains no syntax element that allows the inclusion of redundant picture header information at the slice level, as is available in MPEG-4's header extension code (see Section 14.7.2).

For a discussion of how the preceding tools can be incorporated in a video codec for video transmission over the Internet and wireless networks, see [81].

14.7.2 Error-Resilience Tools in MPEG-4

One of the primary objectives of the MPEG-4 standard is to enable universal access. Therefore, error resilience is an important issue to consider from the very beginning. Because of overlapping in development time, the error-resilience tools defined in MPEG-4 are quite similar to those in H.263. It incorporates many of the error-isolation, synchronization, data-recovery, and resilient entropy-coding tools described in previous sections.

Resynchronization Tools   MPEG-4 defines several approaches that enable quick resynchronization after the occurrence of a transmission error. The *video packet* approach is very similar to the adaptive slice of MPEG-2 and slice structured mode of H.263. A video packet functions in the same way as the scan-order slice in H.263. The length of a video packet is no longer based on the number of MBs (as in the case of the non–error-resilient mode of MPEG-4 or baseline H.263), but instead on the number of bits contained in that packet (see Figure 14.13). It is aimed at providing periodic resynchronization throughout the bit stream.

A *resynchronization marker* is placed at the start of the header of each video packet. Like other start codes, this marker is distinguishable from all other possible VLC codewords. The header contains the information necessary to restart the decoding process, including the address of the first MB contained in this packet and the QP for the first MB. These are followed by a single bit *header extension code* (HEC). If the HEC is set to 1, then the following information that is already specified in the VOP header is duplicated in this packet header: timing information, temporal reference, VOP prediction type, and certain other information. The header extension feature enables the decoder to correctly utilize data contained in the current packet, even if the packet containing the VOP header is lost. It also enables cross-checking, since all packets in the same VOP should share the same QP, time stamp, and so on.

In addition to using variable-length video packets, a method called *fixed interval synchronization* has also been adopted by MPEG-4. This method requires that the start of a video packet appears only at allowable, fixed-interval locations in the bit stream. This helps to avoid the problems associated with start-code emulation due to transmission errors.

Data Partitioning   To achieve better error isolation in the video packet and fixed interval synchronization approaches, MPEG-4 also uses data partitioning as an error-resilience tool. Data partitioning within a video packet reorganizes the data, such that motion vectors and related syntactic elements are transmitted first, followed by

One video packet

Resync. marker	MB index	Quant. info	HEC and header repetition (if present)	Motion vectors	Motion marker	Texture (DCT)	Resync. marker	...

Figure 14.14   Syntactic structure of video packets in MPEG-4 error-resilient mode with data partitioning. Reprinted from Y. Wang, S. Wenger, J. Wen, and A. G. Katsaggelos, Error resilient video coding techniques, *IEEE Signal Processing Magazine* (July 2000), 17(4):61–82. Copyright 2000 IEEE.

syntactic elements like CBP and the DCT coefficients. Figure 14.14 illustrates the syntactic structure of a video packet with data partitioning. Note that the texture (DCT) partition in the structure may be coded with RVLC.

NEWPRED Mode   This tool is very similar to the H.263 RPS Mode (Annex N) and the Slice Structured Mode (Annex K). When the NEWPRED (ND) mode is turned on, the reference used for temporal prediction by the encoder will be updated adaptively according to the feedback messages from the decoder. These messages indicate which NEWPRED segments (which can be either an entire frame, or—in MPEG-4 language— a VOP, or the content of a packet, typically one slice) have not been successfully decoded. Based on the feedback information, the encoder will use either the most recent ND segment, or a spatially corresponding but older ND segment, for prediction.

Data Recovery in Conjunction with RVLC and Data Partitioning   The MPEG-4 error resilience mode utilizes RVLC for better recovery of DCT data. The usage of RVLC is signaled at the VOL. When RVLC is used for DCT data, the bit stream is decoded in the forward direction first. If no errors are detected, the bit stream is assumed to be valid. If an error is detected, however, two-way decoding is applied, and the portion of the packet between the first MBs in which an error was detected in both the forward and backward directions should not be used.

14.8 SUMMARY

In this chapter, we described various error-control mechanisms that can be used to combat transmisson errors in real-time video communications. We focused on techniques developed for block-based hybrid coders, especially those that have been adopted in the H.263 and MPEG-4 standards. The key features of different categories of error-control techniques are summarized here.

Transport-Level Error Control (Section 14.3)

- Transport-level error control mechanisms are the most important, and guarantees a basic level of quality. Error control can be applied through FEC, interleaved packetization, and when feasible, properly constrained retransmission.
- Different levels of protection can be applied to parts of coded bits that are of varying importance. Such unequal error protection is an effective way of employing a limited amount of redundancy to achieve acceptable quality.

Error-Resilient Encoding (Section 14.4)

- Techniques in this category achieve error resilience by adding a certain amount of redundancy in the coded bit streams at the source coder. Some are aimed at guaranteeing a basic level of quality and providing a graceful degradation upon the occurrence of transmission errors (e.g., layered coding and MDC), some help to prevent error propagation (e.g., error-resilient prediction), and others help to detect or recover from bit errors (e.g., error isolation and robust entropy coding). Some techniques require close interaction between the source coder and transport layer (e.g., layered coding with unequal error protection, error-resilient prediction with interleaved packetization), whereas others assume that different substreams are treated equally in the transport layer (e.g., MDC).

- Error isolation, RVLC, and joint source and channel coding as described in Section 14.4.6 are only useful for combating bit errors, and are not helpful for recovering from packet loss.

Decoder Error Concealment (Section 14.5)

- All the error-concealment techniques for texture recover lost information by making use of the temporal and spatial smoothness property of image and video signals. The MSR technique enforces the smoothness constraint by minimizing the roughness of the reconstructed signal. The motion-compensated temporal interpolation and spatial interpolation methods can be considered special cases of the MSR approach in which only the temporal or spatial difference measure is minimized. The POCS method, on the other hand, iteratively projects the reconstructed image block onto the convex sets determined by the received coefficients and the smoothness constraint determined from the estimated edge direction of the block. Although generally giving more accurate results than the MSR, the POCS method is computationally more intensive, as it requires many iterations.

- Simple and yet effective, motion-compensated temporal interpolation has been employed in commercial systems. More complex methods that exploit spatial correlation (spatial interpolation, MSR, and POCS) can offer additional improvement only when the underlying scene has very complex motion or is experiencing scene changes.

- Recovery of coding mode and motion information is more difficult than texture information, because there is less correlation among the coding modes and motion vectors of adjacent MBs. When possible, such information should be more heavily protected than texture information, through the use of data partitioning and unequal error protection.

Encoder–Decoder Interactive Error Control (Section 14.6)

- Such techniques are applicable only when there is a backward channel from the decoder to the encoder, so that the decoder can inform the encoder which part of the coded information is lost.

- The three techniques presented (reference picture selection, error tracking, and re-transmission without waiting), all aim to stop error propagation after the feedback information is received. Reference picture selection and error tracking avoid using damaged regions for future prediction at the encoder; whereas retransmission without waiting corrects the error based on the retransmitted information.

- When only general information is available about, say, packet loss rate, but not which part of the information is lost, one can try to adjust the coding parameters so as to reduce data loss and suppress the effect of any lost data.

Choices among Techniques

- Transport-level error-control mechanisms are the most important, and guarantees a basic level of QoS. Error-resilient encoding and decoder error concealment are necessary to further improve the quality in very lossy environments (such as the Internet and wireless networks). Depending on channel error characteristics and system configuration and requirements, some techniques are more effective than others.

- The burstiness of transmission errors has a significant impact on the choice of algorithms. For a channel with very long bursts, error-resilient encoding techniques may not be appropriate. This is because the redundancy introduced in such coding methods is wasted when the channel is error-free, and such redundancy is not very helpful when a burst error occurs. Retransmission, when properly constrained, may be more suitable since it only introduces the overhead when needed.

- The existence of a backward channel from the decoder to the encoder also affects the deployment of some schemes. In applications such as broadcast, in which there is no backward channel, none of the interactive error-control techniques can be applied. Similarly, in video streaming applications, videos are usually precompressed and stored in a server, so that it is not possible to adapt encoder operations based on network conditions.

- The error-concealment techniques can be applied in any circumstances. However, the effectiveness of such techniques is limited by the available information. Also, some techniques may either be too complicated for cost-effective implementation, or introduce processing delay unacceptable for real-time applications.

- Aside from the delay and complexity issues, one important criterion for comparing different schemes is the total amount of redundancy required in the source coder and the transport layer, to achieve the same degree of error protection. Obviously, the approach that requires the minimal amount of redundancy is the most efficient in terms of bandwidth usage.

14.9 PROBLEMS

14.1 What are the main differences in the requirements for transporting audio and video versus data files?

14.2 What is the typical delay requirement for an interactive video conferencing application? For video streaming?

14.3 What are the factors contributing to the end-to-end delay between a video source and the displayed video at the receiver?

14.4 We divided the approaches for error control in video communications into four categories. What are they? Describe the advantages and limitations of each.

14.5 What are the major differences between circuit-switching networks and packet-based networks? What would you choose as your error-control mechanisms in each case?

14.6 What are the major differences between LC and MDC? How do they differ in terms of network requirements? Describe network conditions under which LC may be preferred over MDC, and vice versa.

14.7 What would you do differently in the implementation for motion-compensated prediction, if the network changed from reliable (almost error free) to unreliable (with high packet losses)?

14.8 In principle, lost information is unrecoverable. What makes error concealment techniques feasible for video?

14.9 What are the major assumptions in error-concealment techniques for recovering damaged texture data?

14.10 Describe the principles of the MSR and POCS methods for error concealment. What are the advantages and disadvantages of each?

14.11 Consider an Internet video conferencing application. Assume that the compressed video streams are packetized, and that a packet is either lost (due to overlong delay or misdelivery or detected bit errors within a packet) or received intact. Assume that the effective packet loss rate is quite high, about 10%, and that there is no feedback channel available. The average end-to-end delay between the source and destination is about 50 ms. Further assume that the maximum end-to-end delay allowed between the source and destination is 150 ms. What would you propose as mechanisms to control and conceal the effect of packet loss? What approaches described in this chapter would not be applicable or effective?

14.12 Consider again the scenario in Problem 14.11, but assume that the network has a lower packet loss rate, say 1%. Would your solution be different?

14.13 Consider again the scenario in Problem 14.11, but assume this time that a feedback channel is available between the destination and the source, and that it usually takes 50 ms to deliver a feedback message. What are the solutions that you would propose in this case? What would you send on the feedback channel?

14.14 Now consider video transport over a wireless network with high bit-error rates and bursty errors, and narrow bandwidth (say 64 kbps). You have the choice to further examine a packet, even if the network or decoder has detected bit errors within the packet based on the FEC code. What are the solutions that you would propose in this case? Compared to the Internet case, would you use a shorter or longer packet?

14.10 BIBLIOGRAPHY

[1] Aravind, R., M. R. Civanlar, and A. R. Reibman. Packet loss resilience of MPEG-2 scalable video coding algorithms. *IEEE Trans. Circuits Syst. for Video Technology* (Oct. 1996), 6(5):426–35.

[2] Ayanoglu, E., R. Pancha, A. R. Reibman, and S. Talwar. Forward error control for MPEG-2 video transport in a wireless ATM LAN. *ACM/Baltzer Mobile Networks and Applications* (Dec. 1996), 1(3):245–58.

[3] Berger, A. W. ATM networks. In M. T. Sun and A. R. Reibman, eds., *Compressed Video over Networks*. New York: Marcel Dekker, 2001, 139–76.

[4] Blake, S., et al. *An architecture for differentiated services,* IETF RFC2475, Dec. 1998. (Available at http://www.ietf.org/rfc/rfc2475.txt).

[5] Bojkovic, Z. S., and K. R. Rao. *Packet Video Communications over ATM Networks*. Englewood Cliffs, NJ: Prentice Hall, 2000.

[6] Bolot, J. C., H. Crepin, and A. Vega-Garcia. Analysis of audio packet loss in the Internet. In *The 5th International Workshop on Network and Operating System Support for Digital Audio and Video* (April 1995), 163–74.

[7] Bormann, C., et al. *RTP payload format for the 1998 version of ITU-T recommendation H.263 video (H.263+),* IETF RFC2429, Oct. 1998.

[8] Budagavi, M., W. R. Heinzelman, J. Webb, and R. Talluri. Wireless MPEG-4 video communication on DSP chips. *IEEE Signal Processing Magazine* (Jan. 2000), 17:36–53.

[9] Chang, L.-F. Wireless systems and networking. In M. T. Sun and A. R. Reibman, eds., *Compressed Video over Networks*. New York: Marcel Dekker, 2001, 177–214.

[10] Chapplapali, K., et al. The grand alliance system for US HDTV. *Proceedings of the IEEE* (Feb. 1995), 83(2):158–74.

[11] Chung, D., and Y. Wang. Multiple description image coding using signal decomposition and reconstruction based on lapped orthogonal transforms. *IEEE Trans. Circuits Syst. for Video Technology* (Sept. 1999), 9(6):895–908.

[12] Chung, D., and Y. Wang. Lapped orthogonal transform designed for error resilient image coding. In *IEEE Int. Conf. Image Proc. (ICIP2000),* 3:368–71. Vancouver, Canada, Sept. 2000.

[13] Civanlar, M. R. Internet video. In M. T. Sun and A. R. Reibman, eds., *Compressed Video over Networks*. New York: Marcel Dekker, 2001, 433–64.

[14] Civanlar, M. R., and G. L. Cash. A practical system for MPEG-2 based video-on-demand over ATM packet networks and the WWW. *Signal Processing: Image Communications* (1996), 8(3):221–27.

[15] Civanlar, M. R., G. L. Cash, and B. G. Haskell. *RTP payload format for bundled MPEG,* IETF RFC2343, May 1998.

[16] Comer, D. E. *Internetworking with TCP/IP*. Vol. 1: Principles, Protocols, and Architecture. 3rd ed. Englewood Cliffs, NJ: Prentice Hall, 1995.

[17] Côté, G., S. Shirani, and F. Kossentini. Optimal mode selection and synchronization for robust video communications over error prone networks. *IEEE Journal on Selected Areas in Communications* (June 2000), 18(6):952–65.

[18] Steinbach, E., N. Färber, and B. Girod. Standard compatible extension of H.263 for robust video transmission in mobile environments. *IEEE Trans. Circuits Syst. for Video Technology* (Dec. 1997), 7:872–81.

[19] Farvardin, N., and V. Vaishampayan. Optimal quantizer design for noisy channels: An approach to combined source-channel coding. *IEEE Trans. Inform. Theory* (Nov. 1987), 38:827–38.

[20] Fischer, T. R. Joint trellis coded quantization/modulation. *IEEE Trans. Commun.* (Feb. 1991), 39:172–76.

[21] Fleming, M., and M. Effros. Generalized multiple description vector quantization. In *Data Compression Conference (DCC'99),* Salt Lake, UT, March 1999.

[22] Fuldseth, A., and T. A. Ramstad. Combined video coding and multilevel modulation. In *IEEE Int. Conf. Image Proc. (ICIP'96),* 77–80, Lausanne, Switzerland, Sept. 1996.

[23] Ghanbari, M. Postprocessing of late cells for packet video. *IEEE Trans. Circuits Syst. for Video Technology* (Dec. 1996), 6(6):669–78.

[24] Gharavi, H., and S. M. Alamouti. Multipriority video transmission for third-generation wireless communication systems. *Proceedings of the IEEE* (Oct. 1999), 87:1751–63.

[25] Girod, B., and N. Färber. Wireless video. In M. T. Sun and A. R. Reibman, eds., *Compressed Video over Networks.* New York: Marcel Dekker, 2001, 465–512.

[26] Girod, B., and N. Färber. Feedback-based error control for mobile video transmission. *Proceedings of the IEEE, Special Issue on Video for Mobile Multimedia* (Oct. 1999), 87:1707–23.

[27] Goodman, D. J. *Wireless Personal Communications Systems.* Reading, MA: Addison-Wesley, 1997.

[28] Goyal, V. K., J. Kovacevic, R. Arean, and M. Vetterli. Multiple description transform coding of images. In *IEEE Int. Conf. Image Proc. (ICIP'98),* 1:674–78, Chicago, Oct. 1998.

[29] Hagenauer, J. Rate compatible punctured convolutional codes (RCPC codes) and their applications. *IEEE Trans. Commun.* (April 1988), 36:389–99.

[30] Hagenauer, J., and T. Stockhammer. Channel coding and transmission aspects for wireless multimedia. *Proceedings of the IEEE* (April 1988), 87(10):1764–77.

[31] Hanzo, L. Bandwidth efficient wireless multimedia communications. *Proceedings of the IEEE* (July 1998), 86:1342–82.

[32] Hoffman, D., G. Fernando, V. Goyal, and M. R. Civanlar. *RTP payload format for MPEG1/MPEG2 video,* IETF RFC2250, Jan. 1998.

[33] Hong, M. C., L. Kondi, H. Scwab, and A. K. Katsaggelos. Video error concealment techniques. *Signal Processing: Image Communications, Special Issue on Error Resilient Video* (1999), 14:437–92.

[34] Ingle, A., and V. A. Vaishampayan. DPCM system design for diversity systems with applications to packetized speech. *IEEE Trans. Speech and Audio Processing* (Jan. 1995), 3:48–57.

[35] ITU-T. Recommendation H.261: Video codec for audiovisual services at $p \times 64$ kbit/s, 1993.

[36] ITU-T. Recommendation H.324: Terminal for low bitrate multimedia communication, 1995.

[37] ITU-T. Recommendation G.114: International Telephone connections and circuits—General recommendations on the transmission quality for an entire international telephone connection—One way transmission time, 2000.

[38] ITU-T. Recommendation H.223: Multiplexing protocol for low bit rate multimedia communication, 1996.

[39] ITU-T. Recommendation H.221: Frame structure for a 64 to 1920 kbit/s channel in audiovisual teleservices, 1997.

[40] ITU-T. Recommendation H.320: Narrow-band visual telephone systems and terminal equipment, 1997.

[41] ITU-T. Recommendation H.245: Control protocol for multimedia communication, 1998.

[42] ITU-T. Recommendation H.263: Video coding for low bit rate communication, 1998.

[43] ITU-T. Recommendation H.323: Packet-based multimedia communications systems, 1998.

[44] Jafarkhani, H., and V. Tarokh. Multiple description trellis coded quantization. *IEEE Trans. Comm.* (June 1999), 47(6):799–803.

[45] Jiang, W., and A. Ortega. Multiple description coding via polyphase transform and selective quantization. In *SPIE Conf. Image Proc. Visual Comm. (VCIP'99)*, SPIE-3653(1–2):998–1008. San Jose, Jan. 1999.

[46] Kansari, M., et al. Low bit rate video transmission over fading channels for wireless microcellular systems. *IEEE Trans. Circuits Syst. for Video Technology* (Feb. 1996), 6:1–11.

[47] Katsaggelos, A. K., and N. P. Galatsanos, eds. *Signal Recovery Techniques for Image and Video Compression and Transmission.* Boston: Kluwer Academic Publishers, 1998.

[48] Kawahara, T., and S. Adachi. Video transmission technology with effective error protection and tough synchronization for wireless channels. In *IEEE Int. Conf. Image Proc. (ICIP'96)*, 101–04, Lausanne, Switzerland, Nov. 1996.

[49] Kurtenbach, J., and P. A. Wintz. Quantizing for noisy channels. *IEEE Trans. Commun.* (April 1969), 17:291–302.

[50] Lam, W.-M., A. R. Reibman, and B. Liu. Recovery of lost or erroneously received motion vectors. In *IEEE Int. Conf. Acoustics, Speech, Signal Proc. (ICASSP'93)*, Minneapolis, April 1993, 5:417–20.

[51] Lee, S. H., P. J. Lee, and R. Ansari. Cell loss detection and recovery in variable rate video. In *3rd Int. Workshop on Packet Video,* Morriston, NJ, March 1990.

[52] Lin, S., and D. J. Costello. *Error Control Coding: Fundamentals and Applications*. Englewood Cliffs, NJ: Prentice Hall, 1983.

[53] Marasli, R., P. D. Amer, and P. T. Conrad. Retransmission-based partially reliable transport service: An analytic model. In *IEEE INFOCOM'96,* 2:621–29, San Francisco, March 1996.

[54] Maxwell, K. Asymmetric digital subscriber line: interim technology for the next forty years. *IEEE Commun. Mag.* (Oct. 1996), 100–06.

[55] Mead, D. C. *Direct Broadcast Satellite Communications: An MPEG Enabled Service*. Englewood Cliffs, NJ: Prentice Hall, 2000.

[56] Modestino, J. W., and D. G. Daut. Combined source-channel coding of images. *IEEE Trans. Commun.* (Nov. 1979), 7:1644–59.

[57] Modestino, J. W., D. G. Daut, and A. L. Vickers. Combined source-channel coding of images using the block cosine transform. *IEEE Trans. Commun.* (Sept. 1981), 29:1261–73.

[58] Ortega, A., and K. Ramchandran. Rate-distortion methods for image and video compression. *IEEE Signal Processing Magazine* (Nov. 1998), 15:23–50.

[59] Ramamurthy, G., and D. Raychaudhuri. Performance of packet video with combined error recovery and concealment. In *IEEE INFOCOMM'95,* 753–61, 1995.

[60] Redmill, D. W., and N. G. Kingsbury. The EREC: An error resilient technique for coding variable-length blocks of data. *IEEE Trans. Image Proc.* (April 1996), 5(4):565–74.

[61] Reibman, A., H. Jafarkhani, Y. Wang, and M. Orchard. Multiple description coding for video using motion compensated prediction. In *IEEE Int. Conf. Image Proc. (ICIP'99),* 3:837–41. Kobe, Japan, Oct. 1999.

[62] Schulzrinne, H. IP networks. In M. T. Sun and A. R. Reibman, eds., *Compressed Video over Networks*. New York: Marcel Dekker, 2001, 81–138.

[63] Schulzrinne, H., S. Casner, R. Frederick, and V. Jacobson. *RTP: A transport protocol for real-time applications*. IETF RFC 1889, Jan. 1996. (Available from ftp://ftp.isi.edu/in-notes/rfc1889.txt).

[64] Smith, B. C. Implementation Techniques for Continuous Media Systems and Applications. Unpublished Ph.D. thesis, University of California, Berkeley, 1994.

[65] Srinivasan, M., and R. Chellappa. Multiple description subband coding. In *IEEE Int. Conf. Image Proc. (ICIP'98),* 1:684–88. Chicago, IL, Oct. 1998.

[66] Stallings, W. *ISDN and Broadband ISDN*. 2d ed. New York:Macmillan, 1992.

[67] Stuhlmuller, K., N. Färber, M. Link, and B. Girod. Analysis of video transmission over lossy channels. *IEEE Journal on Selected Areas in Communications* (June 2000), 18(6):996–1011.

[68] Sun, H., K. Challapali, and J. Zdepski. Error concealment in digital simulcast AD-HDTV decoder. *IEEE Trans. Consumer Electronics* (Aug. 1992), 38(3):108–17.

[69] Sun, H., and W. Kwok. Concealment of damaged block transform coded images using projections onto convex sets. *IEEE Trans. Image Process.* (April 1995), 4(4):470–77.

[70] Sun, H., and J. Zdepski. Error concealment strategy for picture-header loss in MPEG compressed video. In *SPIE Conf. High-Speed Networking and Multimedia Computing,* 145–52, San Jose, CA, Feb. 1994.

[71] Swann, R., and N. G. Kingsbury. Transcoding of MPEG-2 for enhanced resilience to transmission errors. In *IEEE Int. Conf. Image Proc. (ICIP'96),* 2:813–16. Lausanne, Switzerland, Nov. 1996.

[72] Takishima, Y., M. Wada, and H. Murakami. Reversible variable length codes. *IEEE Trans. Commun.* (Feb. 1995), 43(2):158–62.

[73] Turletti, T., T. Huitema, and C. Huitema. *RTP payload format for H.261 video streams,* IETF RFC2032, Oct. 1996.

[74] Vaishampayan, V. A., Design of multiple description scalar quantizers. *IEEE Trans. Inform. Theory* (May 1993), 39:821–34.

[75] Vaishampayan, V. A., and N. Farvardin. Optimal block cosine transform image coding for noisy channels. *IEEE Trans. Commun.* (March 1990), 38:327–36.

[76] Villasenor, J. D., Y.-Q. Zhang, and J. Wen. Robust video coding algorithms and systems. *Proceedings of the IEEE* (Oct. 1999), 87:1724–33.

[77] Wada, M. Selective recovery of video packet loss using error concealment. *IEEE J. Select. Areas Commun.* (June 1989), 7(5):807–14.

[78] Wang, M., and T. R. Fischer. Trellis coded quantization designed for noisy channels. *IEEE Trans. Inform. Theory* (Nov. 1994), 40:1792–802.

[79] Wang, Y., and D. Chung. Non-hierarchical signal decomposition and maximally smooth reconstruction for wireless video transmission. In D. J. Goodman and D. Raychaudhuri, eds., *Mobile Multimedia Communications.* New York: Plenum Press, 1997, 285–92.

[80] Wang, Y., M. Orchard, V. Vaishampayan, and A. R. Reibman. Multiple description coding using pairwise correlating transform. *IEEE Trans. Image Process.* (March 2001), 10(3):351–66.

[81] Wang, Y., S. Wenger, J. Wen, and A. G. Katsaggelos. Error resilient video coding techniques. *IEEE Signal Processing Magazine* (July 2000), 17(4):61–82.

[82] Wang, Y., and Q.-F. Zhu. Error control and concealment for video communication: A review. *Proceedings of the IEEE* (May 1998), 86:974–97.

[83] Wang, Y., and Q.-F. Zhu. Maximally smooth image recovery in transform coding. *IEEE Trans. Commun.* (Oct. 1993), 41(10):1544–51.

[84] Wen, J., and J. Villasenor. Utilizing soft information in decoding of variable length codes. In *Data Compression Conference (DCC'99),* Snowbird, UT, March 1999, 131–39.

[85] Wen, J., and J. Villasenor. Reversible variable length codes for robust image and video transmission. In *Data Compression Conference (DCC'98),* Snowbird, UT, April 1998, 471–80.

[86] Wen, J., and J. Villasenor. A class of reversible variable length codes for robust image and video coding. In *IEEE Int. Conf. on Image Proc. (ICIP'97)*, 2:25–28. Santa Barbara, CA, Oct. 1997.

[87] Wenger, S., Video redundancy coding in H.263+. In *Workshop on Audio-Visual Services for Packet Networks (AVSPN'97)*, Aberdeen, Scotland, Sept. 1997.

[88] Wenger, S., G. Knorr, J. Ott, and F. Kossentini. Error resilience support in H.263+. *IEEE Trans. Circuits Syst. for Video Technology* (Nov. 1998), 8(6):867–77.

[89] Wu, D., et al. An end-to-end approach for optimal mode selection in Internet video communication: Theory and application. *IEEE Journal on Selected Areas in Communications* (June 2000), 18(6):977–95.

[90] Wu, D., et al. On end-to-end architecture for transporting MPEG-4 video over the Internet. *IEEE Trans. Circuits Syst. for Video Technology* (June 2000), 18(6):977–95.

[91] Yang, X., and K. Ramchandran. Optimal multiple description subband coding. In *IEEE Int. Conf. Image Proc. (ICIP'98)*, 1:684–58, Chicago, IL, Oct. 1998.

[92] Yu, G.-S., M. M.-K. Liu, and M. W. Marcellin. POCS-based error concealment for packet video using multiframe overlap information. *IEEE Trans. Circuits Syst. for Video Technology* (Aug. 1998), 8(4):422–34.

[93] Zeger, K., and A. Gersho. Pseudo-Gray coding. *IEEE Trans. Commun.* (Dec. 1990), 38:2147–56.

[94] Zhang, R., S. L. Regunathan, and K. Rose. Video coding with optimal inter/intra-mode switching for packet loss resilience. *IEEE Journal on Selected Areas in Communications* (June 2000), 18(6):966–76.

[95] Zhang, Y.-Q., Y. J. Liu, and R. L. Pickholtz. Layered image transmission over cellular radio channels. *IEEE Trans. Vehicular Tech.* (Aug. 1994), 43:786–96.

[96] Zheng, H., and K. J. R. Liu. Multimedia services over digital subscriber lines. *IEEE Signal Process. Mag.* (July 2000), 17(4):44–60.

[97] Zhu, Q.-F. Device and method of signal loss recovery for real-time and/or interactive communications. U.S. Patent 5,550,847, Aug. 1996.

[98] Zhu, Q.-F., V. Eyuboglu, and M. Sridhar. Device and method of digital video streaming, U.S. Patent 5,768,527, June 1998.

[99] Zhu, Q.-F., and Y. Wang. Error concealment in visual communications. In M. T. Sun and A. R. Reibman, eds., *Compressed Video over Networks*. New York: Marcel Dekker, 2001, 217–50.

[100] Zhu, Q.-F., Y. Wang, and L. Shaw. Coding and cell loss recovery for DCT-based packet video. *IEEE Trans. Circuits Syst. for Video Technology* (June 1993), 3(3):248–58.

15

Streaming Video over the Internet and Wireless IP Networks

Recent developments in computing technology, compression technology, high-bandwidth storage devices, and high-speed networks have made it feasible to provide real-time multimedia services over the Internet. Real-time multimedia, as the name implies, has timing constraints. For example, audio and video data must be played out continuously. If the data does not arrive in time, the playout process will pause, which is annoying to viewers.

Real-time transport of live or stored video is the predominant part of real-time multimedia. In this chapter, we are concerned with video streaming, which refers to real-time transmission of stored video.* Video streaming typically has bandwidth, delay, and loss requirements. However, the current best-effort Internet does not offer any QoS guarantees for streaming video over the Internet. In addition, the heterogeneity of the Internet makes it difficult to efficiently support video multicast while providing service flexibility to meet a wide range of QoS requirements from users. Furthermore, for streaming video over wireless IP networks, fluctuations of wireless channel conditions tend to greatly degrade video quality. Thus, streaming video over the Internet and wireless IP networks poses many challenges.

To deal with the challenges, extensive efforts have been contributed. With the aim of providing a global view of this field, we cover seven areas regarding streaming video, namely: video compression, application-layer QoS control for streaming video, continuous media distribution services, streaming servers, media synchronization mechanisms, protocols for streaming media, and streaming video over wireless

---

*Video streaming implies that the video content need not be downloaded in full, but is being played out while parts of the content are being received and decoded.

IP networks. For each area, we address the particular issues and review representative approaches and mechanisms.

15.1 ARCHITECTURE FOR VIDEO STREAMING SYSTEMS

A video streaming system typically consists of seven building blocks, as illustrated in Figure 15.1. In the figure, raw video and audio data are precompressed by video and audio compression algorithms and then saved in storage devices. Upon the client's request, a *streaming server* retrieves compressed video/audio data from storage devices, and then the *application-layer QoS control* module adapts the video/audio bit streams according to the network status and QoS requirements. After this adaptation, the transport *protocols* packetize the compressed bit streams and send the video/audio packets to the Internet or wireless IP networks. Packets may be dropped or experience excessive delay on the Internet due to congestion; on wireless IP segments, packets may be damaged by bit errors. To improve the quality of video/audio transmission, *continuous media distribution services* are deployed in the Internet. Packets that are successfully delivered to the receiver first pass through the transport layers and then are processed by the application layer before being decoded at the video/audio decoder. To achieve synchronization between video and audio presentations, *media synchronization mechanisms* are required. From Figure 15.1, it can be seen that the seven areas are closely related, and that they are coherent constituents of the video streaming architecture.

Next, we briefly describe the seven areas, respectively.

1. **Video compression:** Raw video must be compressed before transmission so that efficiency can be achieved. Video compression schemes can be classified into two categories: scalable and nonscalable video coding. Since scalable video is capable

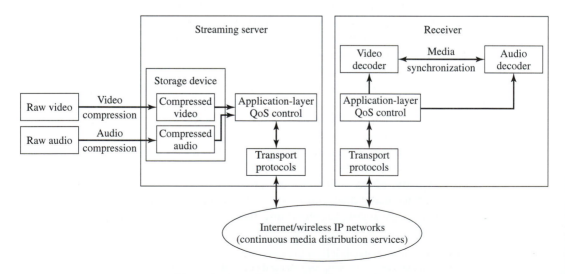

Figure 15.1 Architecture for video streaming.

of gracefully coping with the bandwidth fluctuations of the Internet [45], we are primarily concerned with scalable video coding techniques. We covered this topic in Chapter 11, and thus will provide only a brief review of this topic here.

2. **Application-layer QoS control for streaming video:** To cope with changing network conditions and changing presentation quality requested by users, various application-layer QoS control techniques have been proposed [19, 69, 77]. These techniques include congestion control and error control. Their respective functions are as follows: Congestion control is employed to prevent packet loss and reduce delay. Error control, on the other hand, improves video presentation quality in the presence of packet loss. Techniques for error control for general video communication applications were described in Chapter 14. Here, we will focus on those that are effective for video streaming applications.

3. **Continuous media distribution services:** In order to provide quality multimedia presentations, support from the network is important. This is because network support can reduce transport delay and packet loss ratio. Built on top of the Internet (IP protocol), continuous media distribution services are able to achieve QoS and efficiency for streaming video/audio over the best-effort Internet. Continuous media distribution services include network filtering, application-level multicast, and content replication.

4. **Streaming servers:** Streaming servers play a key role in providing streaming services. To offer quality streaming services, streaming servers are required to process multimedia data under timing constraints and support interactive control operations such as pause/resume, fast forward, and fast backward. Furthermore, streaming servers must retrieve media components in a synchronous fashion. A streaming server typically consists of three subsystems, namely, a communicator (e.g., transport protocols), an operating system, and a storage system.

5. **Media synchronization mechanisms:** Media synchronization is a major feature that distinguishes multimedia applications from other traditional data applications. With media synchronization mechanisms, the application at the receiver side can present various media streams in the same way as they were originally captured. A typical effect of media synchronization is that the movements of the lips of a speaker correspond to the presented audio.

6. **Protocols for streaming media:** Protocols are designed and standardized for communication between clients and streaming servers. Protocols for streaming media provide such services as network addressing, transport, and session control. According to their functionalities, the protocols can be classified in three categories: (1) network-layer protocol such as IP, (2) transport protocol such as User Datagram Protocol (UDP), and (3) session control protocol such as Real-Time Streaming Protocol (RTSP).

7. **Streaming video over wireless IP networks:** Fluctuations of wireless channel conditions pose many challenges to the provision of QoS for video transmission over wireless IP networks. To address this problem, an adaptive framework has been proposed [85]. The adaptive framework includes scalable video representations, network-aware video applications, and adaptive services.

The remainder of this chapter is organized as follows: Section 15.2 discusses video compression techniques. In Section 15.3, we present application-layer QoS control mechanisms for streaming video. Section 15.4 describes continuous media distribution services. In Section 15.5, we discuss key issues of the design of streaming servers. Section 15.6 presents various media synchronization mechanisms. In Section 15.7, we overview the key protocols for streaming video. Section 15.8 presents an adaptive framework for streaming video over wireless IP networks. Finally, a summary is given in Section 15.9.

15.2 VIDEO COMPRESSION

Since raw video consumes large quantities of bandwidth, it must be compressed before transmission so that efficiency can be achieved. Video compression schemes can be classified into two categories: scalable and nonscalable video coding. As described in Chapter 11, a scalable video coding scheme produces a compressed bit stream, parts of which are decodable. Compared with decoding the complete bit stream, decoding part of the compressed bit stream produces pictures with degraded quality, or smaller image size, or smaller frame rate. It has been shown that scalable video is capable of gracefully coping with the bandwidth fluctuations on the Internet [40, 45]. In contrast, nonscalable video is more susceptible to bandwidth variations, since it cannot adapt its video representation to bandwidth fluctuations [45]. Furthermore, scalable video representation is an effective means to achieve efficiency and flexibility for multicasting video over heterogeneous networks (e.g., networks with different access-link bandwidths) [40, 45]. For these reasons, all streaming video services employ scalable video coding techniques. For a more detailed discussion of various scalable coding methods, see Chapter 11. Next, we present the application-layer QoS control mechanisms, which adapt the video bit streams according to the network status and QoS requirements.

15.3 APPLICATION-LAYER QOS CONTROL FOR STREAMING VIDEO

Application-layer QoS control maximizes video quality in the presence of packet loss and changes in available bandwidth. Application-layer QoS control techniques include congestion control and error control. These techniques are employed by the end systems, and do not require QoS support from routers or networks. Section 15.3.1 surveys the approaches for congestion control, and Section 15.3.2 describes mechanisms for error control.

15.3.1 Congestion Control

Bursty loss and excessive delay have devastating effects on video presentation quality, and they are usually caused by network congestion. One way to reduce packet loss and delay is by applying congestion control at the source and/or receiver.

There are two mechanisms for congestion control: rate control and rate shaping. Rate control attempts to minimize network congestion and the amount of packet loss by matching the rate of the video stream to the available network bandwidth. On the other hand, rate shaping is used to force the source to send the video stream at the rate dictated by the rate control algorithm [19].

Rate Control   Rate control is a technique that determines the sending rate of video traffic based on the estimated available bandwidth in the network.* Existing rate-control schemes can be classified into three categories: source-based, receiver-based, and hybrid rate control.

Source-Based Rate Control: Under source-based rate control, the sender is responsible for adapting the video transmission rate. Typically, feedback is employed by source-based rate control mechanisms. Based upon feedback information about the network, the sender can regulate the rate of the video stream. Source-based rate control can be applied to both unicast [81] and multicast [8].

For unicast video, the existing source-based rate control mechanisms follow two approaches: probe-based and model-based.

The probe-based approach is based on probing experiments. Specifically, the source probes for the available network bandwidth by adjusting the sending rate in a way that could maintain the packet loss ratio p below a certain threshold P_{th} [81]. There are two ways to adjust the sending rate: (1) additive increase and multiplicative decrease [81], and (2) multiplicative increase and multiplicative decrease [74].

The model-based approach is based on a throughput model of a TCP connection. Specifically, the throughput of a TCP connection can be characterized by the following formula [22]:

$$\lambda = \frac{1.22 \times \text{MTU}}{\text{RTT} \times \sqrt{p}}, \tag{15.3.1}$$

where λ is the throughput of a TCP connection, MTU (maximum transmission unit) is the packet size used by the connection, RTT is the round-trip time for the connection, and p is the packet-loss ratio experienced by the connection. Under model-based rate control, Equation (15.3.1) is used to determine the sending rate of the video stream. Thus, the video connection can avoid congestion in a way similar to that of TCP, and can compete fairly with TCP flows. For this reason, model-based rate control is also called "TCP-friendly" rate control [22].

For multicast under source-based rate control, the sender uses a single channel to transport video to the receivers. Such multicast is called "single-channel multicast." For single-channel multicast, only probe-based rate control can be employed [8].

Single-channel multicast is efficient, since all the receivers share one channel. However, single-channel multicast is unable to provide flexible services to meet the different demands from receivers with various access-link bandwidths. In contrast, if multicast video were to be delivered through individual unicast streams, the band-width efficiency would be low but the services could be differentiated, since each receiver could negotiate the parameters of the services with the source. Unicast and

---

*In Section 9.3.4, we defined the rate control problem as (1) determining the appropriate encoding rate and (2) adjusting the coding parameters to meet the target rate. Rate control in this section refers to the first task only.

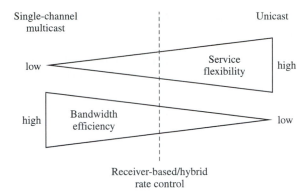

Figure 15.2   Trade-off between efficiency and flexibility.

single-channel multicast are two extreme cases, as shown in Figure 15.2. To achieve good trade-off between bandwidth efficiency and service flexibility for multicast video, receiver-based and hybrid rate control have been proposed.

Receiver-Based Rate Control: Under receiver-based rate control, receivers regulate the receiving rate of video streams by adding and dropping channels, whereas the sender does not participate in rate control. Typically, receiver-based rate control is only applied to layered multicast rather than unicast video.

Similarly to source-based rate control, the existing receiver-based rate control mechanisms follow two approaches: probe-based and model-based. The basic probe-based rate control [45] consists of two parts:

1. When no congestion is detected, a receiver probes for the available bandwidth by joining a layer, resulting in increase of its receiving rate. If no congestion is detected after joining, the join-experiment is successful. Otherwise, the receiver drops the newly added layer.
2. When congestion is detected, a receiver drops a layer, resulting in reduction of its receiving rate.

Unlike the probe-based approach, which implicitly estimates the available network bandwidth through probing experiments, the model-based approach uses explicit estimation for the available network bandwidth. The model-based approach is based on the throughput model of a TCP connection (i.e., Equation (15.3.1)), as described under source-based rate control. Thus, model-based rate control is also "TCP-friendly."

Hybrid Rate Control: Under hybrid rate control, receivers regulate the receiving rate of video streams by adding and dropping channels while the sender also adjusts the transmission rate of each channel based on feedback from the receivers. An example of hybrid rate control is destination set grouping [13].

Rate Shaping   Rate shaping is a technique through which the rate of precompressed video bit streams can be adapted to a target rate constraint. A rate shaper is an interface (or filter) between the compression layer and the network transport layer, or between two network segments, with which the video stream can be matched to the available network bandwidth.

There are many types of rate shapers or filters [87], which include:

1. **Codec filters:** Codec filters decompress and compress a video stream. They are commonly used to perform transcoding between different compression schemes. Depending on the compression scheme used, transcoding can be simplified without full decompression and recompression.

2. **Frame-dropping filters:** Frame-dropping filters can distinguish frame types (e.g., I-, P-, and B-frames in MPEG) and drop frames according to importance. For example, dropping order would be first B-frames, then P-frames, and finally I-frames. Frame-dropping filters are used to reduce the data rate of a video stream by discarding a number of frames. The remaining frames require a lower rate. These filters can be used at the source [90] or in the network (see Section 15.4.1).

3. **Layer-dropping filters:** Layer-dropping filters can distinguish and drop layers according to importance. The dropping order is from the highest enhancement layer down to the base layer.

4. **Frequency filters:** Frequency filters perform operations on the compression layer. Specifically, they operate in the frequency domain (i.e., DCT coefficients). Frequency filtering mechanisms include low-pass, color reduction, and color-to-monochrome filtering.

   Low-pass filtering discards the DCT coefficients of the higher frequencies. A color reduction filter performs the same operation as a low-pass filter, except that it operates only on the chrominance information in the video stream. A color-to-monochrome filter removes all color information from the video stream. In MPEG, this is done by replacing each chrominance block by an empty block.

   Unlike frame-dropping filters, frequency filters reduce the bandwidth without affecting the frame rate. They reduce the presentation quality of the resulting frame.

5. **Requantization filters:** Requantization filters perform operations on the compression layer (i.e., DCT coefficients). The filter first extracts the DCT coefficients from the compressed video stream through dequantization. Then the filter requantizes the DCT coefficients with a larger quantizer stepsize, resulting in rate reduction.

As a summary, the purpose of congestion control is to prevent packet loss. The facts that packet loss is inevitable on the Internet and may have significant impact on perceptual quality prompt the design of mechanisms to maximize video presentation quality in the presence of packet loss. Error control is such a mechanism.

15.3.2 Error Control

Techniques for error control in video communication were discussed in Chapter 14. There, we classified error control techniques into four categories: transport-level error control, including FEC and delay-constrained retransmission; encoder error-resilient coding; decoder error concealment; and encoder–decoder interactive error control. Here, we describe several techniques that are effective for video streaming applications.

FEC   FEC works by adding redundant information in the original message stream, so that the message can be reconstructed in the presence of packet loss. For Internet applications, block codes are typically applied across packets. Specifically, a video stream is first chopped into segments, each of which is packetized into k packets; then, for each segment, a block code (e.g., Tornado code [1]) is applied across the k packets to generate an n-packet block, where $n > k$. To perfectly recover a segment, a user need only receive any k packets in the n-packet block.

Delay-Constrained Retransmission   Retransmission is usually dismissed as a method for transporting real-time video because a retransmitted packet may miss its play-out time. However, for streaming applications, if the one-way trip time is short with respect to the maximum allowable delay, *delay-constrained retransmission* is a viable option for error control (see Section 14.3.3).

For unicast, based on who determines whether to send and/or respond to a retransmission request, three delay-constrained retransmission mechanisms have been proposed: receiver-based, sender-based, and hybrid control.

The objective of receiver-based control is to minimize the requests for retransmissions that will not arrive timely for display. Under receiver-based control, the receiver executes the following algorithm:

When the receiver detects the loss of packet N:

$$\text{if } (T_c + \text{RTT} + D_s < T_d(N))$$

send the request for packet N to the sender

where T_c is the current time, RTT is an estimated round trip time, D_s is a slack term, and $T_d(N)$ is the time when packet N is scheduled for display. The slack term D_s could include tolerance of error in estimating RTT, the sender's response time, and the receiver's decoding delay. The timing diagram for receiver-based control is shown in Figure 15.3, where D_s is only the receiver's decoding delay.

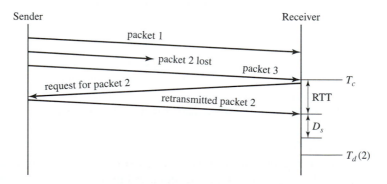

Figure 15.3   Timing diagram for receiver-based control.

The objective of sender-based control is to suppress retransmission of the packets that will miss their display time at receiver. Under sender-based control, the sender executes the following algorithm:

When the sender receives a request for packet N:

if $(T_c + \mathrm{RTT}/2 + D_s < T'_d(N))$

retransmit packet N to the receiver

where $T'_d(N)$ is an estimate of $T_d(N)$.

Hybrid control is a simple combination of sender-based and receiver-based control.

In the multicast case, retransmission must be restricted within closely located multicast members. This is because one-way trip times between these members tend to be small, making retransmissions effective. Typically, a logical tree is configured to limit the number and scope of retransmission requests, and to achieve recovery among closely located multicast members [86]. In addition, to address heterogeneity problems, a receiver-initiated mechanism for error recovery can be adopted [86].

Error-Resilient Encoding   As described in Section 14.4, error-resilient schemes address loss recovery from the compression perspective. Specifically, they attempt to prevent error propagation or limit the scope of the damage (caused by packet losses) in the compression layer. The standardized error-resilient tools include resynchronization marking, data partitioning, and reversible variable length coding [35, 68] (see Section 14.4). These techniques are targeted at bit-error dominant environments such as wireless channels. For Internet video applications, which use packet-based transport, these techniques may not be useful, since a packet loss may cause the loss of all the motion data and its associated shape and texture data. Furthermore, the boundary of a packet already provides a synchronization point in the variable-length coded bit stream at the receiver side, making the resynchronization marking somewhat redundant.

The two techniques that are more promising for robust Internet video transmission are *optimal mode selection* and *multiple description coding*.

Optimal Intra/Inter Mode Selection: The effect of lost packets on video presentation quality depends on the coding scheme used at the source, the network congestion status, and the error-concealment scheme used at the receiver. High-compression coding algorithms usually employ inter-coding (i.e., prediction) to achieve efficiency. With these coding algorithms, loss of a packet may degrade video quality over a large number of frames, until the next intra-coded frame is received. Intra-coding can effectively stop error propagation at the cost of efficiency, whereas inter-coding can achieve compression efficiency at the risk of error propagation. Therefore, a good mechanism for selection between intra- mode and inter- modes should be in place to enhance the robustness of video compressed by both intra- and inter-coding (see Figure 15.4).

For video communication over a network, a coding algorithm such as H.263 or MPEG-4 [35] usually regulates output rate to match the available bandwidth. The objective of rate-regulated compression algorithms is to maximize the video quality

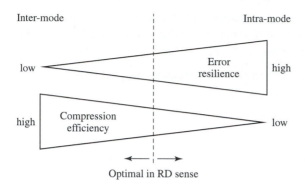

Figure 15.4   Optimal mode selection depends on current network condition.

under the constraint of a given bit budget. This can be achieved by choosing a mode that minimizes the quantization distortion between the original and the reconstructed frame or macroblock under a given bit budget [52], which is the so-called RD-optimized mode selection (see Section 9.3.3). We refer to such RD-optimized mode selection as the classical approach. The classical approach is unable to achieve global optimality in an error-prone environment, since it does not consider network congestion status and receiver behavior.

To address this problem, Wu et al. [83] proposed an end-to-end approach to RD-optimized mode selection. It considers all three factors that have an impact on video presentation quality at the receiver: (1) source behavior, (2) path characteristics, and (3) receiver behavior. Due to the consideration of network congestion status and receiver behavior, this approach is shown to be capable of offering superior performance over the classical approach for Internet video applications [83]. Other approaches for mode selection with the consideration of transmission error effects were introduced in Section 14.4.3.

Multiple-Description Coding: Multiple-description coding (Section 14.4.5) is another way of trading off compression efficiency with robustness to packet loss. Compared to other error-resilient techniques described in Section 14.4, it is more effective at handling packet losses. To carry MDC streams over the Internet, bits from different descriptions should be packetized separately, and preferably transmitted through different routes. Alternatively, the packets containing different descriptions should be properly interleaved, so that traffic congestion will not lead to the loss of both descriptions.

Error Concealment   When packet loss is detected, the receiver can employ error concealment to conceal the lost data and make the presentation more pleasing to the eye. Since viewers can tolerate a certain degree of distortion in video signals, error concealment is a viable technique for handling packet loss [80].

As described in Section 14.5, there are two basic approaches to error concealment, namely, spatial and temporal interpolation. In spatial interpolation, missing pixel values are reconstructed using neighboring spatial information, whereas in temporal interpolation, the lost data is reconstructed from data in the previous frames. Typically, spatial interpolation is used to reconstruct missing data in intra-coded frames, whereas temporal interpolation is used to reconstruct missing data in inter-coded frames.

For streaming applications, which require real-time decoding, most of the techniques described in Section 14.5 are too complex. The most feasible and effective technique is motion-compensated temporal interpolation (see Section 14.5.1), by which the receiver replaces the corrupted block with the block from the previous frame pointed to by a motion vector. The motion vector is copied from its neighboring block when available, otherwise the motion vector is set to zero.

In sum, we have reviewed various application-layer QoS control techniques. These techniques are employed by the end systems and do not require QoS support from routers or networks. If networks support QoS for video streaming, performance will be further enhanced. Next, we present QoS support mechanisms developed for the best-effort Internet.

15.4 CONTINUOUS MEDIA DISTRIBUTION SERVICES

In order to provide quality multimedia presentations, support from the network is important. This is because network support can reduce transport delay and packet loss ratio. Streaming video and audio are classified as continuous media because they consist of a sequence of media quanta (such as audio samples or video frames), which convey meaning only when presented in time. Built on top of the Internet (IP protocol), continuous media distribution services are designed to provide QoS and achieve efficiency for streaming video and audio over the best-effort Internet. Continuous media distribution services include network filtering, application-level multicast, and content replication, which are presented in Sections 15.4.1–3, respectively.

15.4.1 Network Filtering

As a congestion control technique, network filtering is aimed at maximizing video quality during network congestion. As described in Section 15.3.1, the filter at the video server can adapt the rate of video streams according to network congestion status. However, the video server may be too busy to handle the computation required to adapt each unicast video stream. Hence, service providers may wish to place filters in the network [32]. Figure 15.5 illustrates an example of network filtering. In the figure, nodes labeled "R" denote routers that have no knowledge of the format of the media streams and may randomly discard packets. "Filter" nodes receive the client's requests and adapt the stream sent by the server accordingly. This solution allows the service provider to place filters on the nodes that connect to network bottlenecks, and there can be multiple filters along the path from a server to a client.

To illustrate the operations of filters, a system model is presented in Figure 15.6 [32]. The model consists of the server, the client, at least one filter, and two virtual channels between them. Of the two virtual channels, one is for control and the other is for data. The same channels exist between any pair of filters. The control channel is bidirectional, which can be realized by TCP connections. The model shown in

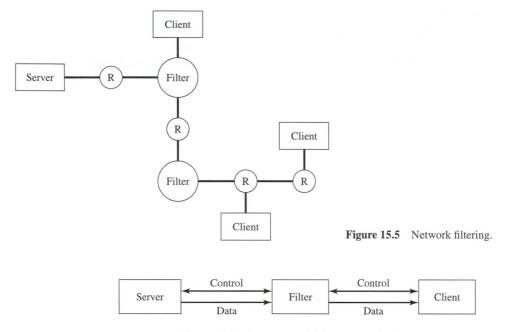

Figure 15.5   Network filtering.

Figure 15.6   A system model for network filtering.

Figure 15.6 allows the client to communicate with only one host (the last filter), which will either forward the requests or act upon them. The operations of a filter on the data plane include: (1) receiving video stream from the server or a previous filter, and (2) sending video to a client or the next filter at the target rate. The operations of a filter on the control plane include: (1) receiving requests from a client or the next filter, (2) acting upon requests, and (3) forwarding requests to the previous filter.

Typically, frame-dropping filters (see Section 15.3.1) are used as network filters. The receiver can change the bandwidth of the media stream by sending requests to the filter to increase or decrease the frame-dropping rate. To facilitate decisions regarding whether the filter should increase or decrease the bandwidth, the receiver continuously measures the packet loss ratio p. Based on the packet loss ratio, a rate-shaping mechanism is designed as follows [32]. If the packet loss ratio is higher than a threshold α, the client will ask the filter to increase the frame-dropping rate. If the packet loss ratio is less than another threshold β ($\beta < \alpha$), the receiver will ask the filter to reduce the frame-dropping rate.

The advantages of using frame-dropping filters inside the network include:

- **Improved video quality.** For example, when a video stream flows from an upstream link with larger available bandwidth to a downstream link with smaller available bandwidth, use of a frame-dropping filter at the connection point (between the links) could help improve video quality. This is because the filter understands the format of the media stream and can drop packets in a way that gracefully degrades the stream's quality instead of corrupting the flow outright.

- **Bandwidth efficiency.** Filtering can help to save network resources by discarding frames that are late or have been corrupted by loss of constituent packets.

Note that network filtering is targeted at unicast (i.e., point-to-point) media streams. Next, we present a multicast mechanism for streaming media.

15.4.2 Application-Level Multicast

The Internet's original design, through well suited for point-to-point applications such as e-mail, file transfer, and Web browsing, fails to effectively support large-scale content delivery such as streaming-media multicast. In an attempt to address this shortcoming, a technology called "IP multicast" was proposed in the early nineties. As an extension to the IP layer, IP multicast is capable of providing efficient multipoint packet delivery. To be specific, efficiency is achieved by ensuring that one and only one copy of an original IP packet (sent by the multicast source) is transported along any physical path in the IP multicast tree. However, with a decade's worth of research and development, IP Multicast is still plagued with concerns pertaining to scalability, network management, deployment, and support for higher-layer functionality such as error, flow, and congestion controls. To address these concerns, an application-level multicast mechanism was proposed [21]. Application-level multicast is aimed at building a multicast service on top of the Internet and traditional IP networks.

The application-level multicast mechanism would allow independent content service providers (CSPs), Internet service providers (ISPs), or enterprises to build their own Internet multicast networks and interconnect them into larger, worldwide "media multicast networks." That is, the media multicast networks could support "peering relationships" at the application level or the streaming-media/content layer, where "content backbones" interconnect service providers. Hence, much as the Internet is built from an interconnection of networks enabled through IP-level peering relationships among ISPs, the media multicast networks can be built from an interconnection of content-distribution networks enabled through application-level peering relationships among various sorts of service providers—for example, traditional ISPs, CSPs, and application service providers (ASPs).

The operations of the media multicast networks are described as follows. In media multicast networks, each multicast-capable node (called a MediaBridge [21]) performs routing at the application layer. In addition, each MediaBridge is interconnected with one or more neighboring MediaBridges through explicit configuration, which defines the application-level overlay topology. Collectively, the MediaBridges in a media multicast network employ a distributed application-level multicast routing algorithm to determine the optimal virtual paths for propagating content throughout the network. When the underlying network fails or becomes overly congested, the media multicast network automatically and dynamically reroutes content via alternate paths according to application-level routing policies. In addition, MediaBridges dynamically subscribe to multicast content when and only when a downstream client requests it. This capability ensures that one and only one copy of the multicast content flows across any physical or virtual path independent of the number of downstream clients, resulting in the conservation of network bandwidth.

The advantage of application-level multicast is that it breaks the "boundaries" between scalability, network management, and support for congestion control, which have prevented ISPs from establishing "IP multicast" peering arrangements.

15.4.3 Content Replication

A fundamental technique for improving scalability of the media delivery system is content/media replication. Content replication takes two forms, namely, caching and mirroring, which are deployed by publishers, CSPs, and ISPs. Both caching and mirroring seek to place content closer to the clients, and both share the following advantages:

- reduced bandwidth consumption on network links,
- reduced load on streaming servers,
- reduced latency for clients, and
- increased availability.

Mirroring places copies of the original multimedia files on several machines scattered around the Internet. That is, the original multimedia files are stored on the main server while copies of these files are placed on duplicate servers. In this way, clients can retrieve multimedia data from the nearest duplicate server, which gives clients the best performance (e.g., lowest latency). Mirroring has several advantages, such as that

- all contents are replicated; and
- the publisher is able to view server access logs and perform user tracking.

On the other hand, there are a number of disadvantages with mirroring. Currently, mechanisms for establishing dedicated mirrors are expensive, ad hoc, and slow. In addition, establishing a mirror on an existing (rather than duplicate) server, while cheaper, is still an ad hoc and administratively complex process. Finally, there is no standard way to make scripts and server setup easily transferable from one server to another.

Caching, which is based on the belief that many different clients will load the same content, makes local copies of content that the clients retrieve. Typically, clients in a single organization retrieve all content from a single local machine, called a cache. The cache retrieves a video file from the streaming server, stores a copy locally, and then passes it on to the client who requests it. If a client asks for a video file that the cache has already stored, the cache will return the copy rather than going all the way back to the streaming server on which the video file resides. In addition, cache sharing and cache hierarchies allow each cache to access files stored at other caches, so that the load on the main server can be reduced and network bottlenecks can be alleviated [11, 20].

Caching has the following advantages:

- caches can form a shared infrastructure that works for all streaming servers; and
- caches are incrementally deployable.

For these reasons, ISPs have been the largest proponent of caching. However, caches cannot supply the services that publishers need: support for quality of service, and security. First, when a cache stores a publisher's document, it makes no promises about how well the document will be treated. The publisher has no way of requesting that its document be delivered within certain time bounds, or that the cache verify the freshness of its copy of the document. Second, today's caches trust everyone, so that attacks such as inserting fraudulent versions of documents into the cache are extremely easy to accomplish. Finally, caches do not provide feedback to publishers, so that publishers are unable to perform user tracking.

Most caching techniques are targeted at generic web objects. Some recent work has demonstrated that caching strategies that are specific to particular types of objects can help improve overall performance [46]. For this reason, a great deal of effort has been contributed in this direction [46, 58, 62, 91]. A trivial extension of caching techniques to video is to store complete video sequences in the cache. However, such an approach may not be applicable due to the large scale of video data size and possibly limited cache space on a proxy server. Instead, it was shown that even a few cached frames can also contribute to significant improvement in performance [46]. Miao and Ortega proposed two video caching strategies, initial caching and selective caching, that store part of the video stream in the cache [46]. In particular, it was shown that selective caching can maximize the robustness of the video stream against network congestion while not violating the limited decoder buffer size.

In this section, we described three network support mechanims for streaming media. Next, we discuss key issues of the design of streaming servers.

15.5 STREAMING SERVERS

Streaming servers play a key role in providing streaming services. To offer quality streaming services, streaming servers are required to process multimedia data under timing constraints in order to prevent artifacts (e.g., jerkiness in video motion and pops in audio) during client playback. In addition, streaming servers must support VCR-like control operations such as stop, pause/resume, fast forward, and fast reverse. Furthermore, streaming servers must retrieve media components in a synchronous fashion. For example, retrieving a lecture presentation requires synchronizing video and audio with lecture slides.

A streaming server typically consists of the following three subsystems:

1. **Communicator:** A communicator involves the application layer and transport protocols implemented on the server (shown in Figure 15.1). Through a communicator, clients can communicate with a server and retrieve multimedia contents in a continuous and synchronous manner. We have addressed the application layer in Section 15.3, and will address transport protocols in Section 15.7.

2. **Operating system:** Different from traditional operating systems, an operating system for streaming services must satisfy real-time requirements for streaming applications.

3. **Storage system:** A storage system for streaming services must support continuous media storage and retrieval.

In this section, we are primarily concerned with operating system support and storage systems for streaming media, which will be presented in Sections 15.5.1 and 15.5.2, respectively.

15.5.1 Real-Time Operating System

The operating system shields computer hardware from all other software. The operating system offers various services related to the essential resources, such as the CPU, main memory, storage, and all input and output devices. In the following sections, we discuss the unique issues of real-time operating systems and review the associated approaches to the problems introduced by streaming services. We first show how process management takes into account the timing requirements imposed by streaming media and applies appropriate scheduling methods; we then describe how to manage resources to accommodate timing requirements; finally, we discuss issues of file management.

Process Management   Process management deals with the main processor resource. The process manager maps single processes onto the CPU resource according to a specified scheduling policy, such that all processes can meet their requirements.

To fulfill the timing requirements of continuous media, the operating system must use real-time scheduling techniques. Most attempts to solve real-time scheduling problems are just variations of two basic algorithms for multimedia systems: earliest deadline first (EDF) [42] and rate-monotonic scheduling [14]. In EDF scheduling, each task is assigned a deadline and the tasks are processed in the order of increasing deadlines. In rate-monotonic scheduling, each task is assigned a static priority according to its request rate.* Specifically, the task with the shortest period (or the highest rate) gets the highest priority, and the task with the longest period (or the lowest rate) gets the lowest priority; then the tasks are processed in the order of priorities.

Both EDF and rate-monotonic scheduling are preemptive—that is, the schedulers can preempt the running task and schedule the new task for the processor based on its deadline or priority. The execution of the interrupted task will resume at a later time. The difference between EDF and rate-monotonic scheduling is as follows. An EDF scheduler is based on a single priority task queue, and the processor runs the task with the earliest deadline. On the other hand, a rate-monotonic scheduler is a static-priority scheduler with multiple-priority task queues; that is, the tasks in the lower-priority queue cannot be executed until all the tasks in the higher-priority queues are served. In the example of Figure 15.7, there are two task sequences. The high-rate sequence comprises Tasks 1–8; the low-rate sequence Tasks A–D. As shown in the figure, in rate-monotonic scheduling, Task 2 preempts Task A, since Task 2 has a higher priority; on the other hand, in EDF, Task 2 does not preempt Task A, since Task A and Task 2 have the same deadlines ($d\text{A} = d2$). It can be seen that a

---

*We assume that each task is periodic.

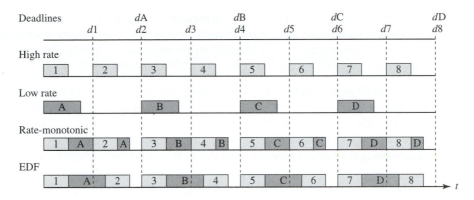

Figure 15.7   EDF versus rate-monotonic schedulers.

rate-monotonic scheduler is more prone to task switching than EDF. In summary, the rate-monotonic algorithm ensures that all deadlines will be met if the processor utilization is under 69 percent [14]; the EDF algorithm can achieve 100 percent utilization of the processor, but may not guarantee the processing of some tasks during overload periods.

Resource Management   Resources in a multimedia server include CPUs, memories, storage devices, and so on. Since resources are limited, a multimedia server can only service a limited number of clients with requested QoS. Therefore, resource management is required to manage resources to accommodate timing requirements. Resource management involves admission control and resource allocation. Specifically, before admitting a new client, a multimedia server must perform an admission control test to decide whether a new connection can be admitted without violating performance guarantees already provided to existing connections. If a connection is accepted, the resource manager allocates resources required to meet the QoS for the new connection.

Admission control algorithms can be classified into two categories: deterministic [24] and statistical [76]. Deterministic admission control algorithms provide hard guarantees to clients, whereas statistical algorithms provide statistical guarantees (i.e., the continuity requirements of at least a fixed percentage of media units are ensured to be met). The advantages of deterministic admission control are simplicity and strict assurance of quality; its limitation is lower utilization of server resources. In contrast, statistical admission control improves the utilization of server resources by exploiting human perceptual tolerances as well as the differences between the average and the worst-case performance characteristics of a multimedia server [76].

Corresponding to admission control algorithms, resource allocation schemes can be either deterministic or statistical. Deterministic admission control algorithms require deterministic resource allocation schemes, whereas statistical admission control algorithms require statistical resource allocation schemes. Deterministic resource allocation schemes make reservations for the worst case—for example, reserving bandwidth for the longest processing time and the highest rate that a task might ever need. On the other hand, statistical resource allocation schemes achieve higher utilization by allowing temporary overload, resulting in a small percentage of QoS violations.

File Management   The file system provides access and control functions for file storage and retrieval. There are two basic approaches to supporting continuous media in file systems. In the first approach, the organization of files on disks remains as it is for discrete data (i.e., a file is not scattered across several disks), with the necessary real-time support provided through special disk-scheduling algorithms and enough buffer capacity to avoid jitter. The second approach is to organize audio and video files on distributed storage, such as disk arrays. Under the second approach, the disk throughput can be improved by scattering or striping each audio/video file across several disks; disk seek times can be reduced by disk-scheduling algorithms.

Traditional disk-scheduling algorithms such as first-come-first-served and SCAN [16, 70] do not provide real-time guarantees. Hence, many disk-scheduling algorithms have been proposed in support of continuous media. These include SCAN-EDF [56], grouped sweeping scheduling (GSS) [88], and dynamic circular SCAN (DC-SCAN) [31], which are described in the following.

- SCAN-EDF combines the seek optimization of the traditional disk-scheduling method SCAN [16] and the real-time guarantees of the EDF mechanism. Note that the EDF mechanism in disk scheduling is nonpreemptive, which is different from the preemptive EDF scheme used in process management.
- GSS divides the set of n streams into g groups; groups can be formed in such a way that all streams belonging to the same group have similar deadlines; individual streams within a group are served according to SCAN.
- DC-SCAN employs a circular SCAN [64] service order to minimize disk seek overhead and variations in interservice time, resulting in high throughput; it reduces start-up delay by dynamically adapting the circular SCAN service order.

As a result, the three algorithms, SCAN-EDF, GSS and DC-SCAN, can improve continuous media data throughput and meet real-time requirements imposed by continuous media.

Another function that must be supported by file management is interactive control, such as pause/resume, fast forward, and fast reverse. Pause/resume operations pose a significant challenge to the design of efficient buffer management schemes, because they interfere with the sharing of a multimedia stream among different viewers. This issue is still under study. Fast-forward and fast-reverse operations can be implemented either by playing back media at a higher rate than normal or by continuing playback at the normal rate while skipping some data. Since the former approach can significantly increase the data rate, its direct implementation is impractical. The latter approach, on the other hand, must be carefully designed if interdata dependencies are present (for example, P- and B-frames depend on I-frames in MPEG) [12]. As a result, for streaming MPEG video, entire GOPs must be skipped during fast-forward operations, and the viewer sees normal resolution video with gaps, which is acceptable.

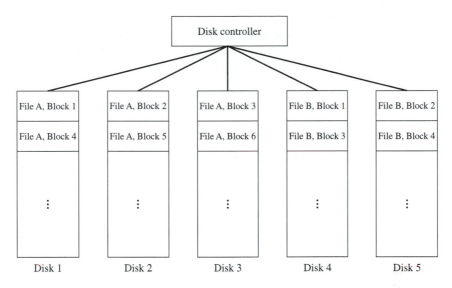

Figure 15.8   Data striped on multiple disks and accessed in parallel.

15.5.2 Storage System

The challenging issues of storage systems design for multimedia are high throughput, large capacity, and fault-tolerance.

Data Striping: A Method to Increase Throughput   If an entire video file is stored on one disk, the number of concurrent accesses to that file are limited by the throughput of that disk. This dictates the number of clients that can view the same video file. To overcome this limitation, data striping was proposed [63]. Under data striping schemes, a multimedia file is scattered across multiple disks and the disk array can be accessed in parallel. An example of data striping is shown in Figure 15.8. As shown in the figure, Blocks 1, 2, and 3 of File A can be read in parallel, resulting in increased throughput. An important issue in the design of a data striping scheme is to balance the load of the most heavily loaded disks to avoid overload situations, while keeping latency low. Designers must trade off load balance with low latency, since they are two conflicting objectives [63]. Note that data striping is different from file replication (an expensive way to increase throughput), in that data striping allows only one copy of a video file to be stored on disks, whereas file replication allows multiple copies of a video file to be stored on disks.

Tertiary and Hierarchical Storage: A Method to Increase Capacity   The introduction of multiple disks can increase storage capacity, as shown in Figure 15.9. However, the cost for large archives (e.g., with 40 terabyte storage requirements) is prohibitively high if a large number of disks are used for storage. To keep the storage cost down, tertiary storage (e.g., an automated tape library or CD-ROM jukebox) must be added.

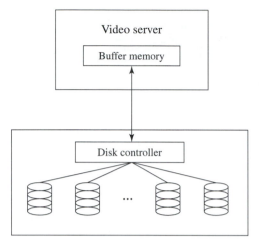

Figure 15.9   Disk-based video storage.

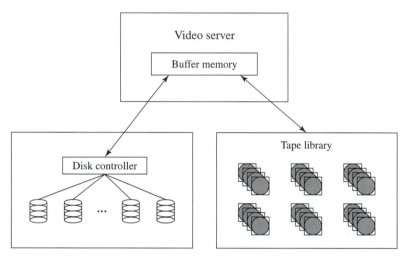

Figure 15.10   Hierarchical storage.

To reduce the overall cost, a hierarchical storage architecture (as shown in Figure 15.10) is typically used. Under the hierarchical storage architecture, only a fraction of the total storage is kept on disks while the major remaining portion is kept on a tertiary tape system. Specifically, frequently requested video files are kept on disks for quick access; the remainder resides in the automated tape library.

To deploy streaming services on a large scale, a storage area network (SAN) architecture must be employed (as shown in Figure 15.11) [17, 29]. A SAN can provide high-speed data pipes between storage devices and hosts at far greater distances than conventional host-attached SCSI (small computer systems interface). The connections in a SAN can be direct links between specific storage devices and individual hosts,

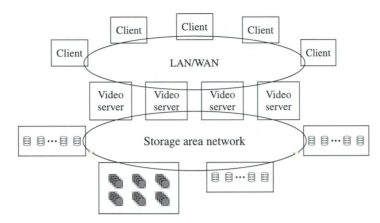

Figure 15.11   A SAN-based server and storage architecture for large-scale deployment.

through fiber channel arbitrated loop (FC-AL) connections; or they can form a matrix through a fiber channel switch. With these high-speed connections, a SAN is able to provide a many-to-many relationship between heterogeneous storage devices (e.g., disk arrays, tape libraries, and optical storage arrays), and multiple servers and storage clients.

Fault Tolerance   In order to ensure uninterrupted service even in the presence of disk failures, a server must be able to reconstruct lost information. This can be achieved by using redundant information. The redundant information could be either parity data generated by error-correcting codes like FEC, or duplicate data on separate disks. That is, there are two fault-tolerant techniques: error-correction coding (i.e., parity-encoding) [4, 53, 72] and mirroring [48]. Parity data adds a small storage overhead, but it requires synchronization of reads and additional processing time to decode lost information. In contrast, mirroring does not require synchronization of reads or additional processing time to decode lost information, which significantly simplifies the design and the implementation of video servers. However, mirroring incurs at least twice as much storage volume as in the non–fault-tolerant case. As a result, there is a trade-off between reliability and complexity (cost). A recent study [23] showed that, for the same degree of reliability, mirroring-based schemes always outperform parity-based schemes in terms of per-stream cost as well as restart latency after disk failure.

As a summary, we have addressed various issues in streaming server design and presented important techniques for efficient, scalable, and reliable storage and retrieval of multimedia files. Next, we discuss synchronization mechanisms for streaming media.

15.6 MEDIA SYNCHRONIZATION

A major feature that distinguishes multimedia applications from other traditional data applications is the integration of various media streams that must be presented in a

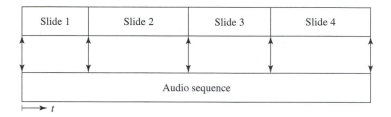

Figure 15.12   Synchronization between slides and a narration audio stream.

synchronized fashion. For example, in distance learning, the presentation of slides should be synchronized with the narration audio stream (see Figure 15.12). Otherwise, the slide currently being displayed on the screen may not correspond to the lecturer's explanation heard by the students, which is problematic. With media synchronization, the application at the receiver can present the media in the same way as they were originally captured.

Media synchronization refers to the preservation of the temporal relationships within a data stream and between various media streams. There are three levels of synchronization, namely, intrastream, interstream and interobject synchronization. The three levels of synchronization correspond to three semantic layers of multimedia data [66], as follows:

1. Intrastream synchronization: The lowest layer of continuous media or time-dependent data (such as video and audio) is the media layer. The unit of the media layer is the logical data unit (LDU), such as a video/audio frame, which adheres to strict temporal constraints to ensure acceptable user perception at playback. Synchronization at this layer is referred to as intrastream synchronization, which maintains the continuity of LDUs. Without intrastream synchronization, the presentation of the stream may be interrupted by pauses or gaps.

2. Interstream synchronization: The second layer of time-dependent data is the stream layer. The unit of the stream layer is a whole stream. Synchronization at this layer is referred to as interstream synchronization, which maintains temporal relationships among different continuous media. Without interstream synchronization, skew between the streams may become intolerable. For example, users could be annoyed if they notice that the movements of the lips of a speaker do not correspond to the presented audio.

3. Interobject synchronization: The highest layer of a multimedia document is the object layer, which integrates streams and time-independent data such as text and still images. Synchronization at this layer is referred to as interobject synchronization. The objective of interobject synchronization is to start and stop the presentation of the time-independant data within a tolerable time interval, if some previously defined points of the presentation of a time-dependent media object are reached. Without interobject synchronization, for example, the audience of a slide show could be annoyed if the audio is referring to one slide while another slide is being presented.

Media streams may lose synchronization after moving from the server to the client. As shown in Figure 15.1, there are many components in the path transporting data from its storage site to the user. Specifically, the server retrieves data from the storage device and sends that data to the network; the network transports the data to the client; the client reads the data from its network interface and presents it to the user; and operating systems and protocols allow these systems to run and do their work. Each of these components on the transport path performs a certain task and affects the data in a different way. They all inevitably introduce delays and delay variations (jitter), in either predictable or unpredictable manners. In particular, the delay introduced by the network is typically unpredictable, due to the best-effort nature of the Internet. The incurred delays and delay variations could disrupt intramedia, intermedia, and interobject synchronization. Therefore, media synchronization mechanisms are required, to ensure proper rendering of the multimedia presentation at the client.

The essential part of any media synchronization mechanism is the specifications of the temporal relations within a medium and between the media. The temporal relations can be specified either automatically or manually. In the case of audio and video recording and playback, the relations are specified automatically by the recording device. In the case of presentations that are composed of independently captured or otherwise created media, the temporal relations must be specified manually (with human intervention). The manual specification can be illustrated by the design of a slide show: the designer selects the appropriate slides, creates an audio object, and defines the segments of the audio stream in which the slides must be presented (see Figure 15.12).

The methods that are used to specify temporal relations include interval-based, axes-based, control flow–based, and event-based specifications [7]. A widely-used method for continuous media is axes-based specification, or time-stamping: at the source, a stream is time-stamped to relate temporal information within the stream and with respect to other streams; at the destination, the application presents the streams according to their temporal relation.

Besides specifying temporal relations, it is desirable that synchronization is supported by each component on the transport path. For example, the servers store large amounts of data in such a way that retrieval is quick and efficient, to reduce delay; the network provides sufficient bandwidth, and delay and jitter introduced by the network are tolerable to the multimedia applications; and the operating systems and applications provide real-time data processing (e.g., retrieval, resynchronization, and display). However, real-time support from the network is not available in the current Internet. Hence, most synchronization mechanisms are based on the end systems. These synchronization mechanisms can be either preventive or corrective [36].

- Preventive mechanisms are designed to minimize synchronization errors as data are transported from the server to the user. In other words, preventive mechanisms attempt to minimize latencies and jitters. These mechanisms involve disk-reading scheduling algorithms, network transport protocols, operating systems, and synchronization schedulers. Disk-reading scheduling is the process of organizing and coordinating the retrieval of data from the storage devices. Network transport protocols provide means for preserving synchronization during data transmission over the Internet. Operating systems achieve the precise control of timing constraints by using EDF or rate-monotonic scheduling. A synchronization scheduler can use the synchronization specifications for a presentation to create a schedule for the delivery of the media streams to the client by the servers (delivery schedule) and the presentation of these media streams to the user by the client application (presentation schedule). This scheduler can be centralized (entirely located at the client) or distributed (the delivery scheduling functionalities are shared among the servers and the client).

- Corrective mechanisms are designed to recover synchronization in the presence of synchronization errors. Synchronization errors are unavoidable, since the Internet introduces random delay. Random delay destroys the continuity of the media stream by producing gaps and jitters during data transmission. Therefore, certain compensations (i.e., corrective mechanisms) at the receiver are necessary when synchronization errors occur.

    An example of a corrective mechanism is the stream synchronization protocol (SSP) [25]. In SSP, the concept of an "intentional delay" is used by the various streams in order to adjust their presentation time to recover from network delay variations. The operations of the SSP are described as follows. At the client side, units that control and monitor the client end of the data connections compare the real arrival times of data with those predicted by the presentation schedule, and notify the scheduler of any discrepancies. These discrepancies are then compensated by the scheduler, which delays the display of data that are "ahead" of other data, allowing the late data to "catch up."

In sum, media synchronization is one of the key issues in the design of media streaming services. A great deal of effort has been contributed in the synchronization area. As an overview of this area, we have described the synchronization concepts, requirements, and approaches. For more information on media synchronization, please refer to [7, 66] and references therein.

15.7 PROTOCOLS FOR STREAMING VIDEO

Protocols are designed and standardized for communication between clients and streaming servers. According to their functionalities, the protocols directly related to streaming video over the Internet can be classified into three categories:

1. **Network-layer protocol:** The network-layer protocol provides basic network service support such as network addressing. The Internet Protocol (IP) serves as the network-layer protocol for Internet video streaming.

2. **Transport protocol:** Transport protocols provide end-to-end network transport functions for streaming applications. These protocols include UDP, TCP, RTP, and RTCP. UDP and TCP are lower-layer transport protocols, whereas RTP and RTCP [59] are upper-layer transport protocols, which are implemented on top of UDP/TCP (see Figure 15.13).

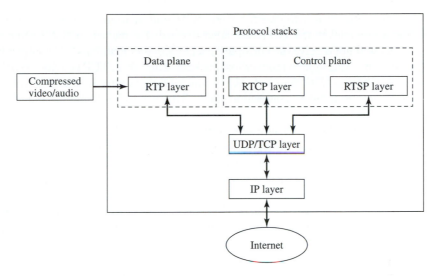

Figure 15.13   Protocol stacks for media streaming.

3. **Session control protocol:** A session control protocol defines the messages and procedures to control the delivery of multimedia data during an established session. RTSP [61] is such a session control protocol.

To illustrate the relations among the three types of protocols, we depict the protocol stacks for media streaming in Figure 15.13. As shown in the figure, at the sending side, the compressed video and audio data are retrieved and packetized at the RTP/RTCP/RTSP layer, to provide timing and synchronization information, as well as sequence numbers for the packets. These RTP-packetized streams are then passed to the UDP/TCP layer and the IP layer. The resulting IP packets are transported over the Internet. At the receiver side, the media streams are processed in reversed manner before their presentations. This is the case for the data plane. For the control plane, RTCP packets and RTSP packets are multiplexed at the UDP/TCP layer and move to the IP layer for transmission over the Internet.

The rest of this section is organized as follows: in Section 15.7.1, we discuss transport protocols for streaming media; Section 15.7.2 describes the session control protocol, that is, RTSP.

15.7.1 Transport Protocols

The transport protocol family for media streaming includes UDP, TCP, RTP, and RTCP protocols. UDP and TCP provide basic transport functions while RTP and RTCP run on top of UDP/TCP.

UDP and TCP protocols support such functions as multiplexing, error control, and flow control. These functions can be briefly described as follows. UDP and TCP can multiplex data streams from different applications running on the same machine with the same IP address. For the purpose of error control, TCP and most UDP implementations employ checksums to detect bit errors. If one or more bit errors are detected in the incoming packet, the TCP/UDP layer discards the packet so that the upper layer (e.g., RTP) will not receive the corrupted packet. On the other hand, in contrast to UDP, TCP uses retransmission to recover lost packets. Therefore, TCP provides reliable transmission, whereas UDP does not. TCP employs flow control to adapt the transmission rate according to network congestion status. This is another feature that distinguishes TCP from UDP.

Since TCP retransmission introduces delays that are unacceptable for streaming applications with stringent delay requirements, UDP is typically employed as the transport protocol for video streams. In addition, since UDP does not guarantee packet delivery, the receiver must rely on an upper layer protocol (i.e., RTP) to detect packet loss.

RTP is an Internet standard protocol designed to provide end-to-end transport functions for supporting real-time applications [59]. RTCP is a companion protocol to RTP. RTCP is designed to provide QoS feedback to the participants of an RTP session. In order words, RTP is a data transfer protocol, whereas RTCP is a control protocol.

RTP does not guarantee QoS or reliable delivery; rather, it provides the following functions in support of media streaming:

- **Time-stamping:** RTP provides time-stamps that can be used to synchronize different media streams. Note that RTP itself is not responsible for the synchronization. This must be done by the applications.

- **Sequence numbering:** Since packets arriving at the receiver may be out of sequence (UDP does not deliver packets in sequence), RTP employs sequence numbering to place the incoming RTP packets in the correct order. The sequence numbers are also used for packet loss detection.

- **Payload type identification:** The type of the payload contained in an RTP packet is indicated by an RTP header field called payload type identifier. The receiver interprets the content of the packet based on the payload type identifier. Certain common payload types such as MPEG-1/2 audio and video have been assigned payload type numbers [60]. For other payloads, this assignment can be done with session control protocols.

- **Source indentification:** The source of each RTP packet is identified by an RTP header field called synchronization source identifier (SSRC). SSRC provides a means for the receiver to distinguish different sources.

RTCP is the control protocol designed to work in conjunction with RTP [59]. In an RTP session, participants periodically send RTCP packets to convey feedback on quality of data delivery and information regarding membership. Basically, RTCP provides the following services:

- **QoS feedback:** This is the primary function of RTCP. RTCP provides feedback to an application about the quality of data distribution. The control information is useful to the senders, the receivers, and third-party monitors. The sender can adjust its transmission rate based on the receiver report feedback (see Section 15.3.1).

The receivers can determine whether a congestion is local, regional, or global. Network managers can evaluate network performance for multicast distribution.

RTCP provides QoS feedback through the use of sender and receiver reports at the source and destination, respectively. These reports can contain information on the quality of reception, such as (1) fraction of lost RTP packets since the last report, (2) cumulative number of lost packets since the beginning of reception, (3) packet interarrival jitter, and (4) delay since receiving the last sender's report.

- **Participant identification:** A source can be identified by the SSRC field in the RTP header. But the SSRC identifier is not convenient for human users. To remedy this problem, RTCP provides a human-friendly mechanism for source identification. Specifically, RTCP SDES (source description) packets contain textual information called canonical names, as globally unique identifiers of the session participants. They may include user's name, telephone number, email address, and other information.

- **Control packet scaling:** To scale the RTCP control packet transmission to the number of participants, a control mechanism is designed as follows. The control mechanism keeps the total control packets to 5% of the total session bandwidth. Among the control packets, 25% are allocated to sender reports and 75% to receiver reports. To prevent control packet starvation, at least one control packet is sent within five seconds at the sender or receiver.

- **Intermedia synchronization:** RTCP sender reports contain an indication of real time and the corresponding RTP time-stamp. This can be used in intermedia synchronization, such as lip synchronization in video.

- **Minimal session control information:** This optional functionality can be used for transporting session information, such as names of the participants.

15.7.2 Session Control Protocol: RTSP

RTSP is a session control protocol for streaming media over the Internet [61]. One of the main functions of RTSP is to support VCR-like control operations such as stop, pause/resume, fast forward, and fast reverse. In addition, RTSP also provides means for choosing delivery channels (e.g., UDP, multicast UDP, or TCP), and delivery mechanisms based upon RTP. RTSP works for multicast as well as unicast.

Another main function of RTSP is to establish and control streams of continuous audio and video media between media servers and clients. Specifically, RTSP provides the following operations:

- **Media retrieval:** The client can request a presentation description, and ask the server to set up a session to send the requested media data.

- **Inviting a media server to a conference:** A media server can be invited to a conference to play back media or to record a presentation.

- **Adding media to an existing session:** A server and a client can notify each other about any additional media that become available to the established session.

RTSP is intended to provide the same services for streamed audio and video that HTTP (hypertext transfer protocol) does for text and graphics. It is designed to have similar syntax and operations, so that most extension mechanisms for HTTP can be added to RTSP.

In RTSP, each presentation and media stream is identified by an RTSP URL (univeral resource locator). The overall presentation and the properties of the media are defined in a presentation description file, which may include the encoding, language, RTSP URLs, destination address, port, and other parameters. The presentation description file can be obtained by the client, using HTTP, email, or other means.

In sum, RTSP is designed to initiate and direct delivery of streaming media data from media servers; RTP is a transport protocol for streaming media data; RTCP is a protocol for monitoring delivery of RTP packets; UDP and TCP are lower-layer transport protocols for RTP/RTCP/RTSP packets; and IP provides a means for delivering UDP/TCP packets over the Internet. The combination of these protocols provides a complete streaming service over the Internet.

15.8 STREAMING VIDEO OVER WIRELESS IP NETWORKS

Recently, the emergence of broadband wireless networks has stimulated great interest in real-time video communications over wireless IP networks. However, delivering quality video over wireless networks in real time is a challenging task. This is primarily because of the following problems.

- **Bandwidth fluctuations:** First, the throughput of a wireless channel may be reduced due to multipath fading, cochannel interference, and noise disturbances. Second, the capacity of a wireless channel may fluctuate with the changing distance between the base station and the mobile host. Third, when a mobile terminal moves between different networks (e.g., from wireless local area network to wireless wide area network), the available bandwidth may vary drastically (e.g., from a few mbps to a few kbps). Finally, when a handoff takes place, a base station may not have enough unused radio resource to meet the demand of a newly joined mobile host. As a result, bandwidth fluctuation is a serious problem for real-time video transmission over wireless networks.

- **High bit-error rate:** Compared with wired links, wireless channels are typically much more noisy and have both small-scale (multipath) and large-scale (shadowing) fades, making the BER very high. The resulting bit errors can have devastating effects on video presentation quality. Therefore, there is a critical need for robust transmission of video over wireless channels.

- **Heterogeneity:** In a multicast scenario, receivers may have different requirements and properties in terms of latency, visual quality, processing capabilities, power limitations (wireless versus wired), and bandwidth limitations. The heterogeneous nature of receivers' requirements and properties makes it difficult to design an efficient multicast mechanism.

It has been shown that scalable video is capable of coping gracefully with the variability of bandwidth (Chapter 11) [2, 45]. A scalable video coding scheme produces a compressed bit stream, parts of which are decodable. Compared with decoding the complete bit stream, decoding part of the compressed bit stream produces pictures with degraded quality, or smaller image size, or smaller frame rate [18]. In contrast, nonscalable video is more susceptible to bandwidth fluctuations, since it cannot adapt its video representation to bandwidth variations [45]. Thus, scalable video is more suitable for use in a wireless environment to cope with the fluctuation of wireless channels. Furthermore, scalable video representation is a good solution to the heterogeneity problem in the multicast case [45].

Recently, application-aware adaptive services have been demonstrated to be able to effectively mitigate fluctuations of resource availability in wireless networks [2]. Scalable video representation naturally fits unequal error protection, which can effectively combat bit errors induced by the wireless medium. This motivates us to present an adaptive framework to support quality video communication over wireless IP networks.

For transporting video over wireless, there have been many proposals of adaptive approaches and services in the literature, which include an "adaptive reserved service" framework [38], an adaptive service based on QoS bounds and revenue [44], an adaptive framework targeted at end-to-end QoS provisioning [49], a utility-fair adaptive service [6], a framework for soft QoS control [57], a teleservice model based on an adaptive QoS paradigm [33], an adaptive QoS management architecture [37], and an adaptive framework for scalable video over wireless IP networks [84].

In this section, we present an adaptive framework [85] for future QoS-enabled wireless IP networks. This adaptive framework consists of (1) scalable video representations, each of which has its own specified QoS requirement, (2) network-aware applications, which are aware of network status, and (3) adaptive services, which make network elements support the QoS requirements of scalable video representations. Under this framework, as wireless channel conditions change, the mobile terminal and network elements can scale the video streams and transport the scaled streams to receivers with acceptable perceptual quality. The adaptive framework has the following key features.

1. *Graceful quality degradation:* In contrast to nonscalable video, scalable video can adapt its video representation to bandwidth variations, and the network can drop packets with awareness of the video representations. As a result, perceptual quality is gracefully degraded under severe channel conditions.

2. *Efficiency:* When there is excess bandwidth (excluding reserved bandwidth), the excess bandwidth will be efficiently used in a way that maximizes the perceptual quality or revenue.

3. *Fairness:* The resources can be shared in either a utility-fair [6] or max-min fair manner [44].

The remainder of this section is organized as follows. Section 15.8.1 describes network-aware applications. In Section 15.8.2, we present the adaptive services for transporting scalable video over wireless IP networks.

15.8.1 Network-Aware Applications

The use of network-aware applications is motivated by the following facts: (1) the bit error rate is very high when channel status is poor, and (2) packet loss is unavoidable if the available bandwidth is less than required. If a sender attempts to transmit each layer without any awareness of the channel status, all layers may become corrupted with equal probability, resulting in very poor picture quality. To address this problem, Wu, Hou, and Zhang [84] proposed to use network-aware applications, which preemptively discard enhancement layers at the sender in an intelligent manner by considering network status.

For the purpose of illustration, we present in Figure 15.14 an architecture that includes a network-aware mobile sender, an application-aware base station, and a receiver. The architecture in the figure is applicable to both live and stored video. At the sender side, the compressed video bit stream is first filtered by the scaler, the operation of which is to select certain video layers to transmit. Then the selected video representation is passed through transport protocols. Before being transmitted to the base station, the bit stream must be modulated by a modem (i.e., modulator/demodulator). Upon receipt of the video packets, the base station transmits them to the destination through the Internet.

Note that a scaler can distinguish video layers, and drop layers according to their significance. The dropping order is from the highest enhancement layer down to the base layer. A scaler performs only two operations: (1) scaling down the received video representation—that is, dropping the enhancement layer(s); and (2) transmitting what is received—that is, not scaling the received video representation.

Under this architecture, a bandwidth manager is maintained at the base station. One function of the bandwidth manager is to notify the sender about the available bandwidth of the wireless channel, through a signaling channel [50]. Upon receiving this information, the rate control module at the sender conveys the bandwidth parameter to the scaler. Then, the scaler regulates the output rate of the video stream so that the transmission rate is less than or equal to the available bandwidth.

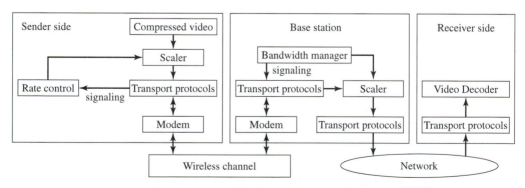

Figure 15.14 An architecture for transporting scalable video from a mobile to a wired terminal.

Another scenario is that the base station notifies the sender about the channel quality (i.e., BER) [3]. Upon receiving this information, the rate control module at the sender commands the scaler to perform the following (supposing that the video is compressed into two layers): if the BER is above a threshold, discard the enhancement layer so that the bandwidth allocated for the enhancement layer can be utilized by forward error correction (FEC) to protect the base layer; otherwise, transmit both layers.

The network-aware application has two advantages. First, by taking the available bandwidth into account, the sender can make the best use of network resources by selectively discarding enhancement layers in order to minimize the likelihood of more significant layers being corrupted, thereby increasing the perceptual quality of the video delivered. Second, by considering the channel error status, the sender can discard the enhancement layers, and FEC can utilize the bandwidth allocated for the enhancement layer to protect the base layer, thereby maximizing the possibility of the base layer being correctly received.

Note that adaptive techniques at the physical/link layer are required to support network-aware applications. Such adaptive techniques include a combination of variable spreading, coding, and code aggregation in code division multiple access (CDMA) systems, adaptive coding and modulation in time division multiple access (TDMA) systems, channel quality estimation, and a measurement feedback channel [50]. In addition, the feedback interval is typically constrained to be on the order of tens to hundreds of milliseconds [50].

15.8.2 Adaptive Service

A scalable video encoder can generate multiple layers or substreams to the network. Adaptive service provides scaling of the substreams based on the resource availability conditions in the fixed and wireless network. Specifically, adaptive service includes the following functions:

- Reserve a minimum bandwidth to meet the demand of the base layer. As a result, perceptual quality can always be achieved acceptable.
- Adapt the enhancement layers based on the available bandwidth and the fair policy. In other words, scale the video streams based on resource availability.

Advantages of using scaling inside the network include:

- *Adaptivity to network heterogeneity.* For example, when an upstream link with larger bandwidth feeds a downstream link with smaller bandwidth, use of a scaler at the connection point could help improve the video quality. This is because the scaler can selectively drop substreams instead of randomly dropping packets.
- *Low latency and low complexity.* Scalable video representations make the operation at a scaler very simple; that is, only discarding enhancement layers. Thus, the processing is fast, compared with processing on non-scalable video.

• *Lower call blocking and handoff dropping probability.* The adaptability of scalable video at base stations can translate into lower call blocking and handoff dropping probability.

Adaptive service can be deployed in the whole network (i.e., end-to-end provisioning) or only at base stations (i.e., local provisioning). Since local provisioning of adaptive service is just a subset of end-to-end provisioning, we will focus on the latter.

The required components of end-to-end adaptive service include: (1) service contract, (2) call admission control and resource reservation, (3) mobile multicast mechanism, (4) substream scaling, (5) substream scheduling, and (6) link-layer error control. These are described in more detail in the following.

Service Contract   A service contract between an application and a network could consist of multiple subcontracts, each of which corresponds to one or more substreams with similar QoS guarantees. Each subcontract must specify traffic characteristics and QoS requirements of the corresponding substream(s). A typical scenario is that a subcontract for the base layer specifies reserved bandwidth, whereas a subcontract for the enhancement layers does not specify any QoS guarantee.

At a video source, substreams must be generated according to subcontracts used by the application and shaped at the network access point. In addition, a substream is assigned a priority according to its significance. For example, the base layer is assigned the highest priority. The priority can be used by routing, scheduling, scaling, and error-control components of the adaptive network.

Call Admission Control and Resource Reservation   Call admission control (CAC) and resource reservation are two of the major components in end-to-end QoS provisioning. The function of CAC is to check whether admitting an incoming connection would reduce the service quality of existing connections, and whether the incoming connection's QoS requirements can be met. If a connection request is accepted, resources must be reserved for this connection, in two parts. First, in order to maintain the specified QoS over a long time scale, the network must reserve resources along the current path of a mobile connection. Second, in order to seamlessly achieve the QoS at a short time scale, some duplication must be made in the transport of the connection to neighboring base stations, so that in the event of a handoff, an outage in the link can be avoided.

The scalable video representation (i.e., substream) concept provides a very flexible and efficient solution to the problem of CAC and resource reservation. First, there is no need to reserve bandwidth for the complete stream, since typically only the base-layer substream requires a QoS guarantee. As a result, CAC is based only on the requirements of the base layer, and resources are reserved only for the base-layer substream. Second, the enhancement-layer substream(s) of multiple connections could share the leftover bandwidth. The enhancement-layer substreams are subject to scaling under bandwidth shortage and/or severe error conditions, which is discussed in subsequent paragraphs.

Mobile Multicast Mechanisms   CAC and resource reservation can provide connection-level QoS guarantee. To guarantee seamless QoS at the packet level, mobile multicast mechanisms must be used. That is, while being transported along its current path, the base-layer stream is also multicast to neighboring base stations, so that QoS at a small time scale can be seamlessly achieved.

To support seamless QoS, the mobile routing protocol must be proactive and anticipatory in order to match the delay, loss, and jitter constraints of a substream. According to the requirements of a substream, multicast paths might need to be established. The multicast paths terminate at base stations that are potential access-point candidates of a mobile terminal. The coverage of such a multicast path depends on the QoS requirements and mobility, as well as handoff characteristics, of a mobile receiver. As a mobile station hands off from one base station to another, new paths are added and old paths are deleted [49].

Substream Scaling   Scaling is employed during bandwidth fluctuations and/or under poor channel conditions. As the available bandwidth on a path decreases due to mobility or fading, lower-priority substreams are dropped by the scaler(s) on the path and substreams with higher priority are transmitted. As more bandwidth becomes available, lower-priority substreams are passed through the scaler, and the perceptual quality at the receivers increases. Figure 15.14 showed an architecture for transporting scalable video from a mobile terminal to a wired terminal. Figure 15.15 depicts an architecture for transporting scalable video from a wired terminal to a mobile terminal. We do not show the case of transporting scalable video from one mobile terminal to another, since it is a combination of Figures 15.14 and 15.15.

The scaling decision is made by a bandwidth manager. When there is no excess bandwidth (excluding reserved bandwidth), the bandwidth manager instructs the scaler to drop the enhancement layers. If there is excess bandwidth, the excess bandwidth can be shared in either a utility-fair [6] or max-min fair manner [44].

Substream Scheduling   The substream scheduler is used in mobile terminals as well as base stations. Its function is to schedule the transmission of packets on the wireless medium according to their substream QoS specifications and priorities.

When a short fading period is observed, a mobile terminal tries to prioritize the transmission of its substreams in order to achieve a minimum QoS. Here, depending

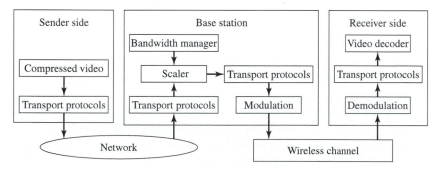

Figure 15.15   An architecture for transporting scalable video from a wired to a mobile terminal.

on channel conditions, a substream might be dropped for a period of time in order to accommodate higher-priority substreams. To determine the transmission time of any packet in a specific substream (or its position in the transmission queue), the scheduler takes two factors into account: the relative importance of the substream compared to other substreams, and wireless channel conditions.

To achieve both QoS (e.g., bounded delay and reserved bandwidth) and fairness, algorithms such as packet fair queueing must be employed [5]. While the existing packet fair queueing algorithms provide both bounded delay and fairness in wired networks, they cannot be applied directly to wireless networks. The key difficulty is that in wireless networks, sessions can experience location-dependent channel errors. This may lead to situations in which one session receives significantly less service than it is supposed to receive, while another receives more. This results in large discrepancies between the sessions' virtual times, making it difficult to provide both delay guarantees and fairness simultaneously.

To apply packet fair queueing algorithms, Ng, Stoica, and Zhang [51] identified a set of properties, called channel-condition independent fair (CIF), that a packet fair queueing algorithm should have in a wireless environment: (1) delay and throughput guarantees for error-free sessions, (2) long-term fairness for error sessions, (3) short-term fairness for error-free sessions, and (4) graceful degradation for sessions that have received excess service time. Then they presented a methodology for adapting packet fair queueing algorithms for wireless networks, and applied the methodology to derive an algorithm based on start-time fair queueing [28], called channel-condition independent packet fair queueing (CIF-Q), that achieves all the preceding properties [51].

As an example, we consider two-layer video. Suppose that a subcontract for the base layer specifies reserved bandwidth while a subcontract for the enhancement layer does not specify any QoS guarantee, which is a typical case. An architecture [84] for substream scheduling is shown in Figure 15.16.

Under this architecture, the buffer pool (i.e., the data memory in Figure 15.16) is divided into two parts: one for base-layer substreams, and one for enhancement-layer substreams. Within the same buffer partition for base or enhancement layer, per-flow queueing for each substream is employed. Furthermore, substreams within the same buffer partition share the buffer pool of that partition, whereas there is no buffer sharing across partitions. It is believed that this approach offers an excellent balance between traffic isolation and buffer sharing [84].

Under this buffering architecture, Wu, Hou, and Zhang [84] designed per-flow–based traffic management algorithms to achieve requested QoS and fairness. The first part of the traffic management is CAC and bandwidth allocation. Video connections are admitted by CAC, based on their base-layer QoS requirements, and bandwidth reservations for the admitted base-layer substreams are made accordingly. For admitted enhancement-layer substreams, their bandwidth will be dynamically allocated by a bandwidth manager, which has been addressed in previous paragraphs. The scaled enhancement layer substreams enter a shared buffer and are scheduled by a first-in-first-out (FIFO) scheduler. The second part of the traffic management is packet scheduling.

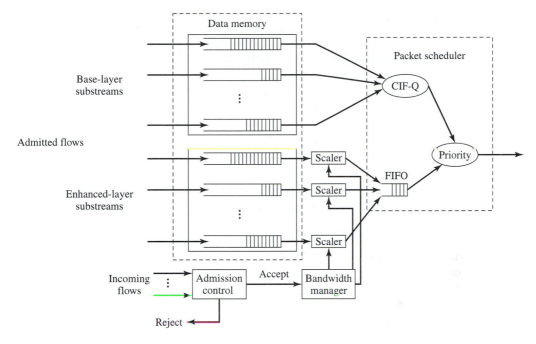

Figure 15.16   An architecture for substream scheduling at a base station.

Shown in Figure 15.16 is a hierarchical packet scheduling architecture, in which a priority link scheduler is shared between a CIF-Q scheduler for base-layer substreams and an FIFO scheduler for enhancement-layer substreams. Service priority is first given to the CIF-Q scheduler and then to the FIFO scheduler.

Link-Layer Error Control   To provide quality video over wireless channels, link-layer error control is required. Basically, there are two kinds of error control mechanisms, namely, forward error correction and automatic retransmission request. The disadvantage of FEC is that it is not adaptive to varying channel conditions, and it works only when the BER is below the FEC code's recovery capability. The disadvantage of ARQ is unbounded delay; that is, in the worst case, a packet may be retransmitted an unlimited number of times to recover from bit errors.

To address the problems associated with FEC and ARQ, truncated type-II hybrid ARQ schemes [41, 89] and delay-constrained hybrid ARQ [82] have been proposed. These hybrid ARQ schemes combine the good features of FEC and ARQ: bounded delay and adaptiveness. On the other hand, unequal error protection [30] naturally fits the hierarchical structure of scalable video. Specifically, the base layer can be better protected against transmission errors than the enhancement layers. This form of unequal error protection is much more desirable than protecting all the substreams. An open issue is how to combine unequal error protection with the hybrid ARQ schemes [85].

15.9 SUMMARY

In this chapter, we surveyed major approaches and mechanisms for Internet video streaming and presented an adaptive framework for video over wireless IP. The objective was not to provide an exhaustive review of existing approaches and mechanisms, but instead to give the reader a perspective on the range of options available and the associated trade-offs among performance, functionality, and complexity.

Video Compression (Section 15.2)

- Scalable video coding is the preferred mode for sending video over wireless and Internet IP networks. Depending on network conditions, the sender, the receiver and intermediate network nodes decide which layers of the video are transmitted, received, or forwarded, respectively.

Application-Layer Quality of Service Control (Section 15.3)

- Techniques in this category limit the influence of packet loss and bandwidth changes on video quality. Congestion control at the sender, in network nodes and at the receiver uses rate shaping and rate control mechanisms to adapt the video stream to the network conditions. Error control, error concealment, FEC and delay constrained retransmission minimize the visual effects of transmission errors.

Continuous Media Distribution Services (Section 15.4)

- Here, aspects of network filtering using video content aware network filters for frame dropping are discussed. For efficient use of bandwidth when serving several clients with the same contents, multicast is used. Application-level multicast enables to connect separate multicast networks. Content replication caches the content on several nodes of the network. The content is then delivered to the client from its closest cache.

Streaming Servers (Section 15.5)

- The streaming server delivers the content to the client across the network. The server processes the media under timing constraints in order to enable the receiver to synchronize different media and to prevent artifacts like jerkiness in video motion. This requires the optimization of the three main components of a server: Operating system, storage system, and communicator, which uses a suite of protocols to interact with clients.

Media Synchronization (Section 15.6)

- Media synchronization enables the client to present different media streams like audio, video, and graphics in a synchronous fashion. The servers add time stamps to every packet they transmit. The receiver uses these time stamps to estimate network jitter and appropriate buffer sizes in order to ensure synchronized and continuous media presentation.

Protocols for Streaming Video (Section 15.7)

- The basic transport protocols TCP and UDP deliver packets with and without sending an acknowledgment to the sender, respectively. RTP enables end-to-end transport for real-time applications, providing functions like time stamps, packet numbering, and payload identification. RTCP is the associated control protocol that mainly provides QoS feedback to the server. RTSP controls the communication session of one or more synchronized media streams. It provides functions like stop, pause/resume, fast forward, and fast reverse.

Streaming Video over Wireless IP Networks (Section 15.8)

- Due to high bandwidth fluctuations and bit-error rates, streaming video over wireless IP networks requires network-aware applications in the sender of a wireless terminal and the base station of the wireless network. They shape the scalable video bit stream according to the currently available bandwidth.

We would like to stress that the seven covered areas are basic building blocks for a streaming video architecture. Such an architecture ties together a broad range of signal processing, networking, and server technologies. Therefore, a thorough understanding of the whole architecture for streaming video is beneficial for the development of signal processing techniques (e.g., scalable video compression) suitable for streaming video. In addition, in-depth knowledge of both signal processing and networking helps in the effective design and use of application-layer QoS control, continuous media distribution services, protocols, and services for video over wireless IP. Furthermore, such architectural understanding is instrumental in the design of efficient, scalable, and fault-tolerant streaming servers.

15.10 BIBLIOGRAPHY

[1] Albanese, A., J. Blömer, J. Edmonds, M. Luby, and M. Sudan. Priority encoding transmission. *IEEE Trans. on Information Theory (Special Issue on Coding Theory)* (Nov. 1996), 42(6, I):1737–44.

[2] Balachandran, A., A. T. Campbell, and M. E. Kounavis. Active filters: delivering scalable media to mobile devices. *Seventh International Workshop on Network and Operating System Support for Digital Audio and Video (NOSSDAV'97)*, 125–35, St. Louis, MO, May 1997.

[3] Balachandran, K., S. R. Kadaba, and S. Nanda. Channel quality estimation and rate adaptation for cellular mobile radio. *IEEE J. Select. Areas Commun.* (July 1999), 17(7):1244–56.

[4] Berson, S., L. Golubchik, and R. R. Muntz. Fault tolerant design of multimedia servers. *ACM SIGMOD'95,* 364–75, May 1995.

[5] Bharghavan, V., S. Lu, and T. Nandagopal. Fair queuing in wireless networks: issues and approaches. *IEEE Personal Commun. Mag.* (Feb. 1999), 44–53.

[6] Bianchi, G., A. T. Campbell, and R. Liao. On utility-fair adaptive services in wireless networks. *Sixth International Workshop on Quality of Service (IWQoS'98),* Napa Valley, CA, May 1998, 256–67.

[7] Blakowski, G., and R. Steinmetz. A media synchronization survey: reference model, specification, and case studies. *IEEE Journal on Selected Areas in Communications* (Jan. 1996), 14(1):5–35.

[8] Bolot, J.-C., T. Turletti, and I. Wakeman. Scalable feedback control for multicast video distribution in the Internet. *ACM SIGCOMM'94,* 58–67, London, Sept. 1994.

[9] Chang, Y.-C., D. G. Messerschmitt, T. Carney, and S. A. Klein. Delay cognizant video coding: architecture, applications, and quality evaluations. Forthcoming in *IEEE Trans. Image Processing.*

[10] Chang, Y.-C., and D. G. Messerschmitt. Adaptive layered video coding for multitime scale bandwidth fluctuations. Submitted to *IEEE Journal on Selected Areas in Communications.*

[11] Chankhunthod, A., P. B. Danzig, C. Neerdaels, M. F. Schwartz, and K. J. Worrell. A hierarchical Internet object cache. *USENIX 1996 Annual Technical Conference,* Jan. 1996, 153–63.

[12] Chen, M., D. D. Kandlur, and P. S. Yu. Support for fully interactive playout in a disk-array-based video server. *ACM Multimedia'94,* San Francisco, CA, Oct. 1994, 391–98.

[13] Cheung, S. Y., M. Ammar, and X. Li. On the use of destination set grouping to improve fairness in multicast video distribution. *IEEE INFOCOM'96,* 2:553–60, March 1996.

[14] Chung, J. Y., J. W. S. Liu, and K. J. Lin. Scheduling periodic jobs that allows imprecise results. *IEEE Transactions on Computers* (Sept. 1990), 19(9): 1156–73.

[15] Davis, G., and J. Danskin. Joint source and channel coding for image transmission over lossy packet networks. *SPIE Conference on Applications of Digital Image Processing XIX,* SPIE-2847:376–87, Denver, Aug. 1996.

[16] Denning, P. J. Effects of scheduling on file memory operations. *AFIPS Spring Joint Computer Conference,* 9–21, Reston, VA, 1967.

[17] Du, D. H. C., and Y.-J. Lee. Scalable server and storage architectures for video streaming. *IEEE International Conference on Multimedia Computing and Systems* (June 1999), 1:62–67.

[18] Ebrahimi, T., and M. Kunt. Visual data compression for multimedia applications. *Proceedings of the IEEE* (June 1998), 86(6):1109–25.

[19] Eleftheriadis, A., and D. Anastassiou. Meeting arbitrary QoS constraints using dynamic rate shaping of coded digital video. *Fifth International Workshop on Network and Operating System Support for Digital Audio and Video (NOSSDAV'95),* 95–106, April 1995.

[20] Fan, L., P. Cao, J. Almeida, and A. Z. Broder. Summary cache: a scalable wide-area web cache sharing protocol. *IEEE/ACM Trans. on Networking* (June 2000), 8(3):281–93.

[21] FastForward Networks. FastForward Networks' broadcast overlay architecture. http://www.ffnet.com/pdfs/boa-whitepaper.pdf.

[22] Floyd, S., and K. Fall. Promoting the use of end-to-end congestion control in the Internet. *IEEE/ACM Trans. on Networking* (Aug. 1999), 7(4):458–72.

[23] Gafsi, J., and E. W. Biersack. Performance and reliability study for distributed video servers: mirroring or parity? *IEEE International Conference on Multimedia Computing and Systems* (June 1999), 2:628–34.

[24] Gemmell, J., and S. Christodoulakis. Principles of delay sensitive multimedia data storage and retrieval. *ACM Trans. on Information Systems* (Jan. 1992), 10(1):51–90.

[25] Georganas, N. D. Synchronization issues in multimedia presentational and conversational applications. *Pacific Workshop on Distributed Multimedia Systems (DMS'96),* June 1996.

[26] Ghanbari, M. Cell-loss concealment in ATM video codes. *IEEE Trans. Circuits Syst. Video Technol.* (June 1993), 3:238–47.

[27] Girod, B., U. Horn, and B. Belzer. Scalable video coding with multiscale motion compensation and unequal error protection. In Y. Wang et al., eds., *Multimedia Communications and Video Coding.* New York: Plenum Press, 1996, 475–482.

[28] Goyal, P., H. M. Vin, and H. Cheng. Start-time fair queueing: a scheduling algorithm for integrated services packet switching networks. *IEEE ACM Trans. Networking* (Oct. 1997), 5(5):690–704.

[29] Guha, A. The evolution to network storage architectures for multimedia applications. *IEEE International Conference on Multimedia Computing and Systems* (June 1999), 1:68–73.

[30] Hagenauer, J., and T. Stockhammer. Channel coding and transmission aspects for wireless multimedia. *Proceedings of the IEEE* (Oct. 1999), 87(10):1764–77.

[31] Hamidzadeh, B., and J. Tsun-Ping. Dynamic scheduling techniques for interactive hypermedia servers. *IEEE Trans. on Consumer Electronics* (Feb. 1999), 45(1):46–56.

[32] Hemy, M., U. Hengartner, P. Steenkiste, and T. Gross. MPEG system streams in best-effort networks. *Packet Video Workshop,* New York, April 1999.

[33] Iera, A., A. Molinaro, and S. Marano. Wireless broadband applications: The teleservice model and adaptive QoS provisioning. *IEEE Communications Magazine* (Oct. 1999), 71–75.

[34] Illgner, K., and F. Mueller. Spatially scalable video compression employing resolution pyramids. *IEEE J. Select. Areas Commun.* (Dec. 1997), 15(9):1688–703.

[35] ISO/IEC. IS14496-2: Information technology—coding of audio-visual objects, part 2: visual. 1999 (MPEG-4 Video).

[36] Jarmasz, J. P., and N. D. Georganas. Designing a distributed multimedia synchronization scheduler. *IEEE International Conference on Multimedia Computing and Systems* (June 1997), 451–457.

[37] Lataoui, O., T. Rachidi, L. G. Samuel, S. Gruhl, and R.-H., Yan. A QoS management architecture for packet switched 3rd generation mobile systems. *Networld+Interop 2000—Engineers Conference on Broadband Internet Access Technologies Systems & Services,* May 2000.

[38] Lee, K. Adaptive network support for mobile multimedia. *ACM Mobicom'95,* 62–74, Nov. 1995.

[39] Li, J. Visual progressive coding. *SPIE Conf. Visual Communications and Image Processing (VCIP'99),* SPIE-3653, Jan. 1999.

[40] Li, X., S. Paul, and M. H. Ammar. Layered video multicast with retransmissions (LVMR): Evaluation of hierarchical rate control. *IEEE INFOCOM'98,* 3:1062–72, March 1998.

[41] Liu, H., and M. El Zarki. Performance of H.263 video transmission over wireless channels using hybrid ARQ. *IEEE J. on Selected Areas in Communications* (Dec. 1997), 15(9):1775–86.

[42] Liu, C. L., and J. W. Layland. Scheduling algorithms for multiprogramming in a hard real-time environment. *Journal of ACM* (Jan. 1973), 20(1):46–61.

[43] Liu, Y.-J., and Y.-Q. Zhang. Wavelet-coded image transmission over land mobile radio channels. *IEEE GLOBECOM'92,* 1:235–39, Orlando, FL, Dec. 1992.

[44] Lu, S., K.-W. Lee, and V. Bharghavan. Adaptive service in mobile computing environments. *Fifth International Workshop on Quality of Service (IWQOS'97),* 25–36, May 1997.

[45] McCanne, S., V. Jacobson, and M. Vetterli. Receiver-driven layered multicast. *ACM SIGCOMM'96,* 117–30, Aug. 1996.

[46] Miao, Z., and A. Ortega. Proxy caching for efficient video services over the Internet. *Packet Video Workshop,* April 1999.

[47] Moura, J., R. S. Jasinschi, H. Shiojiri, and J.-C. Lin. Video over wireless. *IEEE Personal Communications Magazine* (Feb. 1996), 44–54.

[48] Mourad, A. Doubly-striped disk mirroring: reliable storage for video servers. *Multimedia, Tools and Applications* (May 1996), 2:253–72.

[49] Naghshineh, M., and M. Willebeek-LeMair. End-to-end QoS provisioning in multimedia wireless/mobile networks using an adaptive framework. *IEEE Communications Magazine* (Nov. 1997), 72–81.

[50] Nanda, S., K. Balachandran, and S. Kumar. Adaptation techniques in wireless packet data services. *IEEE Communications Magazine* (Jan. 2000), 54–64.

[51] Ng, T. S. E., I. Stoica, and H. Zhang. Packet fair queueing algorithms for wireless networks with location-dependent errors. *IEEE INFOCOM'98,* 1103–11, March 1998.

[52] Ortega, A., and K. Ramchandran. Rate-distortion methods for image and video compression. *IEEE Signal Processing Magazine* (Nov. 1998), 74–90.

[53] Ozden, B., R. Rastogi, P. Shenoy, and A. Silberschatz. Fault-tolerant architectures for continuous media servers. *ACM SIGMOD'96,* 79–90, June 1996.

[54] Puri, R., K. W. Lee, K. Ramchandran, and V. Bharghavan. Application of FEC based multiple description coding to Internet video streaming and multicast. *Packet Video Workshop,* May 2000.

[55] Radha, H., and Y. Chen. Fine-granular-scalable video for packet networks. *Packet Video Workshop,* April 1999.

[56] Reddy, A. L., and J. Wyllie. Disk scheduling in a multimedia I/O system. *ACM Multimedia'93,* 289–97.

[57] Reininger, D., R. Izmailov, B. Rajagopalan, M. Ott, and D. Raychaudhuri. Soft QoS control in the WATMnet broadband wireless system. *IEEE Personal Communications Magazine* (Feb. 1999), 34–43.

[58] Rexford, J., S. Sen, and A. Basso. A smoothing proxy service for variable-bit-rate streaming video. *IEEE GLOBECOM'99,* 3:1823–29, Dec. 1999.

[59] Schulzrinne, H., S. Casner, R. Frederick, and V. Jacobson. RTP: a transport protocol for real-time applications. IETF, RFC 1889, Jan. 1996.

[60] Schulzrinne, H. RTP profile for audio and video conferences with minimal control. IETF, RFC 1890, Jan. 1996.

[61] Schulzrinne, H., A. Rao, and R. Lanphier. Real time streaming protocol (RTSP). IETF, RFC 2326, April 1998.

[62] Sen, S., J. Rexford, and D. Towsley. Proxy prefix caching for multimedia streams. *IEEE INFOCOM'99,* 3:1310–19, March 1999.

[63] Shenoy, P., and H. M. Vin. Efficient striping techniques for multimedia file servers. *Performance Evaluation* (Dec. 1999), 38(3–4):175–99.

[64] Silberschatz, A., J. Peterson, and P. Galvin. *Operating System Concepts.* 3rd ed., Reading, MA: Addison Wesley, 1991.

[65] Sodagar, I., H.-J. Lee, P. Hatrack, and Y.-Q. Zhang. Scalable wavelet coding for synthetic/natural hybrid images. *IEEE Trans. on Circuits and Systems for Video Technology* (March 1999), 9(2):244–54.

[66] Steinmetz, R., and K. Nahrstedt. *Multimedia: Computing, Communications and Applications.* Englewood Cliffs, NJ: Prentice-Hall, 1995.

[67] Sun, H., and W. Kwok. Concealment of damaged block transform coded images using projection onto convex sets. *IEEE Trans. Image Processing* (April 1995), 4:470–77.

[68] Talluri, R. Error-resilience video coding in the ISO MPEG-4 standard. *IEEE Communications Magazine* (June 1998), 112–19.

[69] Tan, W., and A. Zakhor. Real-time Internet video using error resilient scalable compression and TCP-friendly transport protocol. *IEEE Trans. Multimedia* (June 1999), 1(2):172–86.

[70] Tanenbaum, A. S. *Modern Operating Systems.* Englewood Cliffs, NJ: Prentice-Hall, 1992.

[71] Taubman, D., and A. Zakhor. A common framework for rate and distortion based scaling of highly scalable compressed video. *IEEE Trans. on Circuits and Systems for Video Technology* (Aug. 1996), 6(4):329–54.

[72] Tewari, R., D. M. Dias, W. Kish, and H. Vin. Design and performance trade-offs in clustered video servers. *IEEE International Conference on Multimedia Computing and Systems* (June 1996), 144–50.

[73] Tham, J. Y., S. Ranganath, and A. A. Kassim. Highly scalable wavelet-based video codec for very low bit-rate environment. *IEEE J. Select. Areas Commun.* (Jan. 1998), 16(1):12–27.

[74] Turletti, T., and C. Huitema. Videoconferencing on the Internet. *IEEE/ACM Trans. on Networking* (June 1996), 4(3):340–51.

[75] Vetro, A., H. Sun, and Y. Wang. Object-based transcoding for scalable quality of service. *IEEE ISCAS'2000,* 4:17–20, Geneva, May 28–31, 2000.

[76] Vin, H. M., P. Goyal, A. Goyal, and A. Goyal. A statistical admission control algorithm for multimedia servers. *ACM Multimedia'94,* 33–40, Oct. 1994.

[77] Wang, X., and H. Schulzrinne. Comparison of adaptive Internet multimedia applications. *IEICE Trans. Commun.* (June 1999), E82-B(6):806–18.

[78] Wang, Y., Q.-F. Zhu, and L. Shaw. Maximally smooth image recovery in transform coding. *IEEE Trans. Commun.* (Oct. 1993), 41(10): 1544–51.

[79] Wang, Y., M. T. Orchard, and A. R. Reibman. Multiple description image coding for noisy channels by pairing transform coefficients. *IEEE Workshop on Multimedia Signal Processing,* 419–24, June 1997.

[80] Wang, Y., and Q.-F. Zhu. Error control and concealment for video communication: a review. *Proceedings of the IEEE* (May 1998), 86(5):974–97.

[81] Wu, D., et al. On end-to-end architecture for transporting MPEG-4 video over the Internet. Forthcoming in *IEEE Trans. on Circuits and Systems for Video Technology.*

[82] Wu, D., Y. T. Hou, Y.-Q. Zhang, W. Zhu, and H. J. Chao. Adaptive QoS control for MPEG-4 video communication over wireless channels. *IEEE ISCAS'2000,* 1:48–51, Geneva, May 28–31, 2000.

[83] Wu, D., et al. An end-to-end approach for optimal mode selection in Internet video communication: Theory and application. *IEEE Journal on Selected Areas in Communications* (June 2000), 18(6):977–95.

[84] Wu, D., Y. T. Hou, and Y.-Q. Zhang. Scalable video transport over wireless IP networks, in *IEEE International Symposium on Personal, Indoor and Mobile Radio Communication (PIMRC'2000),* 2:1185–91, London, Sept. 18–21, 2000.

[85] Wu, D., Y. T. Hou, and Y.-Q. Zhang. Scalable video coding and transport over broadband wireless networks. Forthcoming in *Proceedings of the IEEE.*

[86] Xu, X. R., A. C. Myers, H. Zhang, and R. Yavatkar. Resilient multicast support for continuous-media applications. *IEEE 7th Int Workshop on Network and Operating System Support for Digital Audio and Video (NOSSDAV'97),* 183–94.

[87] Yeadon, N., F. Garcia, D. Hutchison, and D. Shepherd. Filters: QoS support mechanisms for multipeer communications. *IEEE Journal on Selected Areas in Communications* (Sept. 1996), 14(7):1245–62.

[88] Yu, P. S., M. S. Chen, and D. D. Kandlur. Grouped sweeping scheduling for DASD-based multimedia storage management. *ACM/Springer Multimedia Systems* (1993), 1(3):99–109.

[89] Zhang, Q., and S. A. Kassam. Hybrid ARQ with selective combining for fading channels. *IEEE Journal on Selected Areas in Communications* (May 1999), 17(5):867–80.

[90] Zhang, Z.-L., S. Nelakuditi, R. Aggarwa, and R. P. Tsang. Efficient selective frame discard algorithms for stored video delivery over resource constrained networks. *IEEE INFOCOM'99,* 2:472–79, March 1999.

[91] Zhang, Z.-L., Y. Wang, D. H. C. Du, and D. Su. Video staging: A proxy-server-based approach to end-to-end video delivery over wide-area networks, *IEEE/ACM Transactions on Networking* (Aug. 2000), 8(4):429–42.

Appendix A

Determination of Spatial–Temporal Gradients

To implement the various gradient descent methods required for many optimization problems presented in this book, one must be able to determine the spatial and/or temporal gradient of the underlying video signal. This appendix presents several methods for approximating the continuous gradient operation using differential operators.

A.1 FIRST- AND SECOND-ORDER GRADIENT

The simplest method for approximating the first-order gradient along a particular direction is to use the difference between two pixel values along that direction. This yields:

$$\frac{\partial \psi}{\partial x}\bigg|_{(x,y,t)} = \psi(x, y, t) - \psi(x - 1, y, t),$$

$$\frac{\partial \psi}{\partial y}\bigg|_{(x,y,t)} = \psi(x, y, t) - \psi(x, y - 1, t),$$

$$\frac{\partial \psi}{\partial t}\bigg|_{(x,y,t)} = \psi(x, y, t) - \psi(x, y, t - 1).$$

Using this approximation, one can further derive the approximations for the second-order spatial gradient operators:

$$\frac{\partial^2 \psi}{\partial x^2}\bigg|_{(x,y,t)} = \frac{\partial \psi}{\partial x}\bigg|_{(x+1,y,t)} - \frac{\partial \psi}{\partial x}\bigg|_{(x,y,t)}$$

$$= \psi(x + 1, y, t) - 2\psi(x, y, t) + \psi(x - 1, y, t),$$

$$\left.\frac{\partial^2 \psi}{\partial y^2}\right|_{(x,y,t)} = \left.\frac{\partial \psi}{\partial x}\right|_{(x,y+1,t)} - \left.\frac{\partial \psi}{\partial x}\right|_{(x,y,t)}$$

$$= \psi(x, y+1, t) - 2\psi(x, y, t) + \psi(x, y-1, t),$$

$$\left.\frac{\partial^2 \psi}{\partial xy}\right|_{(x,y,t)} = \left.\frac{\partial \psi}{\partial x}\right|_{(x+1,y+1,t)} - \left.\frac{\partial \psi}{\partial x}\right|_{(x+1,y,t)}$$

$$= \psi(x+1, y+1, t) - \psi(x+1, y, t) - \psi(x, y+1, t) + \psi(x, y, t),$$

$$\left.\frac{\partial^2 \psi_t}{\partial t^2}\right|_{(x,y,t)} = \left.\frac{\partial \psi_t}{\partial t}\right|_{(x,y,t+1)} - \left.\frac{\partial \psi_t}{\partial t}\right|_{(x,y,t)}$$

$$= \psi(x, y, t+1) - 2\psi(x, y, t) + \psi(x, y, t-1).$$

A.2 SOBEL OPERATOR

The digital approximations presented in the preceding section are very sensitive to noise in the video data. Furthermore, some operators are asymmetric. In image processing, the first-order spatial gradient is more often approximated by the Sobel operator, which performs smoothing along the tangent direction, before taking a central difference in the gradient direction. Specifically,

$$\left.\frac{\partial \psi}{\partial x}\right|_{(x,y,t)} = \psi(x+1, y-1, t) + 2\psi(x+1, y, t) + \psi(x+1, y+1, t)$$

$$- \psi(x-1, y-1, t) - 2\psi(x-1, y, t) - \psi(x-1, y+1, t),$$

$$\left.\frac{\partial \psi}{\partial y}\right|_{(x,y,t)} = \psi(x-1, y+1, t) + 2\psi(x, y+1, t) + \psi(x+1, y+1, t)$$

$$- \psi(x-1, y-1, t) - 2\psi(x, y-1, t) - \psi(x+1, y-1, t).$$

Note that these operations can be simply implemented with a 3×3 filter.

A.3 DIFFERENCE OF GAUSSIAN FILTERS

A more noise-robust implementation of the first (or second) order gradient is to first apply a Gaussian smoothing filter to the original data, and then apply the first (or second)–order differential operator to the smoothed signal. Because the smoothing operator (a convolution operator) and the differential operator are interchangeable, this operation is equivalent to convolving the original signal with a filter that is the first (or second)–order gradient of the Gaussian function. To apply this filter to digital signals, the preceding continuous domain filter must be sampled and truncated. The resulting first-order gradient of the Gaussian (or differential of Gaussian) is known as the DoG filter, while the second-order gradient or Laplacian of the Gaussian is called the LoG filter.

Let $g(x, y) = \exp(-(x^2 + y^2/2\sigma^2))$ represent the Gaussian function. Then,

$$\frac{\partial g}{\partial x} = -\frac{x}{\sigma^2} \exp\left(-\frac{x^2 + y^2}{2\sigma^2}\right),$$

$$\frac{\partial g}{\partial y} = -\frac{y}{\sigma^2} \exp\left(-\frac{x^2 + y^2}{2\sigma^2}\right),$$

$$\frac{\partial^2 g}{\partial x^2} = -\frac{1}{\sigma^2}\left(1 - \frac{x^2}{\sigma^2}\right) \exp\left(-\frac{x^2 + y^2}{2\sigma^2}\right),$$

$$\frac{\partial^2 g}{\partial y^2} = -\frac{1}{\sigma^2}\left(1 - \frac{y^2}{\sigma^2}\right) \exp\left(-\frac{x^2 + y^2}{2\sigma^2}\right),$$

$$\frac{\partial^2 g}{\partial xy} = \frac{xy}{\sigma^4} \exp\left(-\frac{x^2 + y^2}{2\sigma^2}\right).$$

Obviously, all of these filters are separable. To obtain the digital approximation of any one of these functions, one can truncate its length to $2 \sim 3\sigma$, and sample the truncated function at integer samples. The parameter σ should be defined in units of pixels, and should be chosen based on the amount of noise present in the data. A popular approximation is to set $\sigma = 1$ pixel and truncate the function at 2σ. The resulting filters are 5×5 filters, which are given here:

$$[\mathbf{G}_x] = \begin{bmatrix} 0.0366 & 0.0821 & 0 & -0.0821 & -0.0366 \\ 0.1642 & 0.3679 & 0 & -0.3679 & -0.1642 \\ 0.2707 & 0.6065 & 0 & -0.6065 & -0.2707 \\ 0.1642 & 0.3679 & 0 & -0.3679 & -0.1642 \\ 0.0366 & 0.0821 & 0 & -0.0821 & -0.0366 \end{bmatrix}$$

$$[\mathbf{G}_{xx}] = \begin{bmatrix} 0.0549 & 0 & -0.1353 & 0 & 0.0549 \\ 0.2463 & 0 & -0.6065 & 0 & 0.2463 \\ 0.4060 & 0 & -1.0000 & 0 & 0.4060 \\ 0.2463 & 0 & -0.6065 & 0 & 0.2463 \\ 0.0549 & 0 & -0.1353 & 0 & 0.0549 \end{bmatrix}$$

$$[\mathbf{G}_{xy}] = \begin{bmatrix} 0.0733 & 0.1642 & 0 & -0.1642 & -0.0733 \\ 0.1642 & 0.3679 & 0 & -0.3679 & -0.1642 \\ 0 & 0 & 0 & 0 & 0 \\ -0.1642 & -0.3679 & 0 & 0.3679 & 0.1642 \\ -0.0733 & -0.1642 & 0 & 0.1642 & 0.0733 \end{bmatrix}$$

$$[\mathbf{G}_y] = [\mathbf{G}_x]^T; \qquad [\mathbf{G}_{yy}] = [\mathbf{G}_{xx}]^T.$$

Appendix B

Gradient Descent Methods

In this appendix, we review some basic gradient descent methods for minimizing an objective function. We consider the general case when the function is multivariable with dimension K.

Let $J(\mathbf{x})$ represent the objective function, where $\mathbf{x} = [x_1, x_2, \ldots, x_k]^T$ is a K-dimensional vector. Gradient descent methods are iterative methods for determining the point $\mathbf{x}^*$ that achieves the minimum of $J(\mathbf{x})$.

B.1 FIRST-ORDER GRADIENT DESCENT METHOD

Let $\mathbf{x}^{(l)}$ represent the solution at the lth iteration. The first-order gradient descent method updates the solution proportional to the first-order gradient of the objective function, that is,

$$\mathbf{x}^{(l+1)} = \mathbf{x}^{(l)} - \alpha \left. \frac{\partial J}{\partial \mathbf{x}} \right|_{\mathbf{x}^{(l)}}. \tag{B.1.1}$$

The rationale behind the first-order gradient descent method is that $\partial J / \partial \mathbf{x}$ represents the direction along which the function $J(\mathbf{x})$ increases most rapidly and, consequently, $-\partial J / \partial \mathbf{x}$ is the direction along which $J(\mathbf{x})$ decreases the fastest.

The constant α is known as the *stepsize*. It must be chosen properly to guarantee the convergence of this iteration process to the desired minimum $\mathbf{x}^*$. If α is too large, the solution may oscillate around $\mathbf{x}^*$; whereas if α is too small, it may take too many iterations to converge. Obviously, if the function $J(\mathbf{x})$ has multiple local minima, then the solution upon convergence will be the local minimum that is closest to the initial solution. Therefore, it is extremely important to select an appropriate initial solution.

In order to obtain the global minimum, one can also start with several different initial solutions that are sufficiently far apart, and run this iteration scheme starting from each initial solution. By comparing the function values at all the local minima obtained from different initial solutions, one can identify the one with the minimum value. However, there is no guarantee that this will be the global minimum. A more sophisticated approach is *simulated annealing,* which allows the iteration process to jump out of the local minima [1].

B.2 STEEPEST DESCENT METHOD

In the preceding method, the stepsize α is set to a small constant. To speed up convergence, one can also vary α at each new iteration, so that the maximum reduction of $J(\mathbf{x})$ is achieved. This is possible because, for fixed $\mathbf{x}^{(l)}$ and $(\partial J/\partial \mathbf{x})|_{\mathbf{x}^{(l)}}$, the new function value $J(\mathbf{x}^{(l+1)})$ is a scalar function of α. A numerical search method can be used to determine the optimal stepsize at each iteration. The first-order gradient descent method that uses such a variable stepsize is known as the steepest descent method. This method will converge in fewer iterations than the method using a constant stepsize, but each iteration will take more time to determine the optimal stepsize.

B.3 NEWTON'S METHOD

The necessary condition for a function to achieve a local minimum is that its gradient at that point is zero. Given a previous solution $\mathbf{x}^{(l)}$ at which the gradient is not yet zero, one way to determine the update is by requiring the gradient at the updated position to be zero, that is,

$$\frac{\partial J}{\partial \mathbf{x}}\bigg|_{\mathbf{x}^{(l)}+\Delta \mathbf{x}} = 0. \tag{B.3.1}$$

Using Taylor's expansion for the preceding gradient function for up to the first term, we obtain

$$\frac{\partial J}{\partial \mathbf{x}}\bigg|_{\mathbf{x}^{(l)}} + \left[\mathbf{H}(\mathbf{x}^{(l)})\right]\Delta \mathbf{x} = 0, \tag{B.3.2}$$

where

$$[\mathbf{H}(\mathbf{x})] = \frac{\partial^2 J}{\partial \mathbf{x}^2} = \begin{bmatrix} \dfrac{\partial^2 J}{\partial x_1 \partial x_1} & \dfrac{\partial^2 J}{\partial x_1 \partial x_2} & \cdots & \dfrac{\partial^2 J}{\partial x_1 \partial x_K} \\ \dfrac{\partial^2 J}{\partial x_2 \partial x_1} & \dfrac{\partial^2 J}{\partial x_2 \partial x_2} & \cdots & \dfrac{\partial^2 J}{\partial x_2 \partial x_K} \\ \cdots & \cdots & \cdots & \cdots \\ \dfrac{\partial^2 J}{\partial x_K \partial x_1} & \dfrac{\partial^2 J}{\partial x_K \partial x_2} & \cdots & \dfrac{\partial^2 J}{\partial x_K \partial x_K} \end{bmatrix}$$

is the second-order derivative of $J(\mathbf{x})$, known as the Hessian matrix. Solving Equation (B.3.2), we obtain

$$\Delta\mathbf{x} = -\left[\mathbf{H}\!\left(\mathbf{x}^{(l)}\right)\right]^{-1}\frac{\partial J}{\partial\mathbf{x}}\bigg|_{\mathbf{x}^{(l)}},$$

or

$$\mathbf{x}^{(l+1)} = \mathbf{x}^{(l)} - \left[\mathbf{H}\!\left(\mathbf{x}^{(l)}\right)\right]^{-1}\frac{\partial J}{\partial\mathbf{x}}\bigg|_{\mathbf{x}^{(l)}}.$$

The preceding iterative algorithm is known as Newton's method, which requires the second-order derivative, in addition to the first-order derivative. Newton's method can converge much faster, but each iteration requires significantly more calculation, and the algorithm is not guaranteed to converge. The use of the second-order derivative also makes this algorithm more prone to noise and numerical errors.

In practice, the following update is more often used:

$$\mathbf{x}^{(l+1)} = \mathbf{x}^{(l)} - \alpha\left[\mathbf{H}\!\left(\mathbf{x}^{(l)}\right)\right]^{-1}\frac{\partial J}{\partial\mathbf{x}}\bigg|_{\mathbf{x}^{(l)}}, \tag{B.3.3}$$

where the stepsize α is a constant smaller than 1, which must be chosen properly to reach a good compromise between guaranteeing convergence and the convergence rate.

B.4 NEWTON-RALPHSON METHOD

When the objective function is a sum of squared errors, that is,

$$J(\mathbf{x}) = \frac{1}{2}\sum_k e_k^2(\mathbf{x}), \tag{B.4.1}$$

we have

$$\frac{\partial J}{\partial\mathbf{x}} = \sum \frac{\partial e_k}{\partial\mathbf{x}} e_k(\mathbf{x}), \tag{B.4.2}$$

and

$$[\mathbf{H}] = \frac{\partial^2 J}{\partial\mathbf{x}^2} = \sum \frac{\partial e_k}{\partial\mathbf{x}}\left(\frac{\partial e_k}{\partial\mathbf{x}}\right)^T + \frac{\partial^2 e_k}{\partial\mathbf{x}^2}e_k(\mathbf{x}) \approx \sum \frac{\partial e_k}{\partial\mathbf{x}}\left(\frac{\partial e_k}{\partial\mathbf{x}}\right)^T. \tag{B.4.3}$$

The last approximation is obtained by assuming that the term with the second-order derivative is significantly smaller than that with the first-order derivative only. Using this approximation in Equation (B.3.3) yields the Newton-Ralphson method. Because almost all of the error minimization problems encountered in practice use a sum of error squares as an objective function, the Newton-Ralphson method is widely used.

B.5 BIBLIOGRAPHY

[1] Van Laarhoven, P. J. M., and E. H. Aarts. *Simulated Annealing: Theory and Applications.* Dordrecht, The Netherlands: Reidel, 1987.

Appendix C

Glossary of Acronyms

2-D	two-dimensional
3-D	three-dimensional
3DTV	three-dimensional TV
K-D	K-dimensional
AAC	advanced audio coder (the MPEG-2 audio coder)
ABR	available bit rate (a service type in ATM networks)
AC	alternating current (refers to transformed coefficients corresponding to nonzero frequencies)
ACE	advanced coding efficiency (an MPEG-4 video coding profile)
A/D	analog-to-digital
ADSL	asymmetric digital subscriber loop
ALF	application layer framing
ARQ	automatic retransmission request
ART	angular radial transform
ARTS	advanced real-time simple (an MPEG-4 video coding profile)
ASP	application service provider
AU	action unit (used in FACS)
AV	audiovisual
ATM	asynchronous transfer mode
BAB	binary alpha block (an MPEG-4 shape coding unit)
BDP	body definition parameters (used in MPEG-4)
BAP	body animation parameters (used in MPEG-4)
BER	bit-error rate

BIFS	binary format for scenes (the MPEG-4 scene modeling language)
B-ISDN	broadband ISDN
BMA	block-matching algorithm
bpp	bits per pixel
bps	bits per second
BT.601	a digital video format defined by ITU-R, formerly known as CCIR601
CAC	call admission control
CBP	coded block pattern (specified for each MB in MPEG-1/2/4 standards)
CBR	constant bit rate (a service type in ATM networks)
CCD	charge-coupled device
CCIF	International Telephone Consultative Committee
CCIR	International Radio Consultative Committee
CCIR601	see BT.601.
CCIT	International Telegraph Consultative Committee
CCITT	International Telephone and Telegraph Consultative Committee (renamed ITU-T)
CDMA	code division multiple access
CELP	code excited linear prediction (a speech coding method)
CIE	Commission Internationale de L'Eclairage, an international body of color scientists
CIF	common intermediate format (a digital video format defined by ITU-T)
CIFQ	channel-condition independent fair queueing
CLR	cell loss rate (in ATM networks)
CMY	color coordinate using cyan, magenta, and yellow as primary colors
CMYK	color coordinate using cyan, magenta, yellow, and black as primary colors
codec	coder and decoder
cpd	cycles per degree, the unit of the spatial frequency in either horizontal or vertical directions
CPS	constrained parameter set
CRT	cathode ray tube
CSFT	continuous-space Fourier transform
CSP	content service provider
CSS	curvature scale space
DAI	delivery application interface
DAVIC	Digital Audio Visual Council
DBMA	deformable BMA
DBS	direct broadcast satellite
DC	direct current (refers to the transformed coefficient corresponding to zero frequency)
DCP	disparity-compensated prediction
DCT	discrete cosine transform

DFD	displaced frame difference
DFT	discrete Fourier transform
DSFT	discrete-space Fourier transform
DTV	digital television
DFD	displaced frame difference
DMIF	delivery multimedia integration framework
DMV	differential motion vector (a motion correction vector defined in MPEG-2 video)
DPCM	differential pulse coded modulation
DSL	digital subscriber loop
DSP	digital signal processor
DV	disparity vector
DVD	digital video disk
DVTR	digital video tape recorder
DWT	discrete wavelet transform
EBMA	exhaustive-search BMA
EDF	earliest deadline first
EDGE	enhanced data rates for GSM evolution
EG	exp-Golomb (a type of probability distribution)
EOB	end of block (a symbol used in coding a DCT block)
EREC	error-resilient entropy coding
EZW	embedded zero-tree wavelet (a wavelet-based coding method)
F2D	flexible 2-D
F3D	flexible 3-D
FACS	facial action coding system
FAP	facial animation parameter (used in MPEG-4)
FC-AL	fiber channel arbitrated loop
FDDI	fiber distributed data interface
FDP	facial definition parameter (used in MPEG-4)
FEC	forward error correction
FEM	finite element method
FFT	fast Fourier transform
FGS	fine-granularity scalability
FIFO	first-in first-out
fps	frames per second, the unit of the frame rate or temporal sampling frequency (also represented as Hz)
GGD	generalized Gaussian distribution
GMF	Gauss-Markov field
GOB	group of blocks (a terminology in video coding standards)
GOP	group of pictures (a terminology in video coding standards)
GPRS	general packet radio service
GR	Golomb-Rice (a type of probability distribution)
GSM	Global System for Mobile Communication

GSS	grouped sweeping scheduling
GVOP	group of video object planes (a terminology in MPEG-4)
HBMA	hierarchical BMA
HDTV	high-definition television
HFM-ME	hierarchical feature-matching motion estimation scheme
HSI	color coordinate using Hue, Saturation, and Intensity (also known as HSV, where V stands for Value)
HTTP	HyperText Transfer Protocol
HVS	human visual system
I3D	incomplete 3-D
IAR	Image aspect ratio
IEC	International Electrotechnical Commission
IETF	Internet Engineering Task Force
i.i.d	independent and identically distributed
IP	Internet Protocol
ISD	independent segment decoding
ISDN	integrated service digital network (usually refers to narrow-band ISDN)
ISA	International Federation of the National Standardizing Associations
ISO	International Standards Organization
ISP	internet service provider
ITU	International Telecommunications Union
ITU-R	International Telecommunications Union—Radio Sector
ITU-T	International Telecommunications Union—Telecommunications Sector
JPEG	Joint Photographic Expert Group
KBASC	knowledge-based analysis-synthesis coding
kbps	kilobits per second
KLT	Karhuner-Loeve Transform
LAN	local area network
LC	layered coding
LCD	liquid crystal display
LDU	logical data unit
LMDS	local multipoint distribution service
LMMSE	linear minimum mean-squared error
lpf	lines per frame
LSI	linear shift-invariant
LTI	linear time-invariant
MAD	mean absolute difference
MAN	metropolitan area network
MAP	maximum a posteriori
MB	macroblock (the coding unit in ITU and MPEG video coding standards)

mbps	megabits per second
MC	model compliance
MCI	motion-compensated interpolation
MCP	motion-compensated prediction
MCU	multipoint control unit
MDC	multiple description coding
MDL	minimum description length (a source coding method)
MDTC	multiple description transform coding
MF	model failure
MMDS	multichannel multipoint distribution service
MMSE	minimum mean-squared error
MOS	mean opinion score
MPEG	Motion Picture Expert Group
MQUANT	quantizer stepsize over an MB in video coding standards
ms	millisecond
MSB	most significant bit
MSE	mean squared error
MSR	maximally smooth recovery (an error concealment method)
MTF	modulation transfer function
MTYPE	coding mode for an MB in video coding standards
MTU	maximum transmission unit (the largest packet size for data delivery in the Internet)
MV	motion vector
MVD	differentially coded motion vector in video coding standards
MZTE	multiscale ZTE (a wavelet-based coding method)
nm	nanometer
NTSC	analog TV system used mostly in North America, defined by the American National Television System Committee
OBASC	object-based analysis-synthesis coding
OBMC	overlapped block motion compensation
OD	object descriptor (an MPEG-4 entity)
OF	optical flow
OTS	object-based temporal scalability
PAL	Phase Alternating Line, the analog TV system used mostly in Western Europe
PAR	pixel aspect ratio
PCF	phase correlation function
PCM	pulse coded modulation
PCT	pairwise correlating transform
pdf	probability density function
pmf	probability mass function
POCS	projection onto convex sets
POTS	plain old telephone system

PS	program stream
PSD	power spectral density
PSNR	peak signal-to-noise ratio
PSTN	public switched telephone network
PTS	presentation time stamps
QAM	quadrature amplitude modulation
QCIF	quarter CIF
QoS	quality of service
QP	quantization parameter (a parameter that controls quantizer stepsizes)
R3D	rigid 3-D
RCPC	rate-compatible punctured convolutional code
RD	rate distortion
READ	relative element address designate
RGB	color coordinate using red, green, and blue as primary colors
RISC	reduced instruction set computer
RPS	reference picture selection
RS	Reed-Solomon (a type of channel error corrector code)
RTCP	real-time control protocol
RTP	real-time transport protocol
RTT	round trip time
RTSP	real-time streaming protocol
RV	random variable
RVLC	reversible variable length coding
SA-DCT	shape-adaptive DCT
SA-DWT	shape-adaptive DWT
SA-ZTE	shape-adaptive ZTE
SAN	storage area network
SAP	session announcement protocol
SCSI	small computer serial interface
SDP	session description protocol
SDTV	standard-definition television
SECAM	Sequential Couleur avec Memoire, the analog TV system used mostly in Eastern Europe
SG	study group (a unit under ITU)
SIP	session initiation protocol
SL	synchronization layer
SMPTE	Society of Motion Picture and Television Engineers
SNR	signal-to-noise ratio
SPIHT	set partitioning in hierarchical trees (a wavelet-based coding method)
SQ	scalar quantization
SSFT	sampled-space Fourier transform
SSP	stream synchronization protocol

SSRC	synchronization source identifier
STC	system time clock
STF	sign-truncated feature (the feature used in HFM-ME)
SVD	singular value decomposition
S-VHS	super-VHS, an enhanced version of VHS
TC	transform coding
TCP	transmission control protocol
TCQ	trellis-coded quantization
TDMA	time division multiple access
TFT	thin-film transistor
TS	transport stream
TTS	text-to-speech
TTSI	TTS Interface
TV	television
UBR	unspecified bit rate (a service type in ATM networks)
UDP	user datagram protocol
URL	univeral resource locator
UMTS	Universal Mobile Telecommunications System
VBR	variable bit rate (a service type in ATM networks)
VCD	video compact disc
VCR	video cassette recorder
VHS	analog videotape recording format
VLC	variable-length coding
VLD	variable-length decoding
VO	video object (terminology in MPEG-4)
VOL	video object layer (terminology in MPEG-4)
VOP	video object plane (terminology in MPEG-4)
VRML	virtual reality modeling language
VQ	vector quantization
VSB	vestigial sideband modulation
VTR	video tape recorder
WAN	wide area network
WP	working party (a unit under ITU)
YCbCr	color coordinate used in most digital video formats, consisting of luminance (Y) and two color difference signals (Cb and Cr)
YIQ	color coordinate used by the NTSC system, consisting of luminance (Y) and two chrominance components I and Q
YUV	color coordinate used by the PAL and SECAM systems, consisting of luminance (Y) and two chrominance components U and V
ZTE	zero tree entropy coding (a wavelet-based coding method)

Appendix D

Answers to Selected Problems

CHAPTER 1

1.2: For illuminating light, red + green + blue = white, red + green = yellow.

1.3: For dyes, red + green + blue = black, red + green = blue.

1.4: Calculate using Equations (1.4.4) and (1.4.2).
For YIQ: (a) (1,0,0), (b) (0.587, −0.274, −0.523), (c) (0.886, 0.321, −0.311), (d) (0.701, −0.596, −0.211).
For YUV: (a) (1,0,0), (b) (0.587, −0.289, −0.515), (c) (0.886, −0.436, 0.100), (d) (0.701, 0.147, −0.615).

1.5: Calculate using Equation (1.5.3). (a) (235, 128, 128), (b) (144, 54, 34), (c) (210, 16, 146), (d) (170, 166, 16).

1.9: The luminance bandwidth is 5.04 MHz for both PAL and SECAM.

CHAPTER 2

2.1:

$$H(f_x, f_y, f_t) = \frac{\sin(\pi f_x T_x)}{\pi f_x T_x} \frac{\sin(\pi f_y T_y)}{\pi f_y T_y} \frac{\sin(\pi f_t \Delta_e)}{\pi f_t \Delta_e} \exp(-j\pi f_t \Delta_e).$$

2.2: The CSFT of the projected image is

$$\Psi(f_x, f_y, f_t) = B^2 \frac{\sin(\pi f_x B)}{\pi f_x B} \frac{\sin(\pi f_y B)}{\pi f_y B} \delta(f_t + f_x v_x).$$

The CSFT of the captured signal is

$$\Psi_c(f_x, f_y, f_t) = \Psi(f_x, f_y, f_t)H(f_x, f_y, f_t)$$

$$= B^2 \frac{\sin(\pi f_x B)}{\pi f_x B} \frac{\sin(\pi f_y B)}{\pi f_y B} \frac{\sin(\pi f_x T_x)}{\pi f_x T_x} \frac{\sin(\pi f_y T_y)}{\pi f_y T_y} \frac{\sin(\pi f_x v_x \Delta_e)}{\pi f_x v_x \Delta_e}$$

$$\times \exp(j\pi f_x v_x \Delta_e)\delta(f_t + f_x v_x).$$

Perceptually, the captured image is a blurred square moving horizontally.

2.5: The angular frequencies are $f_\theta = 5.29, 1.88, 8.75$ cpd, for viewing distances $d = 3, 1, 5$ meters, respectively. The HVS is most sensitive for f_θ at $3 \sim 5$ cpd. So $d = 3$ meters is better than the other two arrangements.

2.6: The temporal frequencies are $f_t = 7, 0, 12, 4$ Hz for $(v_x, v_y) = (1, 1), (4, -3)$, $(4, 0), (0, 1)$, respectively.

2.7: The temporal frequency at the retina is $f'_t = 0$ if the eyes track the object motion exactly. For the fixed eye speed of $(2, 2)$, $f'_t = 7, 14, 2, 10$ Hz, respectively.

CHAPTER 3

3.1: (a) The Voronoi cell of lattice A is a hexagon, with density $1/2\sqrt{3}$. The Voronoi cell of lattice B has a diamond shape, with density $1/4$. (b) The basis vectors of the reciprocal of lattice A are $[1/\sqrt{3}, 0], [-1/2\sqrt{3}, 1/2]$. Its Voronoi cell is again a hexagon, but in a different orientation from the original lattice. The sampling density is $2\sqrt{3}$. The basis vectors of the reciprocal of lattice B are $[\sqrt{2}/4, -\sqrt{2}/4], [0, \sqrt{2}/2]$. Its Voronoi cell is again a diamond. The sampling density is 4. (c) Let r_{max} represent the maximum radius of the spectrum that can be accommodated by sampling without causing aliasing, $r_{A,max} = 1/2\sqrt{3}$, $r_{B,max} = 1/4$. Because $r_{A,max} > r_{B,max}$, lattice A is better.

3.7: (a) $\Psi_c(f_1, f_2) = \sqrt{2\pi} \exp(-2\pi^2(f_1^2 + f_2^2))$. (b) $\alpha = 1.43$.
(c)

$$\psi_s(n_1, n_2) = \frac{1}{\sqrt{2\pi}} \exp\left(-(n_1^2 + n_1 n_2 + n_2^2)\alpha^2/2\right).$$

$$\Psi_s(f_1, f_2) = \frac{4\pi}{\sqrt{3}\alpha} \sum_{m_1, m_2 \in \mathcal{Z}} \exp\left(-2\pi^2\left(\left(f_1 - \frac{2m_1 - m_2}{\sqrt{3}\alpha}\right)^2 \right.\right.$$

$$\left.\left. + \left(f_2 - \frac{m_2}{\alpha}\right)^2\right)\right).$$

For part (d) and (e), see MATLAB script prob3_7.m (available at www.prenhall.com/wang).

3.9: The spatial sampling rates are $f_{s,x} = f_{s,y} = 4$ samples/m in all cases. The temporal sampling rates $f_{s,t}$ in samples/s or Hz are (a) 0, (b) 8, and (c) 4.

3.10: The apparent spatial frequency is $f_x = 2$, $f_y = -2$, with orientation in the off-diagonal direction. The bar appears stationary.

3.13:

$$H_p(f_x, f_y, f_t) = \frac{\sin(\pi f_x T_x)}{\pi f_x T_x} \frac{\sin(\pi f_y T_y)}{\pi f_y T_y} \frac{\sin(\pi f_t \Delta_e)}{\pi f_t \Delta_e} \exp(-j\pi f_t \Delta_e).$$

CHAPTER 4

4.1: The filter used for deinterlacing the top field is

$$h_1(x, y, t) = \begin{cases} 1/2, & t = 0, y = \pm\Delta y; \\ 1/4, & t = -\Delta_t, y = 0; \\ -1/8, & t = -\Delta_t, y = \pm2\Delta y; \\ 0, & \text{otherwise.} \end{cases}$$

The SSFT is $H_1(f_y, f_t) = \cos(2\pi f_y \Delta_y) + \frac{1}{4}(1 - \cos(4\pi f_y \Delta_y))e^{j2\pi f_t \Delta_t}$. The filter for deterlacing the bottom field is $h_2(x, y, t) = h_1(x, y, -t)$, $H_2(f_y, f_t) = H_1(f_y, -f_t)$.

4.2: The filters over the deinterlaced grid are

$$h[n\Delta_y, 0]$$

$$= \left[-\frac{1}{32}, 0, \frac{9}{32}, 1, \frac{9}{32}, 0, -\frac{1}{32} \right], n = -3, \ldots, 3, \; h[0, \Delta_t] = \frac{1}{2},$$

with $\Delta_t = 1/60$ s, $\Delta_y = \frac{1}{525}$ picture-height. The DSFT is

$$H_{\text{DSFT}}(f_y, f_t) = 1 + \frac{18}{32} \cos 2\pi f_y - \frac{2}{32} \cos 6\pi f_y + \frac{1}{2} \exp(-j2\pi f_t).$$

The SSFT is $H_{\text{SSFT}}(f_y, f_t) = H_{\text{DSFT}}(\Delta_y f_y, \Delta_t f_t)$.

4.5: (b) The equivalent filter over the intermediate lattice (120 fps, progressive) is $h(0, m\Delta_t) = 1$, $m = 0, 1, 2, 3, 4$. The SSFT is $H_{\text{SSFT}}(f_y, f_t) = (1 + 2\cos 2\pi \Delta_t f_t + 2\cos 4\pi \Delta_t f_t) \exp(-j4\pi \Delta_t f_t)$.

4.6: (b) The prefilter in field 1 is (vertical direction)

$$h_1[n2\Delta_y] = \frac{1}{138}[-29, 0, 88, 138, 88, 0, -29], n = -3, \ldots, 3.$$

Its DSFT is $H_1(f_y) = 1 + \frac{176}{138} \cos 2\pi f_y - \frac{58}{138} \cos 6\pi f_y$. The interpolation filter in field 2 is

$$h_2[n\delta_y] = \frac{1}{16}[1, 0, 7, 16, 7, 0, 1], n = -3, \ldots, 3,$$

with DSFT $H_2(f_y) = 1 + \frac{14}{16} \cos 2\pi f_y + \frac{2}{16} \cos 6\pi f_y$. The overall vertical-temporal filter is

$$h_3(n\Delta_y, 0) = \left[-\frac{29}{138}, 0, 0, \frac{1}{16}, \frac{88}{138}, \frac{7}{16}, 1, \frac{7}{16}, \frac{88}{138}, \frac{1}{16}, 0, 0, -\frac{29}{138} \right],$$

$n = -6, \ldots, 6$. The DSFT is $H_3(f_y, f_t) = 1 + \frac{14}{16} \cos 2\pi f_y + \frac{176}{138} \cos 4\pi f_y + \frac{2}{16} \cos 6\pi f_y - \frac{58}{138} \cos 12\pi f_y$. The corresponding SSFTs are $H_{1,\text{SSFT}}(f_y) = H_1(2\Delta_y f_y)$, $H_{2,\text{SSFT}}(f_y) = H_2(\Delta_y f_y)$, and $H_{3,\text{SSFT}}(f_y, f_t) = H_3(\Delta_y f_y, \Delta_t f_t)$.

4.7: (b) For field 1,

$$h_1[n2\Delta_y] = [1/2, 1, 1/2], n = -1, 0, 1.$$

For field 2,

$$h_2[n\Delta_y] = [\frac{1}{4}, 0, \frac{3}{4}, 1, \frac{3}{4}, 0, \frac{1}{4}], n = -3, \ldots, 3.$$

The overall filter is

$$h_3[n\Delta_y, 0] = [\frac{1}{4}, \frac{1}{2}, \frac{3}{4}, 1, \frac{3}{4}, \frac{1}{2}, \frac{1}{4}], n = -3, \ldots, 3.$$

The SSFT is

$$H_3(f_y, f_t) = 1 + \frac{3}{2} \cos 2\pi \Delta_y f_y + \cos 4\pi \Delta_y f_y + \frac{1}{2} \cos 6\pi \Delta_y f_y.$$

CHAPTER 5

5.1: The movement in the x direction due to the 3-D movement (X, Y, Z) to (X, Y, Z') is $\Delta x = FX(Z - Z')/ZZ'$. The width of each image pel is $\Delta W = W/N_x$, with $W = 1.33''$, $N_x = 352$. Letting $\Delta x = \Delta W$ yields $\Delta Z_x = Z' - Z = -Z/(1 + XN_x/W)$. Similarly, $\Delta Z_y = -Z/(1 + YN_y/H)$.

5.5: The affine parameters can be obtained by solving:

$$\begin{bmatrix} 1 & x_1 & y_1 \\ 1 & x_2 & y_2 \\ 1 & x_3 & y_3 \end{bmatrix} \begin{bmatrix} a_0 \\ a_1 \\ a_2 \end{bmatrix} = \begin{bmatrix} d_{x,1} \\ d_{x,2} \\ d_{x,3} \end{bmatrix}, \quad \begin{bmatrix} 1 & x_1 & y_1 \\ 1 & x_2 & y_2 \\ 1 & x_3 & y_3 \end{bmatrix} \begin{bmatrix} b_0 \\ b_1 \\ b_2 \end{bmatrix} = \begin{bmatrix} d_{y,1} \\ d_{y,2} \\ d_{y,3} \end{bmatrix}.$$

CHAPTER 6

6.5: The MV $\mathbf{d}_m$ for a block $\mathcal{B}_m$ can be obtained from Equation (6.2.7), where the sum should be over $\mathcal{B}_m$. This solution is valid only if the true MV is small, or the equation is applied iteratively and the initial estimate is close to the true MV.

6.6:

$$\mathbf{a} = \left(\sum_{\mathbf{x} \in \Lambda'} [\mathbf{A}(\mathbf{x})]^T \nabla \psi_1(\mathbf{x}) \nabla \psi_1(\mathbf{x})^T [\mathbf{A}(\mathbf{x})] \right)^{-1}$$

$$\times \left(\sum_{\mathbf{x} \in \Lambda'} (\psi_1(\mathbf{x}) - \psi_2(\mathbf{x}))[\mathbf{A}(\mathbf{x})]^T \nabla \psi_1(\mathbf{x}) \right).$$

6.8: For an image of size $M \times M$, the total number of operations is $(2R + 1)^2 M^2$, independent of the block size.

6.9: $(M/2^{L-1})^2((2R/2^{L-1}) + 1)^2 + 18M^2(1 - (1/2)^{L-1})$.

6.10: For EBMA, $720 \times 480 \times 65^2 = 1.46\text{E}+9$; For HBMA with two layers, with a search range of 16 for the top layer and 1 for the bottom layer, $360 \times 240 \times 33^2 + 720 \times 480 \times 3^2 = 9.72\text{E}+7$.

CHAPTER 8

8.1: (a) $\log_2 L$; (b) 0.

8.6: (a) $H_1 = 1.5850$, $H_2 = 3.0850$, $H_{c,1} = 1.5$, $\bar{H} = 1.5$. (b) $R_1 = 1.6667$, $R_2 = 3.1667$, $R_{c,1} = 1.5$. (c) $R_{min} = \bar{H}$, achievable with first-order conditional Huffman coding.

8.11: (a) $b = 1/\lambda + a$; (b) $a = b/2$; (c) $a = 1/\lambda$, $b = 2/\lambda$, MSE $= (2/\lambda^2)(1 - \frac{2}{e})$.

8.12: (b) $b_l = (g_l + g_{l+1})/2$.

8.14: MSE $= 9/16$ per sample.

CHAPTER 9

9.4: $\sigma_{t,k}^2 = \{1 + 3\rho, 1 - \rho, 1 - \rho, 1 - \rho\}\sigma_s^2 = \{3.85, 0.05, 0.05, 0.05\}$, $\sigma_t^2 = 0.1481$, $G_{TC} = 6.75$. For $R = 2$, $D_{TC} = 0.0093\epsilon_G^2$, $R_k = \{4.35, 1.22, 1.22, 1.22\}$.

9.5: The eigenvalues are $\lambda_k = \{1 + 3\rho, 1 - \rho, 1 - \rho, 1 - \rho\}\sigma_s^2 = \{3.85, 0.05, 0.05, 0.05\}$. The KLT produces the same transform coefficient variances as the DCT for this case.

9.6: $a_k = \rho/(2\rho+1)$, $k = 1, 2, 3$. $\sigma_p^2 = (1 - 3\rho^2/(2\rho+1))\sigma_s^2 = 0.06638$. $G_{DPCM} = 15.06$. For $R = 2$, $D_{DPCM} = 0.0041\epsilon_G^2$.

9.7: $a_k = \rho/(\rho+1)$, $k = 1, 2$. $\sigma_p^2 = \sigma_s^2(1 + \rho - 2\rho^2)/(1 + \rho) = 0.0744$. $G_{DPCM} = 13.45$.

CHAPTER 10

10.5: Denote the four directions by "0", "1", "2", "−1", starting from the direction "East", in clockwise order. The direct chain code with four-neighbor is

$$-1, 2, -1, 2, 2, 2, 0, -1, -1, -1, 0, 1, 1, 0, 0, -1, -1, -1,$$

$$1, 1, 0, 1, 0, 1, 1, 2, 2, 1.$$

The differential code is

$$-1, -1, 1, -1, 0, 0, 2, -1, 0, 0, 1, 1, 0, -1, 0, -1, 0, 0, 2,$$

$$0, -1, 1, -1, 1, 0, 1, 0, -1.$$

10.6: The code tree is: "0111101101111111". The shape tree is "22132233131", where "1" represents the region with the lightest shade, "2" the intermediate shade, and "3" the darkest shade.

Index

1-D DCT, 271
1-D DFT, 267
2-D DCT, 271
2-D DFT, 269
2-D meshes, 394
2-D motion, 142
2-D motion estimation, 141
2-D motion field, 128
2-D motion vector, 128
2-D scene model, 127
2-D silhouette, 220
2.5-D scene model, 127
3-D meshes, 394 (*see* wireframe)
3-D motion vector, 128
3-D sampling lattice, 67, 73
   ALI, 73
   BCO, 73
   FCO, 73
   ORT, 73
3-D VLC, 417, 421, 444
4:1:1, 72
4:2:0, 72
4:2:2, 72
4:2:2 profile, 436
4:4:4, 72

a priori knowledge, 152
absolute entropy, 226
AC coefficients, 271

AC prediction, 443
action units, 338
adaptive arithmetic coding, 363, 365
admission control, 535
advanced coding efficiency, 453
advanced prediction mode, 418
advanced real-time simple (ARTS)
     profile, 452
advanced simple profile, 453
affine function, 172
affine mapping, 132, 177, 196, 387
affine motion, 136, 166
alias components, 62, 63, 76
alias-free condition, 62, 63, 86
aliasing, 62, 63, 67, 69, 73, 78, 162
aliasing versus blurring, 77, 79
alpha-maps, 314, 445, 457
alternate horizontal scan, 443
alternate scan, 430, 431, 443
alternate vertical scan (*see* alternate scan)
ambient light source, 118
ambiguity in motion estimation, 143
analog TV systems, 48
analog video, 221
analog video recording, 21
analog videotape formats, 22
anchor frame, 145
angular frequency, 45
angular radial transform, 463

aperture problem, 144
application layer framing (ALF), 486
application-layer QoS control, 520
application-level multicast, 531
arbitrarily shaped object, 442
arithmetic coding, 235, 238, 283, 447, 450
    bit rate, 241
    comparison with Huffman coding, 241
    context-based, 317, 445
    finite precision, 241
ARQ, 487, 553
asymmetric digital subscriber loop, 479
ATM networks, 480, 488
    ATM packet-switching, 480
    available bit rate (ABR), 480
    cell loss rate (CLR), 480
    constant bit rate (CBR), 480
    unspecified bit rate (UBR), 480
    variable bit rate (VBR), 480
AV document, 460

B-frame, 295
B-mode, 295
B-picture, 418
B-spline, 321
background object, 177
    newly covered, 126
    uncovered, 126, 130
backward channel, 502
backward compatibility, 410
backward motion compensation, 145, 292
backward motion estimation, 145
base layer, 350
baseline distance, 379
Bayesian criterion, 150
BCH code, 485, 506
behavior, 221
BER, 549
best-effort network, 481
bidirectional temporal prediction, 292
BIFS, 437, 440
bilinear functions, 172
bilinear interpolation, 157, 331, 332, 417, 421
bilinear motion, 136, 166
binary alpha block (BAB), 445, 457
binary encoding, 234–240, 258, 490
binary feature matching, 163
    complexity, 165
    HFM-ME, 163
    sign-truncated feature (STF), 163
binary representation, 234
binary tree search, 246

biquadratic mapping, 137
bit allocation, 301
bit rate, 227
bit-plane coder, 453
bitmap-based shape coder, 315
Bloch's law, 44
block, 457
block-based hybrid coding, 220, 293–308,
    339, 341
block-based motion estimation, 154–168
block-based motion representation, 147, 155
block-based transform coding, 263 (see also
    transform coding)
block-matching algorithm (BMA), 154–161
    blocking artifact, 159
    EBMA, 155–157
    fast algorithms, 159
    half-pel search, 157, 301
    integer-pel search, 156
    variable size, 187
    VLSI implementation, 161
blocking artifact, 176, 327, 418
blockwise translational motion model,
    155, 159
blockwise wavelet transform, 362
blue-screen, 326
body animation, 450
bound for
    conditional lossless coding, 229
    lossless coding, 225, 227
    lossy coding, 225, 229, 231
    scalar lossless coding, 227
    vector lossless coding, 228
boundary blocks, 317, 445
bounding box, 324
broadband ISDN (B-ISDN)
    (see ATM networks)
broadcast channels, 484
broadcast, 478, 484, 481
BT.601, 23, 24, 26, 28, 71, 426, 478
    4:2:0, 26
    4:2:2, 26
    4:4:4, 26
    525/50, 23
    525/60, 23
    chrominance subsampling, 25
    color coordinate, 24
    MPEG-2, 26
    NTSC, 23
    PAL, 23
    PAR, 23
    raw data rate, 25

BT.601 (*continued*)
   sampling rate, 23
   SECAM, 23

CAHV camera model, 114
call admission control (CAC), 550
camera, 7–10
   exposure time, 77
   filtering, 76
   frequency response, 77
   impulse response, 77
   Kell factor, 19, 79
   modulation transfer function
      (MTF), 77
   motion picture cameras, 10
   spatial prefilter, 77
   spectral absorption function, 7
   temporal prefilter, 77
camera apertures, 76
   combined, 77
   spatial, 77
   temporal, 76
camera calibration, 385
camera model, 112–116
   CAHV, 114
   extrinsic parameter, 115
   intrinsic parameter, 115
camera motions, 116, 130, 177, 463
   boom, 116
   dolly, 116
   pan, 116, 131, 132
   roll, 116, 132
   tilt, 116, 131, 132
   track, 116
   zoom, 116, 131, 132
Candide, 336, 341
CCD, 9, 22
CCIR, 407
CCITT, 407
CD-ROMs, 27
CDMA, 549
cell loss rate (CLR), 480
centroid, 180, 245
centroid condition, 245, 253, 258
chain coding, 318
   differential, 318
   direct, 318
changed region, 126
channel coding, 227
charge-coupled devices (CCDs), 9
checksum, 481
chirping effect, 134, 136

chroma-keying, 420
chrominance, 4
   hue, 4
   saturation, 4
chrominance subsampling, 24, 25, 72
   4:1:4, 25, 72
   4:2:0, 25, 72
   4:2:2, 25, 72
   4:4:4, 25, 72
CIF, 416
circuit switching, 479
clique, 151
closed-loop prediction, 286
coarse granularity, 349
code excited linear predictive coding, 441
codebook, 234, 249
coded block pattern (CBP), 414, 417, 421
coded frame, 291
codewords, 234, 249
coding parameter selection, 299, 528
   rate-distortion optimization, 300, 528
coding-parameter adaptation, 503
color coordinate, 17, 24
   chromaticity values, 5
   CIE RGB primary, 6
   CIE XYZ primary, 6
   HSI, 7
   PAL, 17
   tristimulus values, 5
   YCbCr, 24
   YDbDr, 19
   YIQ, 18
   YUV, 17
color coordinate conversion, 6
color matching functions, 5
color mixing
   CMY primary, 5
   CMYK primary, 5
   primary color, 4
   RGB primary, 5
   trichromatic theory, 4
color perception
   cones, 3
   frequency responses, 3
   opponent color model, 4
   photoreceptors, 3
   relative absorption functions, 3
   relative luminous efficiency function, 4
   rods, 3
   tri-receptor theory of color vision, 3
   visual cortex, 3
color TV systems, 16, 17 (*see* TV systems)

communication channel, 217
conditional Huffman coding, 238
conditional lossless coding, 228
conformance, 412
congestion control, 522
constant intensity assumption, 40, 120, 143,
        149, 152, 188
constrained parameters set, 425
content replication, 532
content-dependent coding, 217, 220,
        314–343
contents caching mirroring, 532
context, 228
context-based arithmetic coding, 317, 445
continuous frequency, 66
continuous media distribution
        services, 521
continuous RV, 221
continuous source, 221
continuous space signal, 33
continuous-space continuous-amplitude
        random process, 221
continuous-space Fourier transform (CSFT),
        34, 50, 162
contour coding, 318
contrast sensitivity, 43, 376
control grid (*see* mesh-basal motion
        representation)
control points, 121
control protocol, 543
convergence angle, 381
converging camera configuration, 381
conversion between arbitrary lattices, 89
    1D case, 89
    ideal sampling rate conversion
        filter, 89
conversion between PAL and NTSC
        signals, 98
convolution, 34, 50
convolution theorem
    continuous space, 35
    discrete space, 37
    sampled space, 61
core profile, 453
core studio profile, 453
cost functions, 211
critical flicker fusion frequency, 17, 45, 51
cubic lattice, 65
cubic spline, 92
current frame, 291
curvature scale space, 463
cut-off frequency, 43, 51, 67

data communications, 481
data partition (*see* frequency scalability)
data partitioning, 356, 431, 490, 507, 508
data striping, 537
DAVIC, 413
DC coefficient, 271
DCT-based image coders, 281–284
deblocking filter, 419
decimation (*see* down-conversion)
deformable block-matching
        algorithms (DBMA), 165
deinterlacing, 92, 93
    field averaging, 97
    field merging, 96
    ideal interpolation filter, 95
    line and field averaging, 97
    line averaging, 96
delay variation (*see* jitter)
delay-constrained retransmission,
        487, 526
delta function, 34
demultiplexing
    comb filter, 21
    cross-color, 21
    cross-luminance, 21
depth perception, 375, 400
    asymmetric spatial resolution, 376
    asynchronous fusion of binocular
        views, 377
    visual sensitivity thresholds, 375
depth-first tree traversal, 317
description definition language, 459
description scheme, 459
descriptor, 458
    localization, 464
    motion, 463
    shape, 463
differential entropy, 226
    Gaussian source, 226
differential pulse coded modulation
        (DPCM), 286
differentiated services, 488
diffuse reflection, 116
digital filters, 79
digital TV, 27, 68, 478, 484
digital video, 22, 221
    data rate, 22
    frame rate, 22
    line number, 22
    samples per line, 22
    sampling rates, 22
digital video disks (DVD), 26, 27, 412

digital video formats, 26 (*see also* video formats)
digital video recording, 27
   digital video tape formats, 28
   DVD, 28
   hard-disk-based, 28
   tape-based, 28
   VCD, 28
digital video tape formats, 28
   betacam SX, 28
   D1, 28
   D2, 28
   D3, 28
   D5, 28
   digital betacam, 28
   DVCPRO50, 28
digital video tape recorder (DVTR), 27
discrete-time convolution, 37
discrete cosine transform (DCT), 219
   1-D, 271
   2-D, 271
   comparison with KLT, 281
   field, 430
discrete delta function, 37
discrete frequency, 66
discrete RV, 221
discrete source, 221
discrete space signal, 36
discrete wavelet transform (DWT), 361, 363 (*see also* wavelet transform)
discrete-space continuous-amplitude, 221
discrete-space discrete-amplitude, 221
discrete-space Fourier transform (DFST), 37, 50, 60, 162
   periodicity, 37
discrete-space system, 38
disparity (*see* disparity vector)
disparity estimation, 385, 401
   block-based approach, 388, 390
   comparison with motion estimation, 386
   constraints, 386
   dynamic programming, 391
   feature-based approach, 386
   mesh-based approach, 388, 390
   ordering constraint, 387
   unidirectionality, 387
disparity field
   model, 387
disparity quantization, 376
disparity vector, 379
disparity-compensated interpolation, 394

disparity-compensated prediction (DCP), 388, 396, 397
displaced frame difference (DFD), 147, 209
   error, 148
   image, 148
displacement vector field (*see* motion field)
display
   cathode ray tube (CRT), 10
   filtering, 76
   Kell factor, 19
   liquid crystal display (LCD), 10
   temporal filtering, 79
   vertical filtering, 79
display apertures, 79
distortion-rate bound, 231
dominant motion, 181
down-conversion, 84, 87
   ideal prefilter, 87
downward compatibility, 411
drift, 431, 432
DSP, 405
dual prime for P-pictures, 429
DVD consortium, 413
dynamic programming, 300, 391

E-matrix, 198, 214
earliest deadline first (EDF), 534
elementary stream, 437
embedded bit stream, 350, 364
embedded coding, 358
embedded zero-tree wavelet (EZW) coding, 363
end of block (EOB), 416
end-to-end delay, 303, 476, 478
energy, 267
enhancement layers, 350
entropy, 222–227
   conditional, 223–224
   differential, 226
   discrete RV, 222
   discrete source, 224
   joint, 223
entropy coding (*see* variable length coding)
entropy-constrained MMSE quantizer, 257
entropy-constrained optimal quantizer, 255
entropy power, 232
entropy rate, 224, 241
epipolar constraint, 383, 386
epipolar geometry, 386
epipolar line, 198, 199, 203, 214
epipolar plane, 383
epipole, 383

erasure errors, 473
error concealment, 475, 527, 528
    maximally smooth recovery (MSR), 500
    motion-compensated temporal
        interpolation, 500
    projection onto convex sets (POCS), 501
    recovery of coding modes, 501
    recovery of MVs, 501
    recovery of texture information, 500
    spatial interpolation, 500
    syntax-based repair, 502
error control, 472, 525, 543
error control link-layer, 553
error isolation, 489, 508
error propagation, 291, 492, 527
error protection, 483
error resilience, 444
error resilience tools, 505
    in H.263, 505–508
    in MPEG-4, 508–509
error-resilient encoding, 475, 489–498, 527
    intra-coded MBs, 492
    packetization and multiplexing, 486
    prediction, 492, 507
error-resilient entropy coding (EREC), 491
error tracking, 504
essential matrix, 198
essential parameters, 198
event tree, 460
exhaustive block-matching algorithm
        (EBMA), 155–162, 169, 184, 187
    complexity, 156
    search stepsize, 156
exhaustive search, 151, 155, 175
exp-Golomb (EG), 491

face animation, 338, 449
face model, 336
facial action coding system, 338
facial animation parameters (FAPs), 338,
        448, 449
fast algorithms for motion estimation
    2D-log search, 159
    three-step search, 159
fast-Ethernet, 481
fault tolerance, 539
feature points, 204
feature-based approach, 195, 386
feedback channel, 475
feedback information, 502
fiber distributed data interface, 481
field DCT, 430

field picture, 428
field prediction for field pictures, 429
field prediction for frame pictures, 429
field sequential display, 377
file management, 536
    DC-SCAN, 536
    GSS, 536
    SCAN-EDF, 536
filter concatenation, 332
filter design, 66
filter implementation, 91
fine granularity scalability, 350, 357
fine granular scalability profile, 453
finite element method (FEM), 172
fixed interval synchronization, 508
fixed-length coding, 235
flexible objects, 122
    flexible 2-D (F2D), 330
    flexible 3-D (F3D), 332
flicker, 43–45, 68, 71
flow control, 543
flow vectors, 130
focal center, 112
focal length, 112
foreground object, 177
forward compatibility, 410
forward error correction (FEC), 475, 526, 539,
        549, 553
    BCH code, 485
    RS code, 485
forward motion compensation, 145, 292
forward motion estimation, 145
forward transform, 265
Fourier descriptors, 322
Fourier series, 322
fractional accuracy search, 157
frame picture, 428
frame rate, 43, 68
frame-dropping filters, 530
frequency response, 36, 50, 77
frequency scalability, 350, 356
frequency space scaling, 69, 73
fundamental matrix, 384
fundamental parallelepiped, 56
fundamental period, 59
fusion of stereo views, 375

G.723.1, 423
G4, 315
gamma correction, 12
Gamma source, 246
Gauss-Markov field (GMF), 222, 234

Gauss-Markov process, 222
Gaussian filter, 183
Gaussian process (*see* Gaussian source)
Gaussian source, 232, 246, 257, 277, 226
   rate-distortion bounds, 232, 246
   successive refinement property, 371
generalized block-matching algorithm (*see* deformable block-matching algorithm)
generalized centroid, 245, 253
generalized Gaussian distribution (GGD), 491
generalized Lloyd algorithm, 254, 258
generalized Nyquist sampling theorem, 61
generating matrix, 54
geometric distortions, 329, 333
geometric mapping, 132
geometric transformation, 177
Gibbs effect, 79
Gibbs random field, 151
GIF89a, 326
Gilge transform, 325
global minimum, 152, 168
global motion compensation, 444
global motion estimation, 177
   direct estimation, 178
   indirect estimation, 178
global motion representation, 147
global motion, 146
Golomb-Rice code, 491
gradient-based search (*see* gradient-descent method)
gradient descent methods, 152, 175, 565–567
group of blocks (GoBs), 295, 457
groups of pictures (GOPs), 295, 425, 456
guaranteed quality of service, 421, 422

H.221, 479
H.223, 483, 487
   mobile extensions, 483
H.225.0, 421
H.245, 421
H.261, 26, 27, 412, 413–416, 421, 485
H.263, 416–421, 485
H.263+, 416
H.263++, 416, 506
H.320, 416, 479
H.323, 421
H.324, 422
Haar transform, 370
half-pel accuracy search, 157
Hamming distance, 260
harmonic vector excitation coding, 441
HDTV, 27, 68, 84, 413, 478, 484

header, 455
header extension code (HEC), 508
header repetition, 508
hexagonal lattice, 54, 55, 58, 63–65, 251
hierarchical approach (*see* multiresolution approach)
hierarchical block matching algorithm (HBMA), 183, 184
   complexity, 186
hierarchical feature-matching motion estimation (HFM-ME), 163
hierarchical storage, 537
highest confidence first, 175
histogram
   motion vector, 163
homogeneous coordinate, 388
Horn and Schunck method, 153
HTTP, 546
Huffman coding, 235–238, 256, 258
   conditional, 238
   comparison with arithmetic coding, 241
   scalar, 236
   vector, 236
human face, 221
human visual system (HVS), 4, 72, 79, 327, 424
   frequency response, 42–50
hybrid coding (*see* block-based hybrid coding)

ideal interpolation filter, 86
ideal prefilter, 62, 76, 77
IEC, 408
IETF, 408
I-frame (*see* intra frame)
illuminating source
   additive rule, 2
illumination, 210
illumination model, 116–119
image aspect ratio (IAR), 15, 23, 68
image coding, 281–284, 447
image disparity, 379
image noise, 209
image recovery, 499
image signal models, 204
imaging plane, 112
implicit shape coding, 326
impulse function (*see* delta function)
impulse response, 36, 50
incident irradiance, 117
incomplete 3-D (I3D) representation, 398
independent and identically distributed (i.i.d.) sources, 222

independent segment decoding (ISD), 493, 507
indexed faceset, 121
information theory, 227, 475
initial solution, 152, 168
inlier, 181
inner product, 266, 269
instantaneous decodability, 234
integer-pel accuracy search, 156
integrated service digital network, 479
inter-frame, 295
interlaced scan, 9, 13, 67–71, 80, 84, 426–428
    2:1 interlace, 14
    alias components, 69
    bottom field, 13
    comparison with progressive scan, 69–71
    even field, 13
    K:1 interlace, 14
    odd field, 13
    sampling lattices, 69
    top field, 13
    zig-zag artifacts, 13
interlacing artifacts, 71
    interline flicker, 71
    line crawl, 71
intermediate view synthesis, 401
inter-mode, 295
Internet, 480, 488, 519
    header size, 481
    packet loss rate, 481
Internet protocol (IP), 480, 542
Internet telephony, 481
interobject synchronization, 540
interpolation
    cubic spline, 92
    ideal interpolation filter, 86
    interpolation formula, 66
    interpolation kernel, 166, 172
    least-squares fitting, 92
    spline functions, 92
interstream synchronization, 540
intra-frame, 295
intra-mode, 295
inverse transform, 265
ISDN, 27, 412
    basic rate interface, 479
ISO, 26, 408
ISO 11172, 423
ISO/IEC, 408
ISO-MPEG, 27
iterative motion estimation, 212
ITU, 406

ITU-R, 23, 407
ITU-T, 407

JBIG, 315
jitter, 476, 478, 541, 551
joint region segmentation and motion estimation, 181
joint source and channel coding, 410, 475, 498
JPEG, 366, 497
    baseline, 366
JTC 1, 408

Kalman filter, 211
Karhunen-Loeve transform (KLT), 219, 265, 279–280
Kell factor, 19, 79
keystone distortion, 383
keystone effect, 134
K-means algorithm, 254
knowledge-based analysis-synthesis coding (KBASC), 336
knowledge-based coding, 220, 336–338

Lagrangian multiplier method, 257, 275, 297, 300
Lambertian surfaces, 117
Laplacian of Gaussian, 46
Laplacian source, 246
latency (*see* end-to-end delay)
lattice, 54
    convolution, 61
    intersection, 58
    partition, 56
    sum, 59
lattice vector quantizer, 251
layered coding, 339, 483, 488, 493
    with unequal error protection, 493
least-squares method, 200, 210, 322
Levinson-Durbin algorithm, 288
LBG algorithm (*see* generalized Lloyd algorithm)
light, 2
    achromatic, 2
    flux, 2
    illuminating source, 2
    radiant intensity, 2
    reflecting source, 2
    spectral color, 2
linear and shift-invariant (LSI), 36
linear minimum mean square error (LMMSE) predictor, 287

linear motion, 73, 75
linear predictor, 287
linear signal model, 204
linear temporal prediction, 291
linear transform, 266
line flicker, 48
line number, 15
line of sight, 113, 126, 127
line rate, 23, 43
link layer error control, 553
Lloyd algorithm, 247, 258
local area network, 480
local deformation, 334
local minimum, 152, 168, 175
local motion, 124
local multipoint distribution service
        (LMDS), 483
logical data unit, 540
loop filtering, 305–308, 413, 414
    optimal filter, 305
lossless coding (*see* binary encoding)
low-pass filter, 62, 76, 77
low-pass texture extrapolation, 325
luminance, 3, 72
LZW coding, 235, 326

macroblock (MB), 295, 413, 457
macroblock type (MTYPE), 414, 415, 417,
        421, 457
main profile, 435
mapping function, 128, 133, 145
Markov process, 222
master element, 173
maximally smooth recovery (MSR), 500
maximum transmission unit (MTU), 486
Mbone, 478
MC-object, 329
mean absolute difference (MAD), 148, 156
mean absolute error (MAE), 260
mean brightness, 43, 51
mean opinion score (MOS), 454
mean square error (MSE), 148, 230
media synchronization, 521, 593
median, 260
memoryless (*see* i.i.d. sources)
mesh animation, 447
mesh-based disparity estimation, 388
mesh-based motion estimation, 169–171
mesh-based motion representation, 147,
        171–173
    creation of new, 176
    node deletion, 176

mesh coding, 447
meta-information, 458
metropolitan area network (MAN), 480
MF-object, 329, 333
midtread quantizer, 413
minimum mean square error (MMSE)
        quantizer, 245–248
    asymptotic performance, 246
    binary tree search, 246
    centroid condition, 245
    Llyod algorithm, 247
    nearest-neighbor condition, 245
    vector quantizer, 253
mirroring, 539
mixed aliases, 71
mixed-resolution coding, 398
mode selection (*see* coding parameter
        selection)
model-based coding (*see* knowledge-based
        coding)
model compliance (MC), 329
model failure (MF), 210, 329, 333
modified READ code, 315
motion compensation (*see*
        motion-compensated prediction)
motion estimation, 2-D, 141–189, 386
    block-based, 154–169
    criteria, 147, 188
    for video coding, 187
    feature-based, 145, 195
    global motion estimation, 177–179
    intensity-based, 145
    maximum a posteriori (MAP) estimate, 150
    mesh-based, 169–177
    minimum description length (MDL)
        estimate, 151
    multiresolution approach, 182–187
    pixel-based, 152–154
    region-based, 179–182
    regularization approach, 150
motion estimation, 3-D, 194–215
    direct method, 203–212
    eight parameter, 197
    feature-based, 195–203
    six parameter, 195–198, 206, 207
motion field, 145
motion model, 122
    2-D, 128–137
    3-D, 122–125
motion parameter concatenation, 212
motion pictures, 68
    camera, 67

effective playback rate, 68
frame rate, 68
IAR, 68
motion representation, 146, 188
block-based, 147, 155
global, 147
mesh-based, 169, 172
multiresolution, 182, 184
node-based, 166, 172
object-based, 147
pixel-based, 146
region-based, 147
motion smoothness constraints, 150, 161
motion tracking, 174
motion-based region segmentation, 180
clustering, 180
layering, 181
motion-compensated interpolation, 98
motion-compensated prediction (MCP),
291, 332
backward, 145, 292
forward, 145, 292
half-pel, 416
OBMC, 296–299
quarter-pel, 444
motion-compensated processing, 141
motion-compensated temporal
interpolation, 500
movie theater, 68
moving flexible 3-D (F3D) objects, 334
MP3, 423, 426
MPEG, 26, 408, 412, 413
MPEG-1, 27, 28, 423–426
audio, 423
systems, 423
video, 424
MPEG-2, 26–28, 127, 426–437, 478, 484
AAC, 426, 442
audio, 426
HDTV, 27
high profile, 27
level, 436
profile, 435–436
scalability, 430–435
systems, 426
video, 427
MPEG-4, 127, 317, 338, 437–454
audio, 441
profiles, 451–454
systems, 437
visual, 442–447
MPEG-7, 458–465

multicast, 484, 523, 545, 546
mobile, 550
networks, 531
multichannel multipoint distribution service
(MMDS), 483
multimedia services, 519
multipath fading, 546
multiple description coding (MDC), 494,
527, 528
comparison with layered coding, 495
multiple-description transform coding
(MDTC), 496
multiplex errors, 484
multiplexing, 19, 20, 486, 543
error-resilient, 486
NTSC systems, 19–21
multiresolution approach, 152, 161, 182–187
benefits, 183
multiresolution motion estimation
(see multiresolution approach)
multiresolution representation, 152, 184
multiresolution/multifrequency
decomposition, 362
multiscale ZTE (MZTE) coding, 364–366
multiview profile, 436
mutual information, 222–225

nearest-neighbor condition, 245, 253, 258
nearest-neighbor interpolation, 166
nearest-neighbor quantizer, 249
neighborhood, 150
eight connectivity, 150
four connectivity, 150
network filtering, 529
network-aware applications, 521, 548
network-layer protocol, 542
NEWPRED, 509
Newton-Ralphson method, 152, 154, 168, 567
node-based block motion model (see
node-based motion representation)
node-based motion representation, 166, 172
noninterlaced scan (see progressive scan)
norm, 266
normalization, 282
normalized vector, 267
NTSC TV systems, 16, 18, 19, 68, 84
audio subcarrier frequency, 21
bandwidth, 19
chrominance, 19
color subcarrier frequency, 20
conversion to PAL, 98
luminance, 19

NTSC TV systems (*continued*)
  monochrome TV system, 21
  multiplexing, 19
  picture carrier frequency, 21
  RGB primary, 18
  spectral composition, 21
  YIQ, 18
Nyquist sampling theorem
  1-D, 62
  generalized, 59, 61

object descriptor, 437
object model, 120
object motion, 177
object tracking, 207
object-based analysis-synthesis coding
    (OBASC), 220, 328, 339
object-based motion representation, 147
object-based scalability, 350, 359
object-based temporal scalability (OTS), 359
object-based video coding, 328
object recognition, 336
observation points, 203, 209, 211
occlusion, 125
opaque object, 314
operational distortion-rate function, 231
operational rate-distortion function, 231
optical flow, 142
optical flow equation, 143, 149, 152, 188, 204
  discrete version, 149
optimal bit allocation, 275
optimal scalar quantizer, 244
optimal vector quantizer, 253
optimization methods, 142, 151, 181, 189
oriented-smoothness constraint, 153
orthogonal, 266
orthogonality principle, 287
orthographic camera model, 195
orthographic projection, 113
orthonormal, 267
outlier, 181, 210
overlapped block motion compensation,
    296–299, 418, 444

P-frame, 295
P-mode, 295
packetization, 486
  error-resilient, 486
  interleaved, 486
packet loss, 480, 548
padding, 445
PAL TV system, 16, 84

parallax, 379
parallel projection, 113
parametric models, 111
partition, 56
partition regions, 249
payload, 544
PB pictures, 418
pel, 8
pel-recursive methods, 154
pelwise shape coding, 317
perceptual quality, 549
perfect reconstruction, 62
periodicity, 59
  lattice, 59
  matrix, 59
persistence of vision, 43–44
personal computers, 405
perspective camera model, 196, 197, 206
perspective projection, 112, 196
phase correlation function, 162
phase correlation method, 162
Phong shading model, 119
picture reordering, 424
pinhole camera model, 112
pixel, 8
pixel aspect ratio (PAR), 22
pixel-based motion estimation, 152
planar objects, 197
play-out delay, 476
point light source, 118
polygon-based shape coding, 320
polynomial-based motion representation,
    167, 169
postprocessing, 181
predicted frame, 220, 291
predictive coding, 219, 285
  error propagation effect, 291
  error-resilient prediction, 492
  example, 290
  gain over PCM, 288
  lossless, 286
  lossy, 286
prediction coefficients, 287
prediction error image, 220
predictor order, 287
prefilter, 62, 63, 65
prefix code, 234
probability density function (pdf), 222
probability mass function (pmf), 222
profiles, 428, 435, 451
  level, 436
program stream, 426

progressive scan, 9, 27, 67–69, 84
  alias components, 69
  sampling lattices, 69
projection onto convex sets (POCS), 501
projective mapping, 177
pseudo-perspective mapping, 137
PSNR, 29
public switched telephone network, 479
pulse coded modulation (PCM), 219, 286
pure parameters, 197
pyramid structure, 183

QCIF, 416
quad-tree coding, 317
quadratic image signal model, 205
quadrature-amplitude modulation (QAM),
    19–20
quadrilateral element, 172
quality factor, 282
quality measure, 28
  mean absolute difference (MAD), 29
  mean square error (MSE), 29
  peak signal-to-noise ratio (PSNR), 29
  perceptual distortion, 29
quality of service (QoS), 421, 422,
    453, 519
quality scalability, 350
quantization, 218, 241–257
  artifacts, 336
  parameter, 282
  stepsize, 243, 414
  table, 282
quantizer, 258
  dead zone, 414
  stepsize, 414
Questions, 407

random access, 295
random bit errors, 473
random process, 221
random variable (RV), 221
random vector, 221
raster scan (*see* video raster)
rate-compatible punctured convolutional
    (RCPC) code, 498
rate control, 523
rate-distortion bound, 231, 258
rate-distortion function (*see* rate-distortion
    bound)
rate-distortion optimization, 300, 301
  equal-slope condition, 301
rate shaping, 524

rate update interval, 303
rate-monotonic scheduling, 534
real-time applications, 544
real-time control protocol (RTCP), 422,
    481, 522
real-time operating system, 534
real-time services, 481
real-time streaming protocol (RTSP), 481,
    521, 543
real-time transport, 519
real-time transport protocol (RTP), 422, 481,
    488, 542
  payload format, 486
receiver-based rate control, 524
reciprocal lattice, 58, 60
reconstruction filter, 62, 65
recovery of coding modes and motion
    vectors, 501
rectangular lattice, 54, 55, 57, 58, 63, 64, 251
  sampling efficiency, 65
rectification, 385
redundancy, 475
Reed-Solomon (RS) code, 485
reference frame, 145, 187, 291
reference model, 412
reference picture selection, 503, 506
reflectance function, 117
reflected radiance, 117
reflection coefficient, 117
reflecting source
  subtractive rule, 2
region segmentation, 179–181
  ISODATA method, 180
  K-means, 180
  postprocessing, 180
  spatial connectivity, 180
region-based motion estimation, 179–182
  motion-first, 180
  region-first, 180
region-based motion representation, 147
region-based video coding, 327
regularization, 149
relation between Fourier transforms, 66
relative luminous efficiency function, 8
relative projective mapping, 135
resource management, 535
resynchronization markers, 474, 489, 508
retinal disparity, 375
retransmission, 472, 479, 487, 544
  delay-constrained, 487
  prioritized, 488
  without waiting, 504

reverse water filling, 234
reversible variable-length coding (RVLC),
        490, 509
rigid object, 122
rigid components, 124, 333
rigid 3-D model (R3D) objects, 332
ringing artifacts, 92
RISC, 405
robust estimator, 177, 181
    hard-threshold, 178
    inliers, 178
    outliers, 178
    soft-threshold, 178
robust motion estimation, 209
rotation matrix, 122
routers, 480, 529
round trip delay, 477
run-length coding, 416

saccadic eye movement, 46
sampled linear shift-invariant systems, 61
sampled-space Fourier transform (SSFT),
        60, 80
    fundamental period, 60
    periodicity, 60
sampling a raster scan, 71
sampling density, 58, 63
sampling efficiency, 63, 65
    alternative definition, 65
sampling matrix (see generating matrix)
sampling of video, 67, 80
sampling over lattices, 59
sampling rate, 22, 46, 67, 68
sampling rate conversion, 84
scalability, 349–360, 427, 430, 507
scalable video coding, 349–360, 547–550
scalar Huffman coding, 236
scalar lossless coding, 227
scalar quantization, 241, 257, 258
    boundary values, 242
    high-rate approximation, 246
    MMSE quantizer, 245–248
    optimal quantizer, 244
    partition regions, 242,
    reconstruction values, 242
scaled frequency plane, 98
scene composition, 437
scene model, 125–127
scenegraph, 440, 448
scheduling, 534, 541
SCSI, 538
SC 24, 408

SC 29, 408
SDTV, 84, 484
search range, 156, 167
search stepsize, 167
SECAM, 16, 19
second generation image and video
        coding, 327
segment tree, 460
semantic video coding, 221, 338
separable transform, 270
sequence header, 455
session announcement protocol (SAP), 481
session control protocol, 521, 545
session description protocol, 481
session initiation protocol (SIP), 481
settop boxes, 405
Shannon information theory, 475
Shannon lower bound, 232
shape, 341, 463
shape-adaptive DCT, 325, 445
shape-adaptive DWT, 365
shape-adaptive wavelet coding, 364
shape coding, 314–323, 453
shape distortion
    area error, 323
    Euclidean distances, 320
    peak deviation, 323
shape estimation, 202
shape functions, 166, 172
shape model, 121
shape topology, 323
SIF, 28, 423
simple profile, 452
simple studio profile, 453
simulation model, 412
simulcast, 350, 397, 434
singular value decomposition (SVD), 198–201
slice, 295, 457, 506
smooth pursuit eye movement, 48, 49
smoothing buffer, 303, 478
smoothness constraints, 152, 153
smoothness property, 499
SMPTE, 27
SNR scalability (see quality scalability)
soft-decoding, 502
source coding, 221, 227
source model, 218, 219, 314, 330, 334, 339
    F2D, 330
    F3D, 330
    moving known object, 219
    moving known object with known
            behavior, 219

moving unknown objects, 219
   R3D, 332
   statistically dependent pels, 219
   statistically independent pels, 219
   translationally moving blocks, 219
source-based rate control, 523
spatial connectivity, 181
spatial frequency, 38, 50
   angular frequency, 39
spatial frequency response, 45, 46
spatial interpolation, 499, 500
spatial resolution, 68
spatial scalability, 350, 353, 432
spatio temporal frequency plane, 41
spatiotemporal frequency response, 46, 47, 49
   interlaced scan, 48
   reciprocal relation, 48
spectral flatness measure, 289
specular reflection, 116
sphere covering, 65
spherical support region, 63
SPIHT, 364
spline functions, 65, 92, 321
spline-based shape approximation, 321
sprite, 437, 444
standardization requirements, 409
start code, 455
stationary processes (*see* stationary sources)
stationary sources, 222
statistically dependent pixels, 220
steepest gradient descent, 152
stereo and multiview
   capture and display, 374
   coding, 396–400, 403
stereo imaging, 377–384, 401
   arbitrary camera configuration, 377
   converging configuration, 381–383
   image rectification, 387
   parallel configuration, 379–381
stereo sequence coding, 396–400
   block-based coding, 396
   MPEG-2 multiview profile, 396
stereo triangulation, 379
stereopsis, 375
stereoscopic video transmission, 435
storage area network, 538
storage system, 537
streaming servers, 520, 521, 533
structure and motion estimation, 392, 400
structured audio, 442
subband decomposition (*see* discrete wavelet transform)

subjective video quality, 454
support region, 61, 63
SVGA display, 68
sync layer (SL) packets, 438
synchronization marker, 506
synchronization protocol (SSP), 542
synchronization source identifier (SSRC), 544

target frame, 145, 291
Taylor expansion, 204
TCP/IP protocol suite, 480
TDMA, 549
teleconferencing, 412, 477
telephone lines, 217
template, 317
temporal frequency, 40, 50
   linear motion, 40
   of a moving object, 145
temporal frequency response, 43
temporal interpolation, 499
temporal scalability, 350, 356, 396, 434
temporal summation, 44
test model, 412
   long term, 412
   near term, 412
text-to-speech (TTS) synthesizer, 441, 450, 451
texture coding, 447
texture extrapolation, 324
texture mapping, 447
texture maps, 333, 448
texture memory, 331, 332
texture model, 120
texture padding, 324
three-dimensional model-based coding, 400
three-dimensional object-based coding, 399
three-step search, 159
tiling (*see* partition)
time-stamps, 541, 544
transform basis, 264, 265
transform coding, 219, 301
   example, 277, 280
   gain over PCM, 276
   optimal bit allocation, 275
   threshold coding, 310
   zonal coding, 310
transform design criterion, 265
translation vector, 122
translational motion, 166, 177, 179
translationally moving blocks, 220

transmission control protocol (TCP), 481,
        523, 542
transparent object, 315
transport protocol, 542
transport stream, 426
triangular elements, 172
tristimulus values, 4, 5
Tukey algorithm, 212
TV cameras, 67
TV system
    bandwidth, 19
    color coordinate, 17
    color TV system, 16, 17
    digital TV (DTV), 27
    HDTV, 27
    monochrome TV system, 16
    multiplexing, 19
    NTSC, 16
    PAL, 16
    SECAM, 16

UMTS, 408
uncovered region, 116, 126
unequal error protection, 488, 493, 498
unicast, 478, 523, 545
unidirectional temporal prediction, 291
uniform distribution, 222
uniform quantizer, 243
uniform sampling, 59
uniform source, 246
uniquely decodable, 227
unit cell, 56
    fundamental parallelepiped, 56
    volume, 58
    Voronoi cell, 57
unitary matrix, 267
    1-D, 266
    2-D, 269
unitary transform, 267
unrestricted motion vectors, 417, 444
up-conversion, (see interpolation)
up-sampling, 85
upward compatibility, 410
user datagram protocol (UDP), 480, 521, 542

V.34, 422
variable-length coding (VLC), 235, 246, 256,
        258, 473
    3-D VLC, 417, 421, 444
VCD, 27
VCR, 545
vector Huffman coding, 236

vector Lossless coding, 228
vector quantization (VQ), 219, 248, 258, 265
    binary tree search, 250
    complexity, 249
    MMSE quantizer, 253
    optimal quantizer, 253
    reconstruction vector, 249
vector transform coding, 284
vector wavelet transform, 362
velocity vector, 143
verification model, 412
vestigial sideband modulation (VSB), 21
video bit stream syntax, 454
video coding, 217, 442
    general framework, 218, 257
video downloading, 478
video formats
    BT.601, 23
    CIF, 26, 27
    component video, 11
    composite video, 11
    QCIF, 26, 27
    S-video, 11
    SIF, 27
    SIF-I, 27
video indexing, 458
video packet, 508
video packet header, 457
video phone, 217, 405, 477
video raster, 12, 69
    active lines, 14
    bandwidth, 19
    frame rate, 14
    horizontal retrace, 14
    interlaced scan, 13
    line number, 14
    line rate, 14
    progressive scan, 12
    spectrum, 14
    vertical retrace, 14
video redundancy coding, 493, 495, 507
video signal, 2, 7
video streaming, 477, 519
    play-out delay, 478
video-on-demand (VOD), 26
video object plane (VOP), 442, 448, 456
view angle quantization, 376
view synthesis, 393
virtual reality, 393, 399
visual thresholds (see cut-off
        frequency)
Voronoi cell, 55–57, 60, 63, 251

voxel, 121
VRML, 408, 440

waveform-based coding, 217, 219, 263–309
wavelet transform
    arbitrary shape, 362
    self-similarity, 363
    time localization, 370
wavelet-based coding, 350, 361–370
    comparision with DCT, 365, 370
    image coding, 362–367
    motion compensation, 362
    video coding, 367–370
wide area network, 480
wireframe, 121, 220, 332, 334, 394
wireless networks, 481–484, 488, 546
    broadband wireless IP networks, 482

cellular networks, 482
GPRS (general packet radio
    service), 482
IEEE 802.11, 482
mobile IP, 482
packet sizes, 483
third generation (3G) wireless
    systems, 482
wireless ATM, 482
wireless data network, 482
wireless LAN, 482

XML, 459

zero-tree, 447
zero-tree entropy (ZTE) coding, 364
zigzag scan, 282, 416, 430, 431